Mineral Engineering Conference (MEC 2016)

E3S Web of Conferences Volume 8 (2016)

Swieradow-Zdroj, Poland
25 - 28 September 2016

Editors:

Przemyslaw B. Kowalczuk
Jan Drzymala

ISBN: 978-1-5108-3017-2

Printed from e-media with permission by:

Curran Associates, Inc.
57 Morehouse Lane
Red Hook, NY 12571

Some format issues inherent in the e-media version may also appear in this print version.

Printed by Curran Associates, Inc. (2016)

For additional information, please contact EDP Sciences – Web of Conferences at the address below.

EDP Sciences – Web of Conferences
17, Avenue du Hoggar
Parc d'Activité de Courtabœuf
BP 112
F-91944 Les Ulis Cedex A
France

Phone: +33 (0) 1 69 18 75 75
Fax: +33 (0) 1 69 28 84 91

contact-edps@webofconferences.org

Additional copies of this publication are available from:

Curran Associates, Inc.
57 Morehouse Lane
Red Hook, NY 12571 USA
Phone: 845-758-0400
Fax: 845-758-2633
Email: curran@proceedings.com
Web: www.proceedings.com

TABLE OF CONTENTS

E3S Web of Conferences
Open access conference proceedings
in Environment, Energy & Earth Sciences
A flexible, fast and efficient
publication service,
for an open access recording
of your event
and a wide dissemination of
your proceedings
www.e3s-conferences.org
edp open
Open Access Publishing
from EDP Sciences

Aims & scope

E3S Web of Conferences is an open-access journal devoted to the publishing of conference proceedings dealing with Environment, Energy and Earth Sciences.

Coverage includes technological and scientific aspects as well as social and economic matters. Major disciplines include: soil science, hydrology, oceanography, climatology, geology, geography, energy engineering (production, distribution and storage), renewable energy, sustainable development, natural resources management, etc.

The journal aims to make conference proceedings visible, citable and widely usable by the scientific community. It offers outstanding services for the publication and dissemination of contents in order to ensure long-term archiving and maximum exposure of scientific events.

Why publish with E3S Web of Conferences?

- **Open access:** conference papers available worldwide.
- **Fast publication:** less than 3 months.
- **Total flexibility:** no page restrictions, free unlimited use of colour in the online version, any kind of scientific document welcomed (traditional articles, posters, abstracts and slideshows).
- **High visibility:** all documents indexed in main databases and easily found online.
- **Fair author rights:** CC-BY Creative Commons license.
- **Range of additional publishing options:** printed proceedings, CD-ROMs, copy-editing, etc.

Contact: Isaline Augusto
e3s@webofconferences.org

ISSN (electronic edition): 2267-1242

www.e3s-conferences.org

edp open Open Access Publishing from EDP Sciences

edp sciences

E3S Web of Conferences

Statement of Peer review

In submitting conference proceedings to E3S Web of Conferences, I certify to the Publisher that I adhere to the **Policy on Publishing Integrity** of the journal in order to safeguard good scientific practice in publishing.

1. All articles have been subjected to peer review administered by the proceedings editors.
2. Reviews have been conducted by expert referees, who have been requested to provide unbiased and constructive comments aimed, whenever possible, at improving the work.
3. Proceedings editors have taken all reasonable steps to ensure the quality of the materials they publish and their decision to accept or reject a paper for publication has been based only on the merits of the work and the relevance to the journal.

Title, date and place of the conference

Mineral Engineering Conference MEC2016, September 25-28, 2016, Swieradow-Zdroj, Poland

Proceedings editor(s):

Przemyslaw B. Kowalczuk, Jan Drzymala

June 12, 2016

Date and editor's signature

Optimization of operating variables for production of ultra-fine talc in a stirred mill. Specific surface area investigations

Oner Yusuf Toraman[1,a]

[1]*Nigde University, Faculty of Engineering, Mining Engineering Department, 51240, Nigde, Turkey*

Abstract. Due to its properties such as chemical inertness, softness, whiteness, high thermal conductivity, low electrical conductivity and adsorption properties talc has wide industrial applications in paper, cosmetics, paints, polymer, ceramics, refractory materials and pharmaceutical. The demand for ultra-fine talc is emerging which drives the mineral industry to produce value added products. In this study, it was investigated how certain grinding parameters such as mill speed, ball filling ratio, powder filling ratio and grinding time of dry stirred mill affect grindability of talc ore (d_{97}=127 µm). A series of laboratory experiments using a 2^4 full factorial design was conducted to determine the optimal operational parameters of a stirred mill in order to minimize the specific surface area. The main and interaction effects on the volume specific surface area (SV, $m^2.cm^{-3}$) of the ground product were evaluated using the Yates analysis. Under the optimal conditions at the stirrer speed of 600 rpm, grinding time of 20 min, sample mass of 5% and ball ratio of 70%, the resulting talc powder had larger volume specific surface area (*i.e.*, 3.48 $m^2.cm^{-3}$) than the starting material (*i.e.*, 1.84 $m^2.cm^{-3}$).

[a] Corresponding author: otoraman@nigde.edu.tr

Determination of optimal flotation conditions of low-grade graphite ore

Özcan Öney[1,a], and Selçuk Samanli[1]

[1] *Usak University, Department of Mining Engineering, 64200 Usak, Turkey*

Abstract. In this study, the enrichment of the Kütahya Altıntas graphite ore by flotation was carried out. Flotation parameters such as particle size, pH, type and dosage of collector, and frother dosage, dosage of depressant, as well as solid ratio were investigated. As a result, a clean graphite concentrate of 77.80% total carbon content was obtained with 46.68% total carbon recovery.

1 Introduction

Graphite is a crystalline, polymorphic form of elementary carbon [1]. Graphite is lustrous black carbon mineral which is relatively soft and greasy with a hardness of 0.5–1.0 on Moh's scale [2, 3]. Depending upon the mode of occurrence and origin, graphite is graded in to three forms: flake-found in metamorphosed rocks as vein deposit, crystalline (lumpy)-found as fissure filled veins, crypto crystalline (amorphous) form in metamorphosed coal beds [4, 5]. Graphite generally occurs as a result of metamorphism of organic matter in sediments. Flake graphite is assumed to be derived from fine-grained sediments rich in the organic matter. As metamorphic grade increases, carbonaceous material converts to amorphous graphite [3, 6]. Flake graphite is classified based on the size of the crystal flakes and graded according to their graphitic carbon content and particle size. Microcrystalline graphite is commercially called amorphous graphite [6].

Graphite, due to its various physical and chemical characteristics, has a unique role in different industries [7]. Graphite is used mainly for foundry facings, refractories, lubricants, pencils, batteries, brake linings, bearings, conductive coatings and crucibles [8, 9]. The specific properties of graphite result in its increasing importance in the refractory industry [1, 10].

2 Background

To obtain good quality of graphite concentrate, beneficiation is essential in order to obtain optimal prices for the finished product. Froth flotation is commonly used for the concentration of low grade graphite ore [7]. Graphite can be easily enriched by flotation because of its high natural hydrophobicity [8, 11, 12, 13].

[a] Corresponding author: ozcan.oney@usak.edu.tr

In graphite flotation, hydrocarbons like kerosene, fuel oil, paraffin and diesel oil or ionic collectors like potassium amyl xanthate, dithiophosphate are generally used as collectors. Pine oil and methyl isobutyl carbinol are used as frothers and sodium silicate, quebracho and starch are used to depress gangue minerals. Optimum pH in graphite flotation is between 8 and 9 [8, 11].The polyethylene oxide type frothers such as polyoxyproplene glycol butyl ether were also studied as a replacement for MIBC in graphite flotation [6, 14]. Flotation is carried out in different stages ranging from two to several, depending on the liberation characteristics of the ore to be treated [2].

3 Materials and methods

3.1 Material

Low grade run-of-mine graphite ore was received from the Kütahya Altıntaş area in Turkey. The ore was crushed in stages and mixed thoroughly. The graphite ore was subjected to x-ray diffraction studies for the mineralogical phase analysis to identify non-graphite minerals. The diffractogram is shown in Fig. 1.

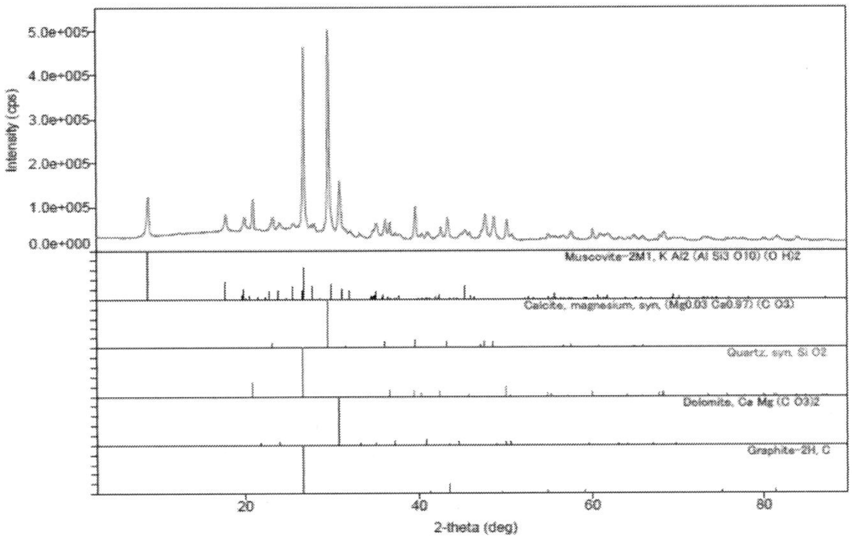

Figure 1. X-ray diffractogram of graphite ore.

The qualitative mineralogical analysis shows that raw graphite consist of muscovite [KAl_2 ($AlSi_3O_{10}$ $(OH)_2$], magnesium calcite [($Mg_{0.03}$ $Ca_{0.97}$) (CO_3)], quartz [SiO_2], dolomite [CaMg $(CO_3)_2$] and graphite. According to the results, raw graphite contains 17.56 % of total carbon.

3.2 Flotation methodology

The aim of the experiments was to obtain high carbon content of froth products. Firstly, various experiments were performed to determine the most proper rougher flotation conditions. The examined parameters for rougher flotation stage were as follows: optimal particle size, effect of pH, collector type, frother type, solid-liquid ratio and effect of depressant. Then, cleaning and scavenging tests were carried out.

The reagents used in the experiments were collectors, frothers, depressants. Collectors used in the experiments were diesel oil, kerosene and n-dodecane. 4 metyl-2 pentanol, 2ethyl-1 hexanol, tri

 DOI: 10.1051/e3sconf/20160801002

(propylene glycol) butyl ether and pine oil were used as frothers and sodium silicate was used as a depressant.

3.3 Experimental methods

Flotation experiments were carried out in a 1 dm^3 Denver laboratory type flotation machine. In the flotation tests, the impeller speed of flotation machine was kept constant at 1400 rpm for both conditioning and flotation. The tests were done gradually as rough and single cleanings. The pulp was mixed for three minutes prior to addition of collector and after an additional mixing of three minutes, the frother was added. Following a three-minute mixing period, air was introduced into the cell and the froth products were collected for five minutes.

4 Materials and methods

4.1 Optimal flotation parameters for rougher concentrate

4.1.1 Effect of optimum particle size fraction

For examining the effect of size fraction for flotation and to determine the optimum size range, the flotation experiments were carried out for the size fractions of -0.300, -0.212, -0.150 and -0.106. The highest total carbon content and total carbon recovery of the concentrate was obtained for -0.106 mm sample as 60.52% and 54.80% respectively (Fig. 2).

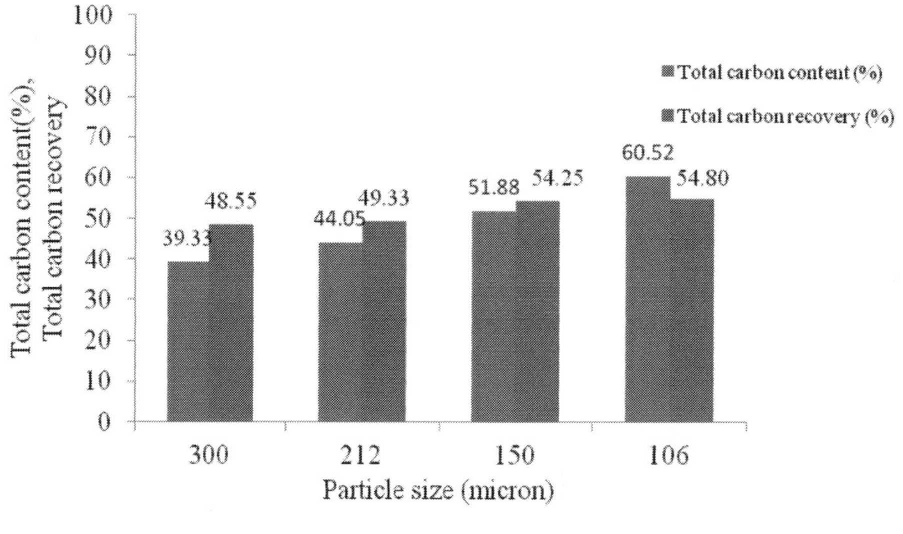

Figure 2. Effect of particle size (diesel oil: 600 g/t, MBIC: 125 g/t, solid ratio: 10%).

As a result the 0.106 mm size fraction was used for further experiment steps. The particle size distribution of the crushed and ground graphite ore is shown in Fig. 3. The calculated d_{80} of this graphite sample was found to be 78 μm.

DOI: 10.1051/e3sconf/20160801002

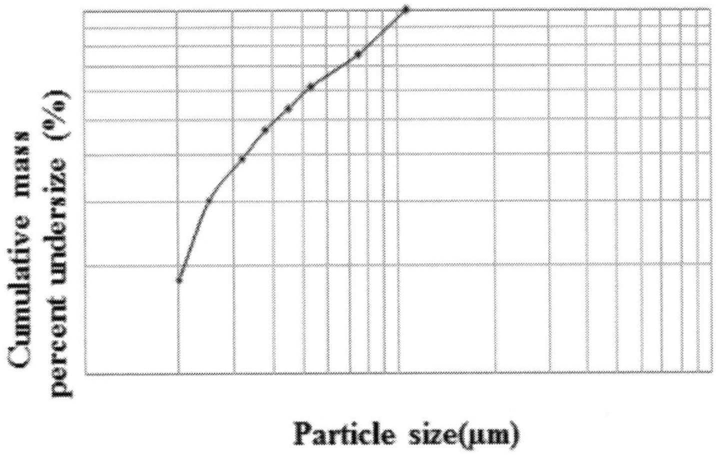

Figure 3. Particle size distribution of graphite sample (d_{80} = 78 μm).

4.1.2 Effect of pH

The flotation trials were conducted at neutral and alkaline (8-11) pH ranges. Na_2CO_3 was used to adjust the pH. The results are shown in Fig. 4. It is clear from Fig. 4 that, while the total carbon content was increasing slightly, the recovery of rougher graphite concentrate showed decreasing trend from pH value of 8 to 11. Hence, the further test were conducted at the natural pH value was equal to 8.

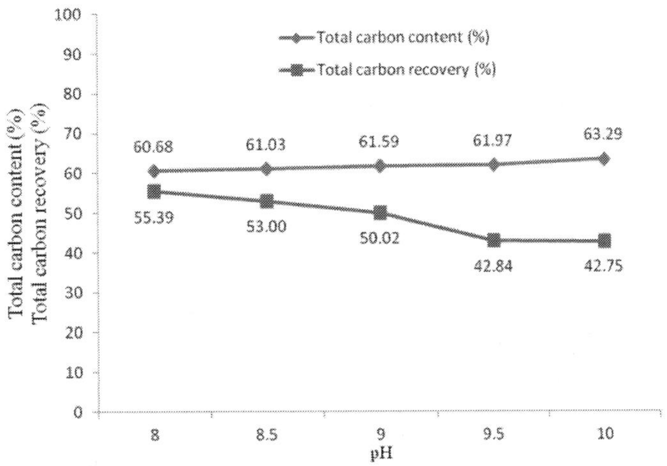

Figure 4. Effect of pH (diesel oil 600 g/t, MBIC: 125 g/t, solid ratio 10%).

4.1.3 Effect of depressant dosage

To determine the effect of depressant dosage, several flotation tests were performed with different dosages of sodium silicate varying from 500 g/t to 2500 g/t. The results obtained are shown in Fig. 5. With increased depressant dosage, the total carbon content remained constant or slightly increased. Total carbon recovery also remained constant or slightly decreased. It was decided to use sodium silicate at 1000 g/t dosage. After this dosage, there was no significant change for total carbon content and the total carbon recovery of the rougher concentrate.

 DOI: 10.1051/e3sconf/20160801002

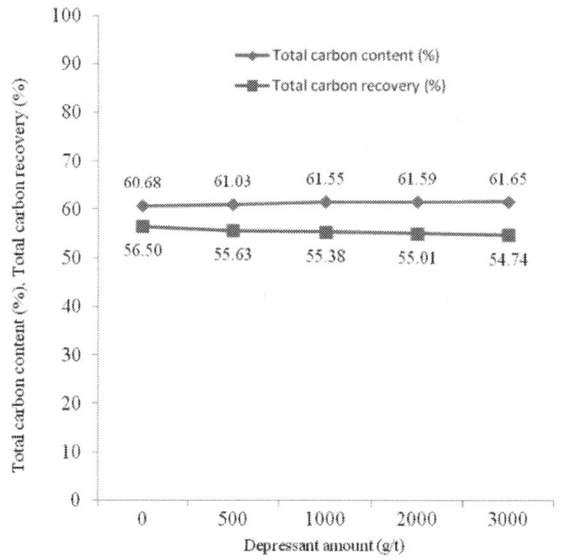

Figure 5. Effect of depressant dosage (diesel oil 600 g/t, MBIC: 125 g/t, solid ratio 10%, Na$_2$SiO3 1000 g/t).

4.1.4 Effect of frother type and dosage

Initially, a few flotation tests were conducted to study the effect of frother type and dosage on the rougher graphite concentrate. Four different types of frother were used in the experiments. The results shown in Fig. 6.

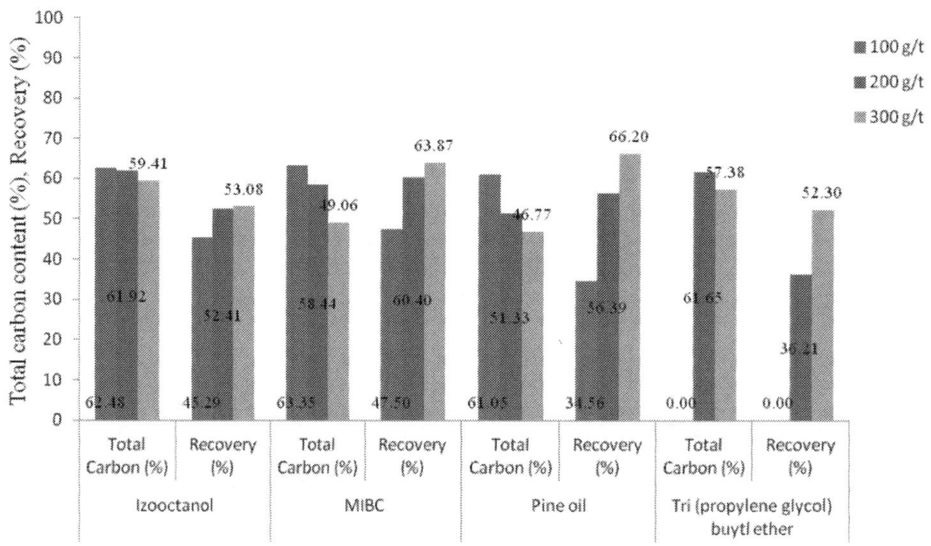

Figure 6. Effect of frother type (diesel oil 600 g/t, solid ratio 10%, pH natural (8), Na$_2$SiO$_3$ 1000 g/t).

The total carbon content of the rougher concentrate decreased continuously with increasing of frother dosage. It was observed that MIBC was the most effective frother. Hence, the next set of trials were performed using various dosages of MIBC (Fig. 7).

 DOI: 10.1051/e3sconf/20160801002

Taking into account the both total carbon content and total carbon recovery of the rougher concentrate, the proper dosage of MIBC was 175 g/ton.

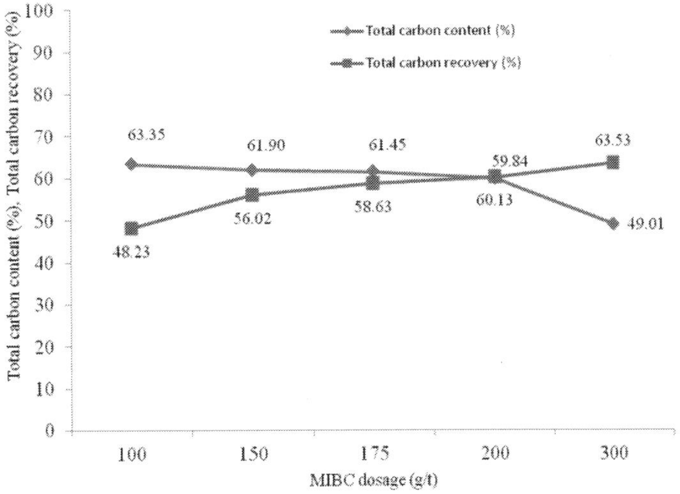

Figure 7. Effect of MIBC dosage (diesel oil 600 g/t, solid ratio 10%, pH= 8, Na_2SiO_3 1000 g/t).

4.1.5 Effect of collector type and dosage

To determine the effect of collector type and dosage on graphite flotation, diesel oil, kerosene and n-dodecane were tested. The results are shown in Fig. 8.

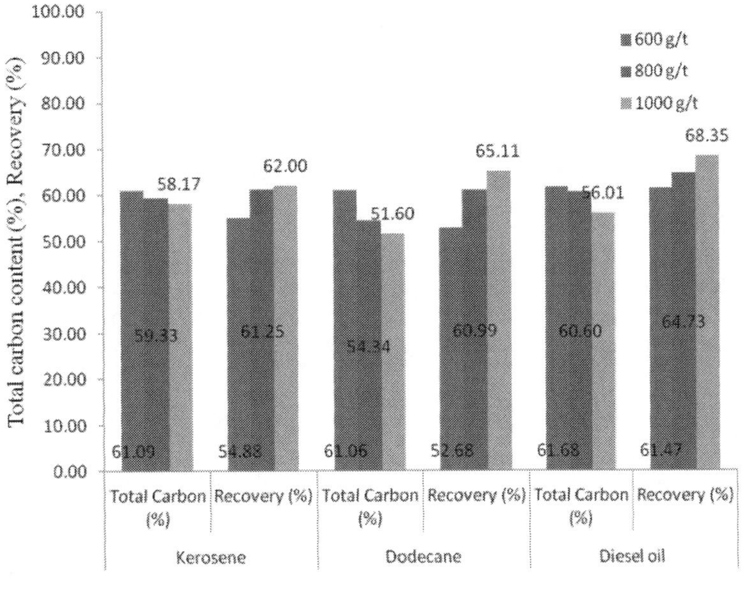

Figure 8. Effect of collector type (type (frother 175g/t, solid ratio 10%, pH = 8, Na_2SiO_3 1000 g/t).

In terms of the total carbon content and total carbon recovery of the rougher concentrate, the best results were obtained with diesel oil. At the dosage of 600 g/t diesel oil, the rougher concentrate with 61.68% total carbon content was obtained with 60.47% total carbon recovery. Diesel oil was selected

DOI: 10.1051/e3sconf/20160801002

as the collector for determining the other steps and the optimal dosage of 800 g/t diesel oil was thought to be appropriate.

4.1.6 Effect of solid ratio

Several experiments were performed to determine the optimal solid rate (Fig. 9.). The total carbon content of 60.60% was obtained with the total carbon recovery of 64.73% at solid ratio of 8%. It can be seen that the solid ratio was determined as 12% (Fig. 9).

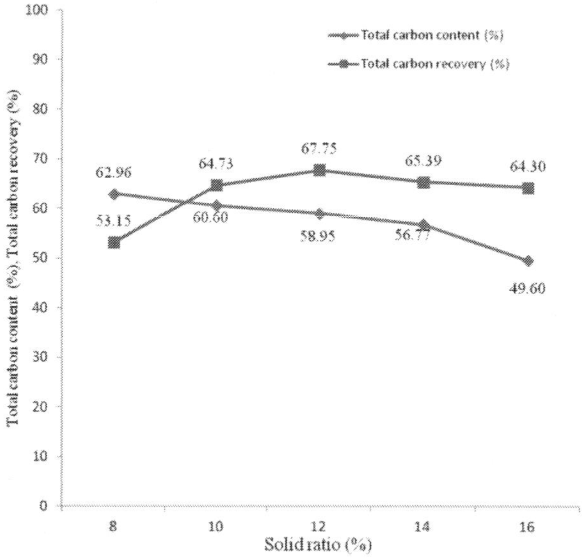

Figure 9. Effect of solid-liquid ratio ((MIBC 175g/t, diesel oil 800 g/t, solid ratio 10%, pH = 8, Na_2SiO_3 1000 g/t).

4.2 Improvement in concentrate grade by cleaning flotation

The flotation tests were carried out using optimum flotation reagents and conditions for rougher flotation followed by cleaner and scavenger flotation. A schematic flowsheet of graphite flotation is shown in Fig. 11. Reagent types and their dosage for each flotation stage are also illustrated in Fig. 10. The graphite concentrate containing 77.80% total carbon can be obtained with the recovery of 46.68%.

5 Conclusions

The optimum particle size was determined as 0.106 mm. While the total carbon content increased slightly, the recovery of rougher graphite concentrate showed a decreasing trend from pH 8 to 11. Hence, the natural pH 8 was kept fixed in the tests. Kerosene, n-dodecane and diesel oil were tested as collectors and among them diesel oil was found to be the most effective. Four different types of frother were examined and the best results were obtained with MIBC. The dosage of sodium silicate had a small influence on the total carbon content and the total carbon recovery of the rougher graphite concentrate.

It was found that rougher, cleaner and scavenger flotation stages should be applied to obtain the high content and recovery of carbon. A clean graphite concentrate of 77.80 % total carbon content was obtained with 46.68% total carbon recovery.

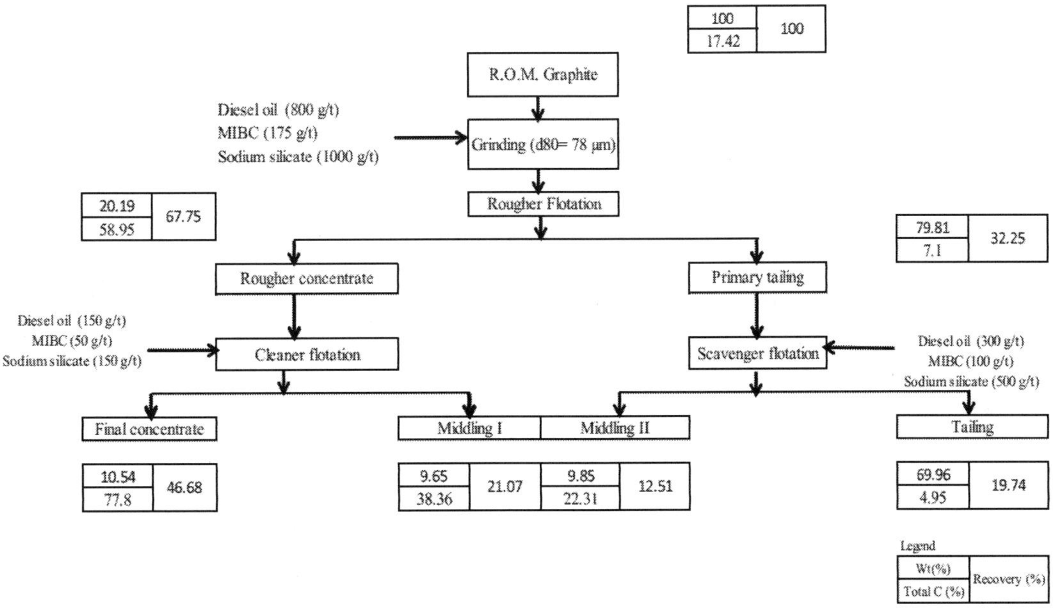

Figure 10. Schematic flowsheet of graphite flotation.

Acknowledgement

This work was supported by Research Fund of Usak University, Usak/Turkey. Project Number: 2014/MF015.

References

1. B. Kwiecinska, H.I. Petersen, *Int. J. Coal Geol.*, **57**, 99 (2004)
2. N. Vasumathi, T. Vijaya Kumar, V. Nayak., S. Subba Rao, S. Prabhakar, G. Bhaskar Raju, *J. Min. Metall.*, **1**, 9 (2014)
3. C.J. Mitchell, British Geological Survey Report No. WG/92/30 (1993)
4. S.K. Nuhu, , *C. J. Appl. Sci.*, **9(2)**, 31 (2014)
5. M.S. Klima, F.F. Aplan, P.T. Luckie, *Elements of Mineral Processing: Lecture Notes* (The Pennsylvania State University, 2010)
6. N. Vasumathi, T.V. Vijaya Kumar, S. Ratchambigal, S. Subba Rao, S. Prabhakar, G. Bhaskar Raju, B.R. Nayak, S. Shiva Kumar, U. Raman, in *Proceedings of the XIII International Seminar on Mineral Processing Technology*,145 (2013)
7. V. Ravichandran, C. Esweraiah, P. ve Manisankar, *Ultra Chemistry*, **8(2)** 159 (2012)
8. O. Kaya, M. Cambazoğlu, *The Journal of Ore Dressing*, **9**, 17 (2007)
9. L.L.Y. Chang, *Industrial Mineralogy Materials, Processes and Uses* (Prentice-Hall Inc., 2002)
10. K. Galos, P. Wyszomirski, *Ceram.-Pol. Ceram. Bull.* **64**, 59 (2001)
11. N. Arbiter, *Flotation, SME Mineral Processing Handbook* (Soc. of Mining Engineers, New York, 1985)
12. P. Crossley, *Ind. Miner.*,**March**, 31 (2000)
13. B.C. Acharya, D.S. Rao, S. Prakash, P.S.R. .Reddy, S.K. Biswal, *Min.Eng.*, **9**, 11 (1996)
14. R.J. Pugh, *Miner. Eng.*, **13**, 2, 151 (2000)

A laboratory study to evaluate the possibility of sulphur and phosphorous removal from iron ore concentrate by leaching

Hassan Rezvani Pour[1,a], Ali Mostafavi[1], Tayebeh Shams Pour[1], Ghobad Ebadi Pour[2], and Amir Haji Zadeh Omran[2]

[1] *Shahid Bahonar University, Department of chemistry, Kerman, Iran*
[2] *Gole-gohar iron ore complex, Sirjan, Iran*

Abstract. Iron ore concentrates with high grade sulfur cause several problems in the steel making process, and hence affect the concentrate price. Environmental issues such as sulfur dioxide emission during the concentrate pelletizing process and effect on the steel quality are other issues. The current study was focused on removal of sulfur from the iron ore concentrate by using the chemical leaching technique. The magnetite iron ore concentrate was chosen for this purpose. The results obtained showed that more than 90% of the total sulfur content was removed from the iron ore concentrate by chemical leaching. Effects of several parameters such as temperature, particle size and use of organic solvent on sulfur removal were investigated by a series of experiments. After optimizing the experimental conditions, it was demonstrated that with addition of sulfur, phosphorus, another important impurity was also removed from the iron ore concentrate. In addition, one of the major advantages of our proposed method was transformation of mineral pyrites to useful by-products such as elemental sulfur.

[a] Corresponding author: rezvani.ch@gmail.com

Flotation of hard coal in saline water

Anna Mlynarczykowska[1,a], Agnieszka Nowak[1], and Klaudia Tupek[1]

[1]*AGH University of Science and Technology, Faculty of Mining and Geoengineering, Department of Environmental Engineering and Mineral Processing, Poland*

Abstract. With the use of saline process water, the ash content in the concentrate was 6% and in the tailings 70%. Of course the upgrading plants which apply flotation for coal slurries still use mine water in technological processes. The water quality, particularly the content and type of salts in water, fundamentally influences the surface properties of the solid, gaseous and liquid phases, which remain in contact during the process. The properties of these phases and interactions between them determine the probability of occurrence of elementary flotation actions that is collision, adhesion and separation, and ultimately the upgrading process effectiveness expressed by technological parameters. The paper presents the results of flotation effectiveness analysis for coal as a model raw material. The object of evaluation was qualitative parameters of the upgrading process products on the basis of flotation tests conducted for variable pulp salinity. The principal goal was to determine the impact of process water quality on the course and the effectiveness of flotation.

1 Introduction

Hard coal is the main source of energy in Poland and its properties result from the condition of formation. The mineral matter of raw coal consists of hydrophilic minerals, mainly clays, quartz, carbonate minerals such as calcite and dolomite, gypsum and pyrite. Flotation is not the basic coal upgrading technology, but it is useful in pyrite removal and is used to reduce the ash content in the concentrate. The use of surfactants in the coal flotation is justified from the physical and chemical point of view. One must also remember that under industrial conditions flotation is performed with the mine water, which contains various well soluble inorganic salts. The types and amounts of salts depend on the mining conditions. The presence of salt in the flotation environment affects the action of flotation reagents, and consequently the process results and quality of concentrate. The use of process water in coal upgrading in processing plants began as early as in the 1980's. However, the main reason to study it was treatment of small coal particles and regulation of water and slurry management in the plants in terms of solids content in concentrators and the quality of concentrate. Since the observation that presence of salts, mostly inorganic, improves the flotation of naturally hydrophobic particles, many theories have been formulated to explain this phenomenon [1, 2, 3]. The research conducted by showed that the flotation recovery reached its maximum at the minimum value of zeta potential. However, Harvey et al. [4] proved that the improvement of coal flotability in the

[a] Corresponding author: mindziu@agh.edu.pl

solutions of electrolytes cannot be explained exclusively by reduction of the zeta potential. Ozdemir et al. [5] revealed that the coal flotation results in brine (1 M NaCl) cannot be fully explained by the physical and chemical surface properties, particularly because the wetting angle for coal in the saline water is the same as in the demineralized water. Hence, explicit determination of mechanisms which improve the coal flotability under conditions of significant salinity is still an open question.

2 Experimental

2.1 Material

The feed used in flotation processes was grade 33 hard coal from one of Polish mines. The material preparation included comminution to obtain the particle size below 0.5 mm. The wet sieve analysis was performed and the following particle size classes were obtained: 0.5-0.4, 0.4-0.315, 0.315-0.2, 0.2-0.1, 0.1-0.063 mm which were tested for the ash content. The tested material was kept in vacuum to prevent coal surface oxidation. All particle size classes were used in the flotation tests, the impact of particle size and density on the upgrading results in the salt-free flotation environment were analysed. On the basis of the results, two particle size classes 0.315-0.2 and 0.1-0.063 mm were selected for flotation tests under conditions of various pulp salinity. The flotation tests were also conducted for the particle size below 0.5 mm.

2.2 Flotation Experiment

The flotation tests were performed in the Denver pneumatic-mechanical laboratory machine. The work space volume was 1dm^3, rotor speed was constant 1850 rpm, and the pulp density was 70 g/dm^3. Using low flotation pulp densities allows to achieve better flotation upgrading results, and the constant rotor speed guarantees the unchanging amount and size of air bubbles [6].

Individual flotation tests for selected narrow grain size classes were performed in distilled water, saline mine water and in brines made by adding NaCl to the mine water. The content of chlorine ions (Cl^-) in saline water was 255.6 mg/dm^3, 23.4 g/dm^3 and 52.6 g/dm^3, and the types of waters were referred to as "mine water I, II, III", respectively. The particles sizes below 0.5 mm were floated in saline water which content of NaCl was 117 mg/dm^3. The products of flotation (concentrates and tailings) were analysed for the ash content in accordance with PN-ISO 1171: 2002.

All flotation experiments were conducted with the procedure described below. The specified coal samples were weighed, then process water with specific salt content was added, and the coal was soaked for 10 minutes. Then, a hexanol frother was added. Hexanol is an aliphatic alcohol and in the flotation process it performs the collecting and frothing properties sine it adsorbs on both solid-liquid and liquid-gas phases. It results in increasing hydrophobicity of coal and reducing the speed at which air bubbles flow to the surface. In addition, it allows to form a durable flotation froth, ensuring stabilization of air bubbles and preventing their coalescence.

At specific pulp density, the dose of flotation reagent was specified as 0.08 g of hexanol per 1 dm^3 of solution, which corresponds to consumption of 1142 g per 1 Mg of dry feed. To ensure reagent adsorption, the flotation machine was started and the solution was mixed for 3 minutes without air [6]. The air was added to the flotation tank after the set time and the fraction flotation was performed. The concentrates were collected at time: 15, 15, 30, 30, 30, every 60 s. The total flotation time was 6 min. The concentrates and tailings were dried at temperature of 60°C. The obtained results allowed calculating the basic indices to estimate the flotation process effectiveness, namely concentrate (γ_k) and tailings (γ_o) yields, ash content in individual upgrading products (β –concentrate, υ – tailings, α– feed), and the recovery of combustibles and volatile matter in the concentrate (ε_k) as well as in the tailing (ε_o).

3 Results and discussion

3.1 Presentation of results

Tables 1 and 2 include the calculation results limited to yield (γ), recovery of combustible and volatile in concentrates (ε_k). The flotation kinetics curves were plotted in order to evaluate the upgrading progress in time. Figures 1 and 2 present the empirical recovery vs time relationships, respectively for particle size classes 0.315-0.2 and 0.1-0.063 mm. The upgrading curves for individual particle size classes were plotted to illustrate the flotation effectiveness (Figures 3 and 4).

Table 1. Selected coal flotation parameters.

Particie size 0,315-0,2 mm												
time	**distilled water**			**Mine water I**			**Mine water II**			**Mine water III**		
t [s]	$\Sigma\gamma\downarrow$ [%]	β[%]	ε [%]	$\Sigma\gamma\downarrow$[%]	β[%]	ε [%]	$\Sigma\gamma\downarrow$ [%]	β[%]	ε [%]	$\Sigma\gamma\downarrow$ [%]	β[%]	ε [%]
0	0.00	0.00	0.00	0.00	0.00	0.00	0.00	0.00	0.00	0.00	0.00	0.00
15	4.90	5.13	4.80	9.54	18.67	9.73	7.48	25.37	7.57	9.77	33.36	8.34
30	8.07	5.07	7.92	11.48	17.99	11.80	10.72	26.86	10.63	14.54	29.12	13.19
60	12.39	4.86	12.18	15.63	14.49	16.75	14.71	24.91	14.98	18.80	26.25	17.74
90	17.00	4.60	16.76	18.12	14.36	19.44	17.96	25.13	18.23	22.56	23.74	22.01
120	19.88	4.67	19.59	20.89	13.94	22.52	21.20	25.61	21.38	25.31	22.50	25.10
180	26.51	4.27	26.23	26.14	13.52	28.33	26.68	25.82	26.84	31.08	20.85	31.47
240	32.28	4.19	31.96	30.29	13.15	32.97	31.42	24.44	32.19	35.34	21.08	35.69
300	38.62	3.96	38.33	34.72	12.70	37.98	35.16	23.44	36.51	39.10	19.91	40.07
360	44.09	4.02	43.74	40.25	12.10	44.33	38.65	22.44	40.65	43.61	18.82	45.30
tailing	100.00	3.24	100.00	100.00	20.20	100.00	100.00	26.26	100.00	100.00	21.85	100.00

Table 2. Selected coal flotation parameters.

Particie size (0,1 -0,063) mm												
time	**distilled water**			**Mine water I**			**Mine water II**			**Mine water III**		
t [s]	$\Sigma\gamma\downarrow$ [%]	β[%]	ε [%]	$\Sigma\gamma\downarrow$[%]	β[%]	ε [%]	$\Sigma\gamma\downarrow$ [%]	β[%]	ε [%]	$\Sigma\gamma\downarrow$ [%]	β[%]	ε [%]
0	0.00	0.00	0.00	0.00	0.00	0.00	0.00	0.00	0.00	0.00	0.00	0.00
15	42.47	15.57	49.77	39.19	17.18	46.97	35.42	16.20	45.38	34.13	15.17	44.42
30	57.05	15.19	67.16	50.90	17.39	60.84	48.19	17.15	61.04	45.44	15.41	58.97
60	67.19	15.66	78.66	59.98	18.06	71.12	57.83	18.05	72.45	55.95	18.43	70.02
90	71.32	16.13	83.02	63.32	18.45	74.73	61.93	18.70	76.97	63.89	21.10	77.34
120	75.75	16.82	87.46	66.19	19.33	77.27	65.30	19.52	80.34	66.47	22.09	79.45
180	78.61	17.29	90.24	68.94	19.96	79.85	71.08	21.19	85.64	74.80	26.11	84.80
240	79.87	17.53	91.43	70.85	20.59	81.41	74.94	22.34	88.97	77.58	26.91	87.00
300	80.03	17.56	91.58	72.16	21.18	82.31	76.87	22.71	90.83	79.76	27.52	88.70
360	81.30	17.93	92.62	73.00	21.42	83.00	78.80	23.19	92.53	81.15	27.96	89.70
tailing	100.00	27.95	100.00	100.00	30.90	100.00	100.00	34.59	100.00	100.00	34.82	100.00

 DOI: 10.1051/e3sconf/20160801004

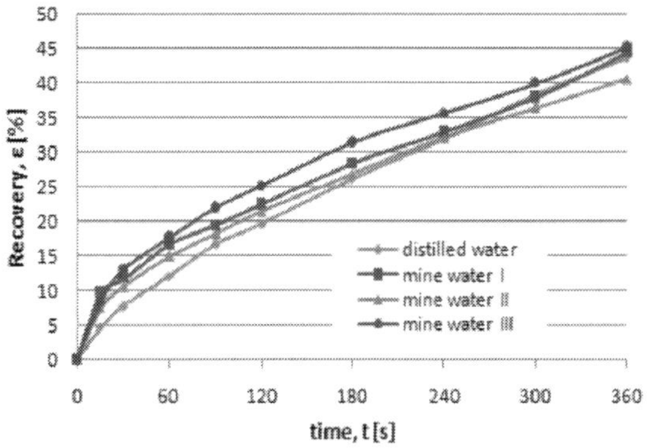

Figure 1. Flotation kinetics curves in waters with different salinity for particle size 0.315-0.2 mm.

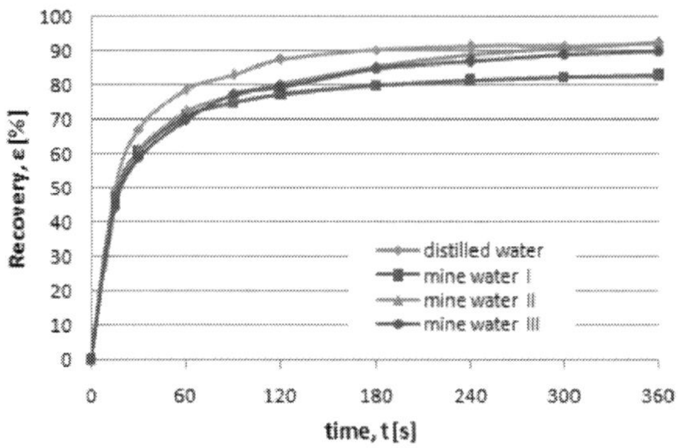

Figure 2. Flotation kinetics curves in waters with different salinity for particle size 0.1-0.063 mm.

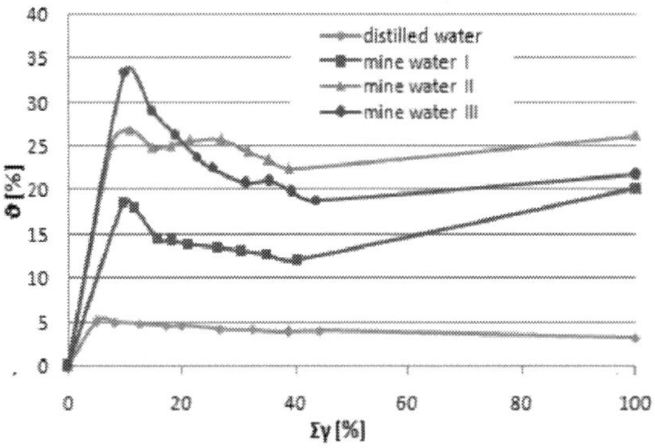

Figure 3. Changes of ash content in concentrate from, yield particle size 0.315-0.2 mm.

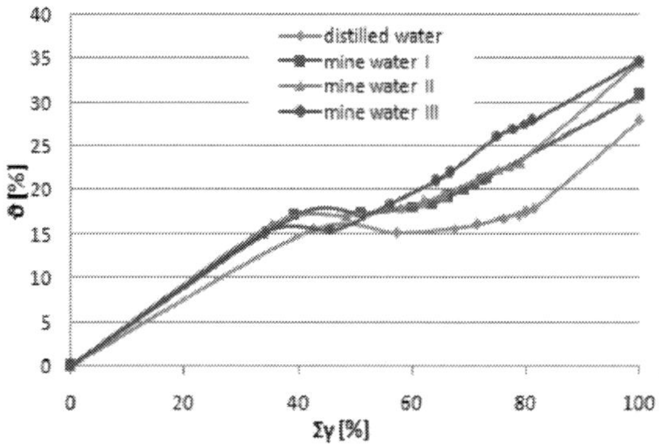

Figure 4. Changes of ash content in concentrate from yield, particle size 0.1-0.063 mm.

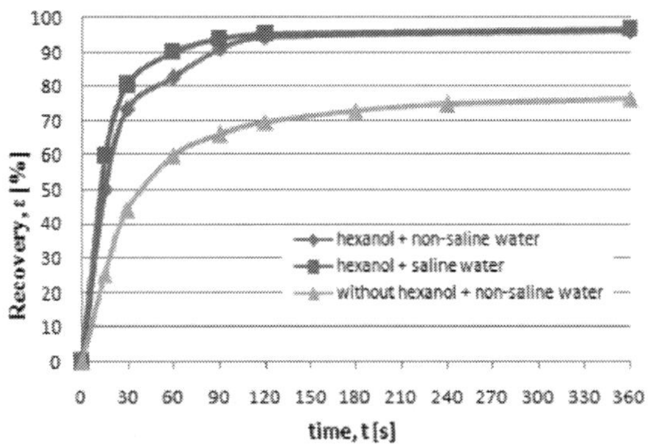

Figure 5. Recovery vs. time.

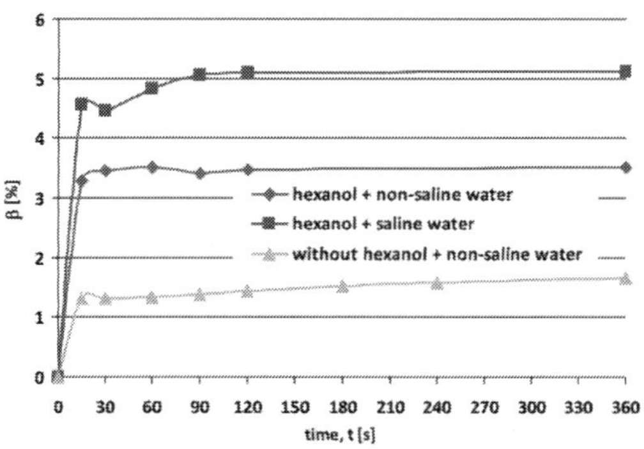

Figure 6. Changes of ash content in concentrate from time.

The graphical interpretation of flotation results for the below 0.5 mm grain size is performed by plotting the flotation kinetics curves (Fig. 5) and the curves indicating the changes of ash content in concentrates (Fig. 6) and in tailings (Fig. 7).

 DOI: 10.1051/e3sconf/20160801004

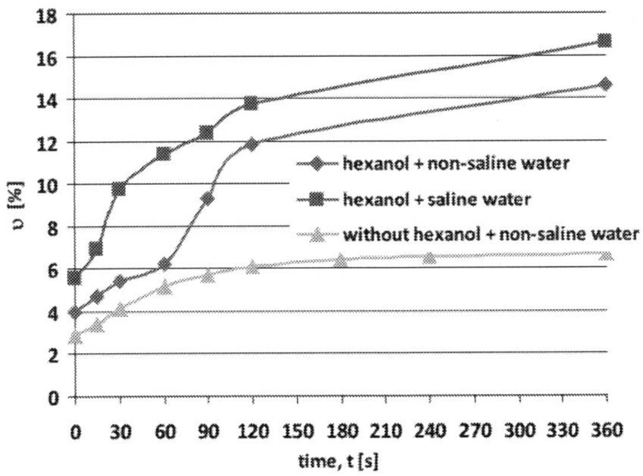

Figure 7. Changes of ash content in tailing with time.

3.2 Discussion

The analysis of results presented in Table 1 and shape of curves in Fig. 1 indicate a constant rising trend for recovery values with time. The shape of kinetics curves suggests that the upgrading process could continue because the balance recovery value (a plateau on the plot) was not reached. The indicated shape of kinetics curves should be expected in the case of highly effective flotation [7]. The course of flotation process is similar for each type of water used as shown by the shape of curves, and the highest intensity is observed in the first 3 minutes of flotation. The review of conditions improving the upgrading effectiveness indicates that in the most saline water (mine water III) flotation reaches the best values of yield, recovery and concentrate parameters. This proves that the presence of NaCl enhances the flotation process.

It is also evident that the concentrate yield in the first 15 seconds of the process is the lowest in thecase of distilled water (4.90%), whereas for the mine water I it reaches 9.54 %, mine water II – 7.48%, and mine water III – 9.77%. The yield increases and also recovery increases by 1/3 or even 1/2 in comparison with flotation in distilled water. It clearly indicates a positive impact of the presents of salt on the properties of solids subjected to flotation and also an improvement of gaseous phase properties under such conditions [7, 8].

Since inder industrial conditions narrow particle size classes are not upgraded, the successive analyses include the results of flotation tests conducted for particle below 0.5 mm. The analysis of results shown in Fig. 6 indicates that the greatest process speed for two flotation process with hexanol was obtained in the first 2 minutes, and the process without hexanol stabilized in the 3rd minute when 69.73% concentrate yield was achieved in flotation without reagent. Almost identical result, with difference of about 1% was detected in flotation in saline and non-saline water with addition of hexanol. It should be noted that both process with hexanol were similar, and the kinetics curves were practically identical. It would suggest similar properties of phases remaining in contact during flotation [6]. Only from 15 to 105 seconds of the process, the obtained recovery results differed by about 10% in each successive concentrate collection. An initial recovery increase was observed for floatation in saline water, but it was later reduced. It is shown in the plot as another convergence of curves in the same point (about 95% concentrate recovery). The final result was the 96% recovery for flotation with hexanol in both the NaCl saline water and non-saline water. The recovery of flotation without reagent was however significantly lower and reached 76.29%. It is evident that hexanol increased the coal flotation effectiveness and at the same time it can be stated that water salinity facilitates the uplift of coal grains. The ash content in the concentrate in the non-saline water without hexanol varies in the 1.31-1.66% range, in the 3.28-3.51% range in non-saline water with hexanol, and in the 4.56-5.12% range in saline water with hexanol.

DOI: 10.1051/e3sconf/20160801004

In order to evaluate the flotability of the tested material, the experimented results were presented in the form of Henry curves (Fig. 9). The ideal upgrading is denoted when on the Henry curve is the relationship between the yield and the content of useful ingredient is a horizontal line [10]. If the relationship is like this, then we obtain only concentrate containing pure useful components. When the upgrading line is vertical and the content of useful components remains constant and equal to α, then there is no upgrading. The analysis of curves presented in Fig. 8 and 9 indicates that the hexanol in this case gives high concentrate yields in both saline and non-saline water. The concentrate yield reached 95% in pure water flotation, and 90% in the process in the saline water. Such yield values substantiate the statement that the coal floated with addition of hexanol becomes easily upgradeable.

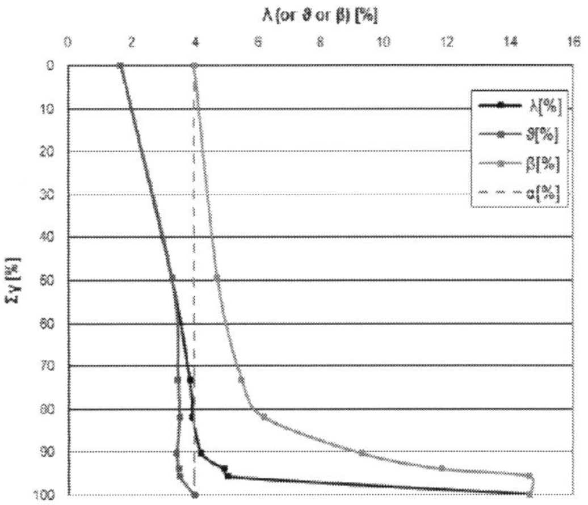

Figure 8. Washability curves in non-saline water.

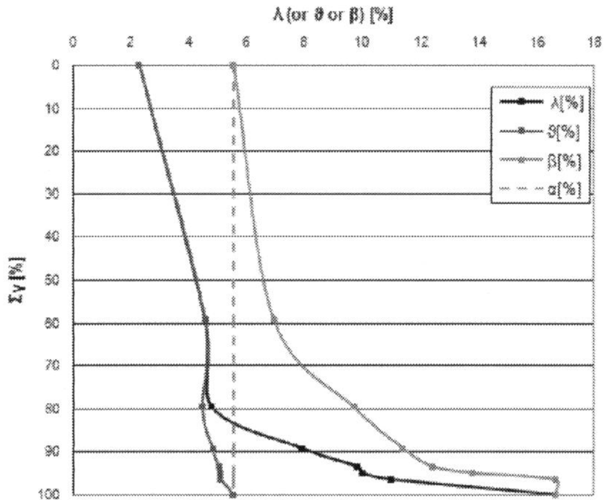

Figure 9. Washability curves in saline water.

3.3 Summary

In the light of literature on the flotation upgrading in the presence of inorganic substances, mainly salt, it seems appropriate to distinguish two terms: salt flotation conducted under conditions of increased surface tension in relation to water, and flotation conducted in the presence of salt in the pulp where the surface tension is reduced. The process conducted in the presence of salt is accompanied by increased froth stability, which rises as the concentration of frothing agent increases.

The froth stability is also very closely related to the bubble size, namely: the smaller the bubbles, the more stable the froth. This conclusion should be verified for each frothing agent used. The froth stability in coal flotation and the increased content of NaCl in process water improves the flotation effectiveness, with maximum recovery equal to 90%. In addition, it should be noted that high flotation effectiveness depends on the feed particle size. Although narrow particle size classes are not upgraded in industrial conditions, a verification of process effectiveness for small particle size ranges allows to forecast their behaviour in real conditions. Hence, on the basis of the presented results one should expect the process effectiveness of 40-90% recovery when upgrading the 0.315-.0063 mm grain size class in saline water. Thus, coal flotation in saline waters brings benefits: *(i)* conscious control of upgrading to increase its effectiveness, and *(ii)* reduced consumption of reagents which means lower costs. And most importantly, it allows the mines to recirculate the process water. The results of laboratory experiments prove that using an aliphatic alcohol, in both non-saline and saline waters, gives the concentrate yield in excess of 90%. This type of reagent improves the effectiveness of coal slurry flotation, and water salinity facilitates floating the coal particles to the froth product. In addition, hexanol adsorbing on the surface of bubbles creates durable flotation aggregates, thus increasing the probability of floating the coal particles. The application of this flotation reagent gives high ash content in tailings, in excess of 16% in saline water, with the ash content in concentrates not exceeding 6%. The purpose of coal upgrading is to maximize the content of combustible and volatile matter and to minimize the ash content in concentrates. Such results are guaranteed by the use of hexanol which has an apolar carbon chain. In addition, it should be noted that:

(i) the process effectiveness is much better in the case of upgrading materials with a wider particle size range than for narrow ranges floated separately. It gives much higher process effectiveness and good quality upgrading products. Wider particle size class and lower content of solids in the flotation pulp allow to reduce the reagent consumption, giving the better upgrading results;

(ii) the excessive salt content in process water causes a transfer of a significant amount of salt to the products of upgrading, thus increasing its yields and simultaneously reducing the technological parameters;

(iii) use of saline mine waters reduces the costs of technological processes by limiting the demand for fresh water and improves the environmental protection by reducing the amount of mine wastewater discharged to rivers.

Acknowledgement

This work was done as a part of the University of Science and Technology Research Program No.11.11.100.276 and was used in the MSc dissertations.

References

1. R.H.Yoon, Min. Cong. J., **68**, 76 (1982)
2. R.H. Yoon, J.B.Sabey, Interfacial Phenomena in Coal Technology, G.D. Botsaris, G.D., Glazman, Y.M. (Eds.) (New York, 87 1989).
3. J.S.Laskowski, Coal Flotation and Fine Coal Utilisation (Elsevier, Amsterdam, 2001) R.H. Yoon, J.B.Sabey, *Interfacial Phenomena in Coal Technology*, G.D. Botsaris, G.D., Glazman, Y.M. (Eds.) (New York, 87 1989).
4. O. Ozdemir, E. Taran, M.A. Hampton, S.I. Karakashev, A.V. Nguyen, Surface Process., **92**, 177 (2009)
5. A. Mlynarczykowska, A. Grabiec, A. Krawczykowska, D. Saramak, Mineral Processing Conference 2013 (Zakopane, 2013)
6. M. Brozek, A. Mlynarczykowska, Kinetics of flotation,(Press. IGSMiE PAN, 2009)A
7. A. Rodrigo, T, Graua, J.S. Laskowski, K. Heiskanen, Int. J. Miner. Process., **76**, 225 (2005)
8. M. Brozek, A. Mlynarczykowska, Physicoch. Prob. Miner. Process., **44**, 23 (2012)
9. J. Drzymala, , Mineral Processing, (Oficyna Wydawnicza Politechniki Wrocławskiej, 2001)
10. L. Wang, X. Qu, Miner. Eng., **26**, 50 (2012)

Dry grinding of talc in a stirred ball mill

Serkan Cayirli[1,a]

[1]*Nigde University, Department of Mining Engineering, 51240 Central Campus, Turkey*

Abstract. The aim of this work was to investigate micro fine size dry grindability of talc in a stirred ball mill. The effects of various parameters such as grinding time, stirrer speed, powder filling ratio and ball filling ratio were investigated. Alumina balls were used as grinding media. Experiments were carried out using the 2^4 full factorial design. The main and interaction effects were evaluated using the Yates method. Test results were evaluated on the basis of product size and surface area.

1 Introduction

Talc, the softest of all minerals with a Mohs hardness of 1, is an organophilic, water repellent and chemically inert mineral. It is characterized as a hydrated magnesium sheet silicate with the formula $Mg_3Si_4O_{10}(OH)_2$. Talc is a functional component (filler) in paper, paints, plastics, rubbers, ceramics, fertilizers, animal feed, cosmetics, pharmaceuticals and other applications [1]. Size specifications for talc are of considerable importance, depending primarily on the particular end purpose, and to a great extent, on the physical properties of the mineral itself [2]. It is therefore pertinent to improve the basic properties of industrial minerals by the grinding method employed. Generally, milling of these minerals can be performed in the conventional tumbling, impact, stirred ball, jet and vibration mills. However, in the fine size range, the economic and process considerations show the last three milling methods to be the most versatile, depending on the product and use [3]. Due to higher power density (density of the energy absorbed by the particles) and better grinding efficiency, high-speed stirred ball mills are considered as an interesting alternative to conventional tumbling mills and vibratory mills for fine grinding and delamination processes in the minerals industry [4-6].

The aim of this study is to determine the operating conditions for the dry grinding of talc to micro fine sizes using a laboratory vertical stirred ball mill.

2 Material and method

2.1 Material

A talc sample (S_v= 1.57 m^2/cm^3) used in the grinding experiments obtained from Mikron'S Company, Nigde, Turkey. The chemical analysis (by X-rayfluorescence) of the sample is given in Table 1.The

[a] Corresponding author: scayirli@nigde.edu.tr

specific density of talc is 2.78 g/cm^3. The typical size distribution of the feed sample is presented in Figure 1. Alumina balls with diameters 2.5-3.5 mm mixtures (33% of each) a density of 3.6 g/cm^3 were used as the grinding media.

Table 1. Chemical analysis of the sample.

SiO_2 (%)	MgO (%)	CaO (%)	Fe_2O_3 (%)	Al_2O_3 (%)
63.15	33.33	0.79	3.12	0.003

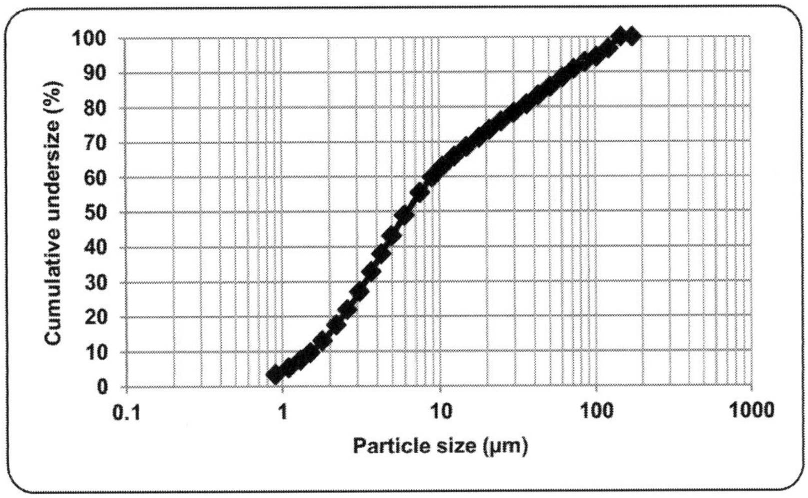

Figure 1. Particle size distribution of a feed sample.

2.1 Method

The grinding tests were performed in a stirred ball mill Standart-01 Model manufactured by Union Process (Fig. 2). It comprises a 595 ml cylindrical alumina (99.9%) grinding tank. The stirrer axis is fitted with four slotted stainless steel shaft and arms. For cooling purposes, the grinding chamber is also equipped with a water jacket for cooling. The shaft run at 100-600 rpm and can be used for either wet or dry grinding.

The grinding tests were carried out as a batch process in which samples were taken from the pot at the determined grinding time intervals. After each test, all of the media and ground samples were removed from the mill and the media were separated from the products by sieving. A series of experiments were carried out to investigate the effects of experimental parameters such as the grinding time, stirrer speed, powder filling ratio and ball filling ratio (J) on the particle size distribution (PSD) and surface area (S_v) of the products. A Sympatec HELOS (H1613) laser sizer was used for analysis of the feed and the ground products. Each test was repeated three times and the values reported are a mean average.

The primary aim of this study was to determine the optimum grinding conditions for producing the maximum surface area. The surface area is one of the basic properties of the powder and is generally represented by the total surface area of all particles contained in a unit mass of the powder [7]. The success of fine grinding depends on the selection of suitable parameters, and optimization of these parameters necessitates many tests. The total number of required experiments can be reduced by employing a factorially designed series, using the Yates technique [8].

DOI: 10.1051/e3sconf/20160801005

Figure 2. A stirred ball mill used in grinding tests.

Within these studies, the main and interaction effects of four variables on fine grinding were investigated. All 2^4 full factorial unreplicated experiments were carried out to evaluate the main and interaction effects of each variable in talc grinding. The Yates notation was used in this work to name each treatment. For example, treatment *abcd* was the experimental run in which the variables *a* (grinding time), *b* (stirrer speed), *c* (powder filling ratio), and *d* (ball filling ratio) were set to their high level. The factorial design of an experiment requires a series of parameter combinations in which all possibilities are considered by the variation of parameters. The number of experiments required is given by the equation $N=2^n$. In this equation, N denotes the number of experiments and n is the number of parameters. The number of parameters in this case is 4, corresponding to a total of 16 experiments. The parameters and the levels of each of the 2^4 designs, along with the actual and coded values of the parameters of the experiments, are presented in Table 2. The higher and lower levels are assigned +1 and −1 values, respectively, and zero represents the base level.

Table 2. Parameter codes and parameters values at different levels for experimental modelling.

Variables	Parameter code	Yates Symbol	Lower level (−1)	Base level (0)	Higher level (+1)
Grinding time (min.)	X_1	*a*	5	10	15
Stirrer speed (rpm)	X_2	*b*	300	450	600
Powder filling ratio (*fc**)	X_3	*c*	0 .03	0 .05	0 .07
Ball filling ratio (*J**)	X_4	*d*	0 .50	0 .60	0 .70

$$fc = \frac{\frac{\text{mass of powder}}{\text{powder density}}}{\text{mill volume}} \cdot \frac{1.0}{0.6} \tag{1}$$

$$J = \frac{mass\ of\ balls / ball\ density}{mill\ volume} \times \frac{1.0}{0.6} \ . \tag{2}$$

According to the basic principles of experimental design, three experiments were carried out at the base level, designated 0, to estimate the error and standard deviation. The tests are listed in Table 3 along with the alternative levels that were used in the experiments.

Table 3. Factor levels for each set of process variables.

Exp. no	Factor levels *a*	*b*	*c*	*d*	Experimental results S_v (m²/cm³)
1	−1	−1	−1	−1	2.14
2	+1	−1	−1	−1	2.62
3	−1	+1	−1	−1	2.27
4	+1	+1	−1	−1	2.98
5	−1	−1	+1	−1	2.01
6	+1	−1	+1	−1	2.28
7	−1	+1	+1	−1	2.06
8	+1	+1	+1	−1	2.48
9	−1	−1	−1	+1	2.28
10	+1	−1	−1	+1	2.89
11	−1	+1	−1	+1	2.50
12	+1	+1	−1	+1	3.34
13	−1	−1	+1	+1	2.13
14	+1	−1	+1	+1	2.53
15	−1	+1	+1	+1	2.22
16	+1	+1	+1	+1	2.82
BLR	0	0	0	0	2.47
BLR	0	0	0	0	2.47
BLR	0	0	0	0	2.50

3 Results and discussion

Within the study, the main and interaction effects of four parameters such as the grinding time, stirrer speed, powder filling ratio and ball filling ratio (J) on the specific surface area (S_v) of the ground product were investigated. In other words, the main goal was to determine the optimal set of variables that could be used in grinding to obtain the maximum surface area.

The Yates technique for the 2^4 experiments (where four is the number of parameters) was used for the statistical design and analysis of the results. The experimental conditions were arranged in the so-called Yates order. Table 4 combines the Yates technique with the analysis of variance method (ANOVA) to simplify the decision-making process on the significance of the parameters investigated.

The procedure for preparing Table 4 is as follows [9, 10].

1. The upper half of column 3 is obtained by adding successive pair responses, and the lower half is obtained by subtracting successive pairs. Columns 4, 5, and 6 are calculated in the same way.
2. Tests are repeated three times at the center points to estimate the error associated with determination of an individual response, which is required to test the statistical significance.
3. The ANOVA procedure is applied.
4. The table value of $F(1.16)$ for α=0.05 is compared with the calculated F value.
5. If $F_{Calculated}>F_{Table}$, then decision is effective. If $F_{Calculated}<F_{Table}$, then decision is ineffective.

Statistical models to predict the surface area can be built up using the Yates data analysis (Table 5). These models can be used to analyze the residues and check the assumption about the experimental error distribution of the factorial designs [11]. The regression equation with interactive terms can be written as follows:

$$Y= b_0 + b_1X_1 + b_2X_2 + b_3X_1X_2 + b_4X_3 + b_5X_1X_3 + b_6X_2X_3 + b_7X_1X_2X_3 + b_8X_4 + b_9X_1X_4 + b_{10}X_2X_4 + b_{11}X_1X_2X_4 + b_{12}X_3X_4 + b_{13}X_1X_3X_4 + b_{14}X_2X_3X_4 + b_{15}X_1X_2X_3X_4 \quad (3)$$

where Y = response (surface area (m^2/cm^3)), X_1 = grinding time (min), X_2 = stirrer speed (rpm), X_3 = powder filling ratio (*fc*) and X_4 = ball filling ratio. The models formed for surface area of talc using the effects of variable significant at either 95% or more confidence level are given below:

$$Y = 2.472 + 0.271X_1 + 0.112 X_2 + 0.051X_1X_2 - 0.156X_3 - 0.059X_1X_3 - 0.033X_2X_3 + 0.117X_4 + 0.036X_1X_4 + 0.019X_2X_4 \ . \quad (4)$$

The coefficient of determination, R^2, was used to check the models ability to predict the response of active ratio accurately. It is determined from the following equation:

$$R^2 = 1 - \left[\left\{ \Sigma\left(y_i - \hat{y}_i\right)^2 \right\} \left\{ \Sigma\left(y_i - \bar{y}_i\right)^2 \right\} \right] \quad (5)$$

where y^ is the predicted response variable and y– is the mean experimental value. If R^2 is 1, then the prediction is nearly perfect. However, if R^2 becomes zero, the model has little value. The empirical models were found to accurately estimate the response variable as indicated by R^2 value (0.9989 for S_v). The residual analyses for volumetric surface area are presented in Fig. 3.

The main effects of each variable on the volumetric surface area (m^2/cm^3) that are significant at the 95% (a<0.05) confidence level are presented in Fig. 4. The order of significance is *a*>*c*>*d*>*b*. The most dominant effect is *a* (grinding time) and its effect is positive. The effect of powder filling ratio (*c*) is the second most significant and it has a negative effect. The stirrer speed (*b*) and ball filling ratio (*d*) have also a positive effect. The *ab*, *ad* and *bd* interactions have positive effects, whereas that of *ac* and *bc* has a negative effect on the surface area of the ground product of grinding of talc. Figure 5 and Table 5 also show the comparison of *a*, *b*, *c*, *d*, *abd* and feed.

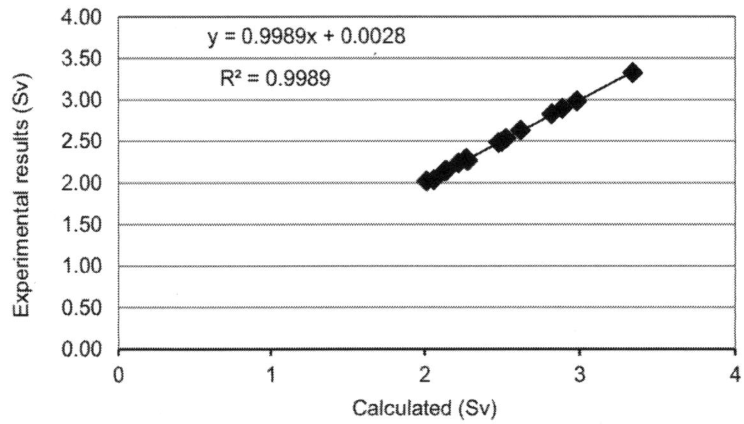

Figure 3. Residual analyses for volumetric surface area (S_v).

Table 5. Comparison of the results for some products.

	S_v (m^2/cm^3)	d_{10} (µm)	d_{50} (µm)	d_{90} (µm)
Feed	1.57	1.54	6.26	69.33
a	2.62	0.99	3.18	14.38
b	2.27	1.13	3.84	17.82
c	2.01	1.28	4.43	26.61
d	2.28	1.09	3.84	27.02
abd	3.34	0.82	2.33	7.09

DOI: 10.1051/e3sconf/20160801005

Table 4. Results of statistical analysis for response for volumetric surface area S_v that is volumetric surface area (m^2/cm^3); *DF* is degree of freedom; *a* is grinding time; *b* is stirrer speed (rpm); *c* is powder filling ratio (*fc*); *d* is ball filling ratio (*J*). Base level responses (*BLR*): (1) 2.47 m^2/cm^3, (2) 2.47 m^2/cm^3, (3) 2.50 m^2/cm^3

$$S_e^2 = \sum_{i=1}^{3} \frac{(BLR_i - BLR_{average})^2}{2} \quad S_e^2 = 0.0003.$$

Yates order	S_v	Column	Column	Column	Total effect [*TE*]	$[TE]^2/16$	*D.F.*	$F_{Calculated}$ $(7)/(8{*}S_e^2)$	F_{Table}	Decision (α=0.05)
1	2	3	4	5	6	7	8	9	10	11
1	2.14	4.76	10.01	18.84	39.55					
a	2.62	5.25	8.83	20.71	4.33	1.17180625	1	3906.020833	4.49	Effective
b	2.27	4.29	11.01	1.88	1.79	0.20025625	1	667.5208333	4.49	Effective
ab	2.98	4.54	9.70	2.45	0.81	0.04100625	1	136.6875	4.49	Effective
c	2.01	5.17	1.19	0.74	-2.49	0.38750625	1	1291.6875	4.49	Effective
ac	2.28	5.84	0.69	1.05	-0.95	0.05640625	1	188.0208333	4.49	Effective
bc	2.06	4.66	1.45	0.38	-0.53	0.01755625	1	58.52083333	4.49	Effective
abc	2.48	5.04	1.00	0.43	-0.11	0.00075625	1	2.520833333	4.49	Ineffective
d	2.28	0.48	0.49	-1.18	1.87	0.21855625	1	728.5208333	4.49	Effective
ad	2.89	0.71	0.25	-1.31	0.57	0.02030625	1	67.6875	4.49	Effective
bd	2.50	0.27	0.67	-0.50	0.31	0.00600625	1	20.02083333	4.49	Effective
abd	3.34	0.42	0.38	-0.45	0.05	0.00015625	1	0.520833333	4.49	Ineffective
cd	2.13	0.61	0.23	-0.24	-0.13	0.00105625	1	3.520833333	4.49	Ineffective
acd	2.53	0.84	0.15	-0.29	0.05	0.00015625	1	0.520833333	4.49	Ineffective
bcd	2.22	0.40	0.23	-0.08	-0.05	0.00015625	1	0.520833333	4.49	Ineffective
abcd	2.82	0.60	0.20	-0.03	0.05	0.00015625	1	0.520833333	4.49	Ineffective

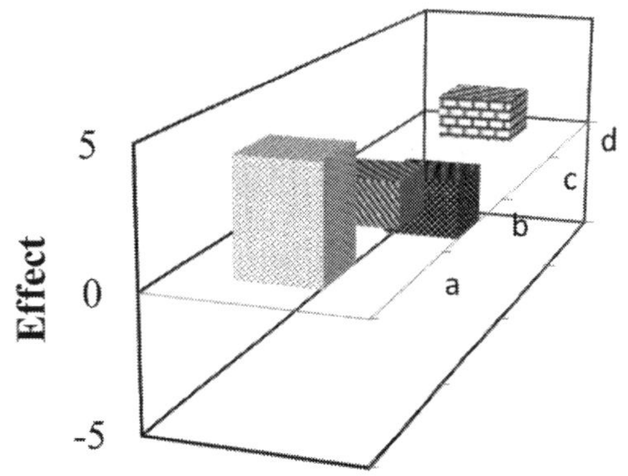

Figure 4. The main effects of each variable on the volumetric surface area (m^2/cm^3): *a* = grinding time (min), *b* = stirrer speed (rpm), *c* = the powder filling ratio (*fc*) and *d* = ball filling ratio.

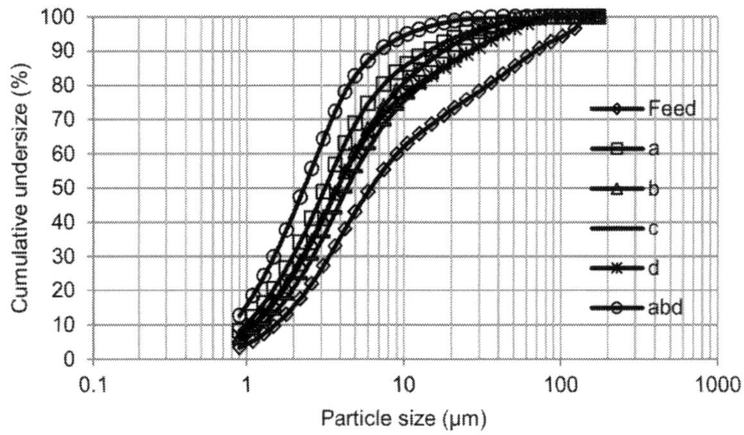

Figure 5. Particle size distributions of some products (*abd*: finest product).

4 Conclusions

As a result of statistically designed experiments performed, the optimum grinding conditions of talc (as a volumetric surface area of product obtained) in a stirred ball mill was determined to be; grinding time 15 min, stirrer speed of 600 rpm, powder filling ratio of 0.03 (*fc*), ball filling ratio of 0.70 (*J*). The most dominant and positive effect was the grinding time The effect of powder filling ratio was the second most significant and it had a negative effect. The stirrer speed and ball filling ratio (*J*) had also a positive effect. The volumetric surface area (S_v) by dry grinding using stirred ball mill was increased in comparison to the feed. The correlation coefficient of the model estimated at 95% confidence level with effective factors had a high value of 0.9989.

Acknowledgments

The author would like to thank MİCRON' S Company (Nigde, Turkey) for providing samples.

References

1. Mondo Minerals, www.mondominerals.com, (2016)
2. R. Sivamohan, P. Vachot, *Powder Technol.*, **61,** 119 (1990)
3. O. A. Orumwense, and E. Forssberg, *Scand. J. Merall.*, **19**, 201 (1990)
4. R. F. Conley, in G. Malghan (ed.), *112th AZME Meeting.* Atlanta, GA, AIME, 37, (1983)
5. N. Stehr, R. K. Mehta and J. A. Herbst, *Coal Prep.*, **4**, 209, (1987)
6. E. G. Davis, J. P. Hansen and G. V. Sullivan, in P. Somasundaran (ed.), *SME*, 74, (1980)
7. O. Y. Toraman, *Part. Sci. Technol.*, **31**, 205, (2013)
8. Yates, F., *Design and analysis of a factorial experiment* (UK: Imperial Bureau of Soil Science, 1976).
9. G. Özbayoğlu, M. Atalay, J. Alloy. Comp., **303–304**, 520 (2000)
10. E. Yoğurtcuoğlu, M. Uçurum, *Powder Technol.*, **214**, 47 (2011)
11. D. C. Montgomery, *Design and analysis of experiments* (Arizona State University, 1981)

Effect of process variables on oxidative pressurized acid leaching of copper electrorefining anode slimes

Sipi Seisko[1,a], Jari Aromaa[1], Petri Latostenmaa[2], Olof Forsen[1], and Mari Lundstrom[1]

[1]*Aalto University, Department of Materials Science and Engineering, Vuorimiehentie 2, 02150 Espoo, Finland*
[2]*Boliden Harjavalta Oy, Teollisuuskatu 1, 29200 Harjavalta, Finland*

Abstract. Extraction of Cu, Ni, As and Te from copper electrorefining anode slimes was studied. Such parameters as sulfuric acid concentration, temperature and oxygen partial pressure had a positive effect on the oxidative pressurized acid leaching in the decreasing order in contrast to the solids concentration. For nickel extraction, temperature, sulfuric acid concentration, combined effect of temperature and sulfuric acid concentration as well as solids concentration and sulfuric acid concentration had a positive effect, in a decreasing order. In the case of arsenic extraction, combined effect of temperature and sulfuric acid concentration, temperature as well as sulfuric acid concentration had positive in the decreasing order. During tellurium extraction, none of the parameters investigated gave the statistically valid response. Basing on the regression analysis, appropriate models for the dissolution of copper and arsenic could be build.

1 Introduction

Anode slimes contain a significant amount of precious metals, therefore, their processing has a substantial economic impact on plant profitability. The first step of the anode slime treatment process is decoppering, in which copper is typically removed from slimes by oxidative pressurized acid leaching [1-3]. Additionally, a major part of nickel is extracted simultaneously [2, 3]. The aim of this research was to define the process variables having the effect on copper and other metal extraction during oxidative pressurized acid leaching.

2 Experiments and materials

The process variables investigated were: temperature (100-160 °C), partial pressure of oxygen (2-8 bar), sulfuric acid concentration (8-16 vol.%) and solids concentration (200-400 g/dm^3). A series of factorial test with 18 experiments, described in Table 1, was performed in a 1.1 dm^3 titanium (Grade 2) autoclave using sulfuric acid media with oxygen purging.

Extraction of the various elements was calculated by comparing the elemental concentration of leach residue to the initial elemental concentration in the anode slime. As the mass of slime decreases during the leaching process due to the dissolution of soluble elements, the mass of the residue can be

[a] Corresponding author: sipi.seisko@aalto.fi

DOI: 10.1051/e3sconf/20160801006

determined by comparing the concentration of insoluble elements left after leaching to the initial concentration of elements in the raw material. For example lead can be used as a marker element since it does not dissolve during oxidative pressurized acid leaching as it is present as insoluble $PbSO_4$ [4]. Only traces of lead were found in the solution samples taken at the end of experiments confirming that lead remained in the residue. The initial composition of copper anode slime is presented in Table 2.

Table 1. Performed pressurized acid leaching experiments (Exp. No. 1-18) for anode slime.

Exp. No.	Temperature [°C]	Partial pressure of O_2 [bar]	Solids concentration [g/l]	Concentration of H_2SO_4 [vol-%]
1	100	2	200	8
2	160	2	200	8
3	100	8	200	8
4	160	8	200	8
5	100	2	400	8
6	160	2	400	8
7	100	8	400	8
8	160	8	400	8
9	100	2	200	16
10	160	2	200	16
11	100	8	200	16
12	160	8	200	16
13	100	2	400	16
14	160	2	400	16
15	100	8	400	16
16	160	8	400	16
17	130	5	300	12
18	130	5	300	12

Table 2. The composition of copper anode slime used as raw material in the leaching experiments.

Bi [mg/kg]	Se [mg/kg]	Te [mg/kg]	Ag [mg/kg]	As [mg/kg]	Ba [mg/kg]	Cu [mg/kg]	Ni [mg/kg]	Pb [mg/kg]	Sb [mg/kg]
36000	73000	26200	159000	62800	26200	188000	15400	106000	21500

3 Results and discussion

Leaching tests showed that certain elements were partially dissolved during autoclave leaching: Cu (67-91%), Ni (5-100%), As (67-89%) and Te (19-65%). The extraction recoveries of Cu, Ni, As and Te are presented in Table 3. The best extractions were achieved in experiment number 12 (i.e. T = 160 °C, $p(O_2)$ = 8 bar, [solids] = 200 g/l and [H_2SO_4] = 16 vol.%, in which the highest temperature, acid concentration and oxygen partial pressure combined with the lowest amount of solids were investigated.

Figure 1a shows the anode slime SEM image prior to leaching. Three main phases of anode slime were Cu-As-Pb oxide phase (spectra 1-3), Cu-As-Pb-Se-Te-Sb-Ag-S oxide phase (spectra 4-9) as well

DOI: 10.1051/e3sconf/20160801006

as Ag-Cu-Se-Te oxide phase (spectra 10-12). Larger Cu-As-Pb-Se-Te-Sb-Ag-S oxide inclusions and smaller Ag-Cu-Se-Te inclusions were surrounded by Cu-As-Pb oxide phase, which included traces of Ag, Sb, Te, S and Si. Additionally, the raw slime contained Cu-As-Bi-Se-Te-Sb-Ag-S oxide phase, silver-rich Cu-Se-Te-Ag phase, Cu-As-Se-Te-Sb-Ag oxide phase and $PbSO_4$ compounds.

Table 3. Extractions of Cu, Ni, As and Te after oxidative pressurized acid leaching experiments (Exp. No. 1-18).

Exp. No.	Cu [%]	Ni [%]	As [%]	Te [%]
1	69	n.a.	73	61
2	72	77	72	22
3	75	7	74	56
4	86	67	71	62
5	67	n.a.	70	n.a.
6	74	13	71	19
7	67	n.a.	68	34
8	74	19	69	52
9	78	n.a.	70	35
10	87	100	84	59
11	83	n.a.	70	49
12	91	100	89	65
13	72	5	68	30
14	85	100	79	47
15	78	n.a.	69	27
16	83	100	77	42
17	82	13	67	59
18	80	16	68	28

*n.a. – not available or analysis failed

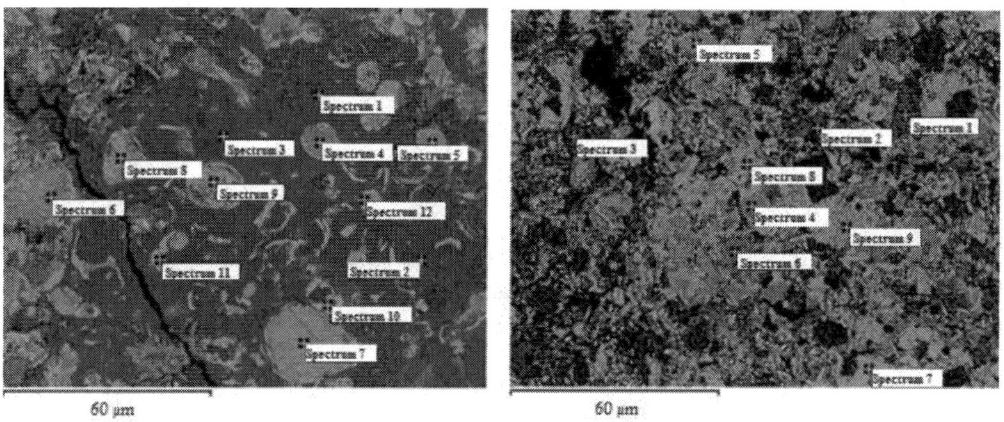

Figure 1. SEM-EDS analysis for copper anode slimes. Figure (a) on the left represents the unleached copper anode slime and (b) on the right represents leached (decopperized) anode slime (test 16).

The general morphology of the leached (decopperized) slime from test 16th is presented in Fig. 1b. For the decopperized anode slime, the main phases were Te-Sb-S oxide phase (spectra 1-3), Bi-As-Se-Te-Sb-Ag oxide phase with traces of Cu (spectra 4-6) and Se-Te-Ag phase with traces of Cu (spectra 7-9). Only traces of copper were shown to remain in the leached anode slime. Furthermore, Cu-As-Bi-Se-Te-Sb-Ag oxide phase, Se-Te-Ag and $PbSO_4$ phase were analyzed. Figures 1a and 1b show that the composition of slime changed, as the base metals were extracted and the nobler elements remained in the slime.

 DOI: 10.1051/e3sconf/20160801006

Several studies have investigated the effect of process variables on copper and nickel dissolution during decoppering [5-7]. In this work, the effect of process variables on extraction of Cu, Ni, As, Te was investigated using the factor and regression analysis. Oxygen partial pressure was shown to have positive effect only on the copper extractions, whereas temperature and sulfuric acid concentration had more significant impact. Figure 2 illustrates the effect of process variables on dissolution of copper. Weichert et al. [6] stated that temperature was the most significant parameter in the autoclave pressure leaching of anode slimes. However, according to this study, sulfuric acid concentration was the parameter having the greatest impact. Furthermore, Weichert et al. [6] suggested that other parameters such as the sulfuric acid concentration, partial pressure of oxygen and other oxidizing agents had only a minor effect on extraction of copper. This was in contradiction with the results of this study, since oxygen partial pressure was shown to significantly affect positively on copper dissolution. Additionally, Zhang and Yan [7] reached the copper extraction near 100% and the partial removal of tellurium and selenium as well as silver under following conditions: T = 99.85 °C, $p(O_2)$ = 8 bar, [solids] = 167 g/l and $[H_2SO_4]$ = 100 g/L with the optimal leaching time of 90 min. The suggested values of oxygen partial pressure and solids concentration were in line with the values that resulted in the best extraction of copper in this study. However, the temperature and sulfuric acid concentration were slightly lower.

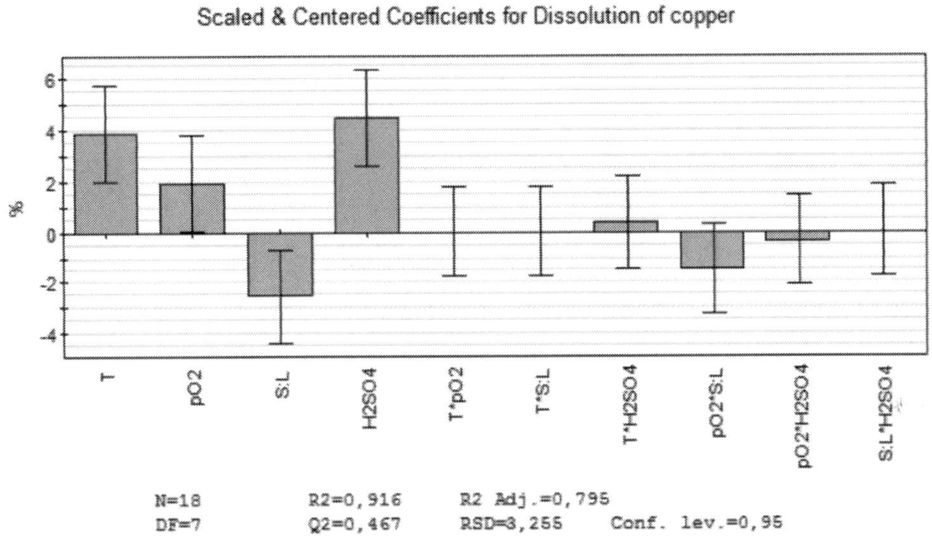

Figure 2. The effect of process variables on the dissolution of copper during oxidative pressurized acid leaching. T represents temperature, pO_2 the partials pressure of oxygen, S:L solids concentration, H_2SO_4 sulfuric acid concentrations, additionally, their combined effects are presented.

The variables affecting nickel extractions were in contradiction to the literature. Sulfuric acid concentration and temperature were found to be the only affecting variables as suggested by [5]. However, Doucet and Stafiej [5] argued that sulfuric acid had a major positive effect on nickel extraction. According to the results of this study, the temperature was shown to be the dominant variable. Weichert et al. [6] suggested that sulfuric acid concentration, partial pressure of oxygen and other oxidizing agents had only a minor effect on nickel dissolution. In the current study, the combined effect of temperature and sulfuric acid concentration as well as solids concentration and sulfuric acid concentration were shown to have a positive impact on nickel extraction.

The combined effect of temperature and sulfuric acid concentration was shown to have a positive action on arsenic extraction. Furthermore, temperature and sulfuric acid concentration affected positively the process. For tellurium extraction, none of the parameters were valid, because the statistical variation exceeded the actual effect. However, despite the major statistical variation, all singular variables, except solids concentration, seemed to have a positive effect. Moreover, based on the regression analysis, valid models could be built only for the dissolution of copper and arsenic.

4 Conclusions

Extraction of Cu, Ni, As and Te from copper electrorefining anode slimes was studied by oxidative pressurized acid leaching. For copper extraction, sulfuric acid concentration, temperature and oxygen partial pressure were shown to have a positive effect in the decreasing order, whereas solids concentration had a negative effect. For nickel extraction, temperature, sulfuric acid concentration, the combined effect of temperature and sulfuric acid concentration as well as solids concentration and sulfuric acid concentration had a positive effect in the decreasing order. For arsenic extraction, the combined effect of temperature and sulfuric acid concentration, temperature as well as sulfuric acid concentration had positive action on arsenic extraction in the decreasing order. For tellurium extraction, none of the parameters investigated gave statistically valid response due to the fact that variation exceeded the actual effect. Basing on the regression analysis, valid models for the dissolution of copper and arsenic could be built.

References

1. K. Pienimäki, H. Virtanen, in *Proceedings Copper* (2007), p. 301
2. J. E. Hoffmann, in *Proceedings EPD Congress* (2000), p. 397
3. B. C. Wesstrom, in *Proceedings EPD Congress* (2000), p. 503
4. T. T. Chen, J. E. Dutrizac,. Can. Metall. Quart. **28(2)**, 127 (1989)
5. M. Doucet, J. Stafiej, in *Proceedings Copper* (2007), p. 173
6. K. Weichert, T. Preiss, H. Kadereit, P. Stantke, G. Leuprecht, in *Proceedings Copper* (2013), p. 339
7. B. Zhang, L. Yan, in *Proceedings EMC* (2013), 969

Effect of process variables on synthesis of MgB_2 by a high energy ball mill

Haldun Kurama[1,a], Savaş Erkuş[1], and Hakan Gaşan[2]

[1] *Eskişehir Osmangazi University, Department of Mining Engineering, 26480, Eskişehir, Turkey*
[2] *Eskişehir Osmangazi University, Department of Metallurgical Engineering, 26480, Eskişehir, Turkey*

Abstract. The discovery of superconductivity of MgB_2 in 2001, with a critical temperature of 39 K, offered the promise of important large-scale applications at around 20 K. Except than the other featured synthesis methods, mechanical activation performed by high energy ball mills, as bulk form synthesis or as a first step of wire and thin film productions, has considered as an effective alternative production route in recent years. The process of mechanical activation (MA) starts with mixing the powders in the right proportion and loading the powder mixture into the mill with the grinding media. The milled powder is then consolidated into a bulk shape and heat-treated to obtain desired microstructure and properties. Thus, the important components of the MA process are the raw materials, mill type and process variables. During the MA process, heavy deformation of particles occure. This is manifested by the presence of a variety of crystal defects such as dislocations, vacancies, stacking faults and increased number of particle boundaries. The presence of this defect structure enhances the diffusivity of solute hence the critical currents and magnetic flux pinning ability of MgB_2 are improved. The aim of the present study is to determine the effects of process variables such as ball-to-powder mass ratio, size of balls, milling time, annealing temperature and contribution of process control agent (toluene) on the product size, morphology and conversion level of precursor powders to MgB_2 after subsequent heat treatment. The morphological analyses of the samples were performed by a high vacuum electron microscope ZEISS SUPRA VP 50. The phase compositions of the samples were performed with an Rigaku-Rint 2200 diffractometer, with nickel filtered Cu $K\alpha$ radiation and conversion level. The MgB_2 phase wt % was calculated by the Rietveld refinement method. The obtained results were discussed according to the process variables to find out their affect on the structure of MgB_2 and to determine the optimum synthesis condition. It was found that the ball-to-powder mass ratio of precursors mixed and the milling time remarkably affect the MgB_2 phase wt % and product size.

[a] Corresponding author: hkurama@ogu.edu.tr

The contributions of combustion of different hard steam coal size grades to overall potential mercury emission in Poland

Ireneusz Pyka[1,a], and Krzysztof Wierzchowski[1]

[1]*Glowny Instytut Gornictwa, Plac Gwarkow 1, 40-166 Katowice, Poland*

Abstract. Different groups of hard coal consumers use various coal size fractions. Therefore, knowing the content of mercury in different coal size fraction one can predict the potential contribution of various coal users to the overall mercury emission in Poland. The article emphasized the potential contribution of mercury emission from residential coal burning to the overall mercury emission in Poland. The base for the discussion is the results of research on mercury content in Polish commercial coal products. All hard steam coal sizes (coal classes) were under consideration. The results confirmed that under Polish conditions the contribution of the residential coal burning is smaller than presented in different papers and inventories concerning global mercury emissions.

1 Introduction

Anthropogenic emissions of mercury create global threats. They result from the both toxicity of compounds of mercury as well as from mobility of mercury in the environment and the lack of natural ways of mercury degradation. Different statistics, inventories and studies have emphasised a huge share of mercury emissions from the residential use of coal in the total anthropogenic mercury emissions for many years. It was estimated that around 43-44% of coal combustion leads to mercury emissions [1, 2, 3]. The latest publications undermine the previous approach. According to the UNEP publication [4], emission of mercury from coal general use of coal, including residential, is a minor share on the global scale but there no data were presented.

Pacyna et al. [3] gives the values of emission factors used for the assessment of mercury emission in 2005. For the residential coal use this factor is 300 μg/kg of coal used. This value can be regarded as relatively high but under Polish conditions one can find the justification for it. In the WHO IAQ Guidelines [5] one can find the values of mercury content in hard coals used in Poland in the residential sector. The range from 140-1780 μg/kg is attributed to such fuels.

All above confirm the need to recognize the real scale of content of mercury in Polish hard coal and the real mass (load) of mercury in the used coal. It will allow assessment of the real environmental risks as well as preventive measures to implement.

The residential sector in Poland is consuming annually 9-11 Mg of hard coal for the last few years. The sector shares around 13% of the total coal use in Poland. Coal secures around 29% of energy needs of residential sector in Poland [6]. It is the highest level among the EU countries (Table 1).

[a] Corresponding author: i.pyka@gig.eu

Table 1. Share of hard coal use in the final energy usage in residential sector in some EU – members (with significant coal use in the residential sector) [6].

Country	Coal use share,%
EU-27	3
Bulgaria	7
Ireland	7
Lithuania	3
Latvia	2
Poland	29
Hungary	2

For European Union member countries, excluding Poland, the share of coal in the final energy use in the residential sector is very small. For Poland alone, with the share of the residential sector in the coal domestic use amounting 13%, the share of this sector in overall mercury emission cannot be like in above mentioned documents [1, 2, 3].

The article presents the assessment of contributions of the combustion of different hard steam coal size classes to the overall mercury emission in Poland. Four groups of hard steam coal size classes are under consideration:

- coarse coal (e.g.: cobbles, coal nuts) - size fraction +25 mm,
- medium coal (pea coals) – size fraction 8-31.5 mm,
- small size fraction -20 (31.5/6) mm
- fine coal – about -1 mm.

Applying rough assumption that the coarse and medium coal classes as well as the finest coals are mainly used in the residential sector and smalls are used mainly in professional and industrial power plants, one can assess the contribution of combustion of hard steam coal split into such two groups of coal users to coal combustion resulting mercury emission in Poland.

2 Materials and methods

In 2014 all coal size grades produced in Polish collieries producing hard steam coal (excluding collieries belonging to Jastrzebska Spolka Weglowa S.A.) were sampled and data on their annual production were collected. It served to assess the weighed mean vales of mercury content in groups of coal size groups as well to assess the mass (load) of mercury “carried” by all coal size groups under consideration.

Gross samples of all coal sizes were collected in portions taken for one to two weeks, depending on local conditions. It was collected at the same time in routine sampling operations in all collieries under consideration. Sampling has conformed to requirements of the following standards:

- Hard coal and brown coal. Methods of sampling and laboratory samples preparation PN-90/G-04502
- Hard coal and coke - Mechanical sampling - Part 2: Coal – Sampling from moving streams PN-ISO 13909-2
- Hard coal and coke - Mechanical sampling - Part 3: Coal – Sampling from stationary lots PN-ISO 13909-3
- Hard coal and coke – Mechanical sampling – Part 4: Coal – Preparation of test samples PN-ISO 13909-4
- Hard coal and coke - Manual sampling PN-ISO 18283.

Mercury content was determined using the certified internal procedure, elaborated in Glowny Instytut Gornictwa No. SC-1.PB.23 applying the Cold-Vapor Atomic Absorption Spectrometry, using analyzer MA-2000 of Nippon Instrument Corporation. It is fully automated measurement system for determination of mercury content in solid materials, gases and liquids through sample combustion or evaporation.

The results of determinations of mercury content in coal in this article are given in μg/kg.

3 Results and discussion

A number of 183 samples of hard steam coal size grades were taken from 31 collieries in 2014. It was split into:

- 44 samples of coarse coal size with weighted mean mercury content: 64 μg/kg,
- 31 samples of medium coal size with weighted mean mercury content: 87 μg/kg,
- 78 samples of small coal with weighted mean mercury content: 81 μg/kg,
- 30 samples the finest coal with weight mean mercury content: 113 μg/kg.

The weighted mean mercury contents in Polish hard steam coal, for analysed coal size groups were relatively small and in the case of the coarse and medium size were very small. The latest could be explained by the fact that these are cleaned coal only. The average mercury content in hard coal in the global scale was assessed to be ca. 100 μg/kg [7]. Even if we assume that the whole mercury content will be emitted to the atmosphere during combustion process, the values of 64, 87 μg/kg, as well as 81 and 113 μg/kg are far much smaller than the value equal to 300 μg/kg.

Using the values of annual production of sampled hard steam coal size groups, the annual load (mass) of mercury for size classes under consideration were calculated. It was done for the production of each colliery separately. The results are shown in Figure 1.

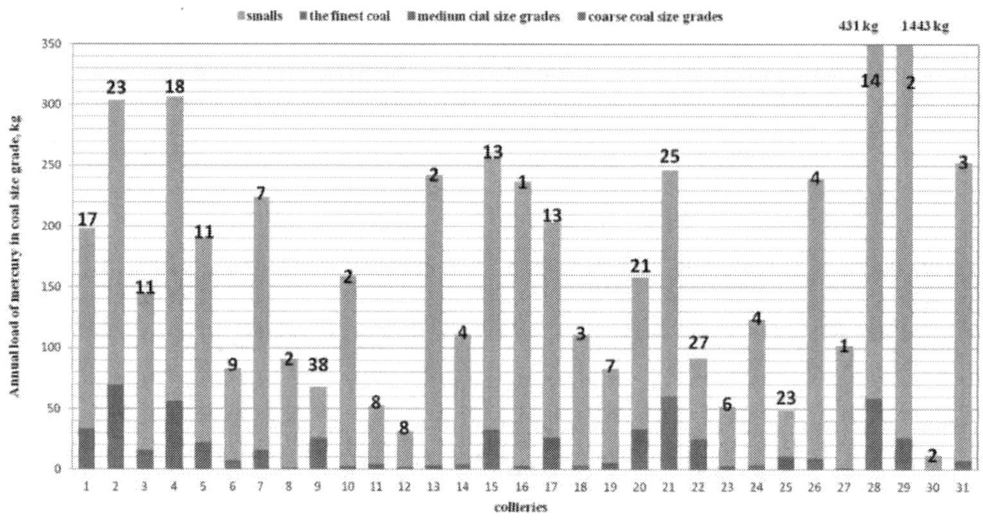

Figure 1. Annual load (mass) of mercury in coal size groups for hard steam coal produced in 31 Polish collieries in 2014.

The numbers in Figure 1 are the values of the share of mercury load in coarse, medium coal size groups and the finest coal being the basic fuel for the residential sector, in total load of mercury in hard steam coal produced in a given colliery. The mean value of this share for the total hard steam coal production in Poland in 2014 was 9%.

The load of mercury in hard steam coal sampled in 31 collieries in 2014 was around 6296 kg and it split into:

- coarse coal: 315 kg,
- medium coal: 198 kg,
- smalls: 5717 kg,
- the finest coal: 66 kg.

Only in the case of one colliery (number 9 in Figure 1) the share of mercury load in coarse, medium coal size groups and the finest coal in total load of mercury in hard steam coal produced in given colliery was high and equal to 38%. For the rest of collieries this share was much smaller and in some case it was equal only to 2%. The mean value as 9% showed that that was no reason, discussing the load of mercury in coal fuel dedicated to the different coal consumers, to foresee that emission of mercury from the residential sector can be comparable with the professional and industrial power plants.

The load of mercury does not mean its emission from coal combustion. It is well known that the mercury emission reduction does not require application of mercury dedicated method. As an accompanying effect, the reduction of mercury emissions occurs in other emissions reduction processes, of ash, SO_x and NO_x. There are many papers dealing with this topic and the effect of this reduction depends on many parameters, including coal properties [7, 8, 9]. The value 80% was taken for further analyses, although such high reduction occurs not in all professional and industrial power plants.

There are not so many works done to assess reduction of mercury emission from small boilers in the residential sector. However, even no environmental protection installations are used there, it is estimated that such reduction occurs and is assessed to be 50% [10].

Using values of mercury load in basic fuels for the residential sector as well as professional and industrial power plants and the above mentioned values of potential mercury emissions reduction, one obtains the possible values of mercury emissions for both sectors: 1143 kg of mercury emitted from professional and industrial power plants and 289 kg of mercury emitted from the residential sector in Poland. The first one is almost 4 times greater than the second. So, in the case of Poland, the country where the share of coal in the final energy usage in the residential sector is the highest among EU member countries, the relation between the emissions from both analyzed groups of coal consumers are far from internationally presented data [1, 2, 3].

4 Conclusions

1. Mercury emission has become the topic of international discussions as well as regulatory works. In many papers a great deal of uncertainty related to the data on mercury emission is emphasized. To await good results of regulatory works, the knowledge on mercury content in coal and mercury emission from coal combustion should be based on direct measurement results but not factors adopted from research results done in single location.
2. Hard steam coal use is not limited to electricity and heat generation in professional and industrial power plants. Coal is also used in the residential sector and combusted in small boilers without installations reducing the impact of such combustion on the environment (mainly the atmosphere). For Poland, where the use of coal in residential sector is significant, it was demonstrated that any inventories on the share of this sector in the global mercury emissions may be overestimated.
3. although hard coal is not the best fuel for the residential sector in Poland, the majority of coal combusted locally is a good quality fuel with low mercury content equal to less than 100 μg/kg.
4. The load of mercury in the coal fuel used in the residential sector was estimated to be around 580 kg per year. It is only 9% of the mercury load in hard steam coal produced annually in Poland. Due to the significant reduction of mercury emissions in the professional and industrial power plants, even no direct methods are applied, the comparison of emission of mercury from the residential sector, professional and industrial power plants will give different result than the comparison of

mercury loads. However, the mercury emission reduction occurs in residential sector and it is difficult to say that the mercury emissions in both sectors are comparable.

Acknowledgements

The results of determination of mercury commercial products (coal size classes) of the Polish hard coal collieries have been obtained from the project founded within the framework of the Applied Research Program of the Polish National Centre for Research and Development entitled *The elaboration of a data base on mercury content in domestic coal, technological guidelines of further mercury reduction, including the defining of benchmarks for domestic indices of mercury emission* – acronym "Hg Base" ("Baza Hg"), (PBS2/A2/14/2013).

References

1. UNEP-United Nations Environmental Programme. *Process Optimization Guidance for Reducing Mercury Emissions from Coal Combustion in Power Plants*, (Division of Technology, Industry and Economics (DTIE) Chemicals Branch, Geneva, Switzerland, 2010)
2. UNEP Chemical Branch. *The Global Atmospheric Mercury Assessment: Sources, Emissions and Transport*, (UNEP-Chemicals, Geneva, 2008)
3. E.G. Pacyna, J.M.Pacyna, K.Sundseth, J.Munthe, K.Kindbom, S.Wilson, F.Steenhuisen, P.Maxson, *Atmos. Environ.*, **44**, 2487 (2010)
4. UNEP-United Nations Environmental Programme. *Global Mercury Assessment 2013: Sources, Emissions, Released Environmental Transport*, (UNEP Chemicals Branch, Geneva, Switzerland, 2013)
5. WHO IAQ Guidelines. *Household fuel combustion* (2014).
6. Central Statistical Office. *Energy consumption in households*. Statistical Information and elaborations (Warsaw, 2014)
7. UNEP Expert Group on BAT/BEP, http://www.mercuryconvention.org/Negotiations/BATBEP ExpertGroup/CommentsonBATBEPguidance/tabid/4545/Default.aspx (2015)
8. J.H. Pavlish, E.A. Sondreal, M.D. Mann, E.S. Olson, K.C. Galbreath, D.L. Laudal, S.A. Benson, *Fuel Process. Technol.*, **82**, 89 (2003).
9. L.L. Sloss, *Legislation, standards and methods for mercury emissions control*, CCC/195 (London, IEA Clean Coal Centre, 2012)
10. S. Hlawiczka, K. Kubica, U. Zielonka, *Sci. Total Environ.*, **312**, 261 (2003)

Separation of copper flotation concentrates into density fractions by means of polytungstate aqueous solution

Andrzej Luszczkiewicz[1,a], Piotr Karwowski[1], Antoni Muszer[2], and Jan Drzymala[1]

[1]*Wroclaw University of Science and Technology, Faculty of Geoengineering, Mining and Geology, Wybrzeze Wyspianskiego 27, 50-370 Wroclaw, Poland*
[2]*Institute of Geological Sciences, University of Wroclaw, Pl. Maksa Borna 9, 50-205 Wroclaw, Poland*

Abstract. Industrial and laboratory flotation copper concentrates were subjected to separation into density fractions by means of heavy liquids in the form of sodium polytungstate aqueous solutions. For two samples, three densities factions were created, however in different density ranges. The density fractions were analyzed to establish the content of copper, lead, silver and organic carbon. The size of particles in both samples was similar (90-95% -0.071 mm). It was found that the lightest density fractions -2.45 and -2.0 g/cm^3 still contained sulfide minerals scattered in the organic carbon bearing particles. Removal of the lightest density fraction (-2.0 g/cm^3) from the industrial concentrate samples led to considerable reduction of organic carbon (92%) and increasing its quality from 13 to 28% Cu. The mineralogical analysis of the heavy liquid separation products showed that most sulfide minerals were evenly dissemination in the heaviest density fractions with the recovery of 95-98%. The lightest density fraction of -2.0 g/cm^3, being the richest in organic carbon, contained approximately 3% of unliberated sulfide minerals.

1 Introduction

One of laboratory separation methods used for mineralogical studies is gravity separation. This method provides density fractions of mineral matter with a high precision [1]. In mineralogical studies using gravity separation, heavy liquids are commonly applied to remove the heavy minerals fraction from the particulate matter as well as ground raw materials. In laboratory tests homogenous heavy liquids are applied. Gravity separation is also used on industrial scale using special suspensions called heavy media and the process is called heavy media separation.

Heavy media separation is especially useful for materials containing components of large differences in densities. The Kupferschiefer copper ore mined in Poland by KGHM Polska Miedz S.A. (shortly called KGHM) from the Lubin-Glogow Copper Basin (LGOM) belongs to this category because it contains both light carbonaceous shale and heavy sulfide minerals. The ore is of sedimentary-hydrothermal origin and contains three lithological fractions of different petrographic character, structure and mineralogical composition. These fractions are associated with three mineral

[a] Corresponding author: andrzej.luszczkiewicz@pwr.edu.pl

layers called sandstone, carbonate and shale. The ore mined by KGHM is a mixture of the three lithological fractions and is processed in concentrating plants by flotation. The final flotation concentrates contain copper-bearing sulfides, gangue minerals (which decrease the quality of concentrates) and organic carbon which increased amount creates many technological problems in metallurgical processing at KGHM. The aim of the present study is to separate the organic carbon bearing particles by means of the gravity method using a heavy liquid to be able to obtain organic carbon concentrates which can be next characterize by chemical and mineralogical analyses. In the present research we wanted to answer the question, whether or not it is possible to remove liberated organic carbon bearing particles from flotation concentrates. It was assumed that the organic material, which is of natural origin and consists of different types of fossil substances, has density below 2.0 g/cm^3, that is much less than the common rock-forming minerals having density above 2.3-2.5 g/cm^3. The density of most metal-bearing minerals is above 3.0 g/cm^3. Such materials in mineralogy and petrography are referred to as heavy minerals. All of the sulfide minerals present in the investigated copper ore have densities above 4.0 g/cm^3 [2].

Recently, organic heavy liquids were commonly used for heavy media separation for research purpose. Presently, the organic heavy liquids are not applied in industry due to their high toxicity and carcinogenic properties. In addition to that, the organic liquids are not suitable for gravity separation of organic substances. Since 1980s the polytungstate salts aqueous solutions, which are non-toxic and non-reactive, have been used for separation, especially for different organic substances present in soils [3, 4]. Also in sedimentology, the polytungstate salts solutions have been used as heavy media for separation of heavy minerals [5, 6].

To separate organic matter present in the copper ores of LGOM, which are mainly associated with black shale, similar methodology, as in the case of separation of organic matter from soil described by Morgun and Makarov [7], can be applied. In this work separation of organic ingredients from the soil required density of the heavy liquid $\geq$ 2.0 g/cm^3.

2 Materials and methods

Two samples of flotation copper concentrates were used in the investigations. Sample 1 was an industrial flotation concentrate from the Division of Concentrators in Lubin of KGHM and sample 2 which was a concentrate separated by laboratory flotation experiments using the copper ore processed at the Division of Concentrators in Lubin. Table 1 shows the particle size distribution of both samples, while the content of Cu, Ag, Pb and C_{org} in the investigated samples is given in Table 2.

Aqueous solutions of sodium polytungstate ($3Na_2WO_4 \cdot 9WO_3 \cdot H_2O$, SPT) were used as a separation medium. This salt is able to provide density up to 3.1 g/cm^3. The solution densities were monitored by using a hydrometer.

The methodology of heavy liquid separation was based on the recommendations of Lewis and McConchie [8] discussed in details elsewhere [2]. Due to fine particle size of the samples and their slow natural sedimentation in a highly viscous polytungstate liquid, separations was supported by using a centrifuge which guarantees greater gravitational acceleration.

The samples were placed in centrifuge glass tubes and after thorough wetting by the heavy liquid, the suspension was centrifuged for 10 minutes at a speed of 3100 rpm. Then, the tubes were stored for 24 hours to achieve complete separation of the floating and sinking fractions. Each separated fraction was cleaned by using distilled water and then dried at 70 °C.

The separation operations started with the highest density liquid, and then the lighter separated fraction was subjected to subsequent separation in a less dense heavy liquid. Sample 1 was separated in two liquids having density of 3.00 and 2.45 g/cm^3, while sample 2 was separated in liquids with density of 2.30 and 2.00 g/cm^3. The obtained fractions were subjected to chemical analysis performed by CBJ sp. z o.o. to determine content of copper, silver, lead, and organic carbon. Additionally, sample 2 was subjected to the mineralogical analysis. The fractions separated in the heavy liquid were

 DOI: 10.1051/e3sconf/20160801009

embedded in a resin and polished. The preparations were subjected to microscopic examination in the reflected light. The content of the sulfide minerals in the density fractions taking into account the size of the particles was determined.

Table 1. Particle size distribution of investigated samples.

Size fraction, mm	Sample 1, %	Sample 2, %
+ 0.071	11.18	6.76
0.040-0.071	25.06	27.30
+0.020-0.040	19.24	24.30
– 0.020	44.53	41.64
Feed	100.00	100.00

Table 2. Content of Cu, Ag, Pb and Corg in investigated samples.

	Cu, %	Ag, ppm	Pb, %	C_{org}, %
Sample 1	12.87	455	5.42	9.44
sample 2	9.93	492	4.93	7.57

3 Results and discussion

Table 3 summarizes the results of gravity separation of the industrial concentrate sample obtained from the Lubin Concentrator (sample 1), while Table 4 shows the results of separation using the concentrate produced by laboratory flotation of the copper ore provided by the Lubin Concentrator (sample 2). The results in Table 3 show that removal of organic carbon from the industrial concentrate by using gravity separation in heavy liquids would allow for a significant increase of its quality with relatively small losses in the yield. More than 88% (as recovery) of copper goes to the heaviest fraction (> 3.0 g/cm^3) having the content of Cu equal to 28%. This faction contains mostly ore minerals with a small amount (1.5%) of organic carbon.

Removal of approximately 54% of organic carbon bearing particles (as recovery, fraction G1, Table 3) and mixing G3 and G2 fractions would improve the grade of concentrate from about 13 to over 17% of Cu with 4% of Cu losses. The fraction enriched in the organic carbon would contain over 18% of C_{org}. The results indicated that for the industrial copper concentrates, despite fine sizes of particles (approximately 65% of the particles below 0.040 mm, Table 1) there was a great number of organic carbon bearing particles containing probably a certain amount of very fine dispersed particles of copper sulfides. However, it can be concluded, the content of copper and organic carbon in the fraction G3 equal to 28% and 1.5%, respectively, indicates the physical possibility of significant improvement of the Lubin flotation concentrates.

The aforementioned findings confirm the results of separation for the laboratory concentrate (sample 2, Table 4) performed in the heavy liquids at lower ranges of density. The main purpose of the experiments was separation of as light as possible fraction, in which one can expect the absence of heavy minerals, such as sulfides. As previously stated, the common rock-forming minerals have a density much larger than 3.0 g/cm^3.

The data on the density fraction -2.0 g/cm^3 show, that despite a significant enrichment in the organic carbon (grade 23% C_{org}), it still contains about 1% of Cu, 150 ppm silver and 1.5% lead. Although this sample particle size distribution was similar to the industrial concentrate (more than 65% of the particles below 0.040 mm, Table 1), the laboratory concentrate contained less coarse particles (+ 0.071 mm). Therefore, it can be concluded that the degree of liberation of sulfide minerals in both samples was similar. The heaviest fraction in this series of tests with density of > 2.3 g/cm^3 contained 11.64% Cu with recovery close to 98% and 4.65% C_{org} with 65% recovery. It indicates that in the density fraction >2.3 g/cm^3 there is a significant amount of unliberated organic carbon bearing particles intergrown with sulfide minerals.

Table 3. Results of gravity separation of industrial copper concentrate (sample 1), γ - weight percent (yield), λ - component content, ε - component recovery.

Density fractions g/cm^3	yield	Cu		Ag		Pb		C_{org}	
	γ, %	λ, %	ε, %	λ, ppm	ε, %	λ, %	ε, %	λ, %	ε, %
G3 (>3.00)	43.03	27.95	88.24	958	77.69	7.74	72.89	1.51	7.72
G2 (2.45-3.00)	32.02	3.40	7.99	224	13.52	2.42	16.96	10.13	38.52
G1 (<2.45)	24.95	2.06	3.77	187	8.79	1.86	10.16	18.14	53.76
calculated feed	100.00	13.63	100.00	531	100.00	4.57	100.00	8.42	100.00
assayed feed		12.87		455		5.42		9.44	

Table 5 summarizes the balance of ore minerals according to mineralogical analyses of the fractions separated by means of polytungstate heavy liquid from the laboratory concentrate (sample 2). Table 5 shows the balance, and independently, the sum of the identified ore minerals and separately major sulfide minerals present in the concentrate. As can be seen from Table 3, almost all copper sulfide mineral particles were transferred, with the recovery of 97-98%, to the heaviest >2.3 g/cm^3 density fraction. A total of about 3% of the ore minerals (recovery) remained in the 2.00-2.30 g/cm^3 density fraction and lightest density fraction < 2.00 g/cm^3. The content of the ore minerals in the lightest fractions was approximately 5%. There is no doubt that this should be attributed to dispersion of the unliberated sulfides particles in the organic carbon bearing particles. In accordance with the previously cited literature, the highest content of organic carbon in the tested concentrates was in the density fractions below 2.00 and below 2.30 g/cm^3.

A small amount of the lightest fraction (Tables 4 and 5) having density of <2.00 g/cm^3 indicates a lack in the tested flotation concentrates particles sulfide minerals liberated from organic carbon bearing particles. Taking into account relatively small particles of the raw investigated samples, separation by means of physical methods gives products which would be enriched in organic carbon and not containing dispersed sulfides seems to be impossible.

4 Conclusions

Gravity separation tests were performed using two samples of copper flotation concentrates with similar particle distributions. More than 65% of particles were -0.040 mm in size. Both samples were subjected to separation into the density fraction by means of heavy liquids prepared as aqueous solutions of sodium polytungstate at different ranges of density. Sample 1 was separated in heavy liquids into heavier densities fractions and the goal was to extract minerals of the highest density. Based on the obtained results, it was concluded that the Cu content equal to 28% (fraction G3 in Table 3) was reached when the content of C_{org} was 1.5%. This indicates on a possible significant improvement of the currently produced Lubin concentrates.

Table 4. Results of gravity separation of laboratory copper concentrate (sample 2), γ - weight percent (yield), λ - metal (component) content, ε - metal (component) recovery.

Density fractions g/cm^3	**γ, %**	**Cu**		**Ag**		**Pb**		**C$_{org}$**	
		λ, %	**ε, %**	**λ, ppm**	**ε, %**	**λ, %**	**ε, %**	**λ, %**	**ε, %**
G3 (>2.30)	82.74	11.64	97.89	558	94.71	5.24	94.30	4.65	65.96
G2 (2.00-2.30)	13.23	1.28	1.72	150	4.07	1.52	4.37	8.04	18.23
G1 (<2.00)	4.03	0.94	0.39	148	1.22	1.51	1.32	22.88	15.82
calculated feed	100.00	9.84	100.00	488	100.00	4.60	100.00	5.83	100.00
assayed feed		9.93		492		4.93		7.57	

Table 5. Balance of separation according to content of ore minerals (sample 2).

Density fractions g/cm^3	**yield**	**ore minerals**		**bornite**		**chalcocite +digenite**		**chalkopyrite**		**covellite**	
	γ, %	**λ, %**	**ε, %**	**λ, %**	**ε, %**	**λ, %**	**ε, %**	**λ, %**	**ε, %**	**λ, %**	**ε, %**
G3 (>2.30)	82.74	35.08	96.92	6.54	97.19	3.18	98.31	10.54	98.71	1.33	96.64
G2 (2.00-2.30)	13.23	5.69	2.51	0.94	2.22	0.30	1.48	0.71	1.07	0.22	2.61
G1 (<2.00)	4.03	4.21	0.57	0.81	0.59	0.14	0.21	0.49	0.23	0.21	0.75
calculated feed	100.00	29.95	100	5.57	100	2.67	100	8.83	100	1.14	100
Density fractions g/cm^3	yield	ore minerals		galena		pyrite + marcasitet		sphalerite		tennatite + tetrahedrite	
	γ, %	λ, %	ε, %	λ, %	ε, %	λ, %	ε, %	λ, %	ε, %	λ, %	ε, %
G3 (>2.30)	82.74	35.08	96.92	5.81	94.10	6.73	96.22	0.37	91.57	0.59	96.11
G2 (2.00-2.30)	13.23	5.69	2.51	1.75	4.52	1.42	3.25	0.20	8.00	0.15	3.89
G1 (<2.00)	4.03	4.21	0.57	1.74	1.38	0.77	0.54	0.04	0.43	0.00	0.00
calculated feed	100	29.95	100	5.11	100	5.79	100	0.33	100	0.51	100

Sample 2 was separated in heavy liquids of lower density, to produce the density fraction <2.0 g/cm^3. This density was considered in the literature as the limit of determination of organic carbon bearing particles. The fraction of this density of the tested flotation concentrates contained mainly particles having a high organic carbon content which exceeded 20%. These particles contained scattered unliberated fine particles of sulfide minerals in the amount of approximately 5%. This fraction represented less than 20% of the total mass of the concentrate.

The gravity separation experiments using the flotation concentrates and heavy liquids indicate that for the used particle size distribution, it is not possible to remove the organic carbon bearing particles without very fine sulfide mineral particles.

Acknowledgement

The authors would like to thank Dr. Przemyslaw B. Kowalczuk for his valuable remarks. Financial support by the Wroclaw University of Technology Statutory Research Grant 0401/0186/15 is also greatly acknowledged.

References

1. G.S. Berger, I.A. Yefimov I.A., *Metody vydelenya monomineralnykh fraktsii. Izd. 2, Gosgeoltekhizdat* (Бергер Г.С., Ефимоф И.А., Методы выделения мономинералных фракции. Изд. 2, Госгеолтехиздат, Москва 1962)
2. A. Luszczkiewicz, *Monograph No. 36* (Prace Naukowe Instytutu Gornictwa PWR, 2002)
3. N.M. Savage, *J. Micropaleontol.*,**7(1)**, 39 (1988)
4. D. Munsterman; S. Kerstholt, *Review of Palaeobotany and Palynology*, **91(1-4)**, 417 (1996)
5. J.A. Callahan, *J. Sedimentary Res.*, **57(4)**, 765 (1987).
6. J.A. Commeau, L.J. Poppe, R.F. Commeau, *U.S. Geolog. Survey Circular*, No. 1071, 1 (1992)
7. E.G. Morgun, M.I. Makarov, *Eurasian Soil Sci.*, **44(4)**, 394 (20110
8. D.W Lewis, D. McConchie, *Analytical Sedimentology* (Chapman and Hall, London, 1994)

Linking impact-related progeny sizes of cement clinker to modes of single-particle breakage

Mahmut Camalan[1,a], and Çetin Hoşten[1]

[1]*Department of Mining Engineering, Middle East Technical University, 06800 Ankara, Turkey*

Abstract. Fragmentation of particulate solids is an important process in many industrial activities, particularly in the mineral, cement and glass industries to comminute the raw materials to a required size. Understanding of processes and mechanisms responsible for fragmentation of particulate solids is a hard task. This may bring great precision for designing and controlling the comminution process. This work aims to contribute to understanding fragmentation of cement clinker particles under impact loading. For this purpose, six narrow-size classes of Portland clinker were fragmented with varying specific impact energy levels in a drop weight test apparatus. Then, breakage probabilities and functions of these size classes were determined from the product size distributions. The calculated breakage functions show evidences for the modes of breakage (meridian cracks to oblique cracks and shattering) given in the literature. Results of this study indicate that self-similarity of progeny size distributions can only be achieved, regardless of particle size and impact energy, if the particles have the same breakage probability, that is the same mode of breakage.

1 Introduction

When a brittle material is subjected to an impact, the resultant mode of breakage is dependent either on the magnitude of the impact energy or the impact velocity. Four modes of breakage have been proposed in this respect: *(i)* the Hertzian crack system [1–4]; *(ii)* meridian (diametrical)/secondary cracks [1, 2, 5–10]; *(iii)* oblique/radial cracks [1, 2, 9]; and *(iv)* shattering or catastrophic failure [1, 2, 11]. The Hertzian crack system are formed preferentially at low-strain rates, involving the cone and undeveloped ring cracks. Additional increase in the either strain-rate or the impact velocity results in an additional inelastic deformation zone, i.e. cone of fines, under contact area. If the rate of strain is sufficient to increase this cone volume to a critical level, tensile stresses prevail around the contact circle, forming and propagating meridian cracks [12] which fracture the solid like slices of orange. It should also be emphasized that either secondary or branching cracks, commonly perpendicular to the meridian fractures, may form just after the completion of meridian cracks [6, 7, 13]. When the applied strain rate is further increased, the cone of fines become, larger and oblique/radial cracks replace meridian cracks. As a result, fine fragments and a coherent large cone formed. The large coherent cone becomes gradually shorter and narrower with increasing strain-rate [1]. If the strain rate is excessive, failure fronts are said to propagate throughout the particle, leading to either intense

[a] Corresponding author: mcamalan@metu.edu.tr

fragmentation or shattering with no large, coherent fragments left [1, 11]. Although the shattering mechanism has not been explicitly defined in the literature, this may be physically similar to failure of brittle solids observed under very high compressive loads where failure waves reduced spall (tensile) and shear strength of brittle material, drastically causing inelastic, shear-induced micro-cracking [14–16]. King [17] defines the breakage mode "shattering" as a series of fracture steps, induced by a rapid application of compressive stress, in which there is a sequential re-breakage of successive generations of daughter fragments until all of the energy available for fracture is dissipated. However, re-breakage events have not been sufficiently elucidated within the context of single-particle studies because of the difficulty in monitoring the fragmentation sequence in an extremely short duration of impact loading. There might still be some indirect evidence from those studies that could be linked to re-breakage. For example, the reason for the secondary crack formation was related to the unbalanced stresses in high-velocity, single-particle impact [6, 7] and to the continued loading on meridian-fracture-related fragments by compression tester [1, 10]. Therefore, despite the limited information, it may be possible that the aforementioned breakage patterns inherently include re-breakage events. Kapur et al. [18] also supported re-breakage of progeny particles and stated that re-breakage at high impact energy levels provide self-similar product size distributions.

Although the abovementioned studies widely explain the breakage phenomena of brittle solids, there is generally lack of clear mathematical relations between any breakage mode and its corresponding fragment distribution. The objective of this paper is to find qualitative evidences for the modes of breakage mentioned above, through impact-related breakage parameters and self-similar progeny distributions. For this purpose, drop-weight tests were performed on six single-size classes (particle sizes ranging from 25.4 to 3.35 mm) of Portland cement clinker. Each size class was stressed with various impact energy levels, and the corresponding product size distributions of each energy level were decomposed to breakage probability and progeny size distributions, both in discrete and cumulative forms, for further evaluation.

1.1 Breakage probability and breakage function

For any given single-size class of particles, two important breakage parameters can be generated from the product size distributions of single-particle impact: (a) breakage probability and (b) breakage function. The former can be defined as the cumulative mass fraction of single-size particles which are broken upon application of a specific impact energy (J/g, kJ/kg or kWh/t). Impact breakage function can be represented in two different forms: the discrete breakage function, denoted as b_{i1}, and the cumulative breakage function, denoted as B_{i1}. The former is the mass percentage (or fraction) of material broken from a single-size feed (indexed as '1') which, upon breakage, goes to a finer size interval '*i*':

$$b_{i1}=p_i/P_2, \text{ where } i>1 \tag{1}$$

where p_i is the mass percentage of product retained in size class '*i*', and P_2 is the cumulative mass percentage of breakage product passing from the upper sieve of size interval '2'. Equation 1 can be graphically shown by plotting b_{i1} values against the geometric mean sizes of the corresponding intervals '*i*'. This function represents the discrete size distribution of progeny particles. The cumulative breakage function, B_{i1}, is the cumulative mass fraction of progeny fragments broken from size 1, which appears less than the upper size of size interval '*i*':

$$B_{i1}=P_i/P_2, \text{ where } i>1 \tag{2}$$

where Pi is the cumulative mass percentage of materials passing from the upper sieve of size interval '*i*'. Obviously, this will yield the cumulative size distribution of progeny particles. Equation 2 can be graphically represented by plotting each B_{i1} value against the corresponding upper sieve size of size interval '*i*'.

2 Materials and methods

Six size fractions (-25.4+19.0, -19.0+12.7, -12.7+9.53, -9.53+6.35, -6.35+4.7 and -4.7+3.35 mm) of Portland cement clinker were used as the experimental material. The oxide contents of each size fractionswere measured by a Spectro IQ X-ray fluorescence (XRF) spectrometer and the results are given in Table 1. These results showed that the chemical compositions of the tested size classes were very similar, inferring material homogeneity in all of the tested size fractions.

Table 1. Chemical composition of the experimental single-size clinker fractions.

Clinker Constituents	Percentage in the size fractions					
	-25.4+19.0 mm	**-19.0+12.7 mm**	**-12.7+9.53 mm**	**-9.53+6.35 mm**	**-6.35+4.7 mm**	**-4.7+3.35 mm**
CaO	66.39	65.51	67.35	64.62	67.78	64.93
SiO_2	19.12	19.77	18.59	20.02	17.35	20.33
Fe_2O_3	3.62	3.62	3.82	3.53	3.63	3.69
Al_2O_3	5.13	5.56	4.62	5.82	5.36	5.75
SO_3	1.59	1.33	1.51	1.31	1.34	1.26
MgO	2.28	2.37	2.03	2.43	1.32	2.43
LOI (Loss-on-Ignition)	0.02	0.01	0.01	0.01	0.02	0.01

The drop-weight test was used to perform single-particle breakage tests. The test consists of dropping a cylindrical steel weight from a known height to a single particle placed on a steel base plate. The specific impact energy E_{is} (kJ/kg) imparted on to a single-size class of particles can be determined by varying the drop weight and the drop height using Eq. 3.

$$Eis = [0.00981 \cdot M \cdot (h_0 - h_f)]/m \quad (3)$$

where M is the mass of the drop weight (kg), h_0 is the initial drop height (mm) measured from the surface of anvil to the bottom of the drop weight, h_f is the offset height (mm) between the bottom of the drop weight and surface of the anvil after impacting on the particle, and m is the average weight of a particle in the set of tested particles (g). For each experiment, 1, 2 or 20 kg drop-weights along with drop-heights ranging from 0.03 to 1 m were used to produce a wide range of E_{is}. The experimental set of specific impact energies and the number of particles stressed for each size class is given in Table 2.

Table 2. E_{is} levels and the number of particles stressed for each sample class.

Fracions (mm)	E_{is} (kJ/kg)	Number of particles stressed
-25.4+19.0	0.04, 0.18, 0.40, 0.79, 1.48, 2.12, 3.17, 4.46, 6.26	20-50
-19.0+12.7	0.11, 0.22, 0.40, 0.79, 1.58, 3.17, 6.44, 8.89	150-180
-12.7+9.53	0.11, 0.40, 0.79, 1.58, 2.23, 3.17, 5.40, 6.12, 9.18	100-200
-9.53+6.35	0.36, 0.79, 1.98, 3.31, 7.96, 15.66	200-250
-6.35+4.7	0.83, 1.62, 3.17, 6.26, 12.24, 18.36, 24.77	250-350
-4.7+3.35	1.94, 3.92, 7.85, 11.95, 17.68, 24.08, 27.40	300-350

3 Results and discussion

Figure 1 shows the distribution of discrete breakage functions for all of the size classes at a range of specific impact energies. The distributions are left-skewed (the long tail on the left) with a distinct major peak located at the coarse end of the breakage product size range. The peak becomes shorter and wider, and shifts towards the finer sizes with an increase in the specific energy input. From the fragmentation pattern viewpoint, the sequential transition from meridian/secondary cracks to oblique/radial cracks and shattering adequately explains this variation in the shape of the distributions since the meridian/secondary cracks forming at low specific energy inputs are expected to give mainly a few coarse fragments of similar size. Whereas oblique/radial cracks and shattering at relatively higher energy inputs are expected to produce many fragments having a wide size range. In fact, a similar variation in the fragment size distribution was also observed by Salman et al. [19] in the transition from low to high velocity impact fragmentation of aluminum oxide particles.

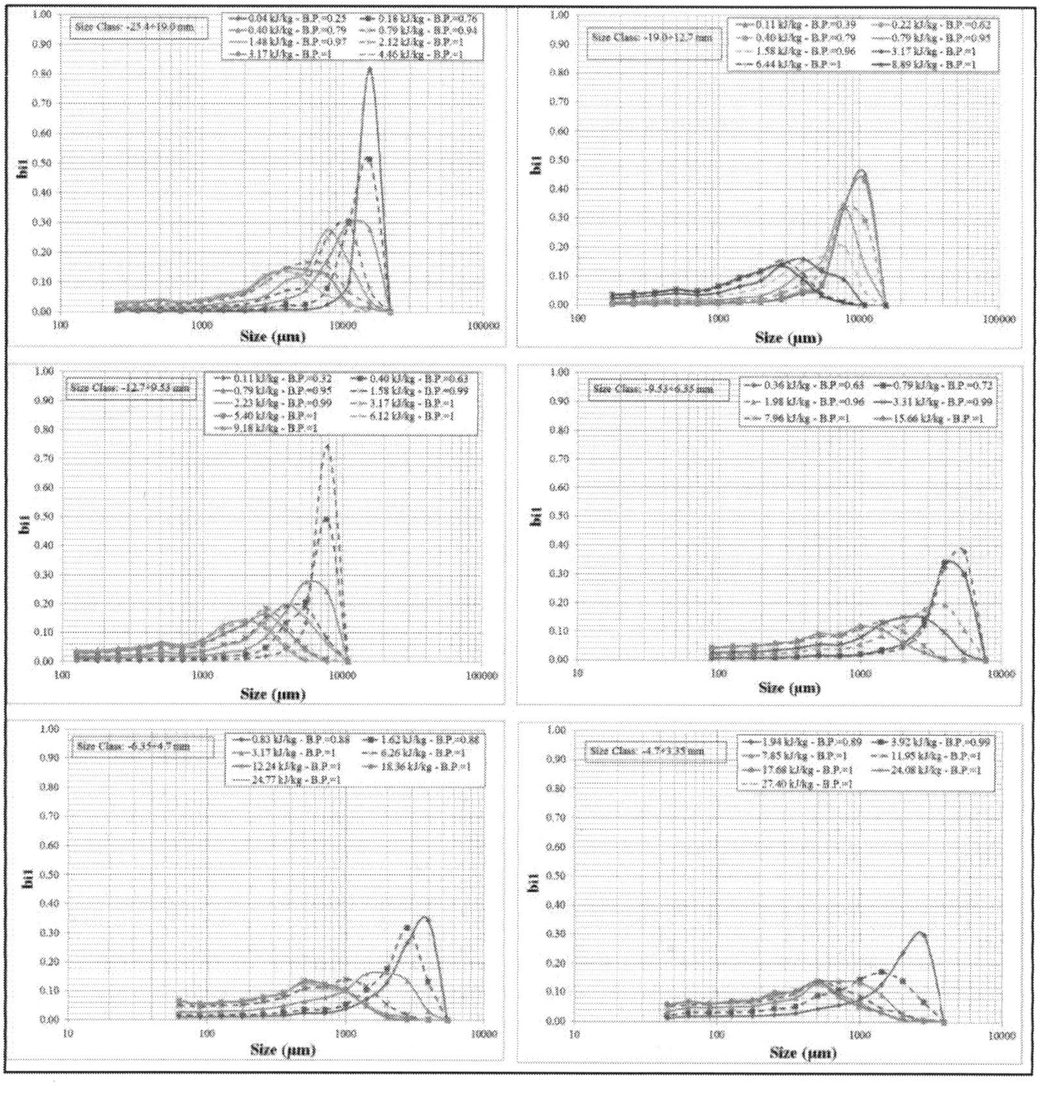

Figure 1. Discrete breakage functions of all size classes. B.P.= breakage probability.

DOI: 10.1051/e3sconf/20160801010

Figure 2 also shows that the discrete breakage functions ($b_{il}s$) approach a limiting distribution around 500 μm, irrespective of the parent particle size if the specific energy input is increased above 12 kJ/kg. The limiting distribution may arise by the following reasons: First, the progeny will become so small that successive fracturing might be difficult. Interestingly, Yashima et al. [20] reported in an earlier study on single-particle impact breakage of natural materials that the specific fracture energies increased rapidly for particles less than 500 μm. Secondly, successive breakage events eventually form a pile of fragments between the flat surfaces of the anvil and the drop weight, such that particle-bed breakage, rather than single-particle breakage, prevails either during the later stages of the primary impact or during the impacts after rebounds of the drop weight.

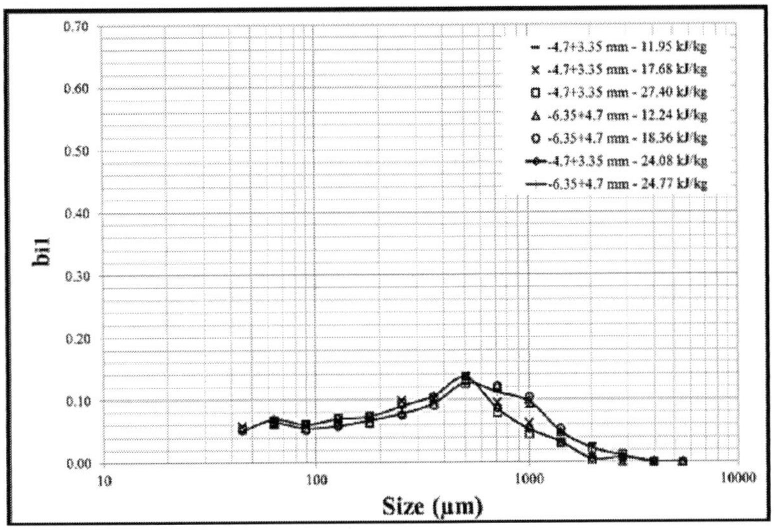

Figure 2. The discrete breakage functions remaining fixed for the specific energy inputs higher than 12 kJ/kg.

As the mode of breakage for a single particle is dependent to a large extent on the applied impact energy, it seems logical that the size distribution of progeny fragments must be non-similar depending on the mode of breakage the mother particle undergoes. Still, it is noteworthy that even though the mother particle size is different, the size distributions of progeny fragments change in the same manner (Figure 1). This brings out the question whether a particle, irrespective of its size, will pass through the same, distinct modes of breakage (from meridian cracking to oblique/radial cracks and shattering) as the breakage energy increases. A graphical means of testing this could be to plot a set of cumulative breakage functions obtained at different levels of the impact energy against a normalized size, that is, a particle size rescaled by the median fragment size (d_{50}) of the breakage function. If these set of normalized curves from such plots fall onto either the same curve or a narrow band, then the distributions are called self-similar reflecting similar breakage patterns [21]. Figure 3 shows semi-logarithmic plots of cumulative breakage functions against the normalized size for the whole set of drop-weight breakage data obtained under various combinations of mother particle size and applied specific impact energy. The plots show that the cumulative breakage functions are non-similar, indicating different breakage patterns depending on the prevailing breakage mode. However, if the data are re-plotted in groups of similar breakage probabilities (Figure 4), we obtain self-similar distributions regardless of mother particle size and the specific impact energy, indicating similar breakage patterns. In fact, this finding was previously supported by Vogel and Peukert [22] such that the similar breakage pattern leads to the same breakage probability for geometrically similar and physically identical particles.

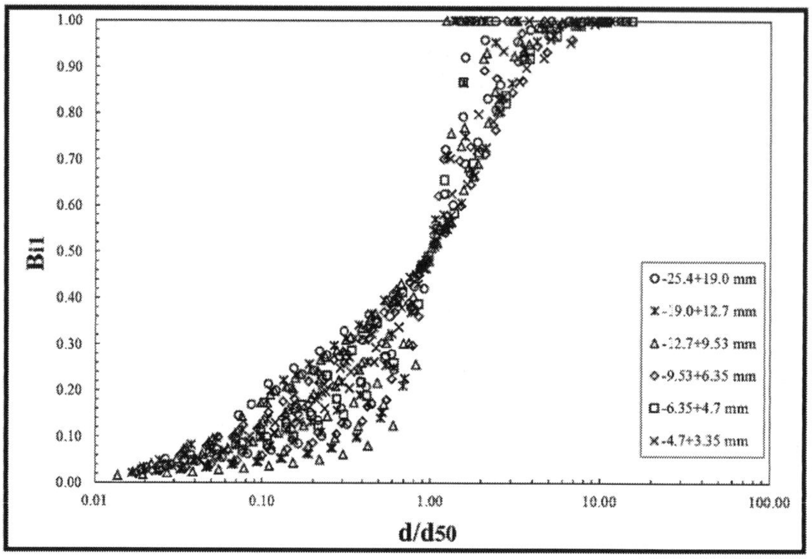

Figure 3. Normalized cumulative breakage functions of all size-classes at different impact energy levels.

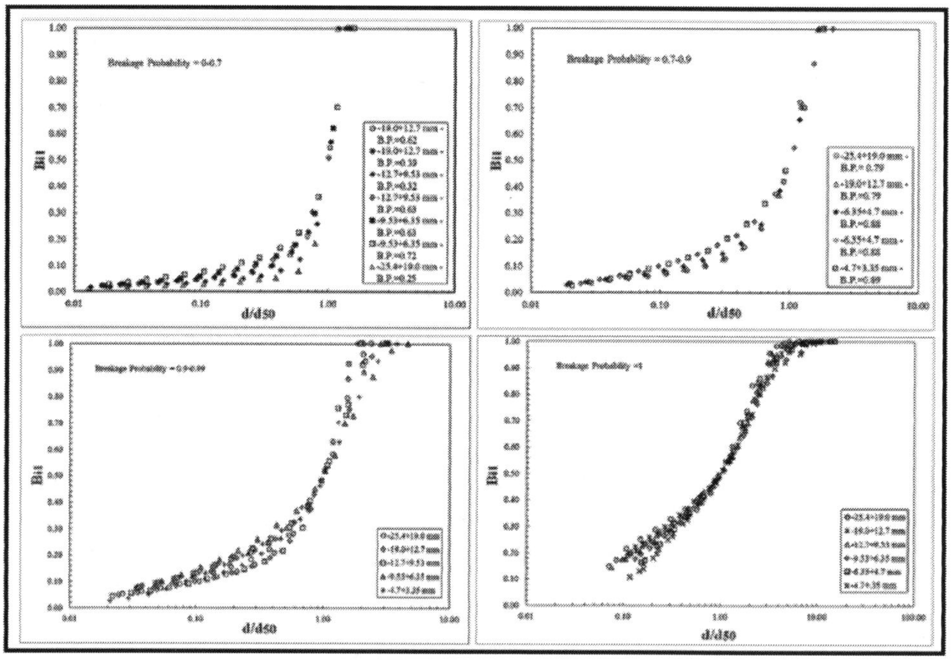

Figure 4. Normalized cumulative breakage functions of all size-classes re-plotted in groups of different breakage probabilities (B.P.= Breakage Probability).

4 Conclusions

Single particle breakage of cement clinker particles produces different progeny distributions or breakage functions where the change in their shapes shows evidence for the distinct modes of breakage (meridian cracks to oblique/radial cracks and shattering). Irrespective of the mother practice size, the shapes of the discrete breakage function distributions approach a limited distribution for specific energies higher than 12 kJ/kg, irrespective of the mother particle size. Self-similar, impact-related progeny size distributions can only be achieved if the particles either have the same breakage probability for any energy input or they are broken in the same mode of breakage.

References

1. D.A. Gorham, A.D. Salman. *Wear* **258**, 580 (2005)
2. D.G. Papadopoulos, *PhD thesis* (University of Surrey,1998)
3. S. Middlemiss, *Int. J. Miner. Process.*, **84**, 207 (2007)
4. D.G. Papadopoulos, M. Ghadiri, *Adv. Powder Technol.*, **7**, 183 (1996)
5. K. Schönert, *Powder Technol.*, **143-144**, 2 (2004)
6. W. Schubert, M. Khanal, J. Tomas, *Int. J. Miner. Process*, **75**, 41 (2005)
7. M. Khanal, W. Schubert, J. Tomas, *Int. J. Miner. Process*, **86**, 104 (2008)
8. S. Antonyuk, J. Tomas, S. Heinrich, L. Mörl, *Chem. Eng. Sci.*, **60**, 4031 (2005)
9. J. Subero, M. Ghadiri, *Powder Technol.*, **120**, 232 (2001)
10. S.Z. Wu, K.T. Chau, T.X. Yu, *Powder Technol.*, **143-144**, 41 (2004)
11. E.W. Andrews, K.-S. Kim, *Mech. Mater.*, **31**, 689 (1999)
12. M.M. Chaudhri, *Powder Technol.*, **143-144**, 31 (2004)
13. S. Antonyuk, M. Khanal, J. Tomas, S. Heinrich, L. Mörl, *Chem. Eng. Prog.*, **45**, 838 (2006)
14. R.J. Clifton, *Appl. Mech. Rev.*, **46,** 540 (1993)
15. N. Bourne, J. Millett, Z. Rosenberg, N. Murray, *J. Mech. Phys. Solids*, **46**, 1887 (1998)
16. D.D. Radford, G.R. Willmott, S.M. Walley, J.E. Field, in *Proceedings J. de Physique IV*, **110**, 687 (2003)
17. R.P., King, *Modeling and Simulation of Mineral Processing Systems* (Oxford: Butterworth-Heinmann Publications, 2001)
18. P.C. Kapur, D. Pande, D.W. Fuerstenau, *Int. J. Miner. Process*, **49**, 223 (1997)
19. A.D. Salman, C.A. Biggs, J. Fu, I. Angyal, M. Szabo, M.J. Hounslow, *Powder Technol.*, **128**, 36 (2002)
20. S. Yashima, Y. Kanda, S. Sano, *Powder Technol.*, **51**, 277 (1987)
21. D.W. Fuerstenau, P.C. Kapur, *Powder Technol.*, **82,** 51 (1995)
22. L. Vogel L, W. Peukert, *Powder Technol.*, **129**, 101 (2003)

Morphology of flotation froth for enrichment of coal in saline water

Anna Mlynarczykowska[1,a], Klaudia Tupek[1], and Konrad Oleksik[1]

[1]*AGH University of Science and Technology, Faculty of Mining and Geoengineering, Department of Environmental Engineering and Mineral Processing, Poland*

Abstract. Gas bubbles have an important influence on the efficiency of flotation. The use of mine saline water in coal preparation plants affects the quality of froth. This paper shows the results of morphological analysis of froth generated in flotation enrichment process performed in technological waters with varying degrees of pulp salinity. The raw material model subjected to flotation tests was coal (type 33 in Polish classification). The appearance of froth was registered and evaluated by using a vision system with a special software. The aim of the study was to determine the size and number of air bubbles formed in the presence of inorganic electrolyte ions in the process water as well as their relationship with the efficiency of enrichment process.

1 Introduction

Flotation is a upgrading method based on physical and chemical differences in surface properties of separated materials [1-2]. In flotation an effective attachment of a particle to a gas bubble occurs due to elementary phenomena that is collision and formation of a stable particle- bubble aggregate [3]. It should be noted that turbulent motions in flotation cell may lead to detachment of particle from a bubble [4-5]. Each aforementioned elementary process has a fundamental impact on the process quality [6].

Gas bubbles play an important role in flotation process as they lift the mineral particles to the pulp surface, forming a froth layer. In the froth zone the raw material is concentrated and separated into useful components and gangue. The froth is a dispersion system [7] which can be formed as a result of either condensation or dispersion of air. In case of froth flotation, it is formed as a result of air dispersion, creating a three-phase system consisting of liquid, gas dispersed in the liquid and solids attached to the gas. Formation of the froth is possible with the presence of surfactants in the pulp which keep the floated particles on the surface, thus giving time to collect them [3-4]. The presence of surfactants in the pulp is the main factor determining formation of a stable froth. They allow to increase the viscosity of solution which reduces liquid leak speed from the froth films (the Plateau theory) and increases the surface flexibility of air bubbles (the Gibbs theory). Nonetheless, it should be noted that the froth stability decreases when an inorganic electrolyte is added to the suspension.

[a] Corresponding author: mindziu@agh.edu.pl

The electrolytes reduce the film stability, and thus the stability is also decreased when oval and large particles are present in the suspension [7].

The pulp is aerated in a pneumo-mechanical flotation machine by a rotating agitator placed at its bottom or by feeding compressed air to the cell [8]. Depending on the liquid content in the froth, the froth can be either dry or wet. The best separation of raw material in flotation is achieved at maximum mineralization of air bubbles [4], thus the wet froth is better since it yields more mineral particles. It should also be noted that the upgrading process effectiveness is related to the gaseous phase properties such as size and distribution of formed bubbles as well as volume and concentration of gas flowing through the flotation cell in a unit of time [9]. The probability of floating particles attached to the gas bubbles increases with decreasing bubble size [10]. Consequently, it can be said that the most advantageous degree of pulp aeration occurs at the optimal saturation of the pulp with gas bubbles [10].

The other factor which affects the flotation effectiveness is the type of process water because the flotation pulp contains 80-85% of water [11]. Water from underground mines is often used in processing plants, and it has a significant concentration of inorganic salts. The stability of formed flotation froth, size of formed air bubbles as well as content of solid phase and water are affected by different physical and chemical properties and hydrodynamic conditions of the process which thus determine the effectiveness of the process [1]. The impact of inorganic electrolytes on the effectiveness of hard coal flotation has been studied for years [12-16]. It is important to understand the influence of salts dissolved in a process water on the flotation results.

One of the elements which allows to effectively control the flotation process is the use of flotation froth image analysis. Traditionally, the flotation process was controlled by operators of flotation machines who controlled the froth existence and stability. However, an effective verification is not easy [18]. There are some papers which provide that the flotation froth image analysis and visualization techniques are effective tools to determine the upgrading efficiency in order to control the industrial processes [18-20]. Banford et al. [19] and Citir et al. [20] verified the possibility of addition a suitable amount of flotation reagent to the hard coal upgrading process by controlling the size and coalescence of air bubbles. On the other hand Jahedsaravani et al. [18] conducted tests to find a relationship between the froth appearance and process efficiency by analysing the images of froth that formed during the coper ore upgrading. The studied parameters included the froth colour, size and velocity of air bubbles. Similar tests were conducted at Divisions of Concentrator of KGHM Polska Miedz S.A. where the dedicated system for technological process control on the basis of froth image analysis was developed and tested. The operation assessment of the visual system used confirmed the improvement of both technological and economical indices at the ore upgrading plant [21-22].

The paper presents the results of flotation processing of hard coal in the presence of process water with varying content of sulfate(VI) ions as well as the analysis of photographs of flotation froth formed during the process. The analysis of images was used to interpret the size and quality of air bubbles in relation to the obtained results.

2 Materials and methods

The tests were conducted on class 33 hard coal which was mechanically ground before flotation to obtain the particle size below 0.5 mm. The flotation tests were performed in a Denver pneumatic-mechanical laboratory machine with 1 dm^3 volume. In each flotation test the rotor speed was constant and equal to 1850 rpm, while the pulp density was 70 g/l [17]. Using the constant rotor speed guaranteed dispersion of air bubbles in the flotation cell, while the low solid content in the pulp allowed to obtain better results [1].

The variable parameter for individual flotations was a degree of pulp salinity expressed as the content of sulfate(VI) ions . The tests were performed on model water solutions of analytical grade sulphuric acid(VI) of various concentrations from 0.063 to 0.5 M [17].

The floatation reagent used during the test was an aqueous solution of hexanol with concentration of 2 mM [17]. In flotation hexanol play a role of both frother and collector. Its adsorption on the solid-liquid and liquid-gas interfaces increases the coal hydrophobicity and prevents the coalescence of air bubbles in order to form a stable flotation froth as well as reduces a velocity of air bubbles to the froth layer [1, 23].

The weighed coal samples were mixed with water for 3 minutes. Then, hexanol was added and mixing continued for another minute to ensure adequate reagent adsorption on the coal surface. After this time the air was introduced to the cell and the flotation process started. The concentrates were collected after 30, 30, 60, 60 and 120 s. The obtained products were dried in a laboratory dryer and weighed in order to determine the yield. However, due to inefficient drying of concentrates caused by increased sorption of sulfates(VI) on the coal surface, the finest fraction below 0.056 mm had to be sieved and the flotation tests had to be repeated. As a result, the flotation time increased to 6 minutes due to improved upgrading effects. The obtained concentrates and tailings were analysed for the ash content with the Polish Standard PN-ISO 1171: 2002.

For each flotation the digital photographs were taken of the formed froth which morphology changed depending on the concentration of sulfuric(VI) acid in flotation pulp. The experimental set-up is presented in Fig. 1.

The pictures with the best quality were analysed. The bubbles were marked manually. Figure 2 shows the flotation froth before marking the measurement objects, while Fig. 3 shows the same image with bubbles marked for analysis.

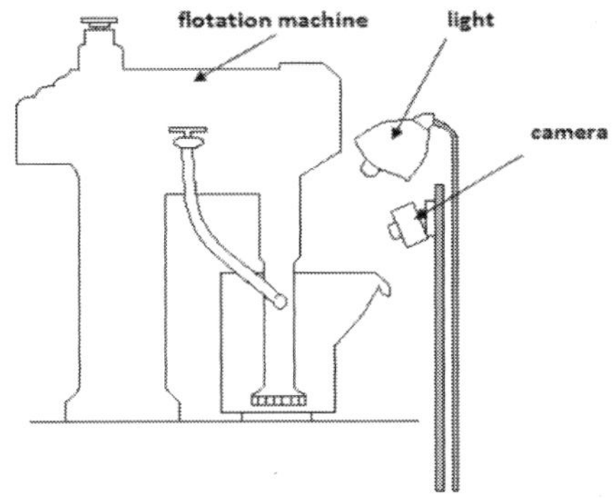

Figure 1. Experimental set-up.

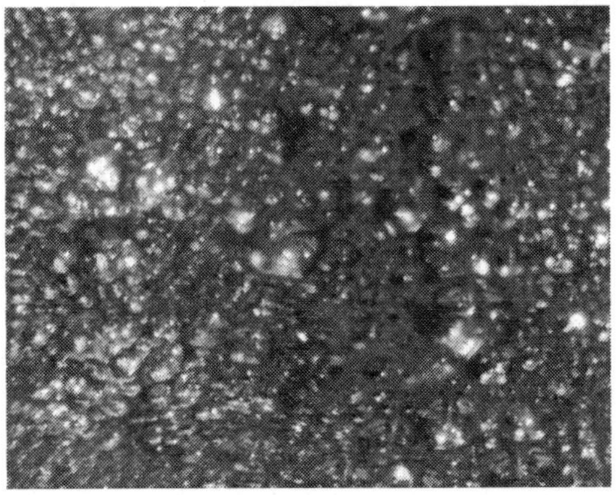

Figure 2. Image of actual flotation froth.

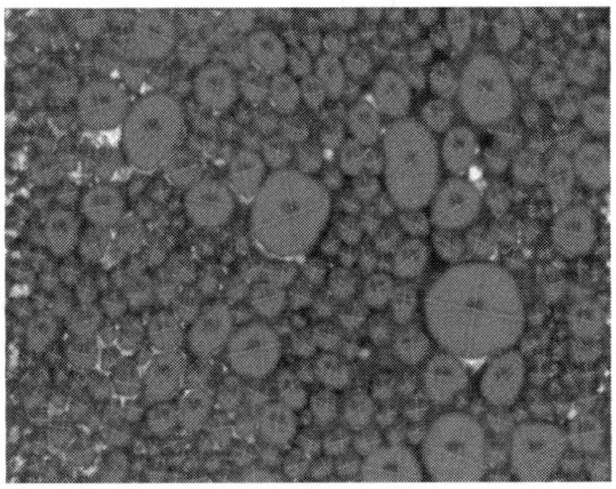

Figure 3. Image of flotation froth with analysis of air bubbles.

3 Results

The analyses allowed to calculate the surface area, perimeter and Feret diameter of individual air bubbles formed during the hard coal upgrading in saline process water. In addition, the software generated statistics for three above-mentioned parameters: mean, standard deviation, standard error, 95 and 99% confidence intervals, and maximum and minimum values of each parameter. Tables 1 and 2 present results for successive flotation with different pulp salinities. Figures 4-8 show the software-processed images of the selected fragment of the froth photograph on which the individual air bubbles were analysed.

The obtained data were used to plot the Feret diameter histograms (Figs. 9-13.) of air bubbles from individual flotation tests along with the selected match of probability distribution. Tables 3-4 present selected coal upgrading parameters from individual flotation tests.

Table 1. Measurements air bubbles for flotation with 0.125 and 0.063 M H_2SO_4.

	salinity of the water 0.063M H_2SO_4			salinity of the water 0.125M H_2SO_4		
	Surface area [pixel^2]	**Ferets diameter [pixel]**	**Perimeter [pixel]**	**Surface area [pixel^2]**	**Feret diameter [pixel]**	**Perimeter [pixel]**
Mean value	947.58	108.86	32.23	932.97	106.25	31.4
Standard deviation	941.09	43.876	12.97	1108.99	48.22	14.22
Standard error	42.65	1.988	0.588	51.15	2.224	0.656
95% confidence interval	83.79	3.907	1.154	100.52	4.371	1.289
99% confidence interval	110.29	5.14	1.519	132.311	5.753	1.697
Measurements number	487	487	487	470	470	470
Min.	94.0	39.79	10.94	78.0	34.042	9.97
Max.	9395.0	372.38	109.37	11923.0	419.75	123.21

Table 2. Measurements air bubbles for flotation with 0.125, 0.375 and 0.5 M H_2SO_4.

	salinity of the water 0.25 M H_2SO_4			salinity of the water 0.375 M H_2SO_4			salinity of the water 0.5 M H_2SO_4		
	Surface area [pixel^2]	**Feret diameter [pixel]**	**Perimeter [pixel]**	**Surface area [pixel^2]**	**Feret diameter [pixel]**	**Perimeter [pixel]**	**Surface area [pixel^2]**	**Feret diameter [pixel]**	**Perimeter [pixel]**
Mean value	1031.13	110.61	32.62	815.01	102.6	29.99	950.20	109.12	32.14
Standard deviation	1312.66	54.14	15.8	792.58	40.32	11.76	956.59	44.80	13.30
Standard error	58.7	2.42	0.707	33.89	1.72	0.503	39.25	1.838	0.546
95% confidence interval	115.34	4.76	1.39	66.57	3.39	0.988	77.09	3.61	1.07
99% confidence interval	151.8	6.26	1.83	87.6	4.46	1.30	101.43	4.75	1.41
Measurements number	500	500	500	547	547	547	594	594	594
Min.	70.0	30.39	9.44	63.0	31.21	8.96	95.0	36.39	10.99
Max.	13903.0	443.20	133.05	8766.0	360.88	105.65	9476.0	370.53	109.84

DOI: 10.1051/e3sconf/20160801011

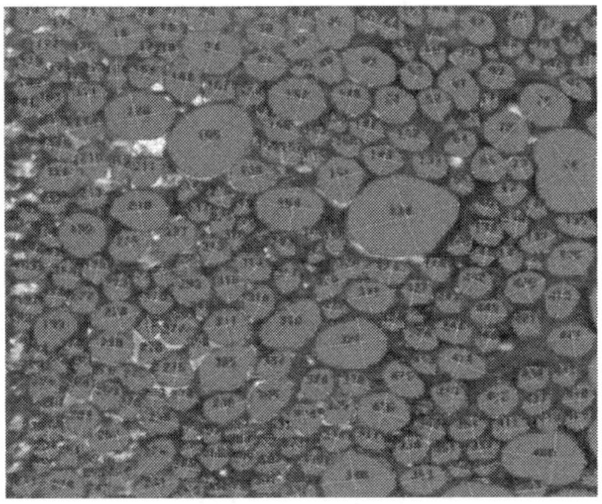

Figure 4. An image of the marked air bubbles for flotation with 0.063 M H_2SO_4.

Figure 5. An image of the marked air bubbles for flotation with 0.125 M H_2SO_4.

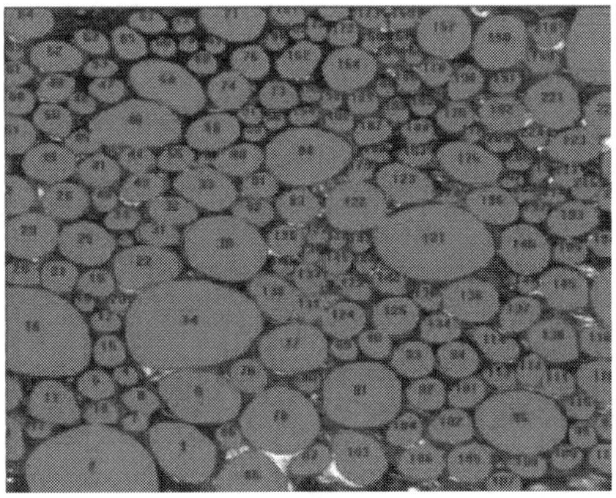

Figure 6. An image of the marked air bubbles for flotation with 0.25 M H_2SO_4.

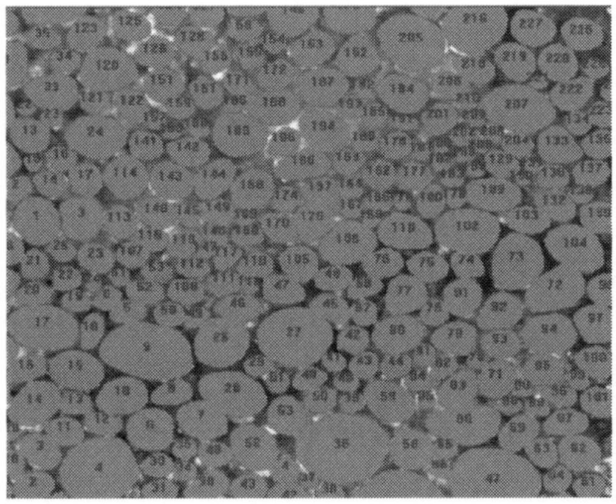

Figure 7. An image of the marked air bubbles for flotation with 0.375 M H_2SO_4.

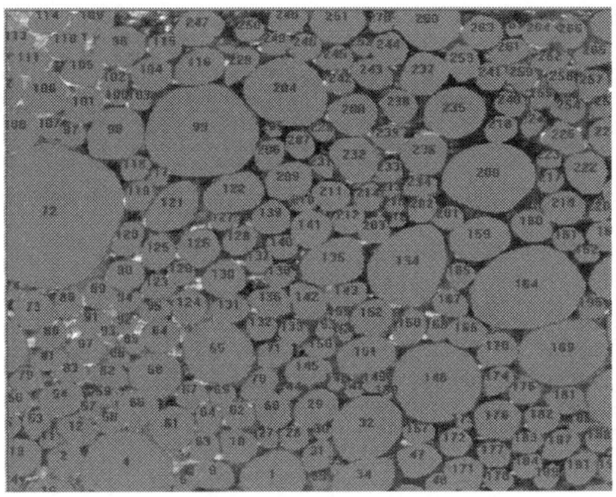

Figure 8. An image of the marked air bubbles for flotation with 0.5 M H_2SO_4.

 DOI: 10.1051/e3sconf/20160801011

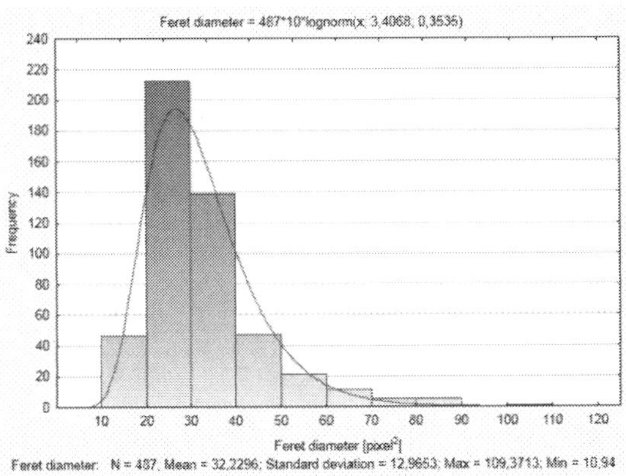

Figure 9. Histogram of Feret diameters for air bubbles together with the adjustment of the log-normal probability distribution for flotation with 0.063 M H_2SO_4.

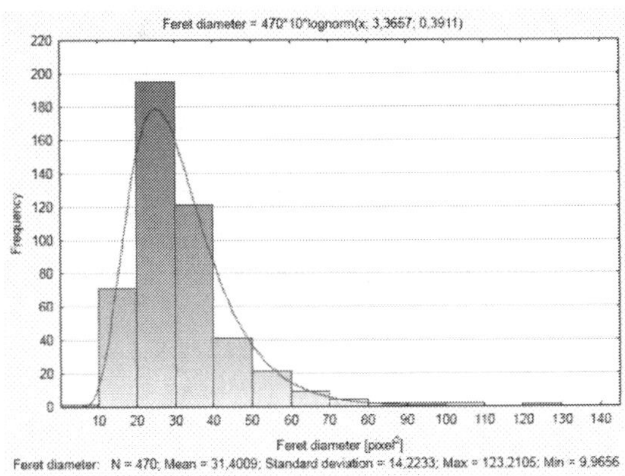

Figure 10. Histogram of Feret diameters for air bubbles together with the adjustment of the log-normal probability distribution for flotation with 0.125 M H_2SO_4.

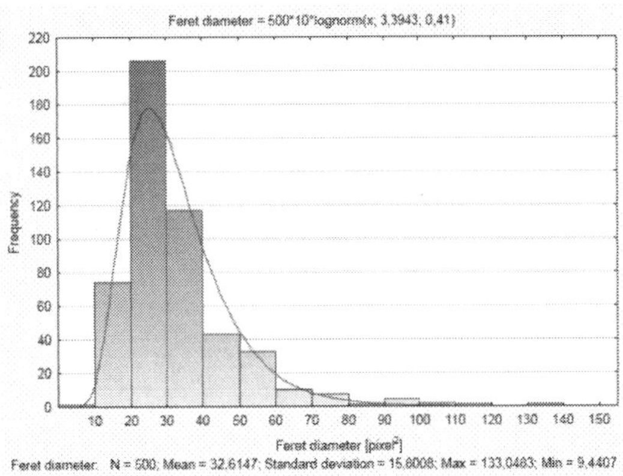

Figure 11. Histogram of Feret diameters for air bubbles together with the adjustment of the log-normal probability distribution for flotation with 0.25 M H_2SO_4.

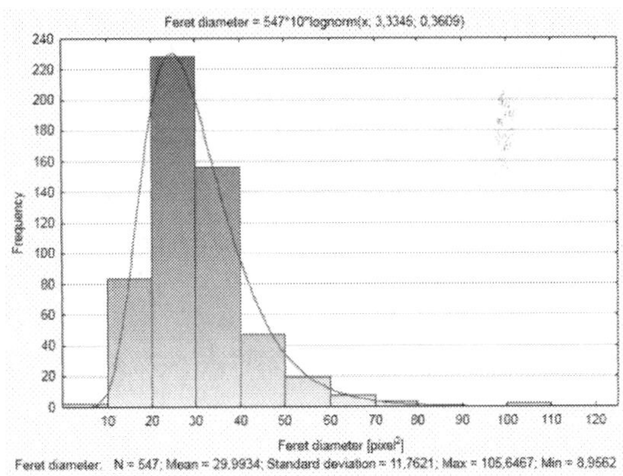

Figure 12. Histogram of Feret diameters for air bubbles together with the adjustment of the log-normal probability distribution for flotation with 0.375 M H_2SO_4.

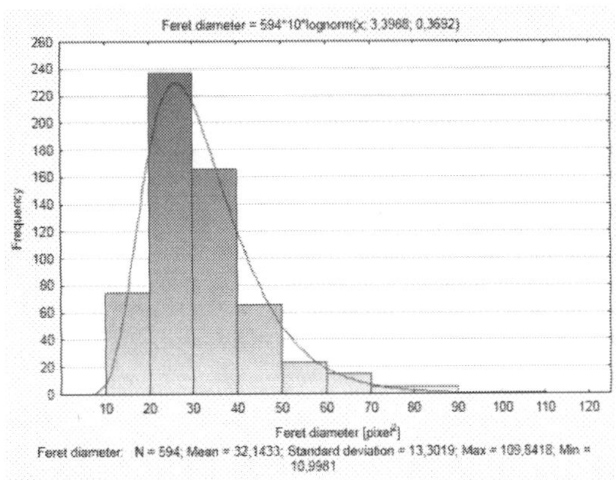

Figure 13. Histogram of Feret diameters for air bubbles together with the adjustment of the log-normal probability distribution for the flotation with 0.5 M H_2SO_4.

Table 3. Selected parameters of coal flotation in the presence of 0.063 and 0.125 M H_2SO_4.

Products	Flotation time t[s]	H_2SO_4 0.063M			H_2SO_4 0.125M		
		yield [%]	grade[%]	recovery[%]	yield [%]	grade [%]	recovery [%]
	0		0.00	0.00		0.00	0.00
K1	30	31.51	4.52	32.51	35.90	4.63	36.81
K2	60	16.44	4.65	49.41	12.82	4.73	49.90
K3	120	13.70	5.01	63.29	12.82	4.86	62.94
K4	180	5.48	5.15	68.81	7.69	4.98	70.72
K5	300	8.22	5.38	77.06	7.69	5.05	78.51
tailing	-	24.66	7.48	100.00	23.08	6.98	100.00

Table 4. Selected parameters of coal flotation for 0.25, 0.375 and 0.5 M of H_2SO_4.

Products	Flotation time t[s]	H_2SO_4 – 0.25M			H_2SO_4 – 0.375M			H_2SO_4 – 0.5M		
		yield [%]	grade[%]	recovery[%]	yield [%]	grade [%]	recovery [%]	yield [%]	grade [%]	recovery [%]
	0		0.00	0.00		0.00	0.00		0.00	0.00
K1	30	21.05	3.15	21.59	20.51	3.05	21.01	30.43	2.76	31.10
K2	60	15.79	1.96	38.25	10.26	3.00	31.53	13.04	2.59	44.50
K3	120	13.16	2.29	51.74	12.82	2.95	44.69	10.87	2.64	55.60
K4	180	7.89	2.46	59.81	5.13	2.96	49.95	4.35	2.61	60.07
K5	300	7.89	2.56	67.89	10.26	3.04	60.42	6.52	2.65	66.72
K6	360	7.89	2.66	75.96	7.69	3.02	68.31	4.35	2.66	71.15
tailing	-	26.32	5.58	100.00	33.33	5.35	100.00	30.43	4.83	100.00

The analysis of experimental results shown in Tables 3-4 indicated that the most advantageous effectiveness of hard coal flotation was achieved when the concentration of H_2SO_4 was 0.25 M, and the tests proved that both high and low concentrations of H_2SO_4 in process water reduce the upgrading effectiveness. In the case of low concentration, the impact of ions concentrations on the obtained concentrate yields can be caused by an increased particles surface charge which reduces the wetting angle, and thus decreases the flotation capability of mineral particles [3, 15]. Higher concentrations, on the other hand, can increase the coal particle hydration, thus worsening the upgrading results [7].

The morphological analysis of concentrates photographs for individual fraction flotation and yields of individual concentrates presented in Tables 3-4 also indicate that more than half of coal particles were lifted to the flotation froth during the first minute of the process. In addition, with flotation time and increased concentration of H_2SO_4 in a model process water, the number of particles floating to the froth slightly decreased. This finally led to smaller coal yields at higher concentrations of inorganic electrolyte. Moreover, it was observed that flotation with inorganic electrolyte ions caused increased dispersion of air bubbles in the flotation cell, leading at the same time to formation of stable and high froth [17]. Smaller gas bubbles caused the reduced probabilities of collision and adhesion, while the larger gas bubble sizes increased the probability of particle-air bubble aggregates being lifted to the flotation froth [24].

The results of image analysis presented in the form of histograms (Figs. 9-13) indicated that with increasing concentration of sulfate(VI) ions in the process water, the number of bubble in the flotation cell increased from 487 to almost 600 with the exception of flotation with 0.125 M salinity where the number of counted air bubbles was 470. In addition, the statistical data suggested that the greatest mean Feret diameter was achieved for 0.25 M concentration of sulfate(VI) ions, and the smallest for 0.375 M. Taking into account the diversity of air bubbles, the biggest ones can be observed at 0.25 M process water salinity as shown by the obtained standard deviation values.

The amount of water in froth for three greatest concentrations of sulfate(VI) ions in process water was tested as an additional element. The reason for studying the water amount in obtained products and its impact on the flotation results was an ineffective drying of concentrates for these tests which was caused by increased sorption of sulfates on the coal surface. This was the reason for retesting after sieving the coal particles below 0.056 mm. During the retests, the wet froth products were weighed before drying and the amount water in the flotation froth was determined. The results are presented in Figs. 14-16.

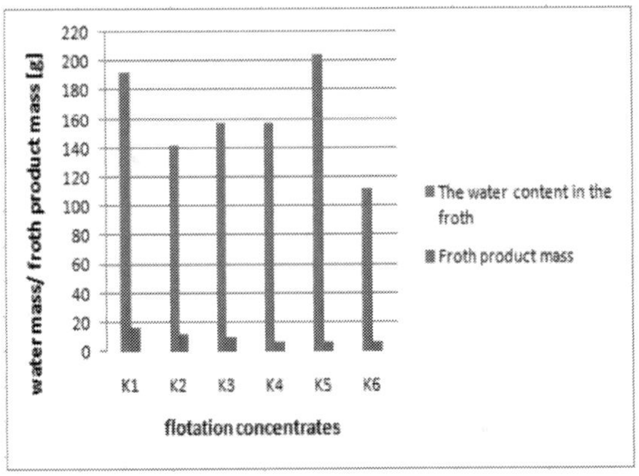

Figure 14. Water and solids yield after flotation with 0.25 M H_2SO_4.

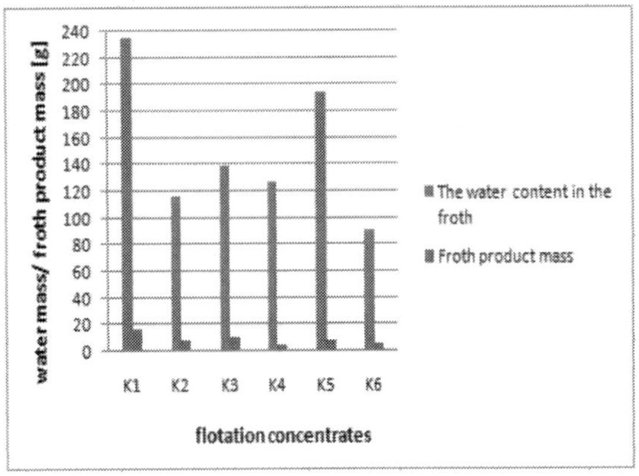

Figure 15. Water and solids yield after flotation with 0.375 M H_2SO_4.

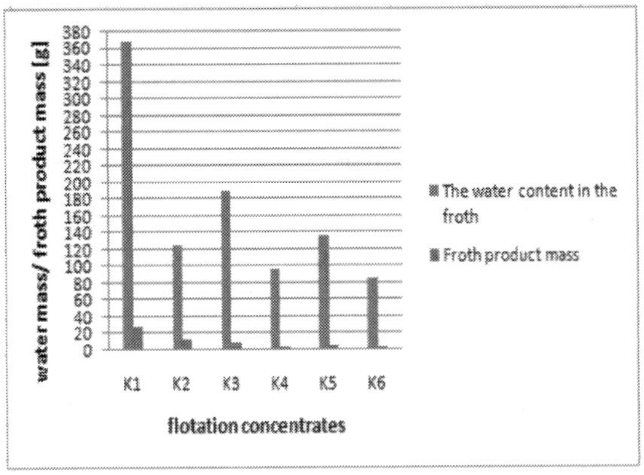

Figure 16. Water and solids yield after flotation with 0.50 M H_2SO_4.

Figures 14-16 shows that with increased concentration of sulfuric(VI) acid in process water, in case of the first concentrate, the amount of water in the froth product increased, but it decreased in concentrates 4 - 6. Simultaneously, in case of all three flotation tests a similar tendency was observed as the flotation time increased. In all flotation tests, the largest amount of water in the froth was obtained in the first 30 seconds of the process when the largest weight of the froth product was obtained. This is linked to water being lifted in the form of thin layers which prolonged both the bubbles and the particles attached to them [25]. Thus, in the last concentrates the water content was the smallest because the product weight was the lowest. Hence, it can be said that the amount of raw material floated to the concentrate depends on the water content in the froth, which is associated with the effectiveness of flotation process. It is believed that similar relationships took place also in case of two lowest salt concentrations in the process water.

4 Summary

As the hard coal processing takes place in water from underground mines, which contains a significant amount of inorganic salts, it is important to study the impact of inorganic electrolytes on the flotation process. Therefore, the presented in this work tests were related to the impact of process water salinity on the size and number of formed air bubbles in the froth and on the upgrading performance. Based on the results the following conclusions can be drawn.

1. The increased presence of inorganic electrolyte ions in a process water resulted in increased dispersion of air bubbles in the flotation cell.
2. The more wet the flotation froth, the larger amounts of froth products were obtained.
3. The greatest mean Feret diameter was achieved for 0.25M concentration of sulfuric acid(VI) where the technological results were the best, and the smallest mean Feret diameter was achieved for 0.375 M sulfuric acid(VI) where the flotation results were the worst.
4. The greater the diversity of air bubble size determined on the basis of standard deviation, the more effective the flotation process.

References

1. A. Grabiec, A. Młynarczykowska, *in Proccedings of Conference of Young Researches ProFuturo* (2014)
2. Z. Blaschke, M. Brożek, E. Mokrzycki, Z. Ociepa, T. Tumidajski, *Mining. częśc V* (Wydawnictwo „AGH", Kraków, 1981)
3. J. Drzymała, *Podstawy mineralurgii,* 2nd ed. (Oficyna Wydawnicza PWR, 2009)
4. S. Blaschke, *Przeróbka mechaniczna kopalin część II* (Wydawnictwo Śląsk, Katowice, 1984)
5. M. Brożek, A. Młynarczykowska, *Physicochem. Probl. Miner. Process.* **44**, 23 (2010)
6. A. Młynarczykowska, A. Turno, *Prace Naukowe Instytutu Górnictwa PWR,* **103**, 211 (2003)
7. E. Małysa, Z. Ociepa, E. Oruba, S. Sanak-Rydlewska, *Ćwiczenia laboratoryjne z flotacji,* 2nd edn. (Wydawnictwo „AGH", Kraków, 1981)
8. A. Nyrek, K. Oleksik, BSc thesis, (2014)
9. A. Młynarczykowska, A. Nyrek, K. Oleksik, *IM,* **Jan.-June,** 181 (2015)
10. S. Blaschke, *Procesy technologiczne w przeróbce kopalin użytecznych* (Wydawnictwo „AGH", Kraków, 1987)
11. E. Muzenda E., *Int. J. Chem., Mol., Nucl. Mater. Metall. Eng.,* **4(9)**, 562 (2010)
12. R.H. Yoon, *Min. Cong. J.,* **68**, 76 (1982)
13. R.H. Yoon, J.B. Sabey, *Interfacial Phenomena in Coal Technology*, (Marcel Dekker, New York, 87, 1989).
14. J.S. Laskowski, *Coal Flotation and Fine Coal Utilisation* (Elsevier, Amsterdam, 2001)
15. J. Drzymała, T. Ratajczak, *Flotacja solna* (Oficyna Wydawnicza PWR, Wrocław, 2003)
16. O. Ozdemir, *Physicochem. Probl. Miner. Process.* **49(2)**, 511 (2013)
17. K. Tupek, *MSc* thesis, (2015)
18. A. Jahedsaravani, M.H. Marhaban, M. Massinaei, *Miner. Eng.* **69**, 137 (2014)
19. AW. Banford, Z. Aktas, E.T. Woodburn, *Powder Techno.* **98(1),** 61 (1998)
20. C. Citir, Z. Aktas, R. Berber, *Comput. Chem. Eng.* **28(5),** 625 (2004)
21. A. Konieczny, W. Pawlos, M. Jach, R. Pępkowski, M. Krzemińska, R. Kaleta, D. Foszcz, *Min. Geol.* **6(2)**, 61 (2011)
22. E. Małysa, A. Iwańska, *Min. Geol,* **34(4/1)**, 165 (2010)
23. D. Tao, *Sep. Sci. Technol.* **39(4)**, 741 (2004)
24. E. Małysa, A. Surowiak, *Miner. Resour. Manag.t* **24(4/1)**, 53 (2008)

Dispersion of air bubbles in a flotation cell under predetermined hydrodynamic conditions

Anna Mlynarczykowska[1,a], Konrad Oleksik[1], and Klaudia Tupek[1]

[1]*AGH University of Science and Technology, Faculty of Mining and Geoengineering, Department of Environmental Engineering and Mineral Processing, Poland*

Abstract. From a practical point of view the most important stage in flotation is a mineralization of gas bubbles, where the bubbles size is an important factor. Numerous studies and models show the relationship between the flotation parameters and gas distribution in a flotation cell. The size of gas bubbles is a random variable with a specific distribution [1] and analysis of this distribution is useful to make mathematical description of the flotation process. The article presents the results of measurements and analysis of the gas phase dispersion by the size distribution of air bubbles in a flotation chamber under fixed hydrodynamic conditions.

References

1. M. Brozek, A. Mlynarczykowska, *Arch. Min. Sci.*, **57(3)**, 729 (2012)

[a] Corresponding author: mindziu@agh.edu.pl

Assessment of soda ash calcination treatment of Turkish trona ore

Sibel Gezer[1,a] and Umit Atalay[2]

[1]*Republic of Turkey Ministry of Energy and Natural Resources, Nuclear Energy Project Implementation Department, Ankara, Turkey*
[2]*Middle East Technical University, Mining Engineering Department, Ankara, Turkey*

Abstract. Trona is relatively rare, non-metallic mineral, $Na_2CO_3 \cdot NaHCO_3 \cdot 2H_2O$. The pure material contains 70.3% sodium carbonate and by calcination the excess CO_2 and water can be driven off, yielding natural soda ash. The terms soda ash and sodium carbonate are used interchangeably. Trona calcining is a key process step in production of soda ash (sodium carbonate anhydrate) from the relatively cheap trona ore. The calcination reaction may proceeds in a sequence of steps. Depending on the conditions, it may result in formation of either sodium carbonate monohydrate ($Na_2CO_3 \cdot H_2O$), sodium sesquicarbonate or weigschederite ($Na_2CO_3 \cdot 3NaHCO_3$). The Beypazarı Turkish trona deposit is the second largest deposit in the world with the content of 84% trona. The decomposition of trona appeared to be a single stage process across the temperature range studied (150-200 °C) with the representative samples of different size fractions in the draught up metallurgical furnace. The optimum particle size and calcination time were -6.35 mm and 30 minutes, respectively, at calcination temperature of 175 °C in a metallurgical furnace. Microwave-induced dry calcination of trona was possible and 5 minutes of calcination time at a power level of 900 was sufficient for complete calcination of -6.35 mm feed. This includes short time calcinations with the goal of improving economics and simplifying the thermal process.

1 Introduction

Soda ash is the common name for sodium carbonate (Na_2CO_3), a compound that has achieved significant economic importance because of its intrinsic role in the manufacturing of glass. The occurrence, preparation and uses of natural soda ash have been known since ancient ages. Soda ash has wide usage areas in chemical, detergent, soap and paper industries, in flue gas desulphurization (FGD) systems and in water treatment systems.

Production of the alkaline materials sodium carbonate and potassium carbonate (which were recognized rather late as chemically distinct) was carried out from ancient times until the 1800s by combustion of marine and land vegetation, followed by calcination at red heat and leaching of ash. The term soda ash originates from this process [1].

[a] Corresponding author: sgezer@enerji.gov.tr

Presently, the well-known sodium carbonate minerals used in production of soda ash are trona ($Na_2CO_3{\cdot}NaHCO_3{\cdot}2H_2O$), thermonatrite ($Na_2CO_3{\cdot}H_2O$) and natrone ($Na_2CO_3{\cdot}10H_2O$). Only trona with 70.3% Na_2CO_3 content is of commercial interest.

Trona deposits are normally mechanically mined. The method uses conventional, either continuous or longwall mining technique or combination of them. Then, the raw trona can be processed by different methods such as sesquicarbonate, monohydrate, alkali extraction and carbonation processes. There are dominantly two, sesquicarbonate and monohydrate processes, used to convert mined trona into either pure soda ash, or sodium carbonate (Na_2CO_3). In the sesquicarbonate process, the trona ore is dissolved in an aqueous solution and after separation of the resulting solution from the insolubles, sodium sesquicarbonate is crystallized from the solution by cooling. The sesquicarbonate crystals are then separated from the mother liquor and finally, calcined to recover soda ash (anhydrous sodium carbonate). In the monohydrate process, the trona ore is first calcined to convert its bicarbonate content to sodium carbonate, and is then dissolved in water. Then the resulting solution is separated from the insoluble and sodium carbonate monohydrate is precipitated to evaporative crystallization. The monohydrate crystals are then separated from the mother liquor and dried to recover soda ash [2].

1.1 Beypazarı Trona deposit in Turkey

The Turkish natural soda (trona) deposit was discovered in Beypazarı by the Mineral Research and Exploration Institute (MTA) in 1979 during coal exploration. The Beypazarı district is a large area of volcano-sedimentary rocks in the interior of central Anatolia, Turkey approximately 100 km northwest of Ankara. The trona deposit is present 130-140 m deep, in layers which are thicker and richer in the quality compared to the ores in Wyoming (USA). The Beypazarı Turkish trona deposit is the second largest deposit in the world (total reserves about 300 million tons) with 84 % trona content.

The Beypazarı trona deposit consists of two lensoidal bodies named as the upper (seam) and lower vein (seam). Due to limited development and preparation work in the mine, the samples named as U3, U4, U5 and U6 were taken from different layers in the upper vein and were mixed to prepare the representative trona ore sample. Characteristics of these layers are given in Table 1.

Table 1. Characteristics of samples taken from different layers.

	U3	U4	U5	U6
Na_2CO_3	34.48	43.06	43.09	43.44
$NaHCO_3$	27.84	37.76	35.45	35.25
Total alkalinity, % Na_2CO_3	52.04	66.88	65.46	65.69
Trona, %	73.53	91.82	91.88	92.63

The dominant mineral in the deposit is trona along with some nahcolite, trace amounts of pirssonite and thermonatrite. Calcite, zeolites (analcime, klinopitolite), feldspar and clays are the most common gangue minerals within the associated rocks of the trona deposit. The trona crystals, which are generally white and occasionally grayish due to the presence of impurities, formed in massive beds and disseminated crystals in claystone and shale.

1.2 Trona calcination

Trona is relatively rare, non-metallic mineral, $Na_2CO_3{\cdot}NaHCO_3{\cdot}2H_2O$. The pure material contains 70.3% sodium carbonate and by calcination the excess CO_2 and water can be driven off, yielding natural soda ash. The terms soda ash and sodium carbonate are used interchangeably. Soda ash is a white, anhydrous, powdered or granular material containing much above 99% sodium carbonate (Na_2CO_3) when shipped. The accepted commercial standard for soda ash is expressed in terms of the total equivalent sodium oxide (Na_2O) content [3].

Trona calcining is a key process step in production of soda ash (sodium carbonate anhydrate) from the relatively cheap trona ore. It is accomplished by heating trona to an appropriate temperature to remove CO_2 and H_2O. The calcination reaction may proceed in a sequence of steps. Depending on the conditions, it may result in formation of sodium carbonate monohydrate ($Na_2CO_3 \cdot H_2O$), sodium sesquicarbonate or weigschederite ($Na_2CO_3 \cdot 3NaHCO_3$). The rate of formation and subsequent decomposition of these compounds are vital in the overall efficiency of the calcinations stage of process. The decomposition reaction occurs according to the following equation:

$$2\,(Na_2CO_3 \cdot NaHCO_3 \cdot 2H_2O) \Rightarrow 3Na_2CO_3 + 5H_2O + 2CO_2. \tag{1}$$

The mechanism is based on thermal treatment of the ore that effects its decomposition and elimination of gangue minerals. To produce the probable main product, dense soda ash, from trona requires the stoichiometry given in Eq. 1.

The calcining step is a simple and exceedingly demanding operation. Trona can be either decomposed or calcined (Eq. 1) at any temperature over 70 °C, but the rate does not become reasonable until temperature is above 120 °C. Above approximately 800 °C the entire ore fuses [4].

Currently, conversion of trona into soda is performed industrially by calcining the raw ore in rotary calciners at temperature of 120 °C or higher. It is postulated that the decomposition reaction according to Eq. 2 occurs only at temperatures beyond 160 °C. At lower temperatures wegscheiderite ($Na_2CO_3 \cdot 3NaHCO_3$) and sodium carbonate monohydrate ($Na_2CO_3 \cdot H_2O$) form as intermediates [5]. Under dry conditions, trona is reported to be stable till ~55-57 °C. Above this temperature trona starts to decompose at a slow rate:

$$3\,(Na_2CO_3 \cdot NaHCO_3 \cdot 2H_2O)\,(s) \Rightarrow (Na_2CO_3 \cdot 3NaHCO_3)(s) + 2Na_2CO_3\,(s) + 6\,H_2O \tag{2}$$

$$2\,(Na_2CO_3 \cdot 3NaHCO_3)\,(s) \Rightarrow 5\,Na_2CO_3\,(s) + 3\,H_2O + 3CO_2. \tag{3}$$

Microwave-induced calcination of trona has the potential to re-evaluate many aspects currently necessary in the conventional processing of trona into soda ash. This includes high temperature calcination, with the goal of improving economics and simplifying the process.

The amount of microwave energy that the trona absorbed is directly related with the operated power of microwave. The measurements of microwave are of input (directly from the generator), reflected (back to the generator from the load) and transmitted powers (through the load). The power absorbed by the trona is then calculated by subtracting the transmitted and reflected power from the input power. The studies in the literature showed that only trona in the solution was capable of absorbing most of the input power. Dry trona did not prove to be a very good absorbent alone. It is interesting to note that pure water only absorbed approximately a tenth of the power that trona in solution absorbed, and when compered the reactor load was near zero. This is conclusive result that trona in a solution is a very good microwave absorbent and an excellent candidate for decomposition in the microwave field [6].

2 Experimental

2.1 Samples

The trona samples used in this study were taken from the trona deposit in Beypazarı, Turkey. The used samples were taken from different layers in the upper vein and were mixed to prepare the representative trona ore samples. The studies showed that trona is present in 130-140 m dep, in layers, which are thicker and richer in quality compared the ones in Wyoming (USA).

Mineralogical analyses show that the trona ore consists of the two different mineralogical structures that is massive and crystalline. The principal sodium carbonate minerals are trona and minor nahcolite, trace amounts of pirssonite and thermonatrite in the Beypazarı ore. Trona and dolomite are associated throughout the trona zone. Calcite, zeolites, feldspar and clays are the most common gangue minerals within the rocks of deposit. The chemical analysis of representative samples is shown in Table 2. The results of the thermogravimetric analysis (TGA) test of representative sample is given in Figure 1.

Table 2. Chemical analysis of representative Beypazarı trona sample.

Component	Weight, %
$NaHCO_3$	33.00
Na_2CO_3	38.59
Total Alkalinity, % Na_2CO_3	59.40
Insolubles	15.30
NaCl	0.025
Na_2SO_4	0.026
Loss of ignition	25.27
TRONA	84.70

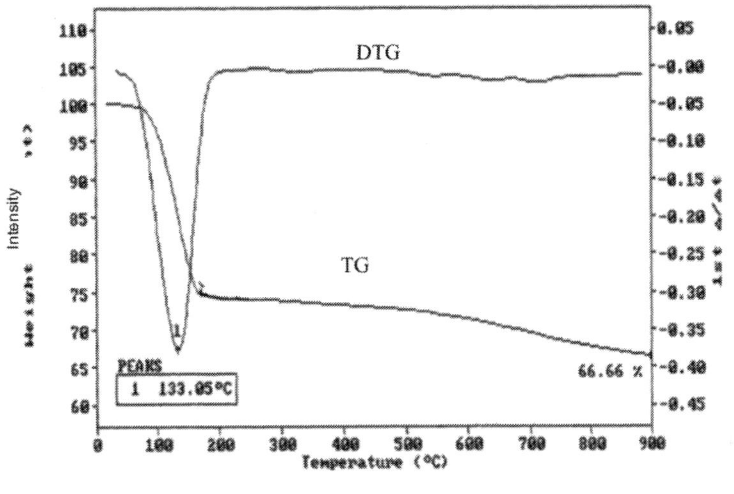

Figure 1. TG and DTG curves of representative Beypazarı trona ore samples.

3 Results and discussion

Calcination process was investigated by using the draught up furnace. The sample solids were contained in a glass vessel and air introduced into the system via a distributor. Experiments were conducted at three different temperatures, 150, 175 and 200 °C. The adjustment of the airflow created some problems. At a low air velocity the solids were in either a packed-bed or fixed-bed state, while solids were thrown away from the heating zone at a high airflow rate. During the tests, attention was given to the adjustment of compressed air flow. At a low airflow rate, the solid samples were in either a packed-bed or fixed bed state and adverse effect of insufficient air flowing was observed noticeably for the small particle size of -3.35 mm. As the gas velocity increased, the lifting force eventually overcame the weight of the particles and eliminated the interparticle forces, and the air with exhausted gases of CO_2 and H_2O found a way to go through the interparticle void space. When the gas flow was too high the tap was thrown away from the top and then the possibility of solid leaking increased. However, under the conditions of good adjustment of air flowing and temperature, trona decomposition was carried out for up draught in shorter time than it was in the muffle furnace.

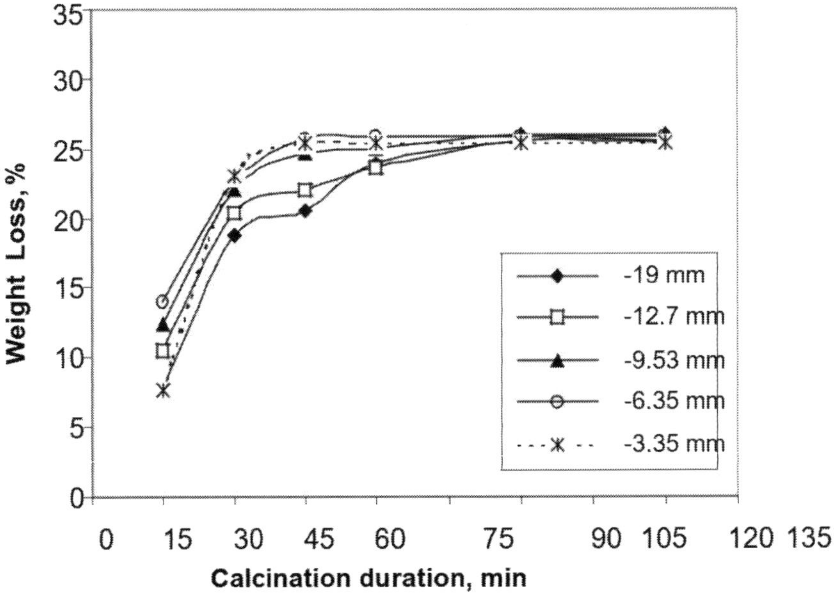

Figure 2. Draught up calcination of different particle size at 150 °C.

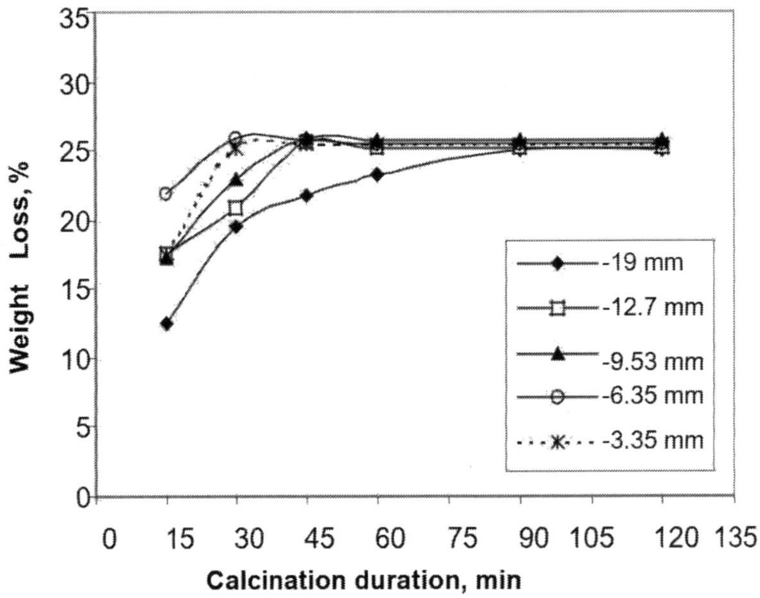

Figure 3. Draught calcination of different particle size at 175 °C.

Decomposition of trona appeared to be a single stage process across the temperature range studied (150-200 °C) and the complete weight loss occurred in a short time. It is possible to interpret that such short time decomposition might relate directly to the heating system and air flowing through the particles. Calcination was not continued to the longer that of 100-120 minutes. It was observed that complete calcination was realized in 80-90 minutes for coarse particle size in draught calcination.

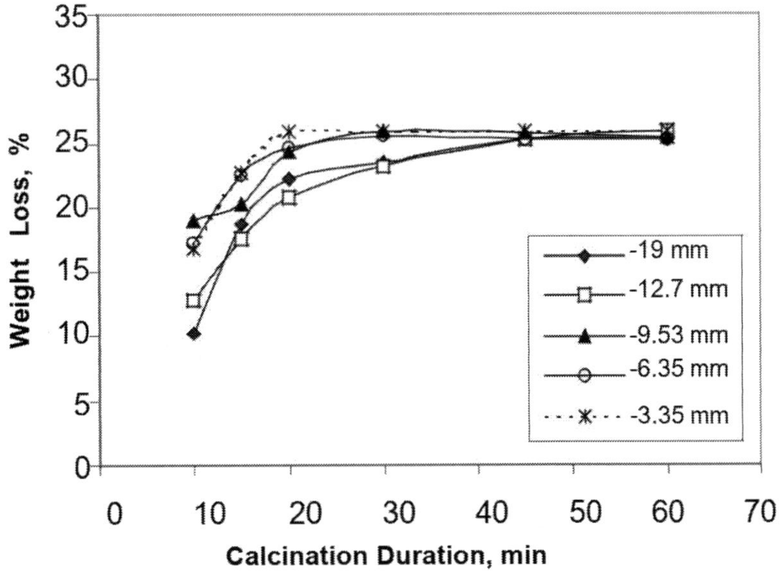

Figure 4. Draught calcination of different particle size at 200 °C.

Microwave-induced calcination of trona has the potential to re-evaluate many of the aspects currently necessary in the conventional processing of trona into soda ash. It includes short time calcination, with the goal of improving economics and simplifying the process.

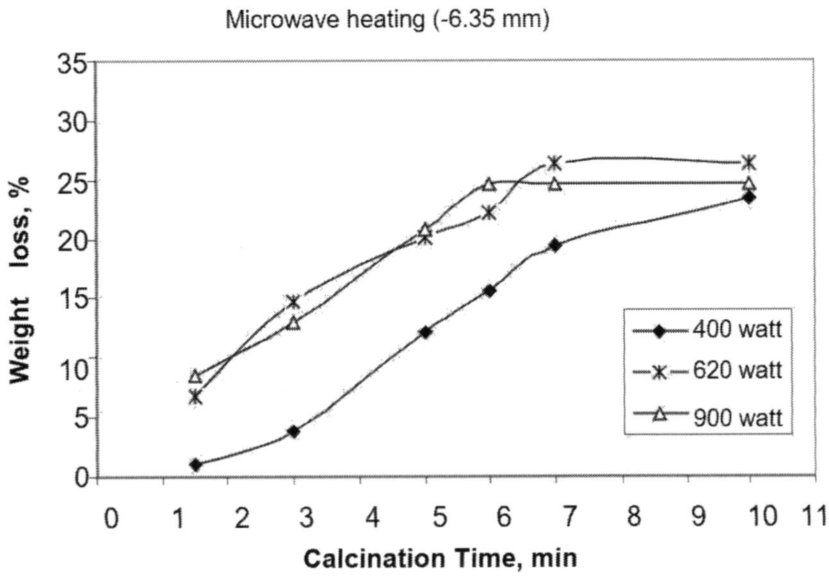

Figure 5. Effect of calcination time on the weight loss of -6.35 mm trona.

The goal of this part of the study were to explore the feasibility of producing soda ash from trona using a microwave process and to investigate the best possible process for converting all of sodium bicarbonate ($NaHCO_3$) content in trona to the required value into either sodium carbonate or soda ash as well as to study the kinetics of reaction. One series of experiment was conducted in microwave using dry samples so the process was named as the dry type microwave calcination [7].

The experiments were intended to compare effects of microwave heating, versus conventional heating. Calcination was completed in 30 minutes in conventional heating, whereas the same weight loss value was obtained in very short calcination time as 6-7 minutes depending on the applied power for the particle size of -6.35 mm. Dry microwave calcination of –6.35 mm particles showed that the weight loss increased with increasing calcination duration and power setting.

The results plotted in Figure 5 showed that both -3.35 and -6.35 mm samples would be completely calcined in 10 minutes, besides -6.35 mm sample would have already completed calcination process in 7 minutes. It shows that the calcination time was not sufficient for complete calcination of -9.53 mm particle samples.

4 Conclusion

Based on the experimental results, the following conclusions can be drawn.

1. Trona can be decomposed at any temperature over 70 °C, but the process rate does not become reasonable below 120 °C.
2. Thermal decomposition of trona at low calcinations temperatures produces sodium carbonate monohydrate ($Na_2CO_3 \cdot H2O$) and wegschederite ($Na_2CO_3 \cdot 3NaHCO_3$) as intermediate products.
3. Calcination of -6.35 mm feed at 175 °C is the muffle furnace can be sufficiently completed after 30 min.
4. The weight loss 25.44 % after complete calcination is explained by removal of crystal water and exposure of CO_2.
5. Microwave-induced dry calcination of trona was possible and 5 minutes of calcination time at a power level of 900 W was sufficient for complete calcination of -6.35 mm feed.

References

1. C. Thieme, *Ullmann's Encyclopedia of Industrial Chemistry,* **A24**, 299 (2000)
2. T. Eggeman, *Kirk-Othmer Encyclopedia of Chemical Technology,* (2011)
3. USGS, Soda Ash, *Mineral Industry Surveys,* (Ann. Rev., U.S. Geolo. Surv.,1, 2002)
4. M. Doğan, C .Güldür, G. Doğu, T. Doğu, in Sprey Dryer., *J.Chem. Tech. Biotechnol.*, **68**, 157 (1997)
5. M.C. Ball, A.N. Strachan, R.M. Strachan, *J.Chem. Soc. Faraday Trans.* **87(12)**, 1911 (1991)
6. R.D. Streuber, C.Y.Cha, in *Proceedings of the First International Soda Ash Conference-Rock Springs* (Wyoming, **2**, 185, 1998)
7. S. Orgul, *Ph.D. Thesis,* METU, Ankara, Turkey, (2003)

Flotation of kaolinite from tailings of kaolin-washing plants by cationic collectors

Kianoush Barani[1 a]

[1]*Department of Mining Engineering, Lorestan University, Khorramabad, Iran*

Abstract. Traditional processing of kaolin is achieved by dispersion of the mined ore and classification by means of multistage hydrocyclones. The inefficiencies inherent to cyclones produce a middling product that is commonly disposed back to the quarry. In this research recovery of kaolinite from tailings of the Zonoupz kaolin washing plant, which is located in Iran, was investigated by cationic flotation. Flotation experiments showed that flotation of kaolinite from tailings was much better in an acidic than in an alkaline medium containing cationic collectors.

1 Introduction

In general, kaolin is upgraded by physical methods such as either water-washing or froth flotation. Kaolin particles are predominantly below 25 μm and are flat shaped particles. Industrial separation is commonly achieved by dispersion of the mined ore and classification using multistage hydrocyclones. Kaolin reports to overflow, which is then reprocessed in other smaller cyclones. Due to inefficiencies associated with hydrocycloning, most plants are not able to produce clean kaolinite products and after either the 4th or 5th hydrocyclone stage the underflow cannot be further separated further. The underflow is either disposed as waste, sold cheaply to the cement industry and brick-manufacturing or used as back fill in the quarry [1-5].

The Zonouz mine, which is located in eastern Azerbaijan Province, has the only kaolin-washing plant in Iran. The plant has a considerable amount of tailings containing about 15% Al_2O_3. Mineralogically, the tailings consist of quartz, kaolin and iron oxide. The tailing has fine particles (100% finer than 100 μm and about 30% finer than 20 μm), which can be easily dispersed by wind and settled down by rain and snow. Without either treatment or recycling, the tailing would cause environmental pollution. It is important to re-evaluate the tailings because of economic and environmental causes. Evaluation is economically feasible since it does not require additional mining costs. Kaolinite can be retrieved from the tailings by using flotation. However, flotation of kaolinite from the tailing is difficult because kaolinite and quartz, the two main minerals occurring in the tailings, have the same flotation behavior.

Both minerals are negatively charged in a wide range of pH, and therefore they can be floated by amine collectors in acidic and alkaline medium [6-8].

[a] Corresponding author: barani.k@lu.ac.ir

Many studies were carried out on kaolinite flotation (pure samples) by cationic collectors such as ether diamine, hydroxyl-alkyl-diethanolamine, cetylpyridinium chloride, cetyltrimethylammonium bromide and dodecylamine. The results of these studies showed that kaolinite flotation by cationic collectors was better under acidic conditions than in alkaline media [7-13]. Yuehua et al. (2004) tried to increase kaolinite recovery in alkaline medium by using a macromolecule as a flocculant. Without polyacrylamide, the kaolinite recovery decreased from 70 to 15%, with increasing the pH from 2 to 10. However, with polyacrylamide as a flocculant, kaolinite recovery increased under both acidic and alkaline conditions [11].

Comparison of flotation of kaolinite and quartz showed that two collectors (dodecylamine and EDA Flotigam) and two quaternary ammonium salts (cetylpyridinium chloride and cetyltrimethylammonium bromide) were the most effective collectors for floating both kaolinite and quartz. However, the time required for kaolinite flotation was 5-10 times longer that for the quartz flotation, depending on the type of collector [10].

The purpose of the present study is to investigate the recovery of kaolinite from tailings of kaolin washing plants using flotation and flotation-flocculation.

2 Experimental

2.1 Materials

Samples used in this study were taken from the tailing dam of the Zonouz kaolin washing plant in East Azerbaijan Province, Iran. The size distribution of the representative sample was carried out by laser diffraction (Fig. 1). It is observed that the percent passing 100 and 50 µm is respectively 100% and 50%, and more than 30% of the particles were smaller than 20 µm (slime particles).

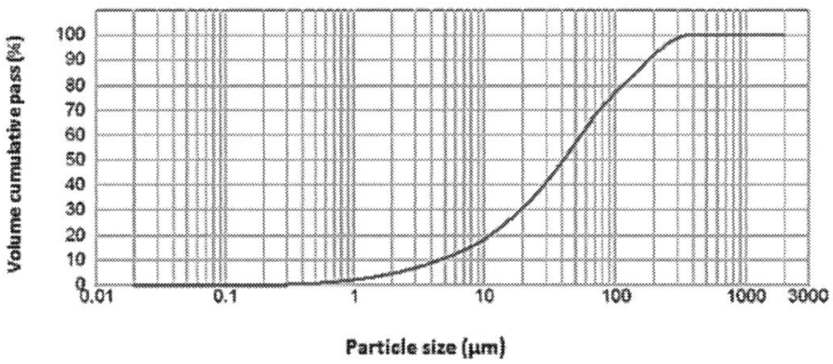

Figure 1. Particle size distribution of the tailing sample by laser diffraction.

According to the mineralogical analyses, it was found that the sample contained quartz, kaolinite and calcite as the main minerals, and muscovite, biotite, montmorillonite and iron-rich epidote as the minor minerals. The chemical composition of the sample is given in Table 1.

Table 1. Chemical composition of a sample.

Composition	**SiO_2**	**Al_2O_3**	**K_2O**	**SO_3**	**Fe_2O_3**
%	74.67	15.35	0.25	0.11	0.33
Composition	CaO	SrO	P_2O_5	MgO	L.O.I
%	1.74	<0.01	<0.01	<0.01	7.54

2.2 Flotation experiments

Flotation experiments were carried out in a 1 dm^3 conventional laboratory cell (Denver model), at 25% solids ratio and agitation rate was 850 rpm. Pine oil was used as frother (150 g/ton), cetylpyridinium chloride (CPCI) and dodecylamine (DDA) as collectors, aluminum chloride as a depressant of quartz; sulfuric acid and sodium hydroxide as pH regulators. All reagents used in this study were of analytical grade.

The experiments were conducted in 5 parts. In each part, the effect of only one parameter was examined, while the other parameters were kept constant.

3 Results and discussion

3.1 Effect of pH with CPCI as collector

The first part of the experimental investigation was conducted to evaluate the effect of pH on the flotation results when CPCI was used as a collector. The results are illustrated in Fig 2. It can be seen, that with increasing pH from 3 to 9 the Al_2O_3 recovery decreased and SiO_2 recovery increased. The assays of Al_2O_3 and SiO_2 were respectively assumed as the kaolinite and quartz indicator recovery.

Figure 3 shows the structure of kaolinite crystal. At low pH (pH<4) the layer to layer interaction between the alumina face surface ($00\bar{1}$) and silica face surface (001) was significant, as well as the interaction between the alumina face surface and edge surface (010). In this case, the cluster structures, and even more complex network structures, were formed by kaolinite particle aggregation. Figure 4 illustrates adsorption of cationic collector by the kaolinite cluster at low pH (pH<4), forming a hydrophobic structure [13].

In alkaline medium, the interaction between the cationic collector and plane (001), which contained oxygen, was stronger than that of the plane ($00\bar{1}$), which contained Al. The plane (001) adsorbed hydrophobic collector, but the plane ($00\bar{1}$) did not adsorb collector due to presence of Al and remained hydrophilic. As a results, collectors of the plane (001) were accumulated and formed structures which the outer surface was hydrophilic (Fig. 5) [9]. Therefore, as indicated in Fig. 2, kaolinite recovery decreased with increasing pH. A pH of 3 was chosen as the optimum condition and was used in for further experiments.

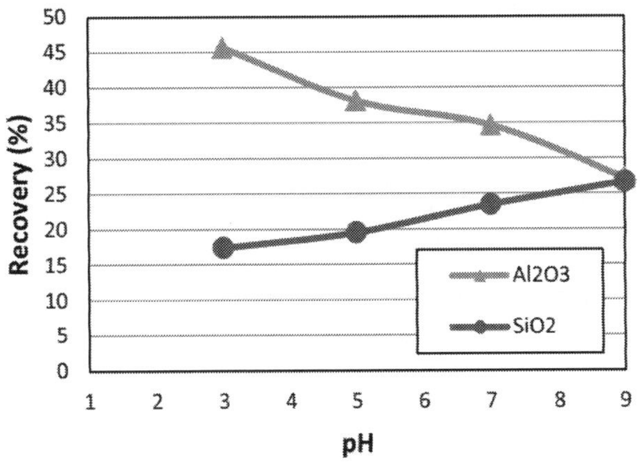

Figure 2. Flotation recovery of Al_2O_3 and SiO_2 as a function of pH with CPCI as collector (CPCI=400 g/t, without depressant and de-slimming).

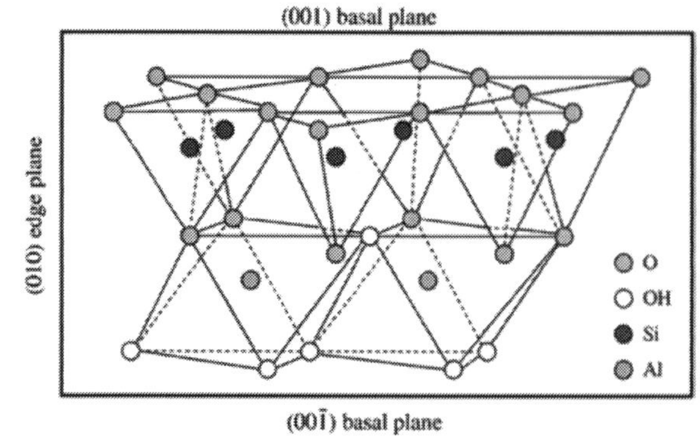

Figure 3. Structure of kaolinite [9].

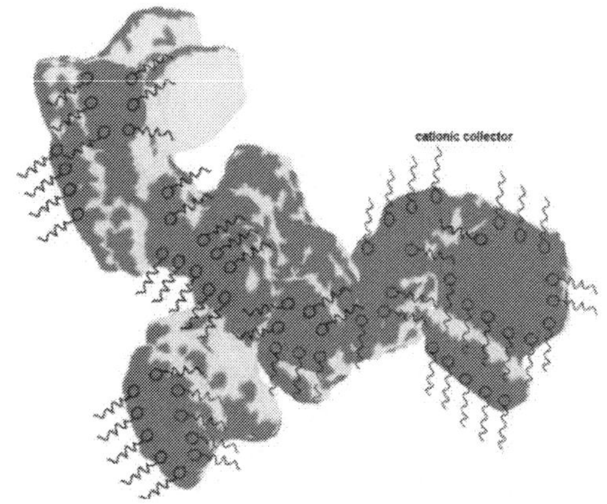

Figure 4. Kaolinite cluster adsorption of cationic collector at low pH. Pink surfaces represent the silica face surfaces of kaolinite particles and blue surfaces represent the alumina face surfaces of kaolinite particles [13].

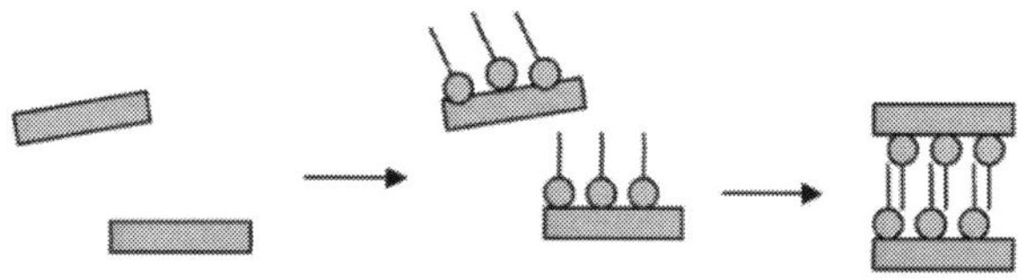

Figure 5. Schematic description of the kaolinite-amines interactions [9].

3.2 Effect of concentration of CPCI collector

The maximum recovery obtained for Al_2O_3 in the first part of investigation (section 3.1) was around 45%, which was significantly low. For increasing the recovery of Al_2O_3, the effect of concentration of CPCI on the flotation results was investigated. Also the slime particles below 38 μm were removed and aluminum chloride was presented used as the depressant.

The results in Figure 6 indicate that the Al_2O_3 recovery increased with increasing concentration of CPCI up to 600 g/t. The Al_2O_3 recovery at a 600 g/t of CPCI exceeded 85%. A further increased of CPCI concentration to 700 g/t, cause that the Al_2O_3 recovery was decreased to 81%. It was probably due to occurrence of the critical because the critical micelle concentration phenomenon was occurring. Although, more research and evidence are needed to prove it. Also, it was observed that the SiO_2 recovery was constant and equal to 20%. It can be attributed to depression of quartz particles by aluminum chloride. The maximum separation efficiency (81%) was achieved at 600 g/t of CPCI.

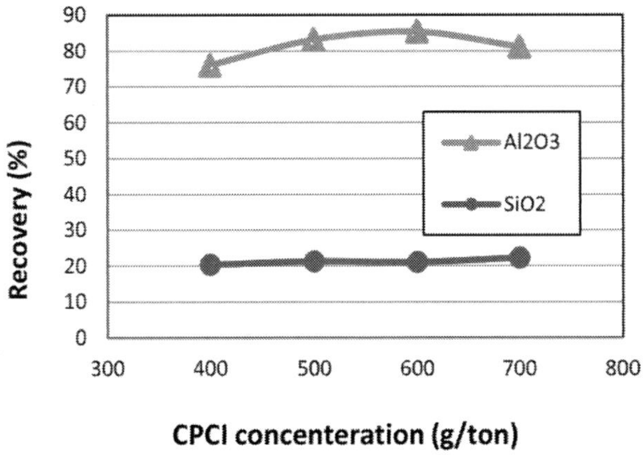

Figure 6. Flotation recovery of Al_2O_3 and SiO_2 as a function of CPCI concentration (pH 3, aluminum chloride depressant 300 g/t, desliming -38 μm).

3.3 Effect of aluminum chloride concentration with CPCI collector

The results of part 2 (section 3.2) showed that the Al_2O_3 recovery and separation efficiency were increased by using aluminum chloride as the quartz depressant. In this part, the effect and role of aluminum chloride was investigated. Figure 7 indicates that with increasing concentration of aluminum chloride from 300 to 600 g/t, the recoveries of Al_2O_3 and SiO_2 did not significantly change. However, the best results were related to aluminum chloride value of 500 g/t, which produced the concentrate with recovery of 86.28% Al_2O_3 and 20.19% SiO_2 and grade of 38.42% Al_2O_3 and 52.96% SiO_2.

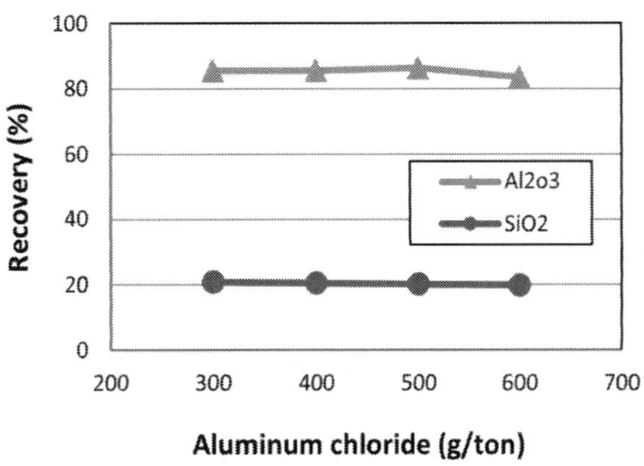

Figure 7. Flotation recovery of Al_2O_3 and SiO_2 as a function of aluminum chloride concentration (pH 3, CPCI 600 g/t, desliming -38μm).

Ions of aluminum chloride (Al^{+3}) play two positive roles during flotation. From one side they act as the depressant of quartz. In fact, these ions deposit on the surface of quartz, which has negative charge and competes with the collector ions that prevent quartz floatation. However, the excessive increased level of Al^{+3} increases the kaolin zeta potential. On the other side, these ions increase the ionic strength of the pulp. In a high ionic strength environment, although the negative charge on the basal the planes $(00\bar{1})$ and (001) is constant, the zeta potential caused by the surface charge on both basal planes and edges is significantly screened and thus aggregation of kaolinite particles can occur (Fig. 8). The surface charge at high ionic strength also weakens self-aggregation between the negatively charged $(00\bar{1})$ planes and positively charged edge planes in the acidic pulp. Thus, the pH dependence of kaolinite flotation becomes less pronounced [9].

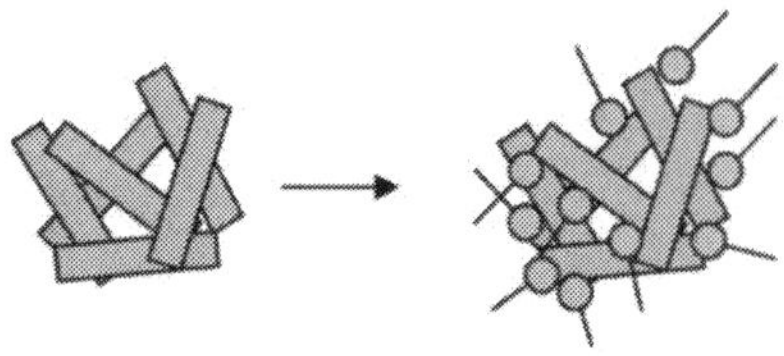

Figure 8. Aggregation of kaolinite particles in a high ionic strength environment [8].

3.4 Effect of pH on with DDA collector

Previous studies showed, dodecylamine (DDA) was the effective collector for flotation of both kaolinite and quartz [10]. In this part, flotation of kaolinite by DDA in the presence of aluminum chloride as quartz depressant was evaluated. Fig. 9 shows that during kaolinite floatation with DDA collector, with increasing pH, both recoveries of Al_2O_3 and SiO_2 increased. The maximum recovery of Al_2O_3 and SiO_2 was obtained at pH=9. It seems that, there are differences between the DDA and CPCI results. As it can be seen in Fig. 2, by using CPCI, with increasing pH, the Al_2O_3 recovery decreased. The effect of pH on selectivity separation is clearly shown in Figure 9. At pH=3 the difference between Al_2O_3 and SiO_2 recoveries was around 28% and a good selectivity was achieved. With increasing pH this difference was reduced and recoveries of Al_2O_3 and SiO_2 were almost equal at pH=9. The best results were obtain at pH=3, where a concentrate with recovery of 66.25% Al_2O_3 and 37.74% SiO_2, respectively, and assays of 20.84% Al_2O_3 and 69.93%SiO_2.

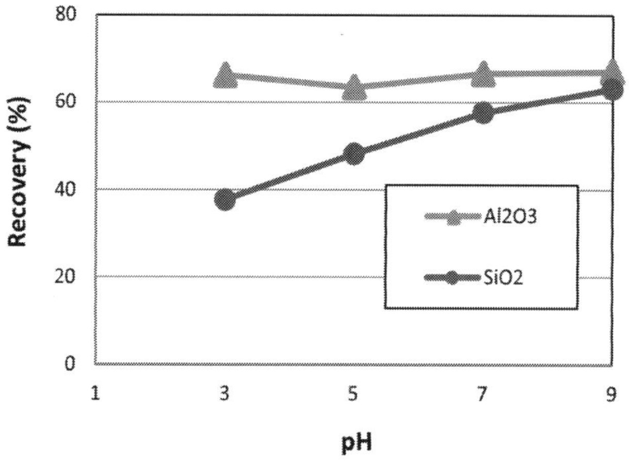

Figure 9. Flotation recovery of Al_2O_3 and SiO_2 as a function of pH with DDA as a collector (DDA 600 g/t, aluminum chloride 300 g/t, desliming -38 μm).

3.5 Effect of concentration of DDA collector

The results of part 4 (section 3.4) showed that satisfactory separation of kaolinite and quartz by the DDA collector in the presence of aluminum chloride can be obtained in acidic pHs, but high dosage of the collector is required. In this part the effect of DDA concentration is studied.

Figure 10 indicates that the maximum recovery of Al_2O_3 at a concentration of 300 g/t was equal to 80% but increasing DDA concentration decreased grade of Al_2O_3. Also, SiO_2 recovery increased from 20 to 38% by increasing the DDA concentration. The best flotation results in term of grade and recovery of Al_2O_3 were achieved for collector concentration of 300 g/t. This condition, the concentrate with a grade of 36% Al_2O_3 and 52% SiO_2 and recovery of 80% Al_2O_3 and 21% SiO_2.

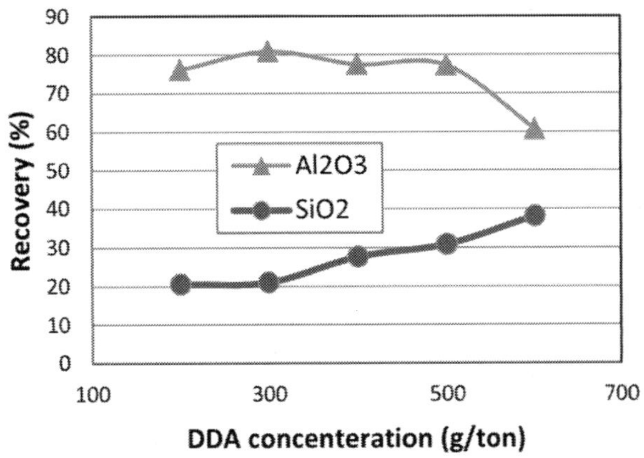

Figure 10. Flotation recovery of Al_2O_3 and SiO_2 as a function of DDA concentration (pH 3, aluminum chloride 300 g/t, desliming -20 μm)

4 Conclusions

This paper showed that for both CPCI and DDA collectors, the best results were obtained at acidic pH. It was shown that the best selectivity was obtained by using CPCI. Aluminum chloride had considerable effect on the floatation performance, which was an effective depressant for quartz. Flotation under acidic condition of pH 3 caused problems such as equipment erosion at industrial scale.

Acknowledgement

The authors of this article would like to thank managers and engineers at Zonouz kaolin-washing plant for providing us with the samples, Lorestan University for providing the required equipment and chemical materials for the experiments and Iranian Mineral Processing Research Center (IMPRC) for performing chemical analysis.

References

1. H. H. Murray, *Applied clay mineralogy: occurrences, processing and applications of kaolins, bentonites, palygorskitesepiolite, and common clays*, vol. *2*. (Elsevier, 2006)
2. M. Valaskova and G. S. Martynková, *InTech* (2012)
3. H. H. Murray, *Int. J. Miner. Process.*, **7(3)**, 263 (1980)

4. R. J. Pruett, *Min. Eng.*, **52(**10), 21 (2000)
5. J. E. Kogel, *SME* (2002)
6. M. C. Fuerstenau, J. D. Miller, and M. C. Kuhn, *Inc., New York*, 132 (1985)
7. R. Koster, B. Schreck, W. von Rybinski, B. Dobiás, R. Kaster, B. Schreck, W. von Rybinski, and B. Dobias, *Miner. Eng.*, **5**(3), 445 (1992)
8. H. Jiang, Z. Sun, L. Xu, Y. Hu, K. Huang, and S. Zhu, *Miner. Eng.*, **65**, 124 (2014)
9. X. Ma, W. J. Bruckard, and R. Holmes, *Int. J. Miner. Process.*, **93**(1), 58 (2009)
10. O. M. S. Rodrigues, A. E. C. Peres, A. H. Martins, and C. A. Pereira, *Miner. Eng.*, **40**, 12 (2013)
11. H. Yuehua, S. Wei, L. Haipu, and Z. Xu, *Miner. Eng.*, **17(9–10)**, 1017 (2004)
12. C. Liu, Y. Hu, and X. Cao, *Miner. Eng.*, **22(** 9–10), 849 (2009)
13. J. Liu, X. Wang, C. L. Lin, and J. D. Miller, *Miner. Eng.*, **78**, 58 (2015)

Value-added materials from the hydrometallurgical processing of jarosite waste

Benjamin P. Wilson[1,a], Petteri Halli[1], Inka Orko[2], Petteri Kangas[2], Mari Lundström[1] and Pertti Koukkari[2]

[1]*Aalto University, School of Chemical Technology, Department of Materials Science and Engineering, P.O. Box 16200, FI-00076 AALTO, Finland*
[2] *VTT – Technical Research Centre of Finland, FI02044-VTT, Finland*

Abstract. Jarosite is a leach residue that can be produced by industrial bulk metal treatment processes and typically has the chemical formula $M_xFe_3(SO_4)_2(OH)_6$, where M normally represents a metal cation. The largest source of jarosite is electrolytic zinc processing [1], which worldwide has an annual production of 11-12 Mt and an associated jarosite waste of 5-6 Mt that can cause important challenges due to its classification as a problem waste. Moreover, as zinc ore typically contains many other commercial/critical metals, the content of valuable materials in this material is significant. An analysis of jarosite from Kokkola, Finland shows that it contained as much metal as many present day commercial ores: ~15% iron, 2% zinc, 3 % lead, 150 g/t silver, 0.5 g/t gold, 100 g/t indium and 40 g/t gallium. Until now, jarosite related research has concentrated on its use in landfill and construction purposes [2], though there is increasing interest in finding methods to efficiently reprocess/recycle jarosite into valuable products [3, 4]. The hydrometallurgical process currently under development by VTT and Aalto University exploits jarosite powdery nature to undertake wet chemical processing. This low cost and energy efficient operation is targeted at the recovery of concentrates which contain the major value-added metals.

References

1. E. Moors, G. Dijkema, *Technol. Forecast. Soc.*, **73**, 250 (2006)
2. E.N. Rathore, M.P. Patil, E. Devendra Dohare, *Int. J. Civil Eng. Technol.*, **5**, 192 (2014)
3. *United Nations Environment Programme*, Report on Metals Recycling (2013)
4. J. Rastas, A. Jarvinen, V. Hintikka, J. Leppinen, *Non-Waste Technol.*, **103(2),** 235 (1989)

[a] Corresponding author: ben.wilson@aalto.fi

E3S Web of Conferences **8**, 01016 (2016) DOI: 10.1051/e3sconf/20160801016
MEC2016

Studies of extractive removal of silver(I) from chloride solutions

Katarzyna Wejman-Gibas[1,a], and Magdalena Pilsniak-Rabiega[1]

[1] *Wroclaw University of Science and Technology, Faculty of Chemistry, Division of Analytical Chemistry and Chemical Metallurgy, 50370 Wroclaw, Poland*

Abstract. For efficient recovery of silver(I) from chloride solutions, amine extractants were used. The effect of the type of diluent, extractant and chloride concentration, contact time and reextraction were studied. It was found that the best diluent for this process was xylene. Extraction of silver depended on concentration of hydrochloric acid, sodium chloride and extractant and additionally on the contact time. Reextraction of Ag(I) ions from the loaded organic phase showed that it can be removed only little using $NH_3 \cdot H_2O$, NaOH, NH_4NO_3, and $NaNO_3$ as the stripping agents.

1 Introduction

Silver has many physical properties which are very difficult to substitute using other metals. It responds greatly to light. It does not corrode. It is a very unusual electricity conductor and has antibacterial properties. Because of these properties many electronics and solar cell manufacturers as well as clothing and pharmaceutical industries have taken an interest in silver. The increased use of this element in production processes has raised its demand. Unfortunately, the amount of available silver is reduced due to irreparable wear. Hence development of an efficient method of obtaining silver is a very important. The most interesting and practical method seems to be hydrometallurgy [1, 2].

Interest in hydrometallurgical methods based on chloride solution leaching of copper sulphide concentrates is due to obvious advantages of these approaches [3-5]. A high solubility of chlorides in comparison to sulphates allows for operation with more concentrated solutions (reduction of liquid volume) and a simultaneous recovery of associated metals like silver, gold and lead. The highly corrosive activity of chloride solutions is no longer a risk due to availability of new construction materials (metals, plastics and ceramics). In addition, chloride leaching is better than other leaching media such as thiourea, cyanide and thiosulphate, due to its low toxicity, low costs and the level of technical "maturity" [6].

Chloride leaching of copper sulphide concentrate under oxidizing conditions changes Ag_2S form of silver into the solution containing chlorocomplexes of silver(I) e.g. $AgCl_2^-$, $AgCl_3^{2-}$, $AgCl_4^{3-}$, and in some cases (such as low concentration of hydrochloric acid or high metal concentration, presence of iron and increased temperature) silver precipitates as AgCl [7].

[a] Corresponding author: katarzyna.wejman@pwr.edu.pl

A low content of silver, a noble metal, in the leach solution makes the process efficient and effective, without losses in solid residues. In addition, all production industries involving silver require purified metals containing as little impurities as possible. One of the best methods of separation is solvent extraction.

Separation of silver(I) ions from the chloride solutions by solvent extraction has been tested and described in several reports [8, 9]. Commonly, the use of Lewis base rule (hard and soft bases and acids) is a guide, where the ligands of the sulphur atoms are considered to be „soft base" and are applied to extract Ag(I) deemed as „soft acids" [10, 11]. For extraction of Ag(I) various organic compounds containing sulphur and phosphorus are used, including Cyanex 471X (triisobutylphosphine sulphide, TIBPS) [10], TBPS (tri-n-butylphosphine sulphide), TOPS (tri-n-octylphosphine sulphide) [12] and bis-O,O-diizobutylophosphorodithioester derivatives [9]. High extraction recoveries of Ag(I) and high selectivity for Fe(III) and Cu(II) were obtained. It was shown that selective extraction of individual metals increases with the length of the alkyl chain between two groups P=S(S), and it had a positive effect on the regeneration process of the loaded organic phase.

Commercial acid extractants such as Cyanex 302 (bis-(2,4,4-trimethylpentyl)-monothiophosphinic acid) and Cyanex 301 (bis-(2,4,4-trimethylpentyl)-dithiophosphinicacid)) [13, 14] also showed a high Ag(I) extraction efficiency. However, neither showed selective extraction of Ag(I) with the co-existing Cu(II) ions in solution and for other noble metals such as Pd(II), Au(III) and Pt(IV).

Ligands containing sulphur and nitrogen (thiourea derivatives) were also checked as effective extractants towards silver(I) ions in concentrated acid and chloride solutions i.e. dodecylthiourea, tetraethylthiuram disulphide, dibutyl- i diphenylthiourea, DPHT and DBT, respectively, as well as other derivatives of thiourea. Dodecyl thiourea was an efficient extractant and allowed for the selective release of Ag(I) from Fe(III) and Cu(II) in strongly acidic chloride solutions [15, 16].

Diphenyl- and di-n-butylothiourea (DPHT and DBT) extractants provided the best extraction efficiency of Ag(I) from the chloride solution containing 2 M HCl and 3 M NaCl at low selectivity ratios in relation to other non-noble metals [16]. When testing the selectivity of Ag(I) extraction from Cu(II) and Fe(III), the separation coefficients of Ag^{+}/Fe^{3+} and Ag^{+}/Cu^{2+} were approximately 3 and 1. In addition, ineffective extraction of the loaded organic phase using sodium thiosulphate was achieved [9].

Extraction of Ag(I) was also conducted using compounds with more complex structure, such as amide derivatives of monomeric calix[4]arene and the compound trialkylthiophosphonate. These extractants systems were well characterized in terms of Ag(I) extraction, but no data are provided in relation to selectivity of Ag(I) extraction to other metals. The resulting extraction recoveries were at low levels, only calix[4]arene obtained an Ag(I) extraction degree of 60% at low acid concentrations [17, 18].

Selectivity was not examined using salts of tri- and tetra alkylammonium, which were used for extraction of Ag(I) to provide a high degree of recovery of Ag(I). These salts caused lower Ag(I) extraction recovery when the concentration of chloride ions in the aqueous solution increased [17].

Despite widespread use in industrial hydrometallurgical processes, phosphorus-containing compounds are considered less environmentally friendly than extractants which have hetero atoms such as either oxygen, nitrogen or sulphur in their structures. For this reason, this study tested the extraction of Ag(I) chloride solutions with amines. This work considers the environmental protection point of view and the lack of reports in the literature about the use of amines. The objective of this study was to examine the process of silver(I) ions extraction from the chloride solutions. The effect of type of diluent, hydrochloric acid concentration, sodium chloride concentration, contact time, extractant concentration and stripping on silver(I) recovery was investigated. Additionally, the process of silver(I) ions extraction was tested from the chloride pregnant leach solutions (PLS), received from atmospheric leaching of the copper flotation concentrate from Lubin Plant of KGHM Polska Miedz S.A..

2 Experimental

2.1 Reagents

Amine extractants were used as organic solvents without further purification. Detailed information on extractants used is presented in Table 1. Xylene (Avantor) and toluene (Avantor) were used as the diluent.

Table 1. Characteristics of amine extractants.

Amine	Sign	Concentration [%]	Supplier
Triethylenetetramine	TETA	97	Sigma-Aldrich
Tris(2-aminoethyl)amine	TAEA	96	Sigma-Aldrich
Diethylenetriamine	DETA	99	Huntsman

2.2 General extraction procedure

Synthetic solutions of silver(I) ions were prepared by dissolving a known amount of silver nitrate ($AgNO_3$) in distilled water with hydrochloric acid and sodium chloride (0.4 – 73.0 g/dm^3 of HCl, 58.5 – 234.0 g/dm^3 of NaCl). The initial Ag(I) concentration for synthetic solutions was 50 mg/dm^3. Chloride leach solutions were generated in atmosphere chloride leaching of a solid residue after sulphate leaching of the commercial flotation sulphide concentrate produced at the Lubin Concentrator (KGHM Polska Miedz S.A.). The solutions after chloride leaching contained ~0.05 g/dm^3 of Ag(I), ~0.50 g/dm^3 of Pb(II), ~0.36 g/dm^3 of Cu(II), ~0.02 g/dm^3 of Co(II), ~0.02 g/dm^3 of Ni(II), and ~0.01 g/dm^3 of Zn(II). The organic phase contained one of three extractants and xylene or toluene as a diluent. Different solutions were used as a stripping agent: NaOH, $NaNO_3$, NH_4NO_3 and $NH_3 \cdot H_2O$. The aqueous and organic phases were shaken at 325 rpm and ambient temperature (25 ± 1°C) using a mechanical shaker ELPAN 357. The concentration of Ag(I) in the aqueous phase before and after extraction and reextraction was determined by means of flame atomic absorption spectrometry (FAAS) using a Varian SpectrAA 20 Plus instrument.

3 Results and discussion

3.1 Effect of type of diluent

The effect of varying diluents such as toluene and xylene in the organic phase (extractant – TETA, DETA, TAEA with concentration 10% (v/v)) on extraction of Ag(I) was studied. The solutions containing 0.05 g/dm^3 of Ag(I), 3.7 g/dm^3 of HCl, 234 g/dm^3 of NaCl were used as the aqueous phase. The results are given in Fig. 1. The best Ag(I) extraction can be observed for TETA extractant, and it can be seen that the extraction percentage is higher when xylene was used as the diluent. Therefore, in the further studies only xylene, as the diluent, was used.

 DOI: 10.1051/e3sconf/20160801016

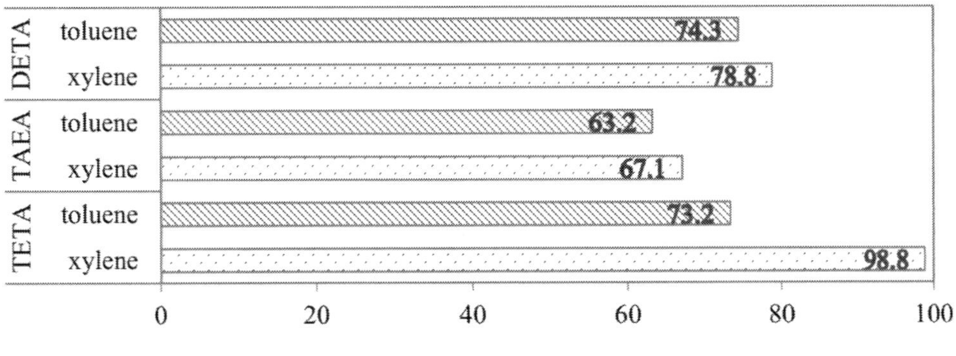

Figure 1. Silver(I) extraction using amine extractants from chloride solutions (extractant: 10% (v/v)).

3.2 Influence of HCl and NaCl concentration on silver(I) extraction

The effect of hydrochloric acid concentration on silver(I) extraction was checked for selected extractants (TETA, TEPA and TAEA) diluted with xylene. The concentration of extractants in the organic phase was equal to 10% (v/v). The solutions containing 0.05 g/dm^3 of Ag(I), 0.4 – 73.0 g/dm^3 of HCl, 58.5 – 234 g/dm^3 of NaCl were used as the aqueous phase. Extraction was performed at ambient temperature. Aqueous (A) and organic phases (O) were contacted for 60 minutes at the phase ratio equal to 1. Figure 2 presents the effect of HCl concentration on extraction of silver with the studied extractants. Extraction of Ag for all extractants decreased with increasing the hydrochloric acid solution concentration. Similar conclusions can be found in [10], where the influence of hydrochloric acid in aqueous phase on silver(I) extraction by using CYANEX 471X was investigated. It was also shown that silver extraction decreased with increasing hydrochloric acid concentration of the aqueous phase. This decrease was attributed to formation of non-extractable anionic silver species in the aqueous phase, which could not be extracted by the organic system.

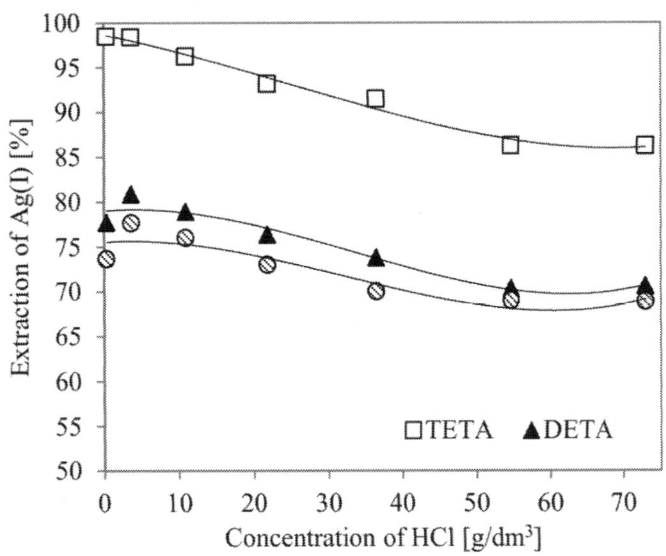

Figure 2. Ag(I) extraction as a function of HCl concentration (extractant: 10% (v/v)).

Figure 3 presents the effect of NaCl concentration on extraction of silver with studied extractants. The extraction efficiency of Ag(I) increased with increasing sodium chloride solution concentration. The highest efficiency of the process for TETA (98%), DETA (83%) and TAEA (78%) was obtained at NaCl concentration of 234 g/dm^3.

The results presented in Figure 2 show that the best Ag(I) extraction efficiency was observed from the solution containing 3.7 g/dm^3 of HCl and 234 g/dm^3 of NaCl in the aqueous phase. The further studies were performed using this composition of the aqueous phase.

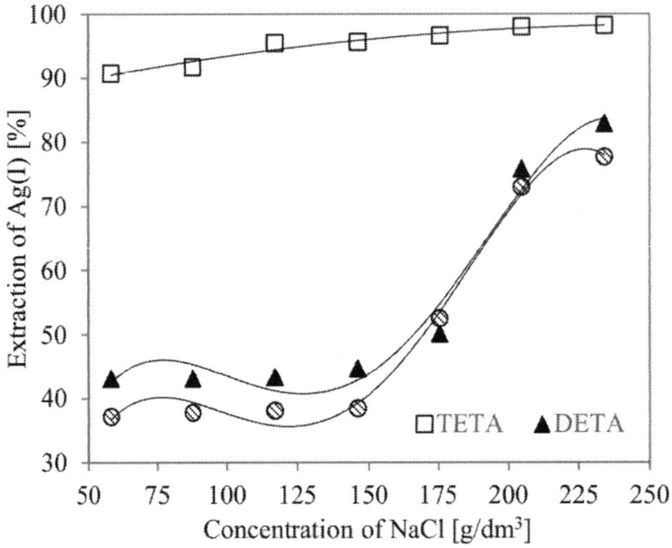

Figure 3. Ag(I) extraction as a function of HCl concentration (extractant: 10% (v/v)).

3.3 Effect of contact time on silver(I) extraction

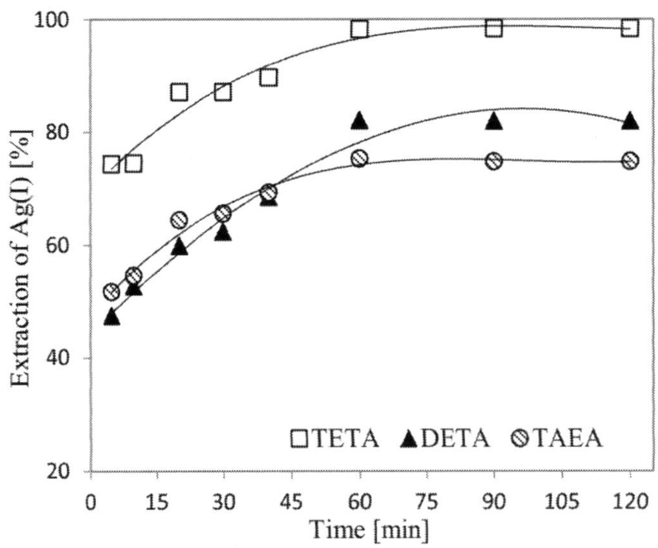

Figure 4. Effect of contact time on silver(I) extraction using TETA, DETA and TAEA as extractants from the chloride solution (extractant: 10% (v/v)).

The effect of contact time on silver extraction is presented in Figure 4, where the extraction efficiency of silver(I) was plotted vs. time. The process was carried out at ambient temperature and the contact time varied in the range of 5 – 120 minutes. The concentration of the extractants (TETA, DETA, TAEA) was equal to 10% (v/v). The phase ratio A/O (aqueous/organic) was equal to 1.

The results show that for extractants tested, the silver(I) ions were removed after 60 min of shaking (Figure 4). The extraction efficiency was equals to 98, 82 and 75% for TETA, DETA and TAEA, respectively. Therefore, the contact time of 60 min was recommended as the time, which enables to achieve the equilibrium of the silver extraction.

3.4 Effect of extractant concentration

To study the effect of extractant concentration the content of TETA, DETA and TAEA in the organic phase varied in the range of 2.5 – 20% (v/v) (Table 2). The aqueous (A) and organic phases (O) were contacted for 60 minutes at the phase ratio A/O equal to 1. Extraction was performed at ambient temperature. It was observed that the extraction efficiency increased with increase of the extractant concentration. The maximum extraction efficiency of silver(I) (96.9, 96.7 and 84.5%, for DETA, TETA and TAEA respectively) was obtained at the highest concentration of extractant equal to 20% (v/v).

This result can be explained in light of increasing number of extractant molecules which leads to the increase of recovery. Applying too high concentration of extractant in the organic phase, while silver concentration in solutions is low, may be unprofitable. Therefore, it was assumed that optimum extractant concentration was 20% (v/v).

Table 2. Silver(I) extraction using amine extractants (TETA, DETA, TAEA) from chloride solutions.

Extractant concentration [% (v/v)]	TETA		TAEA		DETA	
	E [%]	*D*	*E*[%]	*D*	E[%]	*D*
2.5	10.3	0.1	4.4	0.1	3.6	0.04
5	16.9	0.2	43.6	0.8	11.8	0.1
7.5	81.0	4.3	69.8	2.3	59.3	1.5
10	96.5	29.1	75.4	3.1	80.7	4.2
15	96.7	29.0	80.9	4.2	90.6	9.6
20	96.7	27.6	84.5	5.4	96.9	31.8

3.5 Reextraction

A stripping test was performed using freshly loaded and pre-filtered organic phases containing 0.07 g/dm^3 of Ag(I) ions and 0.5 mol/dm^3 of stripping agents ($NH_3{\cdot}H_2O$, NaOH, NH_4NO_3, and $NaNO_3$) in the aqueous phase. Stripping of loaded organic was carried out at one-to-one volume ratios of loaded organic and stripping solutions. After 60 min contact, the phases were allowed to disengage. The aqueous phases were separated and analysed. The results obtained are presented in Figure 6. For all tested extractants low reextraction efficiency of Ag(I) was obtained. The best stripping of Ag(I) ions from the loaded organic phase in the case of TETA and DETA was observed by using 20 g/dm^3 NaOH and the reextraction efficiency was consecutively 22 and 17%. In case of TAEA, the results showed that the metal can be stripped at only 23% efficiency in the one-stage process. Very poor stripping efficiency was obtained using NH_4NO_3, and $NaNO_3$ as the striping agents. The value of reextraction was not higher than 7%.

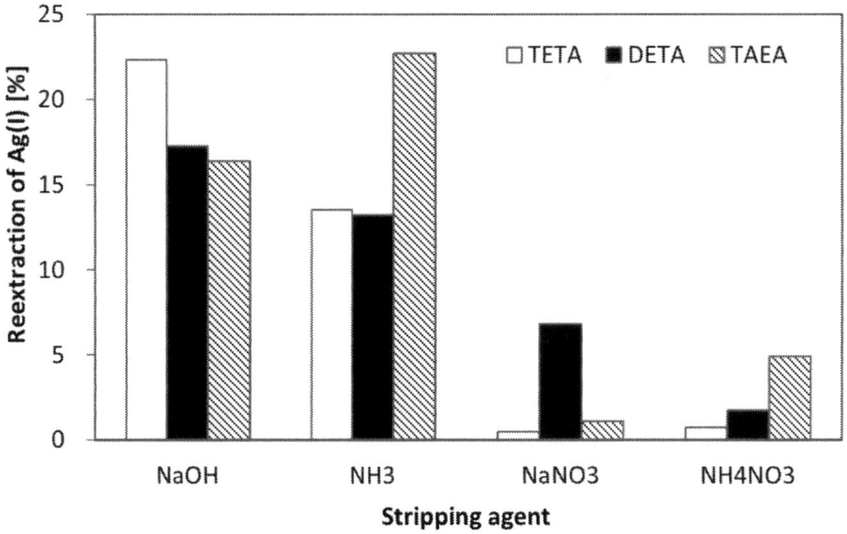

Figure 5. Reextraction of silver(I) for amine extractants (TETA, DETA, TAEA).

3.6 Extraction of silver(I) from multicomponent solutions after chloride leaching

Silver(I) ions were also extracted from the chloride leaching solutions containing lead, copper, cobalt, nickel and zinc. Metals extraction were determined with the organic solution consisting of 10% of extractants (TETA, DETA, TAEA), the A/O ratio of 1:1 and ambient temperature. During extraction, no third phase formation was observed. The extraction efficiency of particular metals and separation factors for silver(I) are shown in Table 3. It can be seen that the metal extraction order was Ag>>Cu>Zn>Ni>Pb>Co for TETA, Ag>>Cu>Co>Zn>Ni>Pb for DETA and Ag>>Cu>Ni> Co>Zn>Pb for TAEA. The highest co-extraction was observed for copper(II), where separation factors (S) were the lowest. In relation to other metals, the most selective extractant for silver(I) was TETA.

Table 3. Silver(I) extraction using amine extractants (TETA, DETA, TAEA) from multicomponent solutions after atmosphere chloride leaching.

Metal	C_{intial} [mg/L]	TETA		DETA		TAEA	
		E [%]	$S^{Ag/Me}$	E [%]	$S^{Ag/Me}$	E [%]	$S^{Ag/Me}$
Ag(I)	46.2	96.6	-	96.8	-	79.5	-
Pb(II)	501.2	0.9	3039.6	0.3	9387.8	1.2	308.7
Cu(II)	364.2	9.7	264.0	20.84	114.9	31.1	8.6
Zn(II)	12.4	2.4	1138.0	4.43	652.7	11.3	30.4
Ni(II)	15.3	1.8	1557.0	3.26	897.1	17.3	18.5
Co(II)	23.7	0.7	4286.2	14.32	181.0	15.1	21.8

4 Conclusions

TETA, DETA and TAEA were used as extractants in solvent extraction of silver(I) ions from the synthetic chloride solutions, and solutions resulting from atmosphere leaching of the copper flotation concentrate. The experimental results showed that Ag(I) can be efficiently extracted from the solutions using amine as an extractant. The best results were obtained by using xylene as a diluent. Additionally, extraction of silver(I) was efficient for 60 minutes of contact time. The process significantly depended on the hydrochloric acid, and sodium chloride concentrations for DETA and TAEA in the aqueous phase, but for TETA the influence of increasing sodium chloride concentration was very small. The most effective results were obtained for the solution containing 3.7 g/dm^3 of HCl and 234 g/dm^3 of NaCl in the aqueous phase. Extraction from the chloride leaching solution showed high Ag(I) ions recovery for all used extractants, and also co-extraction of Cu(II) ions to a small degree. Stripping of Ag(I) ions from the loaded organic phase in the case of TETA, DETA and TAEA was not efficient. The best results for TETA and DETA was observed by using 20 g/dm^3 NaOH (22 and 17% respectively), and for TAEA silver(I) can be stripped with only 23% extraction efficiency.

References

1. Y.J. Park, D.J. Fray, *J. Hazard. Mater.*, **164**, 1152 (2009)
2. S. Lakia, A.A. Shamsabadib, A. Kargaria, *Hydrometallurgy,* **160**, 38 (2006)
3. J.E. Dutrizac, *Hydrometallurgy*, **29**, 1 (1992)
4. J.L. Limpo, J.M. Figueiredo, S. Amer, A. Luis, *Hydrometallurgy,* **28**, 149 (1992)
5. J.L. Limpo, A. Luis, C. Gomez, *Hydrometallurgy*, **28**, 163 (1992)
6. B. Xu, Y. Yang, Q. Li, G. Li, T. Jiang, *Hydrometallurgy,* **147–148**, 79 (2014)
7. O. Dinardo, J.E. Dutrizac, *Hydrometallurgy,* **13**, 345 (1985)
8. A.P. Paiva, *Sep. Sci. Technol.*, **28(4),** 947 (1993)
9. A.P. Paiva, *Solvent Extr. Ion Exch.,* **18(2)**, 223 (2000)
10. Y. Abe, D.S. Flett, *Process Metall.*, **7**, 1127 (1992)
11. P. Malik, A.P. Paiva, N. Neng, J.M.F. Nogueira, *Separ. Sci. Technol.,* **42(6),** 1267 (2007)
12. R.S. Capela, A.P. Paiva, *in Proceedings of ISEC 2002*; K.C. Sole, P.M. Cole, J.S. Preston, and D.J. Robinson, (eds.), Cape Town, South Africa, 335 (2002)
13. A.M. Shafiqul, K. Inoue, K. Yoshizuka, Y. Dong, P. Zhang, *Hydrometallurgy*, **44**, 245 (1997)
14. K.J. Rajesh, H.I. Lee, J.Y. Lee, J.S. Kim, J.S. Sohn, *Sep. Purif. Technol.*, **63**, 184 (2008)
15. G.J. Zuo, M. Muhammed, *Sep. Sci. Technol.* **25**, 13 (1990)
16. A.P. Paiva, *in Proccedings of Extraction and Processing Division Congress*, TMS Annual Meeting (Taylor, P.R., Ed.), The Minerals, Metals & Materials Society, Warrendale, 665 (2001)
17. D. Sevdic, H. Meider, D. Pavkovic, *Solvent Extr. Ion Exch.*, **8(4/5)**, 643 (1990)
18. K. Ohto, H. Yamaga, E. Murakami, K. Inoue, *Talanta,* **44**, 1123 (1997)
19. H.M.N.H. Irving, Damodaran A.D., *Anal. Chim. Acta,* **48(2)**, 67 (1969)

Separation of mercury in industrial processes of Polish hard steam coals cleaning

Krzysztof Wierzchowski[1,a], and Ireneusz Pyka[1]

[1]*Glowny Instytut Gornictwa, 40-166 Katowice, Poland*

Abstract. Coal use is regarded as one of main sources of anthropogenic propagation of mercury in the environment. The coal cleaning is listed among methods of the mercury emission reduction. The article concerns the statistical assessment of mercury separation between coal cleaning products. Two industrial processes employed in the Polish coal preparation plants are analysed: coal cleaning in heavy media vessels and coal cleaning in jigs. It was found that the arithmetic mean mercury content in coarse and medium coal size fractions for clean coal from heavy media vessels, amounts 68.9 μg/kg, and most of the results lay below the mean value, while for rejects it amounts 95.5 μg/kg. It means that it is for around 25 μg/kg greater than in the clean coal. The arithmetic mean mercury content in raw coal smalls amounts around 118 μg/kg. The cleaning of smalls in jigs results in clean coal and steam coal blends characterized by mean mercury content 96.8 μg/kg and rejects with mean mercury content 184.5 μg/kg.

1 Introduction

Coal use is regarded as one of main sources of anthropogenic propagation of mercury in the environment. [1-4]. Determination of the mercury content in raw coal and assessment of potential reduction of mercury in used coal have become the crucial topic for coal dependent economies. Coal cleaning is regarded as one of the methods of mercury in coal reduction. This method is so called pre-combustion [3-9]. Thus, coal cleaning, which reduces the ash in coal and increases the calorific value of the coal fuel, has gained additional significance.

The existing results of research have revealed that cleaning of Polish hard steam coal enables to reduce the mercury content in clean coal, in comparison with raw coal, by 18–79 % [10-12]. More detailed research showed that the result of coal cleaning, in terms of the mercury content reduction, depended on technological characteristics of given coal and on the size fraction of cleaned coal [12]. By the technological characteristics we understand the distribution of ash, mercury and other coal quality parameters as a function of coal grains size and density. The coal cleaning processes were invented to increase the quality parameters of coal fuels, what was quantified first of all by ash reduction and increase of calorific value. The mercury reduction obtained in the coal cleaning processes can be treated as a supporting effect.

[a] Corresponding author: kwierzchowski@gig.eu

The aim of this paper is a statistical assessment of mercury separation between coal cleaning products in industrial processes in the Polish hard steam coal preparation plants. Two coal cleaning processes and results of separation in industrial cleaning devices were analysed:

- coal cleaning of +20 mm size fraction in dense media vessels to produce coarse and medium coal size grades (cobbles, nuts, pea coals)
- smalls cleaning in jigs.

The performed work concerned assessment of results of industrial cleaning devices without analyzing of the effectiveness of their performance. Additionally, in one location the variation of ash and mercury content in the feed (raw coal) and the clean coal was investigated.

2 Material and methods

In 2015 all coal size fractions produced in Polish collieries producing hard steam coal (excluding collieries belonging to Jastrzebska Spolka Weglowa S.A.) were sampled.

Gross samples of all coal size grades were collected from increments taken for from one to two weeks, depending on local conditions. The increments were taken at the same time with those sampled in routine sampling operations in all collieries under consideration. In the case of works concerning the variation of mercury in the feed and the clean coal from a jig, increments taken constituted the gross samples. This sampling lasted one month. One sample of the feed and one sample of the clean coal were taken per day. Sampling conformed to requirements of the following standards:

- hard coal and brown coal: methods of sampling and laboratory samples preparation PN-90/G-04502
- hard coal and coke: mechanical sampling - Part 2: Coal – Sampling from moving streams PN-ISO 13909-2
- hard coal and coke: mechanical sampling - Part 3: Coal – Sampling from stationary lots PN-ISO 13909-3
- hard coal and coke: mechanical sampling – Part 4: Coal – Preparation of test samples PN-ISO 13909-4
- hard coal and coke: manual sampling PN-ISO 18283.

The ash and mercury contents were determined using the following standards and procedures:

- solid fuels: determination of ash by gravimetric method PN-80/G-04512 and PN 80/G-04512Az1
- the certified internal procedure, elaborated in Glowny Instytut Gornictwa No. SC-1.PB.23 applying the Cold-Vapor Atomic Absorption Spectrometry, using the analyzer MA-2000 of Nippon Instrument Corporation. It is a fully automated measurement system for determination of mercury content in solid materials, gases and liquids through sample combustion or evaporation.

In this article the results of determination of ash in coal are given in % (air dried basis) and the mercury content is given in μg/kg (air dried basis).

3 Results

The results of determination of mercury content in the cleaned coal and rejects products of cleaning of +20 mm size fractions in heavy media vessels in 27 Polish coal preparation plants are illustrates in Fig. 1. The statistical distributions of mercury content in both groups of products (cleaned coals and rejects) are presented in Fig. 2 and 3.

DOI: 10.1051/e3sconf/20160801017

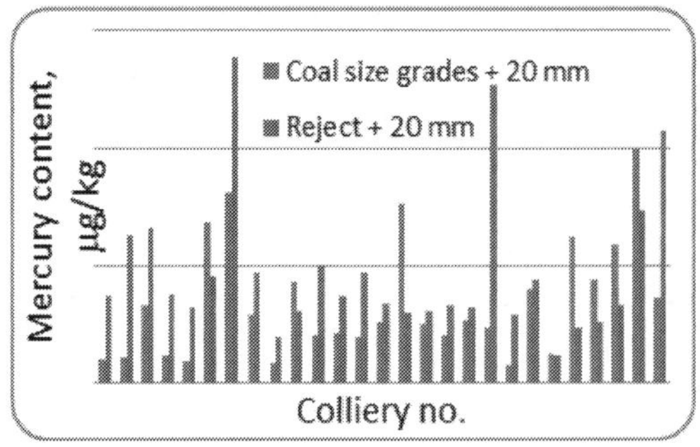

Figure 1. Mercury content in products (cleaned coal and rejects) of cleaning of +20 mm size fractions in heavy media vessels in 27 Polish coal preparation plants.

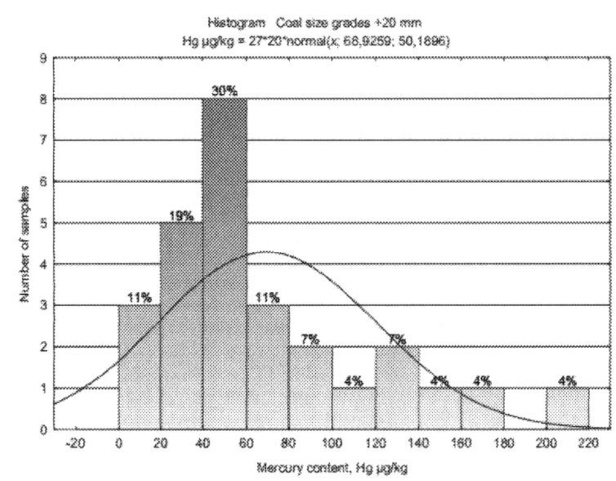

Figure 2. Distribution of mercury content in cleaned coals (coarse and medium coal size fractions – commercial products) from + 20 mm size fraction cleaning.

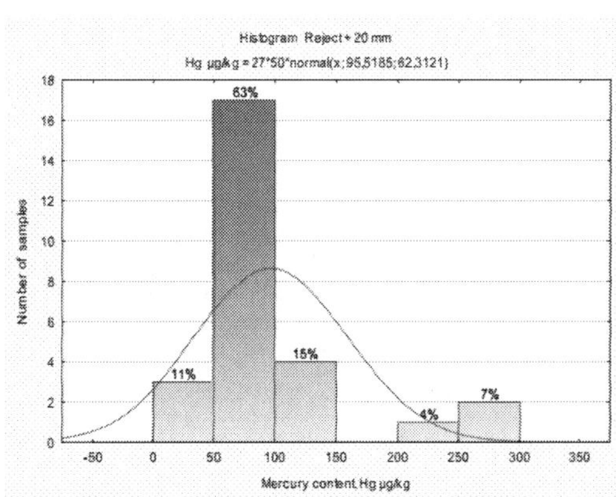

Figure 3. Distribution of mercury content in rejects from + 20 mm size fraction cleaning.

Figures 4, 5 and 6 illustrate distributions of the mercury content in raw smalls and the products of their cleaning in jigs in Polish coal preparation plants. The amounts of feeds and products are not equal to the number of coal preparation plants because in some cases two streams of smalls are treated and they could be analysed separately. The results of analyses of the variation in one month period of ash and mercury in raw smalls (feed) and cleaned coal from a jig are presented in Fig. 7 and 8.

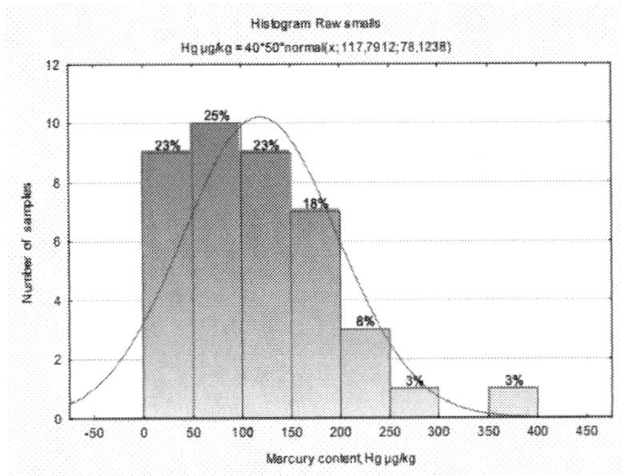

Figure 4. Distribution of mercury content in raw smalls.

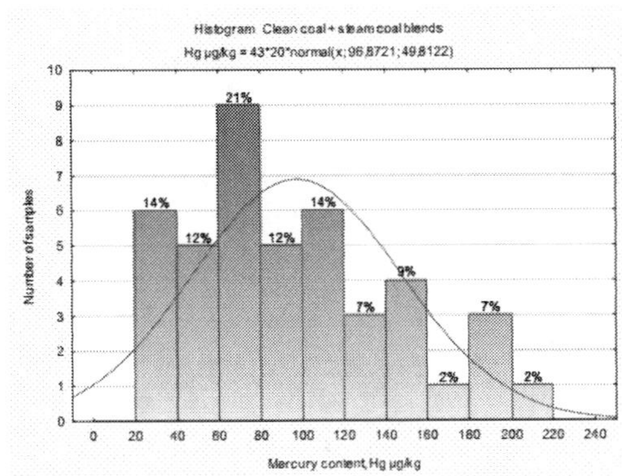

Figure 5. Distribution of mercury content in cleaned coals and steam coal blends.

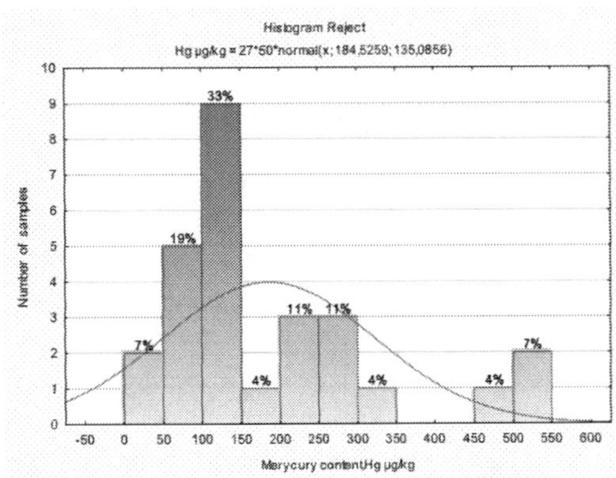

Figure 6. Distribution of mercury content in rejects from smalls cleaning.

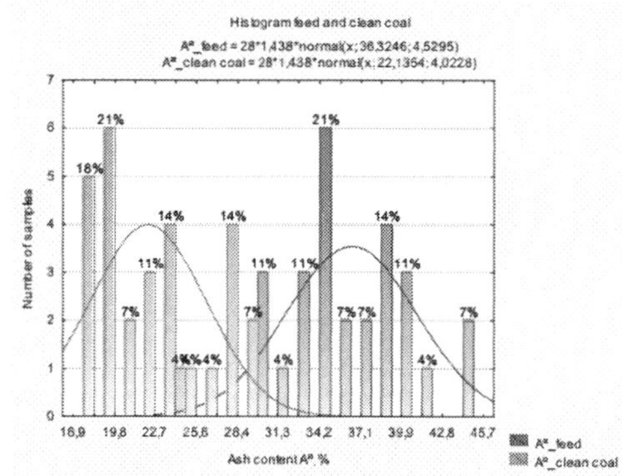

Figure 7. Variations of ash in raw smalls and the clean coal in one of Polish coal preparation plants.

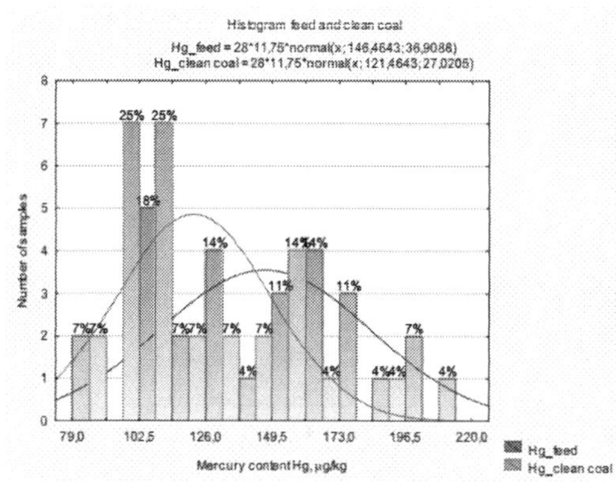

Figure 8. Variations of mercury in raw smalls and the clean coal in one of Polish coal preparation plants.

4 Results and discussion

The mercury content in coarse and medium coal size grades fits the broad interval: 20 – 220 μg/kg. In the most cases mercury in cleaned coals is lower than in rejects from the same process (Fig. 1). In 7 cases only, mercury in cleaned coals is higher than in rejects form the same process. It illustrates differences in technological characteristics of coal cleaned in different coal preparation plants. The densities of heavy media in Polish coal preparation plants amount mainly 1.6-1.7 g/cm^3, although there are exceptions. In the first case mercury concentrates in fraction +1.7 g/cm^3, while in the second case in the fraction -1.7 g/cm^3. The arithmetic mean mercury content in coarse and medium coal size fractions amounts 68.9 μg/kg (Fig. 2). Most of the results lie below the mean value. The characteristic feature of this distribution is the "long tail" values greater than 100 μg/kg. The arithmetic mean mercury content in rejects from the processes of +20 mm size fraction cleaning amounts 95.5 μg/kg (Fig. 3). It means that it is more than 25 μg/kg greater than in clean coals. In the case of rejects the histogram is more concentrated around the mean value but it is also characterized by "the long tail". Three values exceed 200 μg/kg. Taking into consideration that the yield of clean coals form the cleaning +20 mm size fractions amounts 40 - 70 % of the feed mass one can assess that the cleaning of +20 mm coal size fraction in Polish coal preparation plants gives the medium reduction of mercury load in commercial products, in comparison to raw coal by around 50%.

The mercury content in raw smalls fits also the broad interval: 20 – 400 μg/kg, and the arithmetic mean value amounts 117.8 μg/kg (Fig. 4). The histogram is concentrated around the mean value but it is also characterized by "the long tail" and two values exceed 250 μg/kg. From such raw smalls the commercial products from coal cleaning were obtained (clean coals and steam coal blends), for which the mercury distributions are illustrates in Fig. 5. The pure clean coal is sold as a commercial steam coal products only in a few cases, so the analyses are broaden by steam coal blends containing the clean coal. The analysis doesn't assess the whole potential of mercury reduction in processes of smalls cleaning. The mercury content in smalls, commercial products, fits less broad interval: 20 – 220 μg/kg, and the mean value amounts 96.8 μg/kg. The mercury content in rejects from smalls cleaning fits extremely broad interval: 20 – 550 μg/kg, and the mean value amounts 184.5 μg/kg. Most of the results lie below the mean value but the distribution is characterized by the "long tail" In three cases the mercury content is greater than 450 μg/kg.

Unfortunately, the variation of quality parameters of feeds to jigs is significant. Most of Polish coal preparation plants are not equipped in systems for feed quality stabilizing. Fig. 7 and 8 illustrate the statistical attempt to assess dependence of some clean coal quality parameters on the parameters of feed. Although, in the case of ash and mercury, for the whole population of results of one month lasting analyses, the mean values for clean coal are smaller than for raw coal (feed). A huge variation of both parameters can be observed. In the case of ash reduction is round 14% from 36.32% to 22.14%, and in the case of mercury the reduction was around 25 μg/kg, from 146 μg/kg do 121 μg/kg. It is rather impossible to obtain ash in clean coal greater than in raw coal, but in the case of mercury such situations are observed. The distributions of ash in the feed and in the clean coal are more concentrated around the mean values than in the case of the distributions of mercury in the feed and the clean coal. It can be concluded that variation of coal in terms of mercury content is greater than in terms of ash.

5 Conclusions

1. The arithmetic mean mercury content in coarse and medium coal size fractions from the processes of +20 mm size fraction cleaning in Polish coal preparation plants amounted 68.9 μg/kg in 2015. Most of the results lied below the mean value. The arithmetic mean mercury content in rejects amounted 95.5 μg/kg, it means, it was more than 25 μg/kg greater than in clean coals.
2. The arithmetic mean mercury content in raw smalls in Polish coal preparation plants amounted 117.8 μg/kg in 2015. Smalls all commercial products composed using clean coal, were characterised by the medium mercury content 96.8 μg/kg, while the mean mercury content in rejects from smalls cleaning amounted 184.5 μg/kg.
3. Raw small are characterized by huge variation of quality parameters what had to influence the cleaning processes. The statistical analysis of performance of a jig in one of Polish coal preparation plans showed that despite of the fact that the medium mercury content in clean coal was smaller than in raw smalls (feed) in some cases (days) the instantaneous results of mercury determinations in the feed and clean product were inverse.

Acknowledgements

The results of the determination of mercury commercial products (coal size grades) of Polish hard coal collieries have been obtained in the project funded within the framework of the Applied Research Program of the Polish National Centre for Research and Development, entitled: The elaboration of a data base on mercury content in domestic coal, technological guidelines of further mercury reduction, including the defining of benchmarks for domestic indices of mercury emission" – acronym "Hg Base" ("Baza Hg"), (PBS2/A2/14/2013).

References

1. UNEP Chemical Branch. *The Global Atmospheric Mercury Assessment: Sources, Emissions and Transport (* UNEP-Chemicals, Geneva, 2008)
2. A. Glodek, J.M. Pacyna, *Atmos. Environ.*, **43**, 5668 (2009)
3. B. Toole-O'Neil, S.J. Tewalt, R.B. Finkelman, D.J. Akers, *Fuel Process. Technol.*, **78**, 47 (1999)
4. J.H. Pavlish, E.A. Sondreal, M.D. Mann, E.S. Olson, K.C. Galbreath, D.L. Laudal, S.A. Benson, *Fuel Process. Technol.*, **82**, 89 (2003)
5. M. Mastalerz, A. Drobniak, *Int. J. Coal Geol.*, **62**, 223 (2005)
6. L.L. Sloss, *IEA Clean Coal Centre Raport No. CCC/134* (London, UK, 2008)
7. L.L. Sloss, *IEA Clean Coal Centre Raport No. CCC/*195 (London, UK, 2012)
8. UNEP-United Nations Environmental Programme (DTIE, Geneva, Switzerland, 2010)
9. Ya.E. Yudovich, M.P. Ketris, *Int. J. Coal Geol.*, **62,** 135 (2005)
10. T. Dziok, A. Strugala, A. Rozwadowski, J. Gorecki, S. Ziomber, *Energy Pol.*, **17(4)**, 277 (2014)
11. I. Pyka, K. Wierzchowski, *Physicochem. Probl. Miner. Process.* **52(1),** 193 (2016)
12. I. Pyka K. Wierzchowski, *Arch. Min. Sci.*, **55**, 347 (2010)

Spherical agglomeration of acetylsalicylic acid

Izabela Polowczyk[1,a], and Karolina Szczesniak[1]

[1] *Division of Chemical Engineering, Faculty of Chemistry, Wroclaw University of Science and Technology, Wybrzeze Wyspianskiego 27, 50-370 Wroclaw, Poland*

Abstract. In this paper spherical agglomeration of acetylsalicylic acid was described. In the first step, the system of good and poor solvents as well as bridging liquid was selected. As a result of a preliminary study, ethyl alcohol, water and carbon tetrachloride were used as the good solvent, poor one, and bridging liquid, respectively. Then, the amount of acetylsalicylic acid and the ratio of the solvents as well as the volume of the bridging liquid were examined. In the last step, the agglomeration conditions, such as mixing intensity and time, were investigated. The spherical agglomerates obtained under optimum conditions could be subjected to a tableting process afterwards.

1 Introduction

Spherical agglomeration belongs to a group of spherical crystallization processes and has been successfully applied in the pharmaceutical industry in the drugs preparation [1]. In general, in this technique, a particulate substance dispersed in water can be aggregated by immiscible liquids which wet the particles and form bridges between them as well as, in the course of mixing, gather the particles [2]. Under favourable conditions spherical agglomerates can be formed [3]. If the active substance is firstly dissolved in a good solvent, and then poured into a poor solvent in which the crystallization occurs, and then agglomeration of particles is caused by various mechanisms, this technique belongs to crystallization processes.

In an enhanced spherical crystallization technique so-called crystallo-co-agglomeration, the active substance is crystallized and agglomerated with either excipient or another drug, which may or not be crystallized in same systems [4]. A quasi emulsion solvent diffusion method (QESD) is one more modification of spherical crystallization, in which the drug dissolved in a good solvent and dispersed in a poor one producing quasi-emulsion. Then, the good solvent step by step diffuses out of the emulsion droplets into the surrounding phase, and the poor solvent diffuses into the droplets. As a result, the drug crystallizes inside the droplets [5]. In another technique called ammonia diffusion method (AD), the mixture of three partially immiscible solvent can be used as a crystallization system. In this system the ammonia solution acts as a bridging liquid as well as the good solvent. The drug precipitated by the solvent changes without forming ammonium salt. The water immiscible solvent promotes liberation of liquid ammonia [1].

[a] Corresponding author: izabela.polowczyk@mail.pwr.edu.pl

Spherical crystallization has many applications in the pharmaceutical industry such as improving flowability and compressibility of poorly compressible active substances, masking bitter taste of drugs and bettering solubility and disintegration of poorly soluble drugs [5].

In this work, the process of spherical agglomeration of acetylsalicylic acid has been described. Ethyl alcohol, water and carbon tetrachloride were used as a good solvent, a poor one, and a bridging liquid, respectively. The amount of acetylsalicylic acid and the ratio of the solvents as well as the volume of the bridging liquid, were examined. Finally, the agglomeration conditions, such as mixing intensity and time, were investigated, and the optimal conditions of the process were determined.

2 Materials and methods

In this study, acetylsalicylic acid was investigated in terms of a spherical agglomeration process. Acetylsalicylic acid is a medication, often used to treat pain, fever and inflammation. Acidum acetylsalicylicum is an acetyl derivative of salicylic acid. It is obtained in the estrification reaction of salicylic and acetic acids. It appears in a form of long and white crystals.

In order to select proper solvents for agglomeration, solubility of acetylsalicylic acid in ethanol, tetrachloride and water were investigated. Also, a miscibility of all solvents was examined. Acetylsalicylic acid dissolves in ethanol and does not dissolve in water and carbon tetrachloride. Ethyl alcohol is miscible in water in all proportions. On the other hand, carbon tetrachloride can be mixed with ethanol but it does not dissolve in water. Therefore, this system of solvents is recommended for spherical crystallization and agglomeration of acetylsalicylic acid. In this paper the optimum volume of a bridging liquid ranging from 1.5 to 2.0 ml per 2.0 g of drug was examined.

To determine the maximum volume of bridging liquid, resulting in spherical agglomeration but not paste formation, different amounts of carbon tetrachloride was examined. It was found that with 1.5 ml per 2.0 g of acetylsalicylic acid a partial agglomeration started. On the other hand, with 2.0 ml of the bridging liquid, agglomerates started to stick together. In this range, the volume of carbon tetrachloride was changed.

A 2.0 g of acetylsalicylic acid was dissolved at 80 °C in 10 ml of ethanol. Afterwards, a clear solution was poured into 80 ml of water and mixing at 600 rpm was started. After formation of crystals, the bridging liquid was gradually added. Mixing was continued for approximately 15 min. Obtained agglomerates were separated on a sieve and left for drying in the air.

3 Results and discussion

3.1 Optimum conditions of agglomeration

The amounts of solvents were selected on the basis of pictures analysis of obtained agglomerates. It was observed that application of poor solvent (water) led to creation of bigger agglomerates. In addition, the greater volume of ethanol in which acetylsalicylic acid crystals were dissolved, the more difficult became its recrystallization in the poor solvent.

Regarding mixing, crystallization is more effective with prolonged time of mixing before addition of bridging liquid. Additionally, paste formation was observed if the bridging liquid was added in one portion. Thus, carbon tetrachloride was dosed in portions of 0.1 ml.

Too intensive mixing (above 1000 rpm) caused gathering of crystals on the surface of water and obstructed its agglomeration via bridging. Also, prolonged agglomeration time was not recommended, because the formed agglomerates started to break up above 15 minutes.

As soon as agglomeration process was completed, the products were separated on a sieve and agglomerates were left to dry in the air. It was observed that in course of drying, agglomerates were getting harder as a result of surface crystallization of the drug from evaporating solvents.

Analyzing above mentioned conditions, the process parameters were selected to avoid paste formation as well as bigger agglomerates with a broad diameter distribution.

Based on the analysis of agglomerates images, the optimum conditions for spherical agglomeration of the acetylsalicylic acid crystals were established. Those parameters were shown in Table 1. The optimum volume of a bridging liquid was 1.8 ml of carbon tetrachloride for the solid-to-liquid ratio of 2g/10 ml and 80 ml of water. A mixing time before addition of carbon tetrachloride, sufficient for recrystallization, was 5 minutes, while the total time of the process was 15 minutes.

For the optimum parameters of the agglomeration process, the diameter of agglomerates was about 1-3 mm. The agglomerates of this size were expected to show good flowability and compressibility properties [5].

Table 1. Spherical agglomeration process parameters.

Parameter	**Value**
Amount of acetylsalicylic acid	2 [g]
Volume of ethanol	10 [ml]
Volume of water	80 [ml]
Volume of carbon tetrachloride	1.8 [ml]
Mixing intensity	600 [rpm]
Time of mixing of aspirin dissolved in ethanol with water	5 [min]
Total mixing time	15[min]

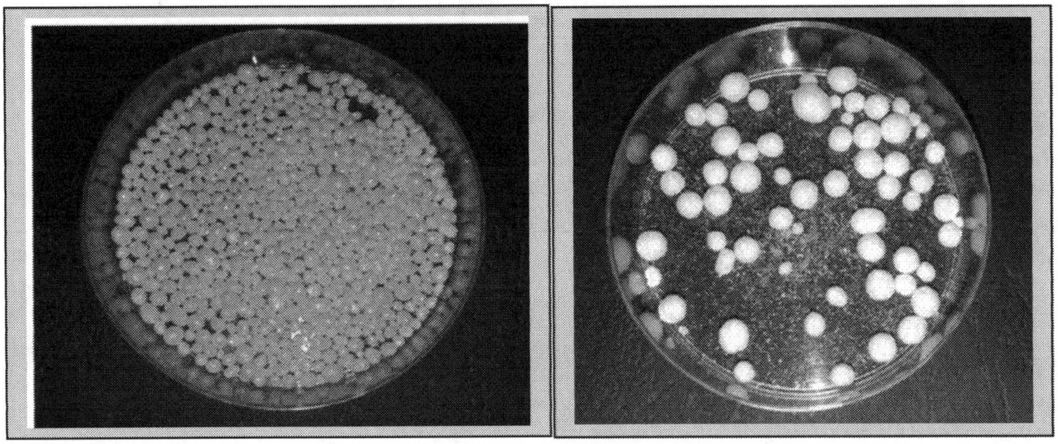

Figure 1. Acetylsalicylic acid crystals agglomerates obtained for the solid-to-liquid ratio of 2 g/10 ml (left) 1 g/10 ml (right) as well as 1.8 ml (left) and 0.8 ml (right) of carbon tetrachloride as the bridging liquid.

The volume of the bridging liquid in the range of 1.5-2.0 ml was used to determine the optimum conditions. The photograph of agglomerates obtained for 1.8 ml of bridging liquid is shown in Fig. 1 (left). Changing a solid-to-liquid ratio to 1 g/10 ml and using 0.8 ml of carbon tetrachloride resulted in coarser agglomerates, as shown in Figure 1 (right).

3.2 Microscopic analysis

The crystals of acetylsalicylic acid before dissolving in ethanol are presented in Fig. 2a, while Fig. 2b shows the crystals from broken agglomerates. Figure 2c shows the example of agglomerate of recrystallized particles by addition of carbon tetrachloride as the bridging oil.

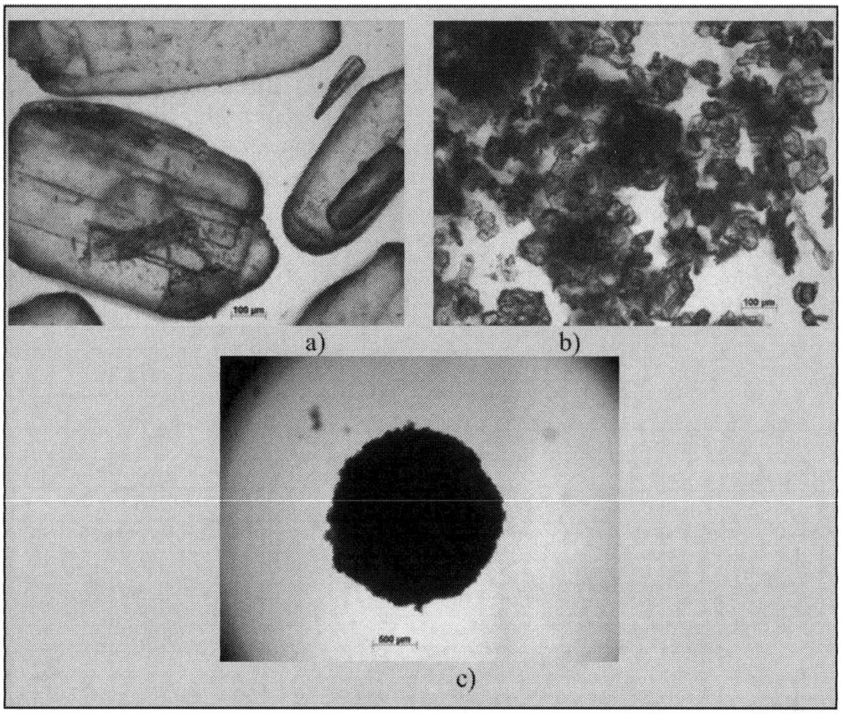

Figure 2. (a) Acetylsalicylic acid crystals before dissolving in ethanol, (b) crystals from broken agglomerates and (c) example of agglomerate of recrystallized particles by addition of carbon tetrachloride as the bridging oil.

From Fig. 2 one can see that dissolved acetylsalicylic acid crystals recrystallized in a course of agglomeration process and the agglomerates were finer than originally used ones. In addition, crystals after spherical crystallization were of irregular shape and interlocked each other, while elongated and well-formed orthogonal particles were used. On the other hand, the agglomerates were found to be close-packed and of spherical shape. This was a result of addition and bridging effect of carbon tetrachloride, where fine crystals were bound together via the oil bridges in the aqueous medium. Acetylsalicylic acid particles were naturally hydrophobic and wetted by the apolar liquid. During mixing, the bridging liquid droplets collided with emerging crystals and spherical agglomerates were formed [2].

4 Conclusions

From the obtained results the following conclusion can be drawn.

1. Ethyl alcohol, water and carbon tetrachloride can be used as a good solvent, a poor one and a bridging liquid, respectively. This system can be appropriate for recrystallization and spherical agglomeration of acetylsalicylic acid.
2. Crystals of acetylsalicylic acid dissolved in ethanol recrystallized after addition to water and were significantly finer in agglomerates.
3. The crystals of acetylsalicylic acid after spherical crystallization were of irregular shape and interlocked each other, when elongated and well-formed orthogonal particles were used.
4. The spherical agglomerates of acetylsalicylic acid were obtained after addition of carbon tetrachloride and were found to be close-packed.
5. The optimum conditions of agglomeration of acetylsalicylic acid were found to be as follows:1.8 ml of carbon tetrachloride as the bridging liquid for the solid-to-liquid ratio of 2g per 10 ml good

solvent and 80 ml of poor solvent (water). In this case, the mean diameter of agglomerates ranged between 1-3 mm.
6. The agglomerates of this size were expected to show the good flowability and compressibility properties and could be subjected to a tableting process afterwards.

References

1. Y. Kawashima, F. Cui, H. Takeuchi, T. Hino, T. Niwa, K. Kiuchi, *Int. J. Pharm.* **119,** 139 (1995)
2. J. Drzymala, *Mineral Processing. Foundations of Theory and Practice of Minerallurgy* (Ofic. Wyd. PWR, Wroclaw, 2007)
3. S. Duzyol and A. Ozkan, *Sep. Pur. Technol.* **72**, 7 (2010)
4. N.B. Rahate, M.M. Bodhankar, P.N. Dhoke, *J. Drug Deliv. Therap.* **3(4)**, 178 (2013)
5. S.V. Patil, S.K. Sahoo, *Der Pharmacia Lettre*, **2 (1)** 421 (2010)

Effect of foamability index of short chain alkyl amines on flotation of quartz

Sabina Szczerkowska[1], Marta Siedlarz[1], Marta Wojcik[1], and Przemyslaw B. Kowalczuk[1,a]

[1] *Wroclaw University of Science and Technology, Faculty of Geoengineering, Mining and Geology, Wybrzeze Wyspianskiego 27, 50-370 Wroclaw, Poland*

Abstract. Amines can be used for flotation of various minerals, especially quartz. The flotation efficiency of quartz depends on the amine type and dose. It was proved that the shorter alkyl amine, higher amine concentration has to be used to recover quartz at the same level. In flotation amines play a role of both collectors and frothers. The ability of a amine to collect particles can be expressed in the form of contact angle, while the foaming properties by different parameters including dynamic foamability index (DFI) and critical coalescence concentration (CCC). Determination of DFI and CCC requires advanced techniques and methods. Therefore, in this paper a rapid and facile method for determination of foaming properties of amines and also other surfactants was used. It was based on measuring the initial foam and froth heights in a conventional flotation machine at different concentrations of surfactants. The foam height-concentration curve was described by utilizing an empirical equation which was based on one-adjustable parameter called the foamability index (*FI*). In this work the foamability index was determined for butylamine ($ButNH_2$), hexylamine ($HexNH_2$) and octylamine ($OctNH_2$) as examples of short chain alkyl amines. The determined foamability indices were 92, 12 and 4 mg/dm^3 for $ButNH_2$, $HexNH_2$ and $OctNH_2$, respectively. It was shown that when the flotation results of quartz were presented in the form of recovery versus normalized amine concentration in relation to the foamability index (*c/FI*), all the experimental data points converged to one curve. It indicates that amines act similarly but at different concentrations expressed as *FI*. The foamability index seems to be a useful parameter for characterizing any flotation frother.

Acknowledgement

This work was financed by the Polish Statutory Research Grant B50199.

[a] Corresponding author: przemyslaw.kowalczuk@pwr.edu.pl

DOI: 10.1051/e3sconf/20160801020

CURRENT STATE, IMPROVEMENTS AND LATEST TRENDS IN COAL PREPARATION IN POLAND

Wiesław S. BLASCHKE[1], Ireneusz BAIC[1], Józef SZAFARCZYK[1]

[1]Institute Mechanised Construction & Rock Mining, Branch in Katowice

INSTITUTE OF MECHANISED CONSTRUCTION & ROCK MINING
BRANCH IN KATOWICE

CURRENT STATE, IMPROVEMENTS AND LATEST TRENDS IN COAL PREPARATION IN POLAND

Wiesław S. BLASCHKE[1], Ireneusz BAIC[1], Józef SZAFARCZYK[1]

[1]Institute of Mechanised Construction & Rock Mining, Branch in Katowice

Mineral Engineering Conference MEC2016

Mineral Engineering Conference
MEC2016
25 -28 September 2016,
Świeradów-Zdrój, Poland

Instytut Mechanizacji Budownictwa i Górnictwa Skalnego

INTRODUCTION

Poland is the largest coal producer in Europe.
In 2015 the coal output of steam coal amounted to 59.21 million tonnes and 12.98 million tonnes of coking coal.

Poland also has the largest coal resources.
They amount to 52.0 billion tonnes (resources supposed economic), but the amount of economic reserve base is deemed to be 3.7 billion tonnes.

Coal can be found in two regions in Poland. These are the Upper Silesian Basin and the Lublin Basin.

2

Current state, improvements and latest trends in coal preparation in Poland

Instytut Mechanizacji Budownictwa i Górnictwa Skalnego

OCCURRENCE OF HARD COAL DEPOSITS IN POLAND

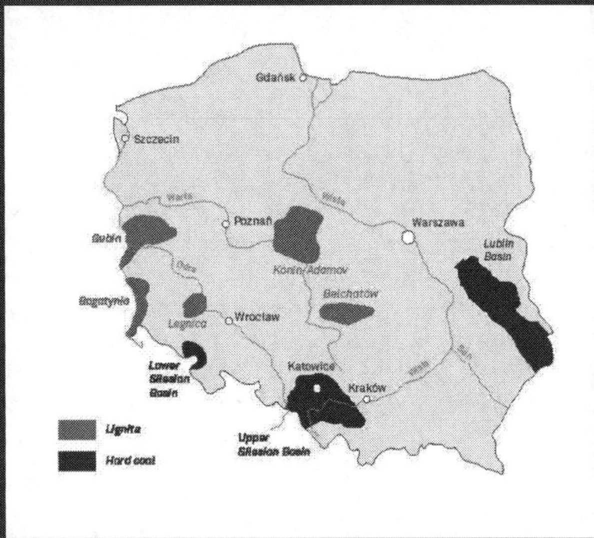

3

Current state, improvements and latest trends in coal preparation in Poland

Instytut Mechanizacji Budownictwa i Górnictwa Skalnego

INTRODUCTION

In the Upper Silesian Coal Basin the following coal companies operate (the status on the mid of 2016):

- Kompania Węglowa S.A. (on 1.05.2016 Polish Mining Group) – operating in 11 coal mines,
- Katowicki Holding Węglowy S.A. – operating in 4 coal mines,
- Węglokoks Kraj Sp. z o.o. - operating in 2 coal mines,
- Jastrzębska Spółka Węglowa S A. – operating in 5 coal mines,
- TURON Wydobycie S.A. – operating in 3 coal mines.

In the Lublin Coal Basin - operating in one mine (LW „Bogdanka" S.A.).

There are also the following small mines:

- PG „Silesia" Sp. z o.o. - property of a Czech coal company,
- Siltech Sp z o.o. – private mine,
- ECO-PLUS Sp. z o.o. – private mine.

4

Current state, improvements and latest trends in coal preparation in Poland

Instytut Mechanizacji Budownictwa i Górnictwa Skalnego

WASHED COAL PRODUCTION

Approximately 60% of the Polish coal production is washed with some form of coal preparation.

The saleable coal has the following quality parameters:

- coking coal - ash content varies from 5.4% to 8.8% (average 6.8%) and sulphur content varies from 0.51% to 0.89% (average 0.67%),
- steam coal for the power industry - the net calorific value ranges from 15.4 to 25.4 kJ/kg, ash content ranges from 9.9 to 30.1% (average 22.4%) and sulphur content from 0.56 to 2.59% (average 0.83%).

In Poland some power stations are adapted to burn raw coal (non-prepared coal) - its net calorific value can be 19.2 MJ/kg, and sometimes even less, whereas the ash content can be up to 26%, and the sulphur content of 1.57%.

5

Current state, improvements and latest trends in coal preparation in Poland

Instytut Mechanizacji Budownictwa i Górnictwa Skalnego

NUMBER OF COAL PREPARATION PLANTS

Coal companies	Number of CPP	Capacity [tph]	Range of size mm	Washed production [million Mg]	Type/Distribution of circuits
Kompania Węglowa S.A. (Polish Mining Group S.A.)	20	600-2.100	20-0 (85 %) >20 (15%)	~14.4	vibration screens, jaw crushers, (20) dense medium washer (18), grain jigger (2) , jig washer (14), dense medium cyclone (1), hydrocyclone (6), spirals separator (4), flotation (7)
Katowicki Holding Węglowy S.A.	5	600-1.600	20-0 (78 %) >20 (22%)	~3.2	vibration screens, jaw crushers (5), dense medium washer (5), jig washer (1),
WĘGLOKOKS KRAJ Sp. z o. o.	2	1.500	20-0 (90 %) >20 (10%)	~0.5	vibration screens, jaw crushers (2), dense medium washer (2), jig washer, Barrel washer (2)
JSW S.A.	8	800-1.600	20-0 (98 %) >20 (2%)	~12.3	Bradford drum crushers, dense medium washer (8), jig washer (8), flotation (8)
TAURON Wydobycie S.A.	2	900	20-0 (80 %) >20 (20%)	~2.9	vibration screens, jaw crushers (2), dense medium washer (2), jig washer (2), spirals separator (2)
LW "Bogdanka" S.A.	1	2.400	20-0 (85 %) >20 (15%)	~7.4	vibration screens, jaw crushers (2), dense medium washer (2), jig washer (2)
PG "Silesia" Sp. z o.o.	1	575	20-0 (82 %) >20 (18%)	~1.2	vibration screens, jaw crushers (1), dense media washer (1), dense medium cyclone (1)

6

Instytut Mechanizacji Budownictwa i Górnictwa Skalnego

Current state, improvements and latest trends in coal preparation in Poland

DEVELOPMENT OF COAL PREPARATION PLANT IN LW „BOGDANKA" (2400 t/h)

Photo: MIFAMA OPA CARBO - MIKOŁÓW

7

Instytut Mechanizacji Budownictwa i Górnictwa Skalnego

Current state, improvements and latest trends in coal preparation in Poland

DEVELOPMENT OF COAL PREPARATION PLANT IN LW „BOGDANKA" (2400 t/h)

Photo: MIFAMA OPA CARBO - MIKOŁÓW

8

Current state, improvements and latest trends in coal preparation in Poland

Instytut Mechanizacji Budownictwa i Górnictwa Skalnego

DEVELOPMENT OF COAL PREPARATION PLANT IN LW „BOGDANKA" (2400 t/h)

Photo: MIFAMA OPA CARBO - MIKOŁÓW

9

Current state, improvements and latest trends in coal preparation in Poland

Instytut Mechanizacji Budownictwa i Górnictwa Skalnego

FLOWSHEET OF STEAM COAL PREPARATION PLANT

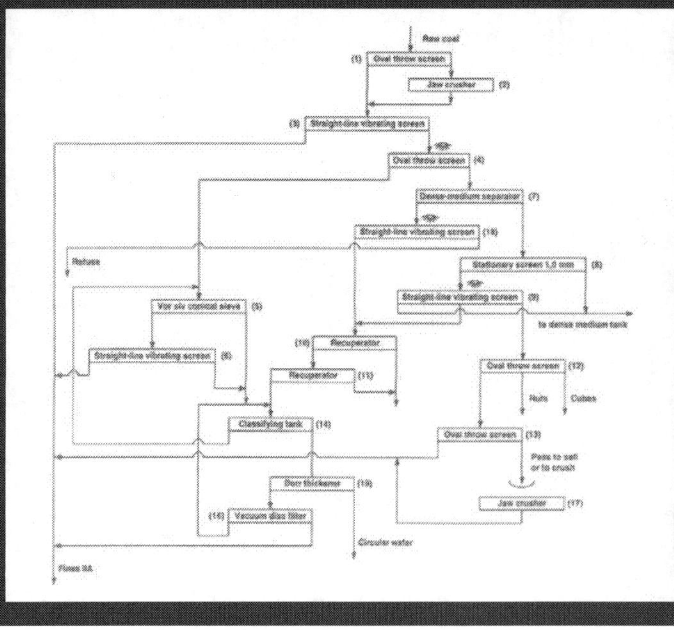

10

Instytut Mechanizacji Budownictwa i Górnictwa Skalnego

Current state, improvements and latest trends in coal preparation in Poland

FLOWSHEET OF COAL FINES PREPARATION AND DESULPHURIZATION PROCESS

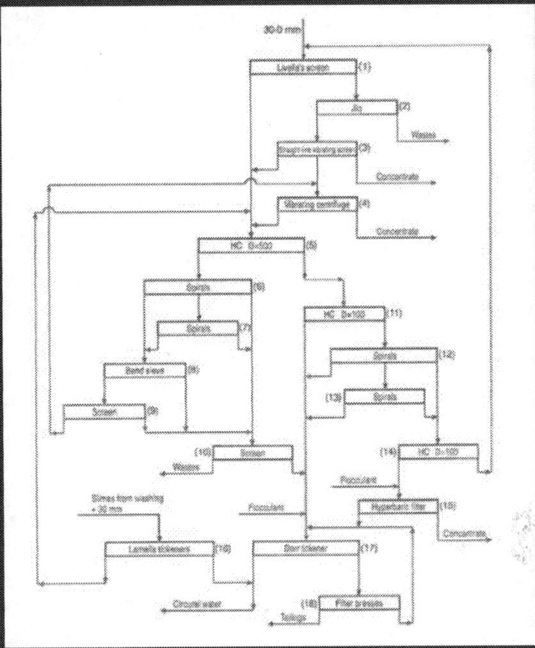

Instytut Mechanizacji Budownictwa i Górnictwa Skalnego

Current state, improvements and latest trends in coal preparation in Poland

FLOWSHEET OF COOKING COAL PREPARATION PLANT

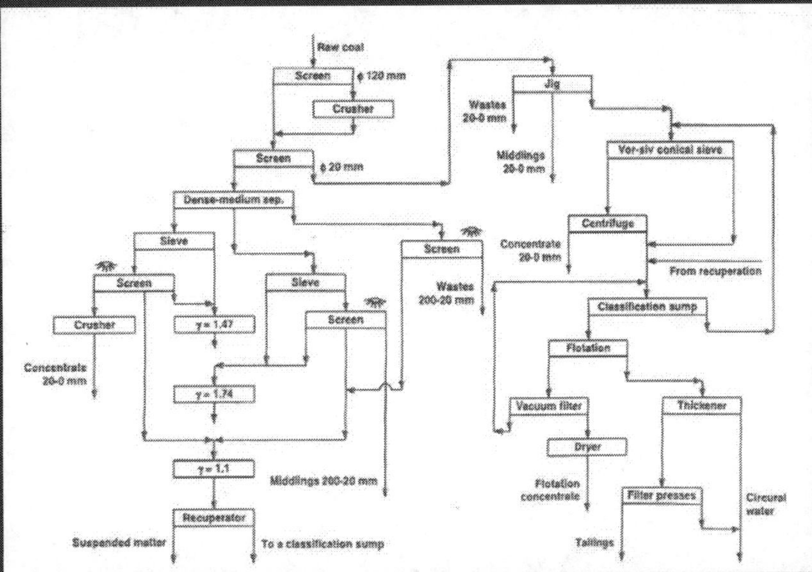

12

Current state, improvements and latest trends in coal preparation in Poland

Instytut Mechanizacji Budownictwa i Górnictwa Skalnego

SIGNIFICANT INDUSTRY CHANGES

The main changes in last 3 years:

- setting in order and improvement of particular process circuits:
  - reconstruction of raw coal preparation station,
  - modernisation of heavy-media separation system,
  - modernisation of jig wash,
  - elimination of flotation concentrate drying plant,
  - construction of flocculate measurement installation,
  - modernisation of dispatcher system,
  - modernisation of dust separation system,
- implementation of more efficient dewatering technologies for fine coal to improve the quality of the products and maximise the reduction of slimes disposal outside the water-slurry circuits,
- modern arrangements for preparation of power mixtures, up to date instrumentation of key technological circuits with electronic monitoring measurement equipment:
  - construction of electronic samplers for saleable coal,
  - construction of electronic analysers for qualification of basic qualitative parameters.

13

Current state, improvements and latest trends in coal preparation in Poland

Instytut Mechanizacji Budownictwa i Górnictwa Skalnego

IMPROVEMENTS

R & D NEEDS

- developing new analyzers of ash, sulphur and moisture contents, which are more accurate for measuring on clean coal, middlings and wastes,
- automation and process control systems for devices of coal preparation technologies to increase productivity and efficiency,
- developing a new method to achieve a quick and accurate data on characterization of coal quality in terms of washability, improving dewatering of finest grains coal (below 0,063 mm) to reduce the load on closed water-slurry systems,
- new alternative utilization waste of mining.

SAFETY

- reduce the risks related to exposure to harmful and dangerous factors by reducing the emission of noise, dust, vibration, etc. derived from the use of machines and other means of production,
- reducing the exposure time of these factors on workers.

14

Instytut Mechanizacji Budownictwa i Górnictwa Skalnego

Current state, improvements and latest trends in coal preparation in Poland

IMPROVEMENTS

WATER USAGE

- further reduction of water consumption by simplifying the water-slurry circuits,
- improving the efficiency of processes especially clarifying water, thickening and dewatering of products,
- reduction of water losses related to the operation of the settling ponds,
- reduction of the duration of wet processes limiting the grains contact with water,
- use of the underground water as the medium for conducting wet processes in a closed water-slurry system,
- implementing of dry separation technology of raw coal.

EFFICIENCY

technological possibilities of reducing the cost of preparation through the implementation of CMMS (Computerised Maintenance Management System) and PIMS (Production Information Management Systems), which includes: area of forecasting the quality of production, planning and integration of the extraction process with the preparation and sales process.

15

Instytut Mechanizacji Budownictwa i Górnictwa Skalnego

Current state, improvements and latest trends in coal preparation in Poland

LATEST TRENDS

- replacement of thermal drying for mechanical dewatering, which reduce the emission of dust and gases into the atmosphere and consumption of coal or other fuels for their own needs or to reduce employment,
- automation of coal preparation process for the regulation parameters enrichment, changing transport system and method for storing (system of selective storage saleable coal),
- increased production of environmentally friendly coal,
- employment optimization,
- automation and visualization of production processes to help reduce employment and cost of processing.

16

Current state, improvements and latest trends in coal preparation in Poland

Instytut Mechanizacji Budownictwa i Górnictwa Skalnego

CONCLUSION

Coal preparation plants existing in the mines fulfil their task in regard to efficiency and technology.

However, they require successive modernisation activities and investments to improve particular process circuits and reduce production costs.

There is need to improve coal quality monitoring and stability of the feed quality and products of coal preparation with particular attention paid to the variety of marketable grades.

17

Current state, improvements and latest trends in coal preparation in Poland

Instytut Mechanizacji Budownictwa i Górnictwa Skalnego

REFERENCES

1. Bilans złóż zasobów złóż kopalin w Polsce, 2015, PIG-PIB, Warszwa,
2. Bilans gospodarki surowcami mineralnymi Polski i świata, 2015, IGSMIE PAN, PIG-PIB, Warszawa,
3. Blaschke W.S., Gawlik L., 2006, Current Situation & Development Prospects of Coal Preparation in Poland. CPSA Journal - The Magazine by the Coal Society of America, vol. 5, no. 2.,
4. Blaschke W.S., Gawlik L., Blaschke S.A, 2010, Coal Preparation Technologies in Poland. CPSA Journal - The magazine by the Coal Preparation Society of America, vol. 9, no. 1,
5. Blaschke W.S., Baic I., 2013, Coal and Lignite Mining in Poland. The Energy Sector will be the Leading Sector of Growth. Turkey Keyword. Ankara. Turkey,
6. Blaschke W.S., Szafarczyk J., 2013, Current Situation of Coal Preparation in Poland. Proceedings of the 17th International Coal Preparation Congress. Istanbul. Turkey,
7. www.mifama.com.pl

18

Current state, improvements and latest trends in coal preparation in Poland
Instytut Mechanizacji Budownictwa i Górnictwa Skalnego
Contact
Prof. PhD. Eng. Wiesław S. BLASCHKE
MSc. Eng. Józef SZAFARCZYK
DSc, PhD. Eng. Ireneusz BAIC, Associate Prof.
Institute of Mechnised Construction & Rock Mining Branch in Katowice
Al. W. Korfantego 193 A
40-157 Katowice
Tel: +48 32 2517-454,
Fax: + 48 32 2517-591
e-mail: i.baic@imbigs.pl
Thank you for your attention !
19

Leaching kinetics of gibbsitic bauxite with sodium hydroxide

El-Sayed A. Abdel-Aal[1,a]

[1]*Central Metallurgical Research and Development Institute (CMRDI), P.O. Box 87 Helwan, Cairo, Egypt*

Abstract. In this paper the results of a leaching kinetics study of bauxite ore with sodium hydroxide are presented. The effect of ore particle size, sodium hydroxide concentration and reaction temperature on the Al_2O_3 extraction rate was determined. The results obtained showed that 99% of Al_2O_3 was leached out using -200+270 mesh ore particle size at a reaction temperature of 105 °C for 60 min reaction time with 250 g/L NaOH. The solid-to-liquid ratio was maintained constant at 1:20. The results indicated that leaching of bauxite is the rate controlling process. The activation energy was determined to be 46.04 kJ/mole, which was characteristic for a chemically controlled process.

1 Introduction

Bauxite is an aluminum ore which consists mainly of aluminum hydroxide. There are three types of aluminum hydroxide minerals occurring in bauxite. The first mineralogical form is gibbsite, alumina trihydrate [$Al_2O_3 \cdot 3H_2O$] or [$Al(OH)_3$], which is the most common occurring constituent of bauxite. The other two mineralogical forms are boehmite and diaspore, which have the same chemical formula of alumina monohydrate [$Al_2O_3 \cdot H_2O$] or aluminum oxyhydroxide [AlOOH] [1].

In 1888, Karl Josef Bayer developed and patented a process, which has become the cornerstone of the aluminum production industry worldwide. The Bayer process is used for refining bauxite to smelting grade alumina, the precursor to aluminum. Typically, depending upon the quality of the ore, between 1.9 and 3.6 tonnes of bauxite is required to produce 1 ton of alumina. The Bayer process involves the digestion of crushed bauxite in the concentrated sodium hydroxide (caustic) solution at temperatures up to 270°C. Under these conditions, the majority of the aluminum containing species in the ore are dissolved, leaving an insoluble residue (called red mud) which is removed. Red mud is composed primarily of iron oxides, quartz, sodium aluminosilicates, calcium carbonate/aluminate and titanium dioxide [2, 3, 4]. Sodium aluminate is separated from the red mud using 7-9 settling tanks with counter-current washing with water. Then, the sodium aluminate solution is diluted and seeded to precipitate aluminum hydroxide. The precipitated aluminum trihydroxide (gibbsite, $Al(OH)_3$) is filtered, counter-current washed with water, dried and calcined at 1000 °C, yielding alumina. The extraction chemical equations can be represented as following:

$$Al(OH)_3(s) + NaOH(aq) \rightarrow NaAl(OH)_4(aq) \quad (1)$$

[a] Corresponding author: eabde2@gmail.com

$$AlO(OH)(s) + NaOH(aq) + H_2O \rightarrow NaAl(OH)_4(aq). \quad (2)$$

Processing of different kinds of bauxite from India, Jamaica, Brazil, Guinea, Australia and Greece on in the laboratory and plots scales was reported. It was found that, the optimum leaching temperatures were ranged from 95 to 230 °C for digestion times of 25 to 120 min depending on the bauxite type [5]. Moreover, bauxite from the Um Bogma area, southwestern Sinai (Egypt), containing mainly gibbsite and kaolinite, was leached using sodium carbonate and lime water [6]. The obtained Al_2O_3 recovery was 92% at reaction temperature of 90 °C, reaction time 15 min, Na_2CO_3 concentration 200 g/L, solid content of 30% and CaO/SiO_2 stoichiometric mole ratio of 2.

The present investigation deals with the kinetics of Indian bauxite leaching with sodium hydroxide. The process conditions studied include the particle size of bauxite, sodium hydroxide concentration, temperature and time.

2 Experimental

2.1 Materials and apparatus

A bauxite ore from India was used. It contains mainly gibbsite [$Al(OH)_3$] and hematite [Fe_2O_3] as major minerals as well as boehmite [AlOOH], anatase [TiO_2], quartz [SiO_2], kaolinite [$Al_2O_3 \cdot 2SiO_2 \cdot 2H_2O$], and alumo-geothite [$Fe_2O_3 \cdot Al_2O_3 \cdot H_2O$] as minor minerals. The ore contains 44.6% Al_2O_3, 25.6% Fe_2O_3, 2.2% TiO_2, 1.4% SiO_2, 0.095% Na_2O, 0.073% MgO, 0.032% CaO and 0.024% K_2O [5, 7]. The ore was ground and sieved. Al_2O_3 contents of the different size fractions was found the same (44.7% ± 0.3%). Sodium hydroxide (purity of 97%) from Fischer Scientific was used in this study. The reaction was performed in 2 liter steel beaker placed in thermostatically controlled oil bath (Fig. 1). Reaction mixture was agitated at a rate of 500 rpm.

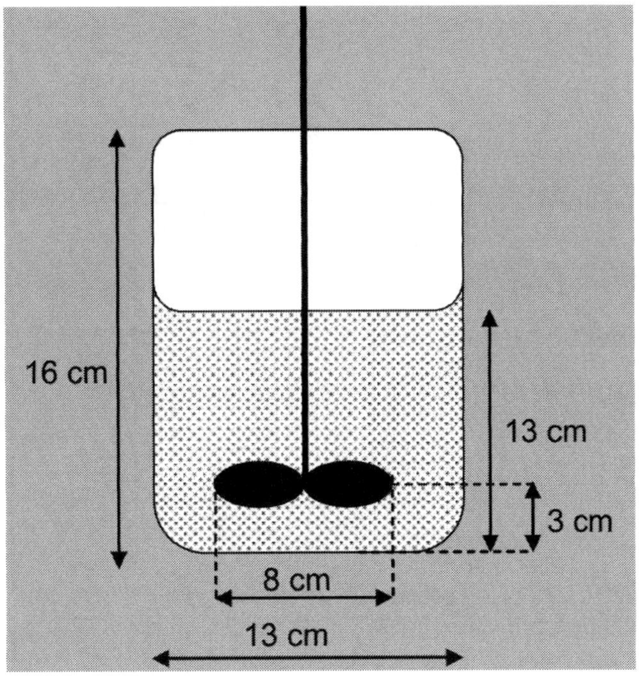

Figure 1. Schematic diagram of reaction vessel.

 DOI: 10.1051/e3sconf/20160801021

2.2. Procedure

A sample of 80 g of ground bauxite ore was added to the agitated sodium hydroxide solution (1600 ml) of required concentration at temperature. At selected time intervals, about 2 ml solution of sample was taken using a syringe filter of 1 μm pore size and the solids corresponding to that volume were discarded. Consequently, the solid-to-liquid ratio was maintained constant at 1:20. The cumulative volume removed by sampling was not significantly compared to the original solution (< 2%). Evaporated water was compensated continuously. The samples were chemically analyzed for determination of Al_2O_3 content from which the percentage extraction of Al_2O_3 was calculated.

3 Results and discussion

3.1 Effect of particle size

The effect of particle size on leaching of bauxite was studied using different size fractions (-100+150 mesh, -150+200 mesh and -200+270 mesh). Within the series of tests the initial sodium hydroxide concentration and leaching temperature were kept constant at 200 g/L and 95 °C, respectively. The solid-to-liquid ratio was 1:20. The obtained results (Fig. 2) show that the particle size has a significant effect on dissolution of bauxite and after 30 minutes of leaching. Depending on the ore particle size, an about 43-58% of Al_2O_3 was extracted. In addition, the results show that about 80% of Al_2O_3 present in the fine fraction (-200+270 mesh) of the ore was extracted after 60 min.

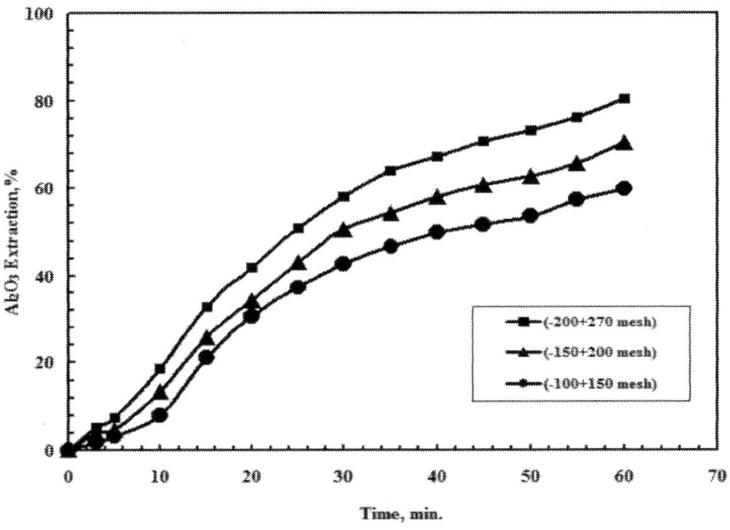

Figure 2. Extraction of Al_2O_3 from different particle sizes of bauxite (at 95 °C, 200 g/L NaOH).

3.2 Effect of sodium hydroxide concentration

A percentage of extracted Al_2O_3 against time is presented in Fig. 3 using fine ore of -200+270 mesh particle size and sodium hydroxide concentration in the range of 150 to 300 g/L at constant temperature of 95 °C. The solid-to-liquid ratio was kept constant at 1:20. It is obvious that the sodium hydroxide concentration also has a pronounced effect on dissolution of Al_2O_3. About 88% of Al_2O_3 present in the fine fraction of the ore was extracted using 250 g/L sodium hydroxide solution after 60 min. The increased sodium hydroxide concentration to 300 g/L had insignificantly increased the Al_2O_3 extraction efficiency.

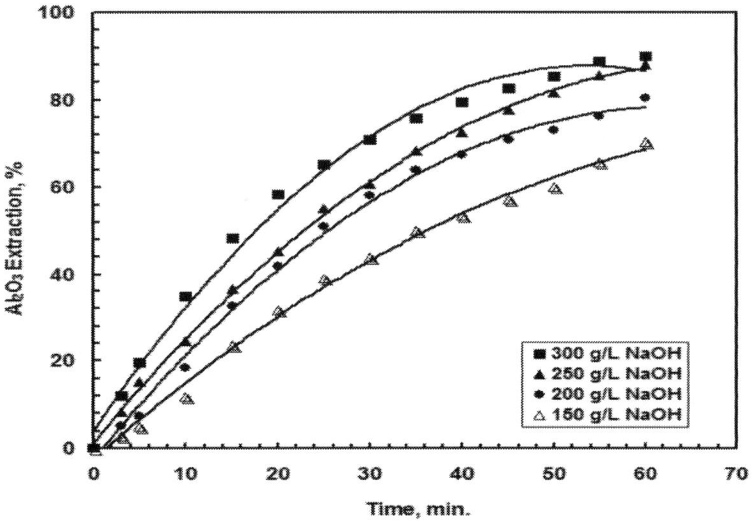

Figure 3. Effect of sodium hydroxide concentrations on extraction of Al_2O_3 (-200+270 mesh particles at 95 °C).

3.3. Effect of reaction temperature

The effect of reaction temperature on extraction of Al_2O_3 at different reaction times is plotted in Fig. 4 using ore of -200+270 mesh particle size and sodium hydroxide concentration of 250 g/L at temperatures in the range of 75 to 105 °C. The solid-to-liquid ratio was kept constant at 1:20. The obtained results show that the studied reaction temperatures had significant effect on dissolution of Al_2O_3. About 99% of Al_2O_3 present in the fine fraction of the ore was extracted after 60 minutes at reaction temperature of 105 °C. At relatively lower temperatures (less than 95 °C), the Al_2O_3 extraction efficiency drastically decreased.

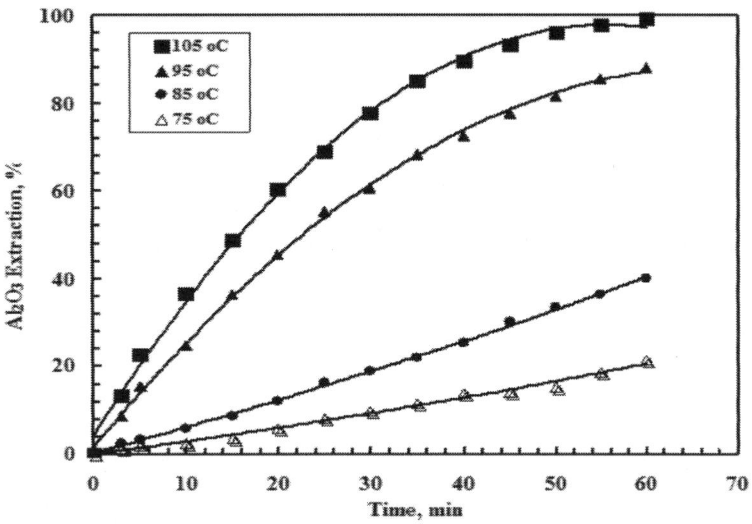

Figure 4. Effect of reaction temperature on extraction of Al_2O_3 (-200+270 mesh particles in 250 g/L NaOH).

3.4 Kinetic aspects

The reaction of bauxite can proceed in a topochemical manner, because the inner core of unreacted particle is decreasing with time. It is clear from Fig. 2 that the rate of reaction decreases with time. This is due to reduction of reactant surface. The rate of reaction is given for models based on control by either chemical reaction at the particle surface, diffusion through the product layer, or combination of both.

The rate control by chemical reaction at the particle surface is [8, 9]:

$$K_c t = 1\text{-}(1\text{-}x)^{1/3} \tag{3}$$

where K_c is the reaction rate constant (min $^{-1}$), t is the time (min), and x is fraction reacted of Al_2O_3 definedas:

$$x = \%\, Al_2O_3\ extraction/100\,. \tag{4}$$

Based on the experimental data presented in Fig. 4, a plot of the right-hand side of Equation (3) versus time is given in Fig. 5. During the whole reaction time, the data in Fig. 5 are linear what indicates that the rate of reaction is controlled by chemical reaction on the surface of bauxite particles. The results are correlated to this model as a straight line and nearly zero point intercept was obtained. In the case of straight lines, the slope equals the rate constant K_c :

$$K_c = MbK_{cc}C/\rho r,\ \text{min}^{-1} \tag{5}$$

where K_{cc} is the chemical rate constant (cm min^{-1}), b is the stoichiometric coefficient (dimensionless), M as molecular weight of the major Al_2O_3 mineral, C concentration of sodium hydroxide (mol/m^3), r radius of unreacted particle (m), ρ as density of bauxite ore (kg/m^3), (pycnometer method = 3660 kg/m^3).

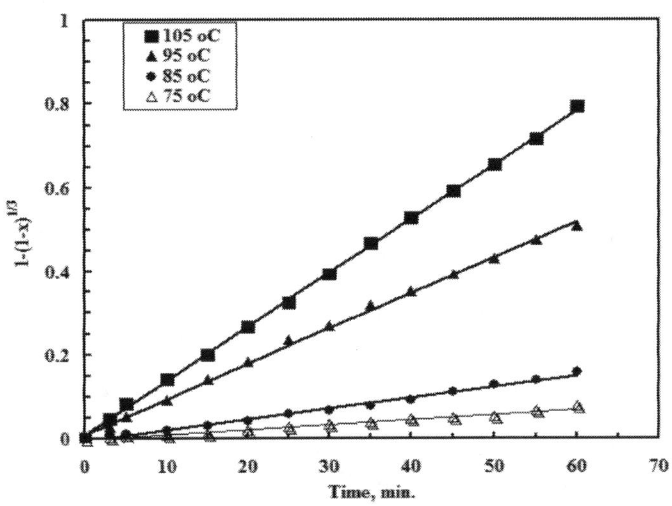

Figure 5. $1\text{-}(1\text{-}x)^{1/3}$ versus time at various reaction temperature (-200+270 mesh particles in 250 g/L NaOH).

When diffusion through the product layer is rate controlling, the kinetics may be correlated graphically using the Valensi equation [8, 9]:

$$K_p t = 1\text{-}\ 2/3x\ \text{-}(1\text{-}x)^{2/3} \tag{6}$$

where K_p = chemical rate constant (cm min^{-1}).

Based on the experimental data in Fig. 4, a plot of the right-hand side of Equation (6) against time is given in Fig. 6. The data cannot be correlated to this model as neither a straight line nor zero point intercept was obtained. In the case of straight lines, the slope is K_p and is equal to:

$$K_p = 2bMDC/\rho r^2 \text{ min}^{-1} \quad (7)$$

where D is the diffusion coefficient of aluminum ions in porous medium (m^2/min).

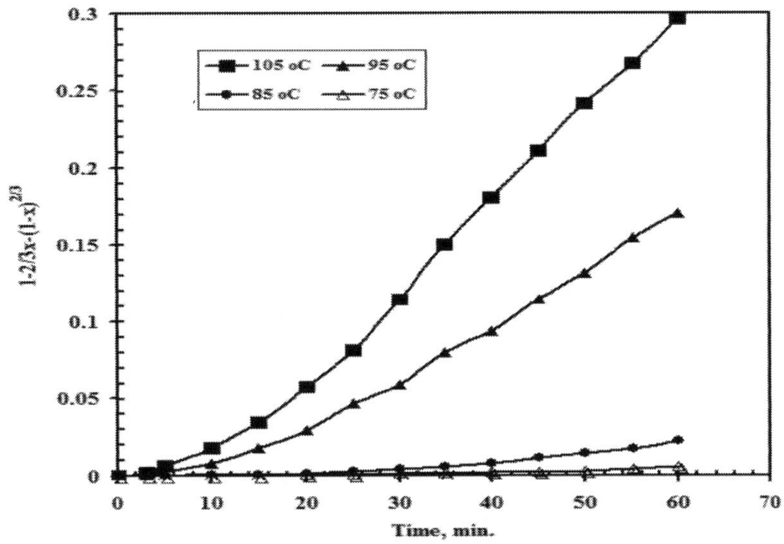

Figure 6. $1\text{-}2/3x\text{-}(1\text{-}x)^{2/3}$ versus time at various reaction temperatures (-200+270 mesh particles in 250 g/L NaOH).

When previous models do not give a straight-line relationship throughout the leaching process, mixed model may be applied. The mixed kinetics equation, in terms of the fraction extracted (x), is [7, 8, 9]:

$$Mbk_{cc}DCt/\rho r^2 = D/r[1-(1-x)^{1/3}] + k_{cc}/2[1-2/3x-(1-x)^{2/3}] \quad (8)$$

where $Kr = Mbk_{cc}C/\rho r$

$$Krt = D/r[1-(1-x)^{1/3}] + k_{cc}/2[1-2/3x-(1-x)^{2/3}] \quad (9)$$

$$krt = 1-(1-x)^{1/3} + B[1-2/3x-(1-x)^{2/3}] \quad (10)$$

where: $B = k_{cc}r/2D$

Two terms on the right side of Eq. (10) represent the chemical and diffusion resistances. To plot Eq. (10), the values of the constant B must be determined, which contains the chemical rate constat k_{cc} and the diffusion coefficient D.

Based on the experimental data from Fig. 4, a plot of the right-hand side of Eq. (10) against time is given in Fig. 7. The data cannot be correlated to this model as neither a straight line nor zero point intercept was obtained.

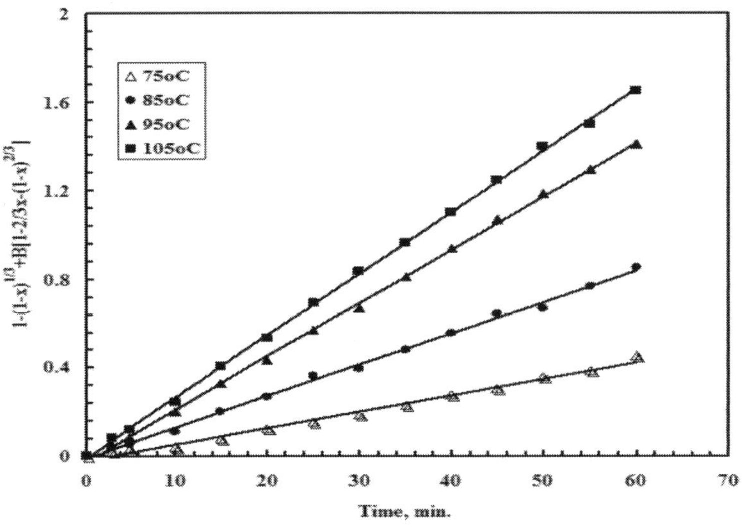

Figure 7. $[1-(1-x)^{1/3}+B[1-2/3x-(1-x)^{2/3}]$ versus time at various reaction temperatures (-200+270 mesh particles in 250 g/L NaOH).

3.5. Calculation of the activation energy

The activation energy of a diffusion controlled process is characterized to be 1 to 3 kcal/mole [8], 2 to 5 kcal/mole [10] or 3 to 6 kcal/mole [11]. In addition, the activation energy for a chemically controlled process is usually greater than 10 kcal/mole [8], or more specifically falls between 10 to 20 kcal/mole [10]. To calculate the activation energy, the values of ln K_c were plotted against $1/T$ (Fig. 8). The activation energy of the overall reaction was calculated to be 11 kcal/mole (45.9 kJ/mole). This value of activation energy confirms that the reaction rate is chemically controlled. This activation energy is in a good agreement with the value of activation energy (48.15 kJ/mole) calculated for oxidative ammonia leaching of sphalerite [12] of chemically rate controlled.

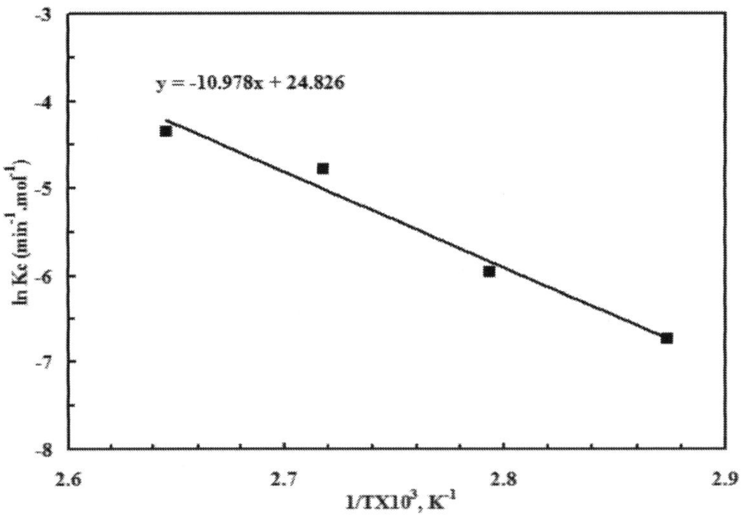

Figure 8. Arrhenius plot for leaching of bauxite ore with sodium hydroxide.

4. Conclusions

The extraction efficiency of about 96% of the Al_2O_3 present in the studied bauxite ore was achieved under the following conditions: particle size -200+270 mesh, temperature 95 °C, reaction time 60 min, sodium hydroxide 250 g/L and solid-to-liquid ratio 1:20. The kinetic study indicated that leaching of Al_2O_3 was the chemically controlled process for all leaching times, and the reaction rate increased with temperature. The activation energy was calculated to be 11 kcal/mole (46.04 kJ/mole), which was consistent with values of activation energies reported for chemically controlled rate.

References

1. D. Panias, P. Asimidis, I. Paspaliaris, *Hydrometallurgy*, **59**, 15 (2001)
2. A.R. Hind, S.K. Bhargava, S.C. Grocott, *Colloids Surf., A*, **146**, 359 (1999)
3. J. Newton, *Extractive Metallurgy* (John Wiley & Sons, Inc., New York, 1959)
4. R.D. Pehlke, *Unit Processes of Extractive Metallurgy* (Elsevier Scientific Publishing Co., 1977)
5. E.A. Abdel-Aal, A. Abdel-Rahman, A Amer, A.K. Ismail, *T. Indian I.Metals*, **49(1-2)**, 79 (1996)
6. A.M. Amer, E.A. Abdel-Aal, *T. Indian I.Metals*, **49(1-2)**, 25 (1996)
7. A.K. Ismail, Report, Academy of Scientific Research and Technology (Cairo, Egypt, 1994)
8. F. Habashi, *Principles of Extractive Metallurgy* (Gordon and Breach, New York, 1996)
9. O. Levenspiel, *Chemical Reaction Engineering*, (Wiley, New York,1986)
10. S. Anand, R.P. Das, *T. Indian I.Metals*, **41**, 335 (1988)
11. L.T. Romankiw, P.L. De Bruyn, *in Proceedings of International Symposium*, M.E. Wadsworth, F.T. Davis,(eds.) (Dallas, Texas, **24**, 62, 1963)
12. M. K. Ghosh, R.P. Das, A.K. Biswas, *Int. J. Miner. Process.*,**70(1-4)**, 55 (2003)

DOI: 10.1051/e3sconf/20160801022

Origin of colour after firing feldspar-quartz raw material from the Sobotka region (Lower Silesia, SW Poland)

Ewa Lewicka[1,a]

[1]*MEERI, Polish Academy of Sciences, 30-261 Krakow, Wybickiego 7, Poland*

Abstract. The article presents the studies of feldspar-quartz raw material from deposits in the Strzegom-Sobotka massif (Lower Silesia, SW Poland). The goal of this paper was to establish the causes of colour variation of samples after firing at 1200°C. The scope of research encompassed among others a detailed chemical analysis of raw material and Mössbauer studies of two selected samples, as well as measurements of colour parameters after firing (L*, a*, b*). The chemical analysis revealed that the samples of darkest colours are characterized by the highest contents of: Fe_2O_3, MnO, Th, U, Ce, Nd. The Mössbauer studies demonstrated the quantitative predominance of Fe^{2+} over Fe^{3+} in the sample of relative darker hue, while its spectra parameters suggest that Fe^{2+} is located in octahedral coordination that can result in cold blue tint. Ions Fe^{3+}, occupying probably the tetrahedral sites, prevail in the other analysed sample, which probably causes its warm, reddish shade. The above-mentioned examinations lead to the finding that colour of analyzed samples after firing originates from ions of transition elements of various valence states and coordination (especially iron), as well as some REE and actinides.

1. Introduction

Colour is one of the primary diagnostic characteristics of minerals. It is also one of key parameters of ceramic raw materials, which determines aesthetic qualities of ceramic bodies obtained in course of their firing. The main components of ceramic mixes used in the domestic ceramic industry are feldspar-quartz raw materials from deposits of granitoids in the Sobotka region (Lower Silesia, SW Poland). The research, are presented in the paper focuses on explanation of the reasons for differentiation of colour of the samples from these deposits after firing at 1200 °C.

Interpretation of colour origin is a complex issue which requires the use of four distinct physical theories, that is crystal field, molecular orbital, and the bands one, as well as physical optics. These theories explain the role of main constituents, admixtures/impurities, structural defects and specific structure of a mineral in creating the visual effect recognized as a certain colour [1-2].

The crystal field theory refers basically to colour of ionic crystals which prevail in the minerals world. In the crystal lattice of these minerals there are ions of transition elements or some REE (lanthanides) and actinides characterized by partially filled *d* or *f* shells. Unpaired electrons present on these shells can interact with visible light that results in absorption and colour. Transition metals such

[a] Corresponding author: lewicka@min-pan.krakow.pl

as Fe, Ti, V, Cr, Mn, Co, Ni, Cu, as well as some REE (Ce, Nd, Pr) and actinides (U, Th) may constitute the major ingredient of a mineral (e.g. either turquoise or malachite being copper compounds) or may be present in the lattice as an impurity (e.g. Fe in citrine, Cr in ruby or emerald). Sometimes a colour can be a result of a presence of colour centres, i.e. lattice defects that can absorb visible light, e.g. smoky quartz or blue Maxixe-type irradiated beryl [1].

In the molecular orbital theory the colours are associated with electron transfer between different ions neighbouring in the lattice of a mineral. This refers to the transfer between differently charged metal ions (Fe^{2+}-Fe^{3+}, Fe^{2+}-Ti^{4+}), metal and non-metal or between two ions of non-metal. The examples are: blue sapphire, crocoite, vivianite. Apart from the valence state, the colour can be influenced by ion coordination. It was found that even small amount of Fe^{3+} tetrahedrally coordinated can result in intense red-brown colour, while its location solely in octahedral position – green shade [1, 3-4]. Minerals containing Fe^{2+} in octahedral coordination also have shades in green and blue.

The band theory applies to a wide range of minerals, especially metals. This theory treats electrons as belonging to the crystal as a whole. They are contributed by each atom in the crystal lattice and create energy bands. These electrons are relatively free to move throughout the whole crystal, which results in the high electrical and thermal conductivities as well as metallic lustre. The colour is connected with electron transfer between energy bands due to absorption of visible light. The examples are native metals being perfect conductors, i.e.: copper, silver, iron, as well as semiconductors such as galena, galena, proustite, pyrite, sulphur. However, the presence of certain impurities can considerably affect both the conductivity as well as the light absorption of a mineral, e.g. a few boron atoms per million carbon atoms in the lattice of diamond can result in a blue colour, while one nitrogen atom for every 100,000 carbon atoms gives its deep-yellow colour.

The colour of some minerals can be explained by physical optics involving scattering, diffraction, interference and dispersion of visible light. These phenomena can be related to either the structural features of mineral formations (regular lamellar intergrowth in labradorite, globular structure of opals, overlapping platelets of aragonite in pearl) or the presence of impurities, i.e. inclusions of coloured minerals (fuchsite in aventurine, hematite in sunstone) or bubbles of fluids or gases (moonstone). One other distinction is the colour deriving from interference in the tarnish film on altered bornite that depends on the film thickness, its refractive index and the nature of incident light.

The detailed investigation of the colour origin often requires highly sensitive methods of examinations as the colour can be caused by more than one of the above-mentioned mechanisms, as well as can be a result of either gamma, X or neutron irradiation, or high temperature. Furthermore, the same element can produce various colours (e.g. Cr - red ruby or green emerald). On the other hand one specific colour may be attributed to different elements/impurities, e.g. blue colour to copper, cobalt or boron (in shattuckite, spinel, and diamond, respectively).

2. Materials and methods

The subject of the research was 19 samples of leucogranites from 4 deposits located in the Strzegom-Sobotka massif of Lower Silesia (SW Poland), i.e.: Pagorki Wschodnie (4 samples), Pagorki Zachodnie (4 samples), Strzeblow I (9 samples) and Stary Lom (2 samples). They represent leucocratic rocks, created in the contact zone of granite magma intrusion with surrounding rocks as a result of alteration (auto metamorphism) of various varieties of granite [5]. This caused an enrichment of their mineral composition in feldspars (plagioclases and microcline-type perthite) and quartz with simultaneous depletion of micas that made them valuable raw material for the ceramic industry [6].

The mineral composition of all samples was determined by transmission-light microscopy in standard thin sections by the use of petrographic microscope Carl Zeiss Jena (Germany). Microstructural and phase observations were performed on fracture surfaces using a scanning electron microscope (HITACHI S-4700, Japan) in both secondary (SE) and back-scattered (BE) electron models after sputtering with a thin layer of carbon in order to ensure electrical conductivity.

Qualitative EDS (NORAN Vantage system) analyses were carried out simultaneously with SEM observations in order to distinguish the various phases containing iron, REE and transition metals. The measurements were taken with the use of accelerating voltage 15-20 kV. Chemical analyses for oxides and trace elements contents were conducted in the Activation Laboratories Ltd. – ACTLABS of Ancaster in Canada by the following methods: inductively coupled plasma (FUS-ICP), by total digestion – inductively coupled plasma (TD-ICP) and instrumental neutron activation analysis (INAA). Two selected samples were examined by the Mössbauer spectroscopy method. The spectra were collected using the RENON MsAa-3 spectrometer at room temperature. The ^{57}Co (RH) isotope was used as a source of radiation. All spectra shifts were reported versus the total shift in α-Fe. From the material of each sample, after grinding (d_{90}=53.3 μm), mixing with water and pressing, the standard pastilles were formed (ϕ37mm x H7 mm), which were fired at 1200°C in laboratory chamber furnace for 2 h 15 min. The colour parameters of fired bodies were measured using UV-Vis spectrophotometer (Minolta 2300D, Japan) and the chromatic coordinates expressed as L*, a*, b*.

The aim of the study was to determine forms and concentrations of colouring elements (chromophores), especially of transition elements (mainly Fe and Ti), some REE and actinides, contents and relative proportions of Fe^{2+} and Fe^{3+} as well as establishing the relation of results of chemical composition and spectroscopic measurements to colour of samples after firing.

3. Results and discussion

The sample material consists of well characterized mineral components [7]. Both observations in transmission-light microscope and examinations using scanning electron microscope techniques (SEM-EDS) performed in back-scattered electrons bean (BSE) revealed the presence of minerals containing colouring components, i.e. transition elements: Fe, Ti, Mn, Ni, V and Cu, as well as some REE (Ce, Nd) and actinides (Th, U). It was also shown by chemical analysis (Table 1). Other colouring elements such as Cr and Co were not detected, while the contents of Ni, Cu and V were very low, and therefore are not mentioned in the Table 1. From the above-mentioned transition elements the dominant colour contributor was iron, the most plentiful in nature (about 5% of the earth crust). Its main mineral phases ascertain in the examined samples were as following: chlorites (containing up to 50% Fe_2O_3), chloritized biotite (up to 40% Fe_2O_3), light micas – muscovite, sericite (up to 53.7% Fe_2O_3), rutile (up to 20.5% Fe_2O_3), and Fe-Mn garnets (up to 31% Fe_2O_3 and 14.4% MnO). It was less present in kaolinite (1.0–4.7% Fe_2O_3), feldspars (occasionally up to 0.3-0.7% Fe_2O_3), monazite (up to 1.23% Fe), sphalerite (up to 1.83% Fe), xenotime (up to 1.37% Fe_2O_3), zircon (up to 0.7% Fe_2O_3) and epidote, occurring in a form of very thin veins in feldspar grains (greenish colour in transmission-light microscope suggested that it can be pistacite – Fe-rich variety of epidote). There were also observed some iron own mineral phases, however in negligible amounts, i.e. pyrite and titanium magnetite or magnetite. Other minerals containing colouring elements detected in the samples were as following: chlorites, micas, garnets (Mn), monazite (Th, U, Ce, and Nd), rutile (Ti, V), zircon (U), and xenotime (Th, REE). Some REE can be also incorporated in feldspars as it was confirmed by Götze et al. [8] in investigations performed by high-resolution spectral CL. Scanning images of selected minerals rich in the above-mentioned elements in the examined samples are presented in Figure 1.

The results of chemical analysis of the examined samples revealed profound differentiation of the contents of colouring oxides, which was reflected by colours of fired samples and significant variation in the values of chromatic coordinates L*, a*, b* (Table 1, Fig. 2).

The values of L* and a* coordinates correlated with the contents of Fe_2O_3 in the majority of samples, i.e. higher Fe_2O_3 content, lower brightness (L*) of the fired body and higher a* value (red tint) with relatively good match factor (R^2) of the trend line to the experimental values (Figs. 3-4). High value of b* (yellow shade) can be correlated with TiO_2 and also Fe_2O_3 contents, although the values of R^2 coefficient are low (Figs. 5-6). When comparing samples characterized by similar content of TiO_2, which differs in Fe_2O_3 contents, it can be noted that b* value is lower in the samples enriched in iron.

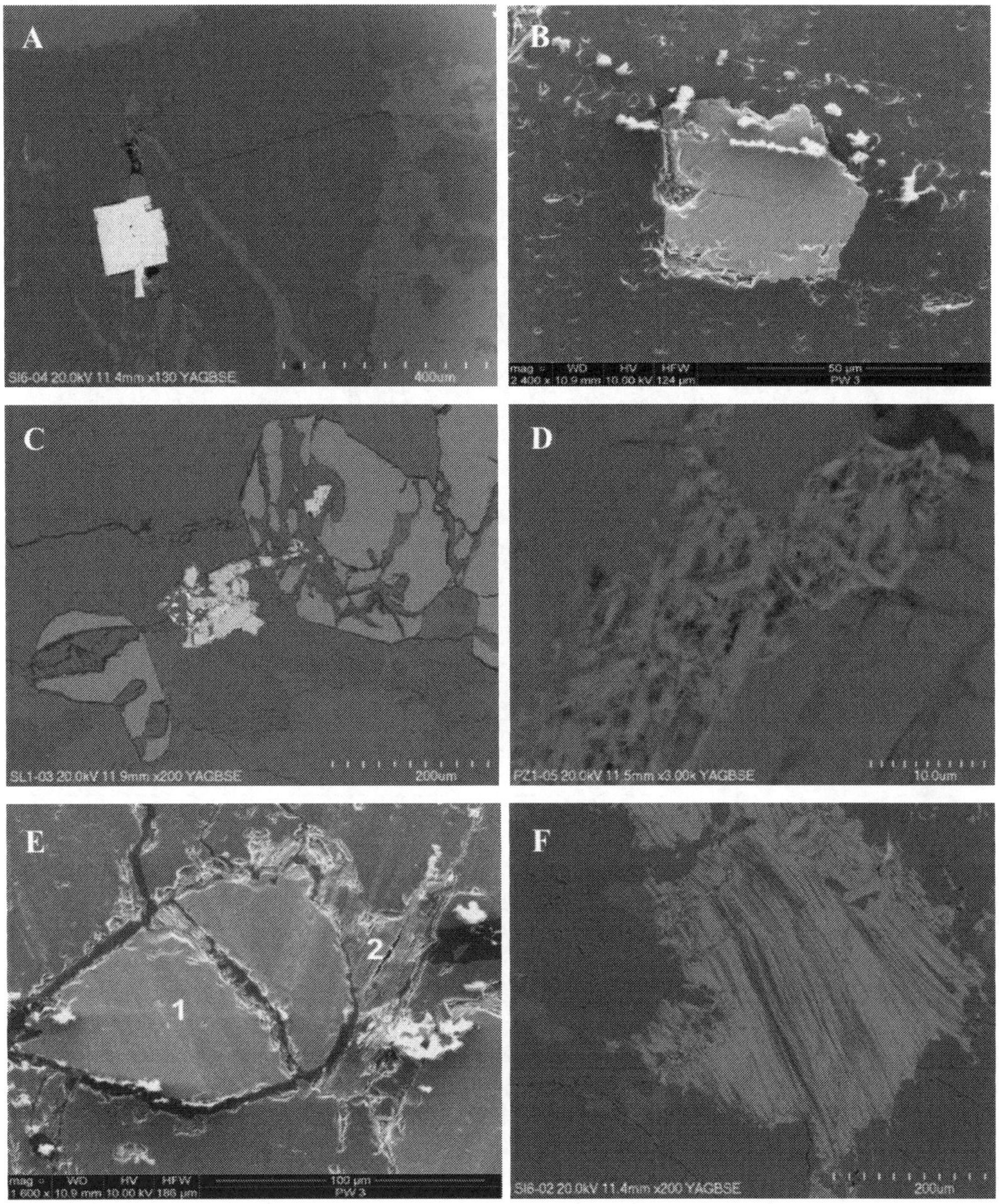

Figure 1. Selected minerals containing colouring elements in granitoids from deposits in the Sobotka region (SEM-EDS): A – pyrite (Strzeblow I), B – monazite (Pagorki Wschodnie), C – Fe-Mn garnets with inclusions of zircon and Th-rich monazite in the centre (Stary Lom), D – rutile (Pagorki Zachodnie), E – Fe-Mn garnet (1), next to chloritized biotite (2) (Pagorki Wschodnie), F – Fe-rich biotite (Strzeblow I)

Table 1. Contents of selected colouring components and values of colour parameters after firing the samples from: Pagorki Zachodnie (PZ), Pagorki Wschodnie (PW), Stary Lom (SL), Strzeblow I (SI) deposits

Sample	Fe_2O_3 [%]	TiO_2 [%]	MnO [%]	Th [ppm]	U [ppm]	Ce [ppm]	Nd [ppm]	L* [%]	a*	b*
PZ-1	0.43	0.052	<0.01	12.2	5.6	25	12	58.68	8.86	7.32
PZ-2	0.33	0.073	<0.01	12.8	1.8	40	24	69.98	5.86	9.27
PZ-3	0.19	0.095	<0.01	12.9	2.5	22	9	82.64	0.68	9.60
PZ-4	0.34	0.048	<0.01	20.2	5.1	33	10	61.37	6.05	5.59
PW-1	0.30	0.027	0.02	15.3	2.4	27	<5	75.39	1.53	6.76
PW-2	0.65	0.021	0.11	9.4	16.2	19	10	57.67	3.52	3.20
PW-3	0.34	0.019	0.03	9.5	0.9	21	10	73.91	2.88	5.66
PW-4	0.22	0.020	<0.01	13.6	1.7	21	9	83.26	1.50	6.35
SL-1	0.26	0.022	0.02	11.0	2.9	28	7	74.31	2.03	7.04
SL-2	0.28	0.019	0.03	17.5	2.6	40	24	66.96	1.22	1.83
SI-1	0.17	0.053	<0.01	14.9	1.1	32	13	78.44	0.11	6.24
SI-2	0.25	0.060	<0.01	12.6	<0.5	24	10	75.93	2.90	7.70
SI-3	0.22	0.048	<0.01	18.7	2.7	30	11	74.49	1.91	7.26
SI-4	0.31	0.050	<0.01	22.3	2.3	29	14	70.99	2.92	6.03
SI-5	0.66	0.054	<0.01	19.1	3.5	23	10	53.08	3.89	2.49
SI-6	0.60	0.056	0.01	14.7	7.5	18	5	52.79	4.92	4.07
SI-7	0.23	0.049	<0.01	14.3	2.6	25	8	78.50	2.06	8.85
SI-8	0.31	0.065	<0.01	21.5	4.9	29	18	75.23	3.88	9.10
SI-9	0.15	0.032	<0.01	9.9	2.4	24	<5	81.73	0.62	6.37

Mössbauer spectroscopy studies were performed for two samples that distinctly differ in respect to Fe_2O_3 contents and colour after firing, i.e. PW-2 and PZ-2 (Tab. 1, Fig. 2). In the sample PW-2, one of the richest in Fe_2O_3, MnO and U among all examined samples (0.65%, 0.11% and 16.2 ppm, respectively), a significant prevalence of the relative content of Fe^{2+} (69%) over Fe^{3+} (31%) was revealed (Table 2). That was probably, along with relatively large amount of these elements – the reason for cold, dark grey/blue shade of that sample (Table 1, Fig. 2). In the sample PZ-2 (containing 0.33% Fe_2O_3) Fe^{3+} cation dominated over Fe^{2+} (78% and 22%, respectively). It was characterized by a relatively high percentage of TiO_2, Ce and Nd (0.073%, 40 and 24 ppm, respectively). The value of isomer shift of Fe^{2+} (1.11 mm/s) in this sample indicated that ferrous ion probably occurred in the octahedral coordination as it was set for this cation in crystalline substances [9]. Simultaneously, the isomer shift of ferric ion in both samples (0.30-0.35 mm/s) can refer to tetrahedral coordination in the crystal lattice which was found by Steffen et al. [10] as well as Spiering and Seifert [11] in their Mössbauer studies of silicate glasses of granitic composition. Therefore, it can be assumed that this ion determines the colour of both samples, however its influence coluld be modified by other elements, especially by titanium. In course of research on the redox equilibria of silica melts rich in Fe and Ti [12] it was proven that Ti^{4+} tended to enhance the stability of Fe^{3+} relative to that of Fe^{2+} due to the structural instability of the Fe^{2+}–O–Ti^{4+} network. In such a configuration, the following transfer of electron via oxygen atom was possible: Fe^{2+}–O–Ti^{4+} $\rightarrow$ Fe^{3+}–O–Ti^{+3} [13-14]. It seems to be consistent with the fact that in the sample PZ-2, in which Fe^{3+} cations dominate, the content of TiO_2 (0.07%) was much higher than in the PW-2 sample (0.02%) marked by a predominance Fe^{2+} cations (Table 2).

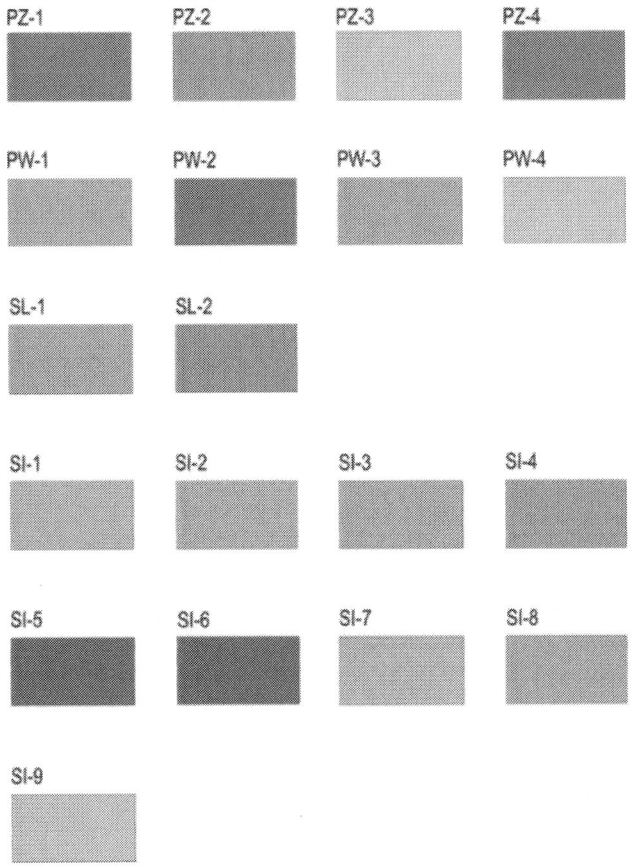

Figure 2. Colour of samples after firing in 1200 °C.

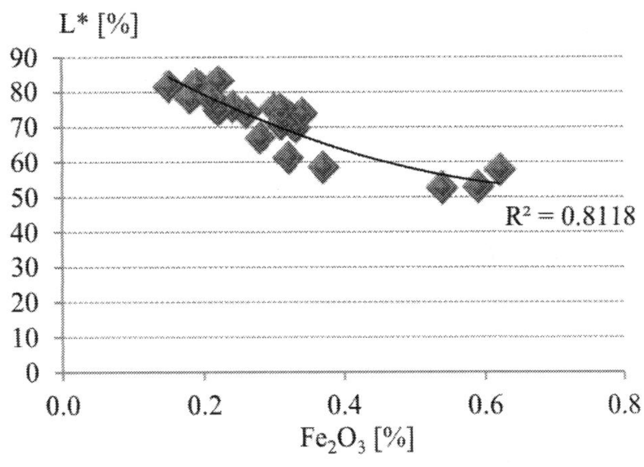

Figure 3. Content of Fe_2O_3 versus brightness of the samples after firing.

DOI: 10.1051/e3sconf/20160801022

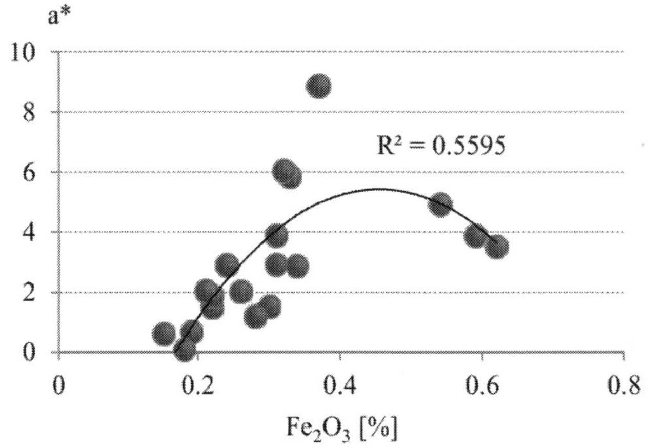

Figure 4. Content of Fe_2O_3 versus a* parameter of the samples after firing.

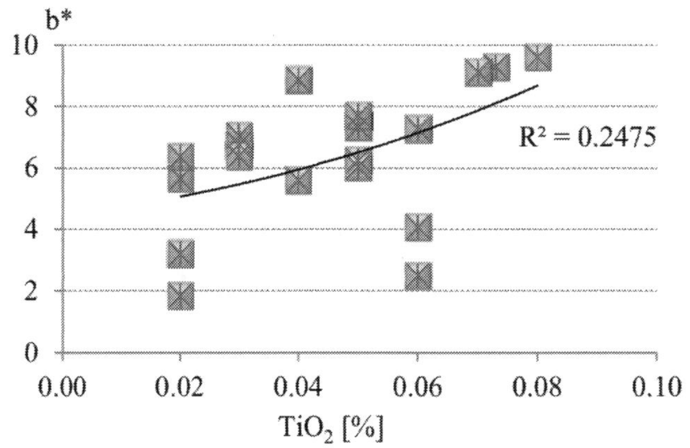

Figure 5. Content of TiO_2 versus b* parameter of the samples after firing.

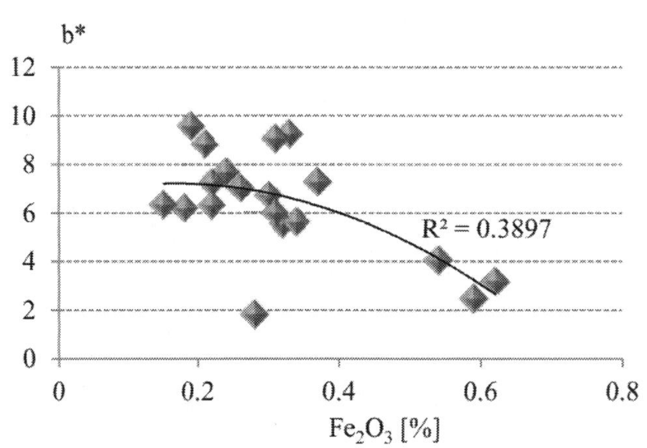

Figure 6. Content of TiO_2 versus b* parameter of the samples after firing.

Table 2. Parameters of the Mössbauer spectra of PW-2 and PZ-2 samples, where IS – isomer shift versus α-Fe, QS – quadrupole splitting, W – relative content of iron of respective valence state.

Sample	Cation	IS [mm/s]	QS [mm/s]	W [% at.]
PW-2	Fe^{2+}	1.44±0.01	3.22±0.01	34
		0.99±0.01	2.90±0.03	35
	Fe^{3+}	0.30±0.05	1.4±0.2	14
		0.35±0.02	0.5±0.1	17
PZ-2	Fe^{2+}	1.11±0.04	2.9±0.1	22
	Fe^{3+}	0.32±0.02	0.64±0.03	78

4. Conclusions

The investigation of granitoids from deposits in the Sobotka region by means of transmission-light microscopy, scanning microscopy, chemical analysis and Mössbauer spectroscopy showed that colour differentiation of the samples after firing can be related to the presence of transition metals, especially Fe and Ti, some REE (Ce, Nd) and actinides (U, Th) in the structure of their mineral components. It seems to be consistent with the crystal field and molecular orbital theories.

The Mössbauer studies indicated that the colour can result from a combined effect of the presence of iron on various valence states and titanium, as in the case of sample with Fe^{3+} prevalence (PZ-2) which is characterized by high content of TiO_2. Apart from the oxidation state, the colour is influenced by the site occupancy of iron, i.e. its either tetrahedral or octahedral coordination in the crystal lattice. The presence of Fe^{3+} tetrahedrally coordinated influences the colour much stronger than its location in the octahedral position, resulting in the red-brown colour, on the other hand exclusive octahedral occupancy of Fe^{3+} can produce the green colour (vide [3]). Ion of Fe^{2+} octahedrally coordinated gives shades in green and blue. The isomer shift of the PZ-2 spectrum suggests tetrahedral coordination of Fe^{3+} that can be attributed to brown-reddish colour of this sample, while another one – PW-2 – showed a predominance of Fe^{2+} in probably octahedral coordination that could result in its cold dark colour with grey/blueish hue.

References

1. K. Nassau, *Am. Miner.* **63**, 219 (1978)
2. A. Bolewski, J. Kubisz, W. Żabiński, *Mineralogia ogolna* (Wyd. Geol., Warszawa, 1981)
3. W. Sikora, *Prace Miner*, **39** (Wyd. Geol., Warszawa, 1974)
4. A. Bolewski, M. Budkiewicz, P. Wyszomirski, *Surowce ceramiczne* (Wyd. Geol., Warszawa, 1991)
5. M. Niec, *Surowce mineralne Polski. Surowce skalne. Surowce krzemionkowe i skaleniowe*, edited by R. Ney, 101 (Wyd. IGSMiE PAN, Krakow, 2005)
6. E. Lewicka, *Surowce mineralne Polski. Surowce skalne. Surowce krzemionkowe i skaleniowe* [ed. R. Ney], 139-155 (Wyd. IGSMiE PAN, Krakow, 2005)
7. E. Lewicka, *Min. Res. Manag.*, **31(1)**, 81 (2013)
8. J. Götze, D. Haberman, R. D. Neuser, D. K. Richter, *Chem. Geol.*, **153**, 81 (1999)
9. G. M. Bancroft, *Mössbauer Spectroscopy* (McGraw Hill, London, 1973)
10. G. Steffen, F. A. Seifert, G. Amthauer, *Am. Miner.*, **69**, 339 (1984)
11. B. Spiering, F. A. Seifert, *Contrib. Mineral. Petrol.*, **90**, 63 (1985).
12. H.V. Alberto, J.M. Gil, N. Ayres DeCampos, B.O. Mysen, *J. Non-Cryst. Solids*, **151(1-2)**, 39 (1992)
13. C. E. Weaver, *Clays Clay Miner.*, **24**, 215 (1976)
14. S. Chyandrasekhar, S. Ramaswamy, *Appl. Clay Sci.*, **21**, 133 (2002)

Evaluation of the effects of coal grinding in terms of coal water slurry preparation

Jolanta Robak[1,a], Karina Ignasiak[1], and Michał Rejdak[1]

[1] *Institute for Chemical Processing of Coal, 41-803 Zabrze, Zamkowa 1 St.*

Abstract. Coal Water Slurry (CWS) is a specific form of solid fuel. It occurs in the form of finely ground coal particles and water. Depending on the use, the content of combustible matter is from 40 to 70% by weight. The attractiveness of the fuel is primarily its properties, i.e. liquid form, high energy efficiency (for water evaporation 4% energy is used – for CSW with 70% concentration of coal), decreased environmental impurities (lower NOx emission) and reduced risk of explosion. The advantages of CWS fuels, the possibility of independence from petrochemical fuels, wide availability of coal and emphasis on the use of cleaner technologies are the driving force for development of slurry fuel technologies. The major parameters characterizing the fuel suspension are: solid phase concentration (share of coal in the slurry expressed as either weight or volume fraction of dry coal), time stability (resistance to delamination and separation of the dispersed phase from the continuous phase) and viscosity, determining the flow of suspension. The mentioned parameters are dependent on the susceptibility of coal for production of aqueous suspensions (slurrability), conditioned by natural properties of coal, such as: coalification degree, petrographic composition and surface properties. They are also dependent on the slurry fuel preparation process: particle size, solid phase concentration, used additives (stabilizing and dispersion agents) and modification of primary coal properties (ash removal, change of surface properties). Preparation of sustainable, high concentrated CWS fuel coal is promoted by the hydrophobic nature of the coal surface, characteristic for coals of higher coalification. A great technological problem is to obtain a highly concentrated coal slurry fuel from less coalified hydrophilic steam coals. The paper presents the results of lab scale research on the CWS prepared from Polish steam coal by wet grinding in mill drum and vibrating. The milling process was carried out in a batch system by varying the composition of the grinding media and varying degrees of working space filling. The feed was pre-ground steam coal mixed with water at the ratio of 1: 1. The changes of particle size, viscosity and stability of suspensions in time were assessed. The milling effectiveness was higher for the vibrating mill. After the same milling time the quality of the obtained product was higher. The obtained milling degree was depended on the share and composition of grinding aids. In the investigated range (10-30% share mill working space), the use of higher share of grinding aids caused higher milling degree of coal. A trend was observed that stronger fragmentation of coal was achieved with the use of polydisperse composition of aids distinguished by smaller diameters. The changes of granulation influenced the viscosity and stability of the obtained product. The lower the mean particle size, the higher the viscosity and stability of the obtained product was particularly noted for the vibrating mill.

[a] Corresponding author: office@ichpw.pl

Flow resistance reduction of coal water slurry through gas phase addition

Jolanta Robak[1,a], and Karina Ignasiak[1]

[1]*Institute for Chemical Processing of Coal, 41-803 Zabrze, Zamkowa 1 St., Poland*

Abstract. One of the main advantages of coal water slurry fuel (CWS) is a physical form that allows, among others, their transfer by pipelines over long distances. For this form of transport actions towards reducing the flow resistance of the transmitted medium are important. One of the treatments leading to reduction in the flow resistance of suspensions is to introduce gas into the stream of flowing slurry. The goal of that action is to either loosen the structure of densely packed grains or increase the velocity of the suspension. The paper presents the flow resistance of CWS in a horizontal pipeline and the effect of addition of the gas phase on the resistance level. The investigation was carried out with the use of a research stand enabling to measure the flow resistance of the multiphase/multicomponent systems. The measured diameter and length of sections were respectively: 0.03 and 2 m. The coal-water slurries (based on steam coals) with concentration of dry coal in the range of 51 do 60% obtained by wet milling in a drum mill were used. During the tests, the following parameters were measured: slurry flow rate, air flow rate, temperature and pressure difference in inlet and outlet of the measured section. The volume flow rate of slurry fuel was in the range of 30 to 110 dm^3/min while the volume flow rate of air was from 0.15 to 4 m^3/h. Based on the obtained results, the slurry flow resistance as a function of the flow rate and share of introduced air was evaluated. The performed research allowed for assessment of flow resistance reduction condition and to determine the pipe flow curves for different temperatures. It was found that the effect of reducing the flow resistance of the coal slurry by introducing gas into the flow tube depended on the volumetric flow rate, and thus the linear velocity of the slurry. Under the experimental condition, this effect only occurred at low flow rates (30 - 50 dm^3/min) and low temperature of the suspension. The share of air needed to flow resistance reduction was in the range of 0.02 to 0.04% by volume.

[a] Corresponding author: office@ichpw.pl

Performance of 500 m^3 TankCell® at Kevitsa Cu-Ni-PGM concentrator

Toni Mattsson[1,a], Tomi Maksimainen[2], Rodrigo Grau[1], and Antti Rinne[1]

[1]*Outotec Beneficiation, Espoo, Finland*
[2]*Kevitsa mine, Finland*

Abstract. Outotec TankCell e500 flotation cell, with 500 m^3 of efficient flotation volume, has been in operation since October 2014 at Kevitsa Cu-Ni-PGM concentrator as the first Cu rougher flotation cell. The 500 m^3 flotation cell has proven to provide metallurgical superiority at very low specific power. On average the cell has recovered 71% of copper contained in the flotation feed. The cell has produced the concentrate with the Cu grade equal to 17% Cu. The typical specific power for the cell is around 0.4 kW/m^3 (blower power not included). After the start-up of the cell the operating parameters have varied. The mixing speed have varied from 4.9 to 7.0 m/s and the superficial gas velocity from 0.3 to 1.5 cm/s. At various operating parameters the mixing, gas dispersion and metallurgical performance of the cell have been evaluated. In this paper a review of the hydrodynamic and metallurgical performance of the cell is presented. The paper focuses on the interactions of mixing intensity, bubble size and metallurgical performance in industrial application.

[a] Corresponding author: toni.mattsson@outotec.com

Pilot-scale recovery of rare earths and scandium from phosphogypsum and uranium leachates

Maxim Mashkovtsev[1], Maxim Botalov[1], Denis Smyshlyaev[1] , Risto Pajarre[2,a] , Petteri Kangas[2], Vladimir Rychkov[1], and Pertti Koukkari[2]

[1]*Institute of Physics and Technology Ural Federal University, Russia*
[2] *VTT – Technical Research Centre of Finland*

Abstract. Ural Federal University (UrFU) and VTT have performed joint research on development of industrial technologies for the extraction of REM and Scandium compounds from phosphogypsum and Uranium ISL leachate solutions. Leaching-absorption experiments at UrFU have been supported with multicomponent solution modelling by VTT. The simulations have been performed with VTT's ChemSheet/Balas program and can be used for speciation calculations in the lixiviant solution. The experimental work combines solvent extraction with advanced ion exchange methodology in a pilot facility capable of treating 5 m3 solution per hour. Currently, the plant produces cerium carbonate, lanthanum oxide, neodymium oxide and concentrate of heavy rare earth metals. A batch of 45 t solids has been processed with the gain of 100 kg's of REM concentrate. A mini-pilot plant with productivity above 50 liters per hour has been applied to recover scandium oxide and REE concentrates from the uranium ISL solution. As the preliminary product contains radioactivity (mainly strontium), an additional decontamination and cleaning of both concentrates by extraction has rendered a necessity. Finally a purified 99% concentrate of scandium oxide as well as 99% rare earth concentrate are received.

1 Introduction

In the processing or Rare Earth Elements (REE) saline solutions containing large amounts of Fe, Al and other elements are typically formed. The extraction of REE and Sc from those is an elaborate task by traditional chemical methods. For example in the industrial scale recovery of REE and Sc from these solutions the traditional method is deposition by ammonia, alkalis and anions such as F^-, $S_2O_4^{2-}$, and PO_4^{3-}. Disadvantages then include large loss of REE (20-25%) [1, 2] as coprecipitating hydroxides of metals such as Fe(III), Al, Zr and Ti. In addition, waste waters caused by the large flow rates, chemicals used for precipitation and low regeneration rates are difficult to process.

Sorption and extraction are currently the most promising method for extracting REEs from dilute solutions [2, 3], since they have a high-performance, and a simple hardware design and are selective and well suitable for construction of closed circulation. In this study of concentrating REE and Sc

[a] Corresponding author: Risto.Pajarre@vtt.fi

from Uranium ISL solutions, solid ion exchange materials, insoluble to aqueous media were used. The received concentrates were subjected to additional purification to achieve decontaminated products.

2 Materials and applied methods

Table 1. Composition of return liquor used in the experiments.

Element	Content, mg/dm^3	Element	Content, mg/dm^3	Element	Content, mg/dm^3
Copper	0.63	Boron	24.6	Dysprosium	0.93
Zinc	23	Barium	0.044	Holmium	0.16
Iron	692	Indium	0.009	Erbium	0.46
Calcium	339	Ytrium	4.68	Thulium	0.075
Magnesium	252	Lanthanum	2.10	Ytterbium	0.51
Aluminium	1111	Cerium	9.66	Lutetium	0.073
Beryllium	0.56	Praseodymium	1.77	Hafnium	0.014
Molybdenum	0.23	Neodymium	6.25	Scandium	1.18
Rhenium	0.005	Samarium	1.24	Thorium	11.7
Manganese	15	Europium	0.34	Uranium	5.7
Cadmium	0.1	Gadolinium	1.03		
Chromium	16	Terbium	0.13		

2.1 Raw materials and research facilities

Russian uranium ore deposits handled by ISL are located usually in sediments alternating with gravel sands, sandstones and siltstones. The major components of the main ores consist of aluminosilicates (with silica and alumina contents 70-85% and 3.5-18.5%, respectively) containing practically no carbonates.

Uranium is present as pitchblende (an average of 40-60%) and coffinite in the intergranular spaces of sandy rocks, sometimes substituting cementitious clay. Uranium minerals tend to have high contents of molybdenum and selenium (up to 0.01 – 0.2%). Content of scandium in permeable lithological precipitate types varies from 4.7 g/t to 10.42 g/t. In average for the permeate side of the section Sc content is 6 g/t.

The content of the lanthanide varies from 123 g/t in the grits to 171-222 g/t in the sandstones, reaching maximum values in clay varieties (up to 447 g/t) in the barren rocks of the ore-bearing layers.

Iron is closely connected with sulfur concentration and varies from 0.4% to 2-2.8 %. The content of sulfates is 2-2.5%, with an average of 0.95%.

The study used returnable solutions ISL Uranium Company "Dalur" after sorption of uranium, which are weakly acidic green solutions. All samples were subjected to prolonged settling and decantation for clarity. This was followed by chemical analysis by ICP mass spectrometry and AAS. The contents of the main impurities are shown in Table 1.

The exact composition of the anionic productive solutions was not fixed in the experiments. The average content were as follows: sulfate ion 10-15 g/dm^3; chloride 0,5-1,2 g/dm^3; acidity of 4-5 (up to 10) g/dm^3; suspended solids up to 27 mg/dm^3.

2.2 Methodology

The behaviour of metal ions in the process of sorption of uranium from ISL solutions was studied in both in static and dynamic conditions. The experimental setup for static conditions for determination of the static exchange capacity and desorption study included a shaking platform with a circular action. Plastic flasks with crew caps were used varying the weight ratios of ion exchanger and ISL uranium solution.

The limiting step during adsorption/desorption is a chemical complex compound formation between the functional group and a rare earth or Sc ion. From previous experience on sorption of rare-earth elements in various ion exchangers it was deduced that 36 hours contact time is required and consequently was also adopted in this study.

After completion of the experiment the ion exchanger was separated from the mother liquor using a ceramic sieve. Calculation of the static exchange capacity and the degree of adsorption or desorption was carried out based on the difference in the concentrations of REE / Sc in solution before and after the experiment and the results of desorption. Reproducibility of results was within 5%. All static tests used completely dry resins in H^{+}/ NH^{4+} -forms.

Static equilibrium exchange capacity for REE/Sc of the dry resin, calculated on the residual content of REE/Sc solution is defined by the formula (1):

$$E_{STAT} = (C_1 - C_2)\frac{V_p}{m_c} \quad (1)$$

where C_1 is concentration of REE/Sc in the initial solution (g/dm^3), Vp volume of the solution taken for tests dm^3, C_2 concentration of REE/Sc in solution after sorption, mc dry sample weight in the resin H^{+}/NH_4^{+} form. To study the dynamics of the process, the dry ion exchanger (H^{+}/NH_4^{+} form) was soaked in distilled water and then placed in a glass column with a diameter of 8 mm. The test solution was passed through the resin bed until equal composition at the inlet and outlet of the column was reached. Under controlled flow rate, the eluate fractions were selected for analysis. The flow rate in the sorption and desorption experiments was set using a peristaltic pump. The volume of the resin in the dynamic tests was approximately 10 cm^3.

Full dynamic exchange capacity of the dry resin for REM/Sc, calculated on the desorption results is determined by the formula (2):

$$E_{DYN} = \frac{CV_pV_{res}}{V_{ionic}} \quad (2)$$

where:
C-REE/Sc concentration of the stripping solution, mg/cm^3
V_p-Volume of desorption solution cm^3
V_{ionic}-Volume of the resin taken for tests
V_{res}-Specific volume of the resin, cm^3/g.

To study the effect of temperature on the static exchange capacity, the reaction mixture in a glass beaker was placed in an incubator. Stirring was performed with an overhead stirrer.

The effect of the temperature on the dynamic exchange capacity was studied with a ion exchanger column placed in a thermostat. The level of liquid was set to completely cover the resin. Moreover, elution solution of the same temperature was fed into the column using a peristaltic pump.

The degree of desorption was determined by the difference between the REE content of the ion exchanger at saturation and after desorption. The REE content was determined by dissolving the resin in boiling nitric acid and analysing the resulting solution.

The study of the influence of various parameters (acidity, solid-to-liquid ratio) on the deposition of REE/Sc concentrates was performed in beaker mixed by an overhead stirrer equipped with a pH-meter.

Precipitation of REE/Sc concentrates was conducted in an Atlas Calorimeter (Syrris). A liquid circulation thermostat was used keep temperature steady during the exothermic and endothermic reactions. The correct pH level was automatically maintained, while the acid consumption was recorded by the gravimetric and volumetric methods.

DOI: 10.1051/e3sconf/20160801026

3 Results

3.1 Sorption of REE from the leaching solutions

Based on the earlier studies [4-6], it was found that the highly crosslinked sulfonated resins adsorb poorly polymeric ions (Zr, Fe (III), etc.), mainly due to the sieve effect. These differences in the absorption and elution properties of different hydrocomplexes on crosslinked ion exchangers allow for enhanced separation of Zr, Fe, Al, Th, and REE ions without adding special complexing agents.

Cations of the same charge those with smaller ionic radii are more prone to hydration and, consequently, have a greater influence on the selectivity with on strong acid cation exchangers. There are comprehensive reviews on the structure of the aqueous ions using the methods of neutron and X-ray diffraction [7-8].

To study the sorption of the REE from ISL uranium solutions commercially available sulfonated resins with different contents of divinylbenzene (DVB) were used. While the increase of DVB does not change the structure of the copolymer, it increases the density of the globular formation (gel). This also decreases the distance between the globular molecular entities from 30Å to 10Å (8% DVB → 20% DVB) in dry cation exchanger.

The increase in volume of the resin gel by water leads to straightening of the polymer chains by, which means that, on average, the distance between the elementary and interglobular formations increase on average of 2-10 times depending on the DVB content.

The specific surface area sorbent polymer, bulk density, total pore volume, pore volume size distribution were determined for the average pore radius. With increasing DVB content 12 → 40% it was found that the specific surface area of the polysorbent increased while the bulk density decreased and the pore volume increased from 1.4 to 2.1 cm^3.

Practical application of macroporous sulfonated resins for the separation of ions revealed the following:

- good operational stability with resins containing no more than 20% of DVB.
- Industrial sulfonated resins can effectively separate due to the sieve effect only ions that can be polymerized in solution (Mo, W, Zr, Ti)
- the copolymerization with DVB does not always ensure a uniform distribution of crosslinks in the polymer network. This leads to the formation of spatial areas in the sorbents difficult for ions to access, which is the reason for the relatively low values of their sorption capacity.

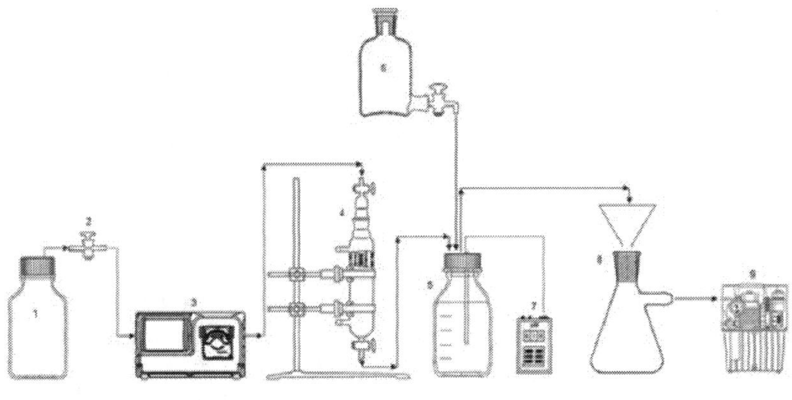

Figure 1. Schematic diagram of the laboratory setup, where 1 reactor feed solutions; 2 stopcock valve; 3 pump; 4 ion-exchange column; 5 reactor precipitator;6 reactant feed (reservoir); 7 pH meter; 8 filter; 9 vacuum pump.

All the above material suggests the possibility of an effective REE separation from sulfuric acid solutions in macroporous sulfonic resins that are commercially available in the DVB content of more than 12%. For further research the macroporous sulphocationite M-31 was chosen. Research for the development of processing parameters was performed in a laboratory setting. Figure 1 gives a schematic diagram of the setup.

In a series of experiments the specific load on the ISL uranium solution in an ion exchange column was investigated. Determining the optimal unit load will further enable the calculation of the optimal size of the ion exchange column for the breakthrough for the given conditions and the degree of extraction of REE.

Macroporous cation exchangers with high internal surface and a large number of pores have a relatively fast sorption kinetics. For the M-31 the optimum load was estimated to be 5 column volumes of solution to unit volume of the cation exchanger per hour.

3.2 Washing impurities from cation exchangers

As mentioned previously, the functional groups of strongly acidic cation exchangers strongly bind REE ions in a wide range of concentrations of sulfuric acid. Therefore, it becomes possible to separate the sulfuric acid adsorbed components according to their distribution coefficients.

To investigate this question, experiments were conducted on the washing of the cation exchanger M-31 saturated ISL uranium solution with sulfuric acid of different concentrations. The studies were conducted in static and dynamic mode. The static value for the washing solution: ion exchanger was 5, while washing for 1 hour. For the dynamic study the transmission rates was 2.5 and 5 volumes of the washing solution to 1 volume of M-31 per hour. The results are shown in Table 2.

Table 2. Results from cation exchanger M-31 washing.

	10	25	50	75	100	125	150	200
	washing, %							
Al	9.02	24.40	42.99	56.18	60.37	73.93	77.40	91.20
Ca	11.85	20.28	27.28	34.95	44.85	53.92	56.17	68.31
Fe	13.24	17.55	51.07	38.12	49.36	56.06	61.75	85.48
Y	2.19	4.20	10.39	17.31	17.84	27.64	30.22	48.76
La	1.82	1.81	4.52	8.16	7.65	12.87	14.16	27.45
Ce	1.44	1.76	4.85	8.54	8.49	14.26	15.49	30.15
Pr	1.64	1.88	5.02	8.84	9.86	16.36	17.76	34.17
Nd	1.53	2.05	5.32	9.52	8.84	14.68	15.76	29.56
Sm	1.30	2.32	5.76	10.56	10.33	16.47	18.01	33.72
Eu	9.92	9.65	14.78	20.27	10.11	18.28	17.32	30.73
Dy	1.29	2.23	5.92	10.56	9.86	15.74	16.88	30.88
Gd	1.10	2.43	6.18	11.41	10.89	17.72	18.99	34.52
Tb	1.04	2.74	7.24	13.13	14.12	21.90	23.38	40.56
Ho	1.41	3.58	9.12	16.42	17.05	26.51	28.28	46.87
Er	1.48	4.08	9.69	17.56	17.30	25.90	28.23	45.58
Tm	1.55	4.66	11.52	19.94	21.35	31.98	34.04	54.12
Yb	1.74	4.55	11.71	19.94	20.08	30.26	32.70	50.32
Lu	2.41	4.74	13.07	21.52	24.07	35.01	37.94	59.27
Th	7.09	3.95	9.42	14.39	25.19	38.39	45.26	93.08
U	32.81	39.73	60.33	75.15	73.43	81.55	80.91	88.62
ΣREE	1.65	2.28	5.82	10.21	14.89	16.20	17.53	32.38

From the results it is evident that the optimum concentration of sulfuric acid to the washing solution is 75 g / dm^3. This washes away 38% of the iron, 58% of aluminum, 35% of calcium, 14% of thorium and 75% of uranium. Improving the washing efficiency by introducing of additional sulfate to the washing solution ions failed because along with the increase in desorption of impurity elements also desorption of REE ions was increased. The optimum load on the column during the washing is 5 volumes of solution to 1 volume of the cation exchanger per hour.

3.3 Desorption of REEs

Desorption was carried out with a solution of ammonium sulfate. The choice of this reagent was dictated by the fact that further processing of the eluates will involve the use of ammonia and ammonium salts to precipitate concentrates. Furthermore, ammonium ions have a good desorbing capability because of the affinity to cation binding functional groups [9].

In a series of experiments we studied the effect of concentration of ammonium sulfate on the degree of desorption and the transmission rate on prewashed M-31. From the data it was evident that concentrations in the range of 200-400 g /dm^3 ammonium sulfate can be used for desorption equally effectively. Recommended concentration of ammonium sulfate is ~300g/dm^3.

The presented data show that the change in the feed rate of the desorbing solution has no strong effect on the dynamics of desorption. Recommended feed rate of 5 times V/V_B per hour stripping solution.

3.4 Deposition of the REE concentrate

Before precipitation of carbonates REE solution concentrate was treated with 10% NH_4OH. Pre-neutralization of the excess acidity of ammonia was necessary, firstly, to help reduce the formation of ammonium carbonate salts, and secondly, to precipitate heavy metal hydroxides of Fe^{3+}, Al^{3+} and Th^{4+} simultaneously sorbed with REE. Precipitation led to pH 4.5-5. This interval is characterized by the end of the precipitation of hydroxides of the above metals. At the same time, substantial hydrolysis of REM is not observed. Coprecipitation of REM does not exceed 1%.

Precipitation of REE as carbonates is a well-studied process. The precipitation is performed by use of ammonium carbonate (130-150 g/dm^3 solution), while use of excess precipitant must be considered to avoid undesired hydrolysis of the carbonates (typical for e.g. the yttrium group elements). In Table 3 the average contents (as elements) of the mother solution and the precipitated REE concentrate are given

Table 3. Content of elements in the mother solution and in the REE carbonate concentrate.

	Element				
	Fe	Al	Ca	ΣREM	Th
mg/dm^3 (solution)	17.2	2.0	319.6	11.0	0.1
w-% (concentrate)	0.13	9.02	0.11	32.89	0.01

4 Extraction of scandium

4.1 Sorption of scandium from the ISL solution

In the recovery of Sc from uranium ISL solutions the kinetic characteristics are important for elucidating the mechanism of sorption process. For example, from the kinetic curves the time to reach equilibrium as well as the maximum operating capacity of the resin at given composition and temperature of the solution can be obtained. Knowledge of the limiting stage allows for refining the extraction process conditions of a particular element in the solution for the optimal choice of ion exchange resins. Furthermore, the kinetic parameters are used in the design of the process hardware.

From the experiments it was determined that at least 10 hours of contact of the ion exchanger and solution was required to approach equilibrium.

Sorption of metal cations strongly depends on their concentration in solution. The nature of the sorption isotherms gives an indication of the selectivity of the selected ions in these conditions. Therefore, it was important to assess the equilibrium sorption process. In figure 2, the achieved equilibrium content of Sc and Th in the ion exchanger ECO-10M by the Russian company AXION as a function of the saturated solution is presented. The bulge on the sorption isotherm confirms the high selectivity of the selected sorbent with respect to scandium ions in sorption from sulphate solutions of complex cation composition. The sorption extraction of scandium is due to the formation of a coordination bond with the phosphonate group of cation exchanger ECO-10M.

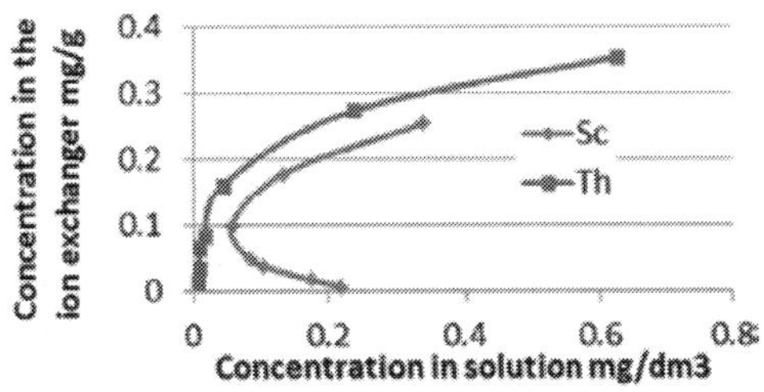

Figure 2. Equilibrium content of Sc and Th in the ECO-10M ion exchanger.

The next series of experiments was conducted in a laboratory setting studying specific load of the ISL uranium solution in an ion exchange column. Determining the optimal unit load allows calculating the optimal size of the ion exchange column for the breakthrough in the given conditions and the degree of extraction of REE.

From the results it was also clear that the increase in the load onto the column by a solution in four times does not affect the capacitive properties cation exchanger. The optimum load is 5 column volumes of solution to 1 volume of the cation exchanger per hour.

4.2 Washing impurities from the cation exchanger ECO-10M

Phosphonic functional groups of the cation exchanger ECO-10M firmly hold scandium ions in a wide range of concentrations of sulfuric acid. Therefore, there is the possibility of separating the sulfuric acid adsorbed components according to their distribution coefficients.

To investigate a cation exchanger ECO-10M saturated with ISL uranium solution was washed with sulfuric acid of different concentrations. The studies were conducted in static and dynamic mode.

The optimum concentration of sulfuric acid in the washing solution was found to be 150 g/dm^3. It was found possible to wash away iron ions to 92% of aluminum ion at 50% of the calcium ion by 21%, of the thorium ion at 14% of the uranium ions to 7.6%. The optimum load on the column during the washing is 5 volumes of solution to 1 volume of the cation exchanger per hour.

4.3 Desorption of scandium

Some of the REE ions, accompanying elements and thorium series radionuclides are largely adsorbed phosphorus-containing ion exchangers from carbonate solutions [10].

At the same time, the imbibition values of the studied elements are largely dependent on the pH and the concentration of carbonate ions. Imbibition is lowest for scandium, thorium, and zirconium ions at ion exchangers in high pH (above 11) and at the solubility limit of alkali metal carbonates and ammonium salts. This allows use concentrated solutions of carbonate ions with high pH for elution of adsorbed ions from the ion exchanger phase after washing.

In light of [11] data to determine the values of the thermodynamic constants of complex ions of thorium, scandium, zirconium, as well as a list of items included in the initial ISL uranium solutions may choose desorption conditions and a carbonate based precipitation reagent. Knowing the thermodynamic properties of the elements enables the separation or concentration of them from the process solutions and avoids formation of deposits in the cation exchanger.

It is known that polynuclear hydrolysis of scandium occurs in weakly acidic solutions, thus resulting hydroxo scandium species. Adding carbonate ions to the solution results in dissolution of scandium (at a ratio of $CO_3^{2-}:Sc^{3+}= 0.75$), followed by its complete precipitation in the form of basic carbonate of scandium, which again is completely dissolved when the molar ratio of $CO_3^{2-}:Sc^{3+}$ is increased to 8.35. Based on the available data of scandium ions in solutions of carbonates the main factor affecting the value of imbibition is the change in charge of complex carbonate ions.

Based on the effectiveness of the stripping scandium cation exchanger ECO-10M carbonate solutions can be arranged in the following order:

$$NaHCO_3 < KHCO_3 < K_2CO_3 < (NH_4)_2CO_3 < NH_4HCO_3 < Na_2CO_3. \quad (3)$$

Studies carried out by stripping under static conditions at a volume ratio solution/cationic ion exchanger = 10. The most efficient solution was found to be sodium carbonate.

The molar ratio in the solution $CO_3^{2-}:Sc^{3+}$ determines the composition and the charge of the scandium complex and thus affects the efficiency of desorption. Thus studies were conducted on the degree of desorption as a function of the ratio of the desorption solution: cationic ion exchanger.

The highest efficiency Sc stripping in sodium carbonate solution was found with a concentration of 200 g/dm^3. Desorption of scandium after washing the cation exchanger ECO-10M with sulfate solution should be carried out in a static mode with constant stirring. The optimum ratio of the desorption solution: cation exchanger equals 5.

The precipitation of scandium concentrate again is a well-known process to be done with an alkaline agent from the desorbed acidified solution (pH 1-2). Precipitation pH for scandium from the concentrate is 5.0-5.5.

5 REE extraction process model

A process model of the rare earth elements (REE) fractionation process was developed, see Figure 1. Model is divided in two parts: i) adsorption-desorption where the respective phenomena are modelled based on the reaction equations, and ii) precipitation where the equilibrium chemistry of aqueous water solution is modelled by applying Pitzer model. The applied tools for modelling are Balas [12] for the mass and energy balandes and ChemSheet/ChemApp [13] for the thermodynamic calculations.

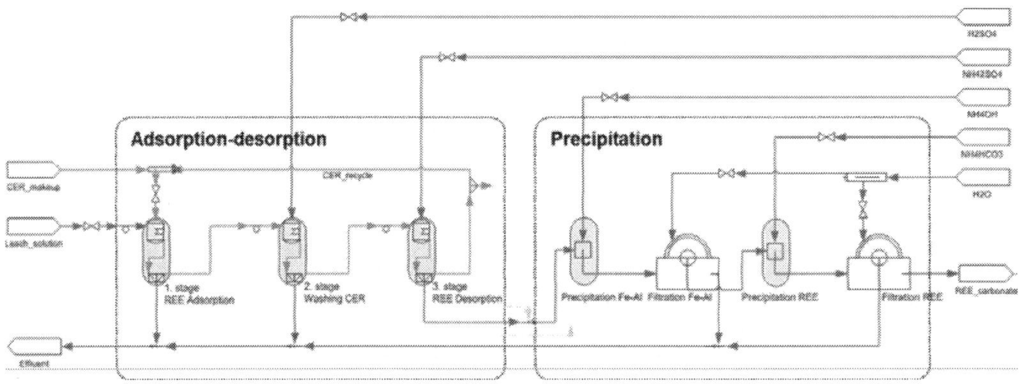

Figure 3. Fractionation of REE from solution. Adsorption-desorption model is based on the reaction equations and Precipitation model is based on the aqueous Pitzer model.

A process model of the rare earth elements (REE) fractionation process was developed, see Figure 1. Model is divided in two parts: i) adsorption-desorption where the respective phenomena are modelled based on the reaction equations, and ii) precipitation where the equilibrium chemistry of aqueous water solution is modelled by applying Pitzer model. The applied tools for modelling are Balas [12] for the mass and energy balandes and ChemSheet/ChemApp [13] for the thermodynamic calculations.

As the original chemical systems included multiple rare earth elements and other species a somewhat reduced chemical system is applied for the modelling purposes. Following elements are included as system components: H, O, S, N, Ca, Al, Fe, Sc, La and Eu. Applied constituents are given in Table 4.

Table 4. Applied chemical system for modelling the fractionation of rare earth elements.

Vapour phase	**Aqueous phase**				**Pure phases**
	Cations		**Anions**	**Neutral**	
$H_2O(g)$	H(+aq)	FeO(+aq)	OH(-aq)	H_2O	$Al_2(SO_4)_3 \cdot 6H_2O$
$NH_3(g)$	Al(+3aq)	FeOH(+2aq)	$Al(OH)_4(-aq)$	$AlO_2H(aq)$	$CaSO_4 \cdot 2H_2O$
	AlO(+aq)	La(+3aq)	$EuO_2(-aq)$	$CaSO_4(aq)$	$CaSO_4$
	AlOH(+2aq)	LaO(+aq)	$FeO_2(-aq)$	$EuO_2H(aq)$	$CaSO_4 \cdot 0.5H_2O$
	Ca(+2aq)	LaOH(+2aq)	$HSO_4(-aq)$	$LaO_2H(aq)$	$(NH_4)_2SO_4$
	Eu(+3aq)	$LaSO_4(+aq)$	$LaO_2(-aq)$	$NH_3(aq)$	$Fe(OH)_3$
	EuO(+aq)	$NH_4(+aq)$	$Sc(SO_4)_2(-aq)$	$Sc_2(SO_4)_3(aq)$	$Eu_2(SO_4)_3 \cdot 8H_2O$
	EuOH(+2aq)	Sc(+3aq)	$Sc_2(OH)_2(+4aq)$		$Al(OH)_3$
	$EuSO_4(+aq)$	$Sc(SO_4)(+aq)$	$ScO_2(-aq)$		$Eu(OH)_3$
	Fe(+3aq)	$Sc_3(OH)_5(+4aq)$	$ScO_2H(aq)$		ScOOH
	$Fe_2(OH)_2(+4aq)$	ScO(+aq)	$SO_4(-2aq)$		$Sc(OH)_3$
	$Fe_3(OH)_4(+5aq)$	ScOH(+2aq)			Sc_2O_3
					$Ca(OH)_2$
					$Fe_2(SO_4)_3$
					CER (as resin)

Some additions of the chemical system were needed: The cationic resin (CER) was modelled as $(C_8H_8SO_3)_2(NH_4)_2$ and respective resin after ion exchange was modelled e.g. $(C_8H_8SO_3)_2Ca$ or $(C_8H_8SO_3)_3La$. The resin was applied during adsorption-desorption process and omitted for the precipitation modelling. As Pitzer parameters are not available for the majority of rare earth element ions respective parameters of known ions were applied, e.g. Pitzer parameters of Fe^{3+} were used for modelling La^{3+} ion.

Adsorption-desorption model is based on the experimental measurements. Precipitation model is parametrized based on the chemical feeds and pH values of the process. The model itself is able to predict the Fe-Al-Sc precipitation and recovering La and Eu at the final stage.

6 Conclusions

In a series of experiments on the extraction of REE from uranium in-situ-leaching solutions concentrates were obtained as REE carbonates containing 50-55% (98-99% of oxides). The content of all impurities was not more than 1%. According to the results of the first experiment was prepared concentrate containing 30-40% Sc. The average specific activity in the concentrates was 1000-1500 Bq/kg. From the experimental procedure the following conclusions could be drawn.

1. REE in sulfuric acid solutions is present in the form of positively charged or neutral complexes, which, obviously, will play a major role in the processes of sorption and extraction. With decreasing pH stability of negatively charged complexes increases.
2. The selectivity of REE and associated elements can be achieved using sulfonated cation exchanger with a DVB content of > 12% at pH = 1.5-2.0.
3. Functional groups of strongly acidic cation exchangers strongly affected REM ions in a wide range of concentrations of sulfuric acid. Therefore, there is the possibility of separating the

sulfuric acid adsorbed components according to their distribution coefficients. We investigated the pre-rinsing sulfonic resins with sulfuric acid. The optimum concentration of sulfuric acid in the washing solution is 75 g/dm^3. Thus, it becomes possible to wash away iron ions to 38% of aluminum ion at 58% of the calcium ion by 35%, of the thorium ion at 14% of the uranium ions to 75%. The optimum load on the column during the washing is 5 volumes of solution to 1 volume of the cation exchanger per hour.

4. Desorption of REEs can be carried out with ammonium sulfate solution. When depositing REE from the eluate a concentrate with an average grade of REE at 32.8% was obtained.
5. The greatest capacity of scandium in the investigated conditions was achieved with a multifunctional cation exchange resin containing sulfonic and phosphonic functional groups. An ion exchanger of this type is under development by the Russian company AXION. Presented test cation exchanger ECO-10M exceeds the capacitive characteristics of known commercial resins.
6. Phosphonic functional groups of the cation exchanger ECO-10M binds scandium ions in a wide range of concentrations of sulfuric acid, enabling separation of adsorbed components according to their distribution coefficients. The optimum concentration of sulfuric acid to the washing solution is 150 g/dm^3. Thus, it becomes possible to wash away iron ions to 92% of aluminum ion at 50% of the calcium ion by 21%, of the thorium ion at 14% and of the uranium ions to 7.6%. The optimum load is 5:1 solution:cation exchanger volume per hour.
7. The highest efficiency in the stripping scandium from ECO-10M appeared in sodium carbonate solution with a concentration of 200 g /dm^3. Desorption of scandium after washing the cation exchanger ECO-10M sulfate solution should be carried out in a static mode with constant stirring . The optimum ratio of the desorption solution: cation exchanger = 5.
8. The specific activity of the primary REE concentrates and scandium was (5.8 ± 0.9)·106 Bq/kg (1.8 ± 0.25)·108 Bq/kg, respectively. Specific activity primary concentrates do not satisfy the technical requirements 95.148-77 TU 1767-009-00545484-2000 (up to 103 Bq / kg) and will require further purification of radionuclides purification factor of at least 6·103 in the case of lanthanide concentrate and 2·105 in the case of concentrate scandium. Radiochemical treatment at 5000-10000 times the level that produces a concentrate Sc with a specific activity of not more than 105 Bq/kg - meets MZUA levels in the workplace. Further decontamination will be carried out in the preparation of Sc oxides.

References

1. L. N. Komissarova et al. Разделение близких по свойствам редких металлов. – Металлургиздат, 130 (1962)
2. K. A. M. Bolshakova, (ed.) Химия и технология редких и рассеянных элементов. Vol. 2, Высшая школа 360 (1976)
3. R. A. Abdulvaliev, L. P. Ni, V. L. Reizman, Райзман Получение скандия из бокситового сырья. Almaty, Гылым 196 (1992)
4. E. I. Kazantsev, A. N. Denisov, *Russ. J. Inorg. Chem.*, **8(9)**, 2189 (1963)
5. E. I. Kazantsev, N. V. Sapogov, *Russ. J. Inorg. Chem.*, **39(8)**, 1973 (1966)
6. E. I. Kazantsev, N. V. Sapogov, *Proc. of Higher Schools, Non-Ferrous Metallurgy*, **6**,103 (1966)
7. S. F. Lincoln, *Inorg. Bioinorg. Mech.* **4,** 217 (1986)
8. G.W. Nestor, J. E. Enderby, *Adv. Inorg. Chem.* **34**, 195 (1989)
9. L. N. Komissarova, Редкоземельные металлы. Сб. статей. М. Химия, 19, 256 (1957)
10. Y. P. Kudryavsky, et al., Известия ВУЗов. Химия и химическая технология. **49,** 1524 (1975)
11. M. P. Noskova et al., *Russ. J. Phys. Chem.* A. **57(3)**, 707 (1983)
12. VTT, Balas, balas.vtt.fi (2007)
13. P. Koukkari, K. Penttilä, K. Hack, S. Petersen, *CHEMSHEET – An Efficient Worksheet Tool for Thermodynamic Process Simulation,* in: Bréchet, Y. (Ed.), (Wiley-VCH, Weinheim, 2005)

Multiphase simulation of mine waters and aqueous leaching processes

Risto Pajarre[1,a], Petteri Kangas[1], Karri Penttilä[1], and Pertti Koukkari[1]

[1]*VTT – Technical Research Centre of Finland, FI02044-VTT, Finland*

Abstract. Managing of large amounts of water in mining and mineral processing sites remains a concern in both actively operated and closed mining areas. When the mining site with its metal or concentrate producing units is operational, the challenge is to find either ways for economical processing with maximum yields, while minimizing the environmental impact of the water usage and waste salt treatments. For safe closure of the site, the environmental control of possible drainage will be needed. For both challenges, the present-day multiphase process simulations tools can be used to provide improved accuracy and better economy in controlling the smooth and environmentally sound operation of the plant. One of the pioneering studies in using the multiphase thermodynamic software in simulation of hydrometallurgical processes was that of Koukkari et al. [1]. The study covered the use of Solgasmix equilibrium software for a number of practical acid digesters. The models were made for sulfuric acid treatments in titania pigment production and in NPK fertilizer manufacturing. During the past two decades the extensive data assessment has taken place particularly in geochemistry and a new versions of geochemical multiphase equilibrium software has been developed. On the other hand, there has been some progress in development of the process simulation software in all the aforementioned fields. Thus, the thermodynamic simulation has become a tool of great importance in development of hydrometallurgical processes. The presentation will cover three example cases of either true pilot or industrial systems including a South African acid mine water drainage treatment, hydrometallurgical extraction of rare earths from uranium leachate in Russia and a multistage process simulation of a Finnish heap leaching mine with its subsequent water treatment system.

References

1. P. Koukkari, H. Sippola, A, Sundquist, *Hydrometallurgy*, **94**, 139 (1994)

[a] Corresponding author: risto.pajarre@vtt.fi

Mathematical models of hydrocyclone performance in various copper ores preparation circuits

Tomasz Niedoba[1,a], Tadeusz Tumidajski[1], and Agnieszka Surowiak[1]

[1]*AGH University of Science and Technology, Faculty of Mining and Geoengineering, Department of Environmental Engineering and Mineral Processing, al. Mickiewicza 30, 30-059 Krakow, Poland*

Abstract. Copper ores located in the Lubin-Glogow Copper Basin contain three main lithological fractions: sandstone, carbonate and shale. This fact is the basic problem of organization and conducting ore preparation to flotation. The existing circuit of feed preparation contains (in ZG Rudna) five classification nodes of hydrocyclones which fit for various purposes. The elaboration of concept of monitoring work of these nodes should be based on appropriate mathematical models of process. It was decided that either regressive or non-dimensional models that is classification according to Svarovsky, and particularly Plitt's, will be suitable, in the aspects of d_{50} and partition sharpness. Errors resulting from determination of the features being part of the equations were also important. In this paper the errors were divided into technical and technological. The issue of experiments organization (mineralogical investigation connected with each classification node) was also addressed. It is obvious that each classification node must obtain its characteristics because of various comminution products directed to classification.

1 Introduction

It can be accepted that the basic problem of beneficiation of Polish copper ores is their complex lithological composition (sandstone, carbonate, shale ores) which is indicated by different comminution course, different size of useful minerals incrustations and other useful components contents. Hydrocyclones decide about the course of qualification (direction) of feed particles to flotation which determine also the amount of material being returned back to the process and amount of additionally ground products. Figure 1 presents the circuit with five hydrocyclone batteries. Each one realizes a slightly different purpose and directs overflows to various flotation processes. The determination of strategy of conducting hydrocyclones work is a complex task and must be based on mathematical models of good quality which describe their work. The additional difficulty is the necessity of conducting modeling under industrial conditions by full acceptable changeability of the feed.

Because of practical applications, the interesting ones are models determining size of cut point d_{50} and so-called partition sharpness s, which can be accepted as the ratio of d_{75}/d_{25}, where d_{75}, d_{50} and d_{25} are the values being determined on the basis of partition curve (k-percentage particles).

[a] Corresponding author: tniedoba@agh.edu.pl

In present paper the problems related to modeling of hydrocyclones work will be considered presented for Lubin-Glogow Copper Basin (LGOM) because of d_{50} and s. Furthermore, the difficulties with including determination of some variables, influence of errors of their determining on quality of models as well ideas on strategy of conducting processes of feed preparation for beneficiation were investigated.

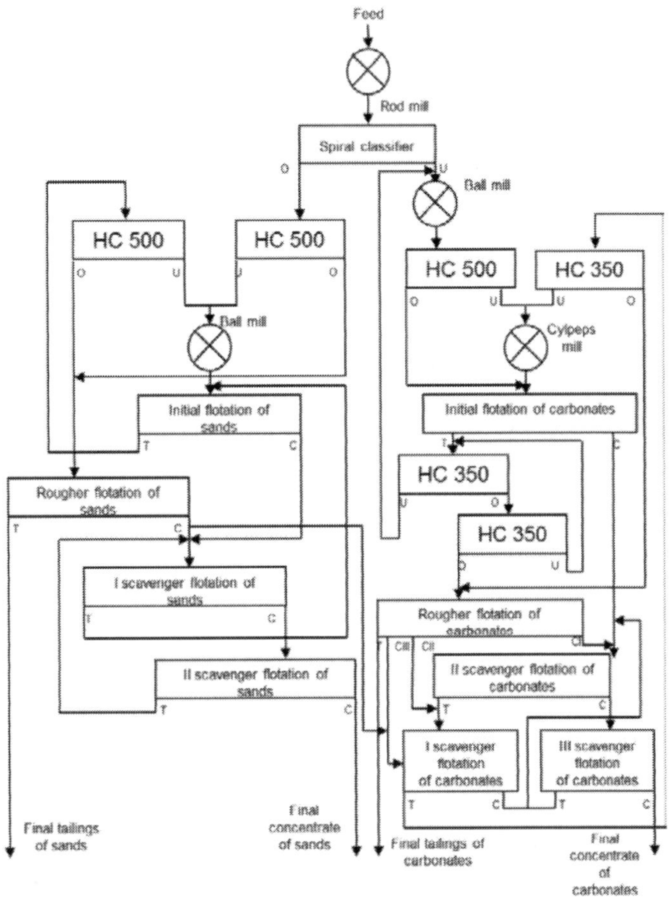

Figure 1. Scheme of beneficiation of material rich in sandstone and carbonate fractions at KGHM Polska Miedz S.A., O/ZWR Region Rudna (U – underflow, O – overflow, T – tailings, C – concentrate) [5].

2 General characteristics of Polish copper ores

As it was mentioned, in area of LGOM, three main copper ores types can be found. Below, the characteristics of individual ores types and their mineralization was discussed [10].

2.1 Sandstone ore

Sandstone ore is created mainly by quartz, sandstones of high changeability of sort of binder in which dissipated type of mineralization with sulfide minerals occurs [26]. It consists usually of quartz (75 to 97% of rock volume) in which size of the material is within 0.04-0.5 mm (most often 0.08-0.3 mm). The particle composition does not change too much in various types of sandstones. However, the composition of binders is differentiated. The clay, carbonate, clay-carbonate and sulfide types of binders can be found, while the most often is calcite [16]. Sandstones of clay and clay-carbonate binders are featured usually with high contents of copper sulfides. The typical concentration of Cu is from 1 to 2%. Sandstones with carbonate binders have high changeability concerning metals sulfides contents, including copper. However, in sandstones with sulfide binders the contents of sulfides oscillates between 0.2 and 1% [1, 9].

Mineralization in the sandstone ore is water-repellent and occurs in spaces between particles. The dominant fraction is 0.098 – 0.5 mm being 75% of weight of ore. Granulation is rather even with the smallest share of the finest particles [11].

2.2 Shale ore

The shale ore is located directly above sandstones. The thickness of the shale fraction in the deposit is equal, on average, to about 30 cm. It consists mainly of clay, dolomite, quartz and also organic matter. As a result of changing proportions between main components, two types of shale: clay-dolomite and clay-organic are created [24]. Clay-organic shale, being about 10% of all shale, contains of 70% clay minerals, 10% organic matter, 10-15% copper sulfide minerals and 5% carbonates and quartz. This type of shale contains the highest contents of additional elements such as Ag, Pb, Co, Zn, Mo, Ni and As in comparison to other ore types [2, 12]. Furthermore, clay-dolomite shales consist of about 60% dolomite, 30% clay minerals, 5% organic matter and 5% copper sulfide minerals. They represent about 90% of the whole amount of shale fraction occurring in the deposit [13, 26].

The ore mineralization is mainly of vein and lens types [16]. Copper shales have very fine granulation because about 94% of particles are below 42 μm in size, that is about 11.5% of the weight. The content of the coarsest fraction (0.14 – 10 mm) is 37.5% of the weight [11].

2.3 Carbonate ore

The areas of occurrence of the sandstone ore are located in northern and eastern parts of Lubin – Malomice and Rudna Regions. Deposits of carbonates dominates mainly in Radwanice East (deposit without sandstone ore) and Polkowice. In the copper deposits located in the Fore-Sudetic Monocline about 140 of minerals were described [25]. The main copper-bearing minerals in the deposit zone are chalcocite, bornite, chalcopyrite, covellite, digenite and tennantite. In the carbonate-shale ores the minerals occur in forms of dissipated particles and aggregates as well as nests and veins. In sandstone, the minerals occur usually in dissipated form being one of the form of either binder or mineral lamination [3, 20]. In all types of ores, the mineralization with chalcocite is dominant. The additional elements are mainly Ag, Au, Pb, Co, Mo and others. They occur mainly in the form of either isomorphic additions or create their own minerals [1]. Because of a lack of mineralization continuity and high irregularity, these concentrations are not like the deposit fulfilling conditions of economic profitability of its single exploitation. A significant part of precious metals contained in material being delivered from mining plants and processed in a beneficiation plant is transferred to a flotation concentrate and is recovered in pyrometallurgical processes.

Mineralization occurring in the carbonate ore is characteristic because of the presence of coarse particles, from 0.14 to 5 mm, which is about 60% of the weight, while 7-28 μm size fraction is only 3% of the weight [11, 12, 26].

3 Structure of models of hydrocyclone work

As it was mentioned, the majority of models describing hydrocyclones work is based on such values as cut point d_{50}, partition sharpness s and, eventually, flowrate. Svarovsky and Svarovsky [31] made the division of theory and models of hydrocyclones work into seven groups:

- simple fundamental theories;

- two-phase flow theory;
- crowding theory;
- regression models;
- dimensionless group models;
- analytical flow models;
- numerical simulations of flow.

Without consideration of rules of this division it is worthy to state that, from practical point of view, it is possible to achieve regression and so-called dimensionless models under industrial conditions. It can be underlined that analytical models of flows and their simulations are future solutions of issues considered in this paper. This idea is in accordance with the one presented in [30, 31].

Currently, for example, simulations CFD (Computational Fluid Dynamics) are applied to determine parameters of equations by means of special software like package FLUENT. Obtaining of certain recommendations for monitoring strategy from these models is very difficult [22, 23, 32]. According to many authors [7, 8, 14, 15, 17, 19, 21, 27, 28, 29, 30, 31, 33] the cut point and other characteristics of hydrocyclone work depend on many technical parameters of device and feed features. Generally, it is accepted that exponential and multiplied models are in use. The generalization of the research is the JKTech model [18]:

$$d_{50} = K_{DO}\left(\frac{D_O}{D_C}\right)^{\alpha_1}\left(\frac{D_U}{D_C}\right)^{\alpha_2}\left(\frac{P}{\rho_{SL} g D_C}\right)^{\alpha_3}\left(\frac{D_I}{D_C}\right)^{\alpha_4}\left(\frac{L_{CYL}}{D_C}\right)^{\alpha_5}\Theta^{\alpha_6} H_S^{\alpha_7} \tag{1}$$

where:
P - feed pressure [kPa]; g –acceleration due to gravity; Θ - cone angle, [°]; H_S – hindered settling factor; K_{DO} – material constant depending on the specific gravity and size of particles in the feed, D_U – underflow diameter [m]; D_I – inlet diameter [m]; D_O – overflow diameter [m]; L_{CYL} – length of cylindrical section [m]; D_C – diameter of the cylindrical section [m]; ρ_{SL} – feed slurry density [Mg/m^3]; α_1, …, α_7 – parameters of equation to be determined.

Working under laboratory conditions, by classification of sands, Plitt [27] elaborated a general model which was several times modified and adjusted:

$$d_{50} = \frac{k_1 D_C^{\beta_1} D_I^{\beta_2} D_O^{\beta_3} \mu^{\beta_4} exp(\beta_5 C_{VS(F)})}{D_U^{\beta_6} L_{VF}^{\beta_7} Q_{V(F)}^{\beta_8} (\rho_S - \rho_L)^{\beta_9}} \tag{2}$$

where:
$Q_{V(F)}$ – volumetric flowrate of the feed [m^3/s]; $C_{VS(F)}$ – % of solids by volume in the feed; μ - liquid viscosity [mPa.s]; ρ_S – density of solid [kg/m^3]; ρ_L – density of liquid [kg/m^3]; k_1 – calibration factor; L_{VF} – free vortex height [m]. The other symbols as in the equation (1).

Assuming that partition sharpness s can be defined as $s = \frac{d_{75}}{d_{25}}$, it can be accepted that the mathematical model for s will assume a general form of equation (1) or (2) with correctly determined parameters.

The key issue in modeling of classification process in hydrocyclones is then the problem of determining partition curve for d_{50}, d_{75} and d_{25}. By determining d_{50} (partition curve) the Grumbrecht method can be applied which allows elimination of eventual anomalies in shape of obtained curve. To do it the following parameters should be accepted:
a_i – share of ith particle fraction in feed
b_i – share of ith particle fraction in underflow
c_i – share of ith particle fraction in overflow.

If the underflow yield is denoted as γ, then

DOI: 10.1051/e3sconf/20160801028

$$a_i = \gamma b_i + (1-\gamma)c_i \, . \tag{3}$$

Because of existing errors of analyses and other disturbances, the sides of Equation (3) will not be equal. By minimizing the differences (squares and sum):

$$S = \sum (a_i - \gamma b_i - (1-\gamma)c_i)^2 \tag{4}$$

according to the rules of determining extreme, it is obtained that

$$\gamma = \frac{\sum (b_i - c_i)(a_i - c_i)}{\sum (c_i - b_i)^2} \, . \tag{5}$$

If $y = AB$, is accepted as simplified version of Plitt's model, where

$$A = \frac{D_I^{\beta_2} D_O^{\beta_3}}{D_U^{\beta_6}} \tag{6}$$

and

$$B = \frac{k_1 D_C^{\beta_1} \mu^{\beta_4} exp(\beta_5 C_{VS(F)})}{L_{VF}^{\beta_7} Q_{V(F)}^{\beta_8} (\rho_S - \rho_L)^{\beta_9}} \tag{7}$$

then the certain numerical simulation can be done. Let us assume that B=const and y=d_{50}=certain value. If these assumptions are accepted, then is possible, by changing values of D_I, D_O and D_U, to show in what range the value of d_{50} can change. Also, the reverse option can be accepted that A=const, d_{50}=certain value and analogical simulation can be conducted on the basis of changes of the values occurring in part B.

In the experiment, the real values of settings for H500 hydrocyclones were accepted and also the limits of the changeability of individual variables in the model were assumed for these, which were qualified to part A. Part B was assumed as constant, calculated on the basis of part A and assumed cut point d_{50}. The values of parameters β_1-β_9 were taken from the original Plitt's model and they were shown in Table 1. The settings and values of cut point occurring from the model were presented in Tables 2 and 3.

Because the recommended cut point in hydrocyclones is particle of size 0.150 mm, such value was accepted as the expected one for the investigated H500hydrocyclone, then the susceptibility of the model to assumed acceptable changes of individual factors was investigated. As the basic cut point sizes, apart from 0.150 mm, the sizes 0.250, 0.500, 0.600, 0.700, 0.800, 0.900, 1.000 and 1.100 mm were accepted. Table 3 shows the changeability of these cut points dependably on accepted minimum and maximum settings for variables from part A, keeping value of B as constant (presented in Table 1). The value of k_1 was assumed to be equal to 1.

Table 1. Values of parameters in initial Plitt's model [7].

Parameter	Value
β_1	0.46
β_2	0.6
β_3	1.21
β_4	0.5
β_5	0.063
β_6	0.71
β_7	0.38
β_8	0.45
β_9	0.5

Table 2. Accepted values of variables in Plitt's model.

Variable	Units	Assumed position	Minimum position	Maximum position
D_I	mm	125	100	150
D_O	mm	175	150	200
D_U	mm	60	50	70

Table 3. Values of cut point.

Position	Cut point d_{50} [mm]								
Assumed position	0.150	0.250	0.500	0.600	0.700	0.800	0.900	1.000	1.100
Minimum position	0.124	0.207	0.414	0.497	0.580	0.662	0.745	0.828	0.911
Maximum position	0.177	0.295	0.589	0.707	0.825	0.942	1.060	1.178	1.296

The susceptibility of the model to individual changes of parameters can be observed. The hydrocyclone operator can then adjust these parameters in the way assuring the final effect (appropriate cut point and partition sharpness) to be satisfying one. For example, to achieve smaller cut points, lower diameters of individual nozzles should be taken. However, it has to be remembered that such hydrocyclones require appropriate feed granulation what causes necessity of its appropriate comminution. Furthermore, the entry of Plitt's models require adaptation to the data obtained on other types of hydrocyclones, what will be the issue of further investigations. Also, Plitt's model concerning partition of slurries volume on products should be noticed [28]:

$$S = \frac{Q_{V(U)}}{Q_{V(O)}} = \frac{k_2\left(\frac{D_U}{D_O}\right)^{\gamma_1} L_{VF}^{\gamma_2}\left(D_U^2+D_O^2\right)^{\gamma_3} exp\left(\gamma_4 C_{VS(F)}\right)\rho_{SL}^{\gamma_7}}{P^{\gamma_5} D_C^{\gamma_6}} \quad (8)$$

where:
$Q_{V(U)}$ - volumetric flow of underflow [m^3/h]; $Q_{V(O)}$ – volumetric flow of overflow [m^3/h]. The rest of marks as in equations (1) and (2).

Furthermore, the determination of appropriate pressure P is important.

$$P = \frac{k_3 Q_{V(F)}^{\delta_1} exp\left(\delta_2 C_{VS(F)}\right)}{D_C^{\delta_3} D_I^{\delta_4} L_{VF}\left(D_U^2+D_O^2\right)^{\delta_5}}. \quad (9)$$

Apart from this, currently it is suitable to use not only one, specific equation, for example, determining cut point size, but a set of equations determining, for example, flowrate, reduced cut point, capacity and partition sharpness. According to [19], the appropriate set of equations gives very good effects of fitting the empirical data.

4 Problems of modeling of hydrocyclones work under industrial conditions

The presented models of hydrocyclones work were obtained under laboratory conditions which ensured mainly stability of feed structure and allowed to organize course of experiment in the way maximally flexible. Under industrial conditions (Polish copper ores processing) momentary changes of quality of processed ore change such features as density of slurries, share of individual fractions, share of lithological layers as well density of solids and useful component grade. By modeling hydrocyclones work, the basic problem is correct determination of partition curve and especially characteristic values like d_{50}, d_{25} and d_{75}. Because of the reasons mentioned above, the collection of representative samples is practically impossible. The application of granulometers, densimeters or other measuring units do not solve the problems in this aspect.

It is worthy to notice that if yields of hydrocyclones products are determined by Grumbrecht's method (complex balance of fractions shares), the possibility of change of fractions shares during realization of separation process must be taken into consideration (blurring of particles during dynamic course of the process) what in connection with analyzes errors may lead to inequalities in balance equations. Errors committed on this stage lead also to errors in determining the value of d_{50} (more generally – course of partition curve).

The determination of the technological values occurring in discussed models is also vitiated by errors and it is obvious that these errors (together with the error of d_{50}) influence the determined values of parameters of regression [4, 6].

The conducted contemplations may be the basis for elaboration of the industrial experiment schedule, which results would be the basis for determination of mathematical model for hydrocyclones.

5 Conclusions

If the purpose is to optimize the classification of hydrocyclones during the beneficiation of the Polish copper ores it is necessary to determine mathematical models of hydrocyclones worked under real conditions. Taking into consideration the eventual planned change of hydrocyclones in the circuit, possibility of selecting their parameters and using the models presented in this paper, the course of industrial experiment was proposed with changes of technical and technological values taken into account. The collected data set, by acceptable change of variables (proportionally), will be the basis to conduct calculations and create model of hydrocyclones work (equivalent of Equation 2). The obtained model will create the base of elaborating concept of monitoring the feed preparation circuit – determining the values of cut points, flowrates (eventual return flows) and amount of processed feed dependably on various shares of lithological layers. For this purpose the circuit work must be characterized for useful minerals liberation – the samples of the feed and separation products must be collected and the chemical and mineralogical analyses for selected particle fractions must be conducted. These investigations, together with the attempt of determining models of the form presented in Equations 8 and 9 will allow to elaborate the concept of monitoring, to determine the feed flowrate and returned material flowrate. The additional grinding of the products in ball mills and especially in cylpebbs mills lead to sufficient liberation of useful minerals, so the amount of the so-called "hard" particles is lower. The appropriate monitoring of classification products yields and adaptation of the process (amount of the feed) to lithological structure of the ore should increase the recovery and quality of flotation concentrates. Such analyses will provide information which can be used in answering the question connected with algorithms of process monitoring.

The presented indications show how to prepare experiments concerning optimization of the hydrocyclones work. It must be remembered that the whole procedure can be different for various ores types to obtain satisfying results.

References

1. M. Banas, P. Kijewski, W. Salamon, J. Pieczonka, A. Piestrzynski, in *Monograph of KGHM PM SA* (KGHM CUPRUM Sp. z o.o. CBR, 214, 2007)
2. M. Banas, W. Mayer, A. Piestrzynski, W. Salomon, *Ann. Socie. Geologorum Poloniae*, **53(1–4)**, 13(1982)

3. A. Bolewski, A. Manecki: *Detailed mineralogy* (Wydawnictwo Polska Agencja Ekologiczna, Warszawa, 1993)
4. J. M. Czaplicki, *Elements of mathematical statistics and their applications in mining and ground works engineering* (Wydawnictwo Politechniki Slaskiej, Gliwice, 2011)
5. J. Drzewiecki, J. Butra, A. Kosior, *Cuprum*, **1**, 185 (2008)
6. W.A. Fuller, *Measurement Error Models* (John Wiley and Sons, New York, 1987)
7. A. Gupta, D.S. Yan, *Mineral Processing Design and Operations: An Introduction* (Elsevier, 2006)
8. K.T. Hsieh, K. Rajamani, *AIChE Journal*, **37(5)**, 735 (1991)
9. W. Kaczmarek, R. Rozek, Miner. Resour. Manag., **24(4)**, 129 (2008)
10. E. Kasinska-Pilut, *PhD Thesis* (AGH Krakow, 2014)
11. P. Kijewski, J. Jarosz, *in Accompanying metals in copper ores deposit – state of research and perspectives of their further applications*, 21, (1987)
12. H. Kucha, W. Mayer, A. Piestrzynski, *Ores and Non-Ferrous Metals*, **6**, 254 (1982)
13. H. Kucha, *Ann. Socie. Geologorum Poloniae*, **46(3)**, 363 (1985)
14. A.J. Lynch, T.C. Rao, *Indian J. Technol.*, **6**, 106 (1968)
15. A.J. Lynch, *Mineral Crushing and Grinding Circuits: Their Simulation, Optimisation, Design and Control* (Elsevier, 1977)
16. W. Mayer, A. Piestrzynski, *Mineralogical Works*, **75,(1)** (1985)
17. K. Nageswararao, *AusIMM Proceedings*, **300(2)**, 21 (1995)
18. T.J. Napier-Munn, S. Morrell, R. Morrison, T. Kojovic, *Mineral Comminution Circuits Their Operation and Optimisation* (JKMRC, 1996)
19. M. Narasimha, A.N. Mainza, P.N. Holtham, M.S. Powell, M.S. Brennan, *Int,. J. Miner. Process.*, **133**, 1 (2014)
20. M. Niec, A. Piestrzynski, *Monograph of KGHM PM SA, Part II*, (Wyd. KGHM CUPRUM Sp. z o.o. CBR, Wroclaw, 167, 2007)
21. A. Nowak, A. Surowiak, *Archives of Min. Sci.*, **58(4)**, 1209 (2013)
22. E. Nowak, Z. Stegowski, B. Tora, M. Kurzac, *in Proceedings of the XXIII International Mineral Processing Congress* (Istanbul, 1743, 2006)
23. E. Nowak, Z. Stegowski, *Tracers and tracing methods* (Warszawa, 62, 2004)
24. S. Oszczepalski, A. Rydzewski, *Metalogenous atlas of Zechstein cupriferous series in Poland*, (Panstwowy Instytut Geologiczny, Warszawa, 1997)
25. J. Pieczonka, A. Piestrzynski, *Miner. Resour. Manag.*, **22(3)**, 187 (2006)
26. A. Piestrzynski, *Monograph of KGHM PM SA, Part II*, (Wyd. KGHM CUPRUM Sp. z o.o. CBR, Wroclaw, 167, 2007)
27. L.R. Plitt, J.A. Finch, B.C. Flintoff, *in Proceedings of European Symposium Particle Technology* (Amsterdam, 790 1980)
28. L.R. Plitt, *CIM Bulletin*, **12**, 114 (1976)
29. C.M. Scheid, L.A. Calcada, R.S.A. Goncalves, G. Massarani, *Brazilian J. Chem. Eng.*, **30(4)** 781 (2013)
30. L. Svarovsky, *Hydrocyclones* (Holt Rinchart & Winston, London, 1984)
31. L. Svarovsky, G. Svarovsky, *Hydrocyclones, volume II* (Svarovsky, 2014)
32. M. Tora, *PhD Thesis* (Ostrava, (2013)
33. B.A. Wills, T.J. Napier-Munn, *An Introduction to the Practical Aspects of Ore Treatment and Mineral Recovery* (Butterworth-Heinemann, 2006)

Bacterial bioleaching of low grade nickel limonite and saprolite ores by mixotrophic bacteria

Siti Khodijah Chaerun[1,2,a], Siti Aisyah Alting[2], Mohammad Zaki Mubarok[1], and Edy Sanwani[1]

[1]*Department of Metallurgical Engineering, Faculty of Mining and Petroleum Engineering, Institut Teknologi Bandung, Ganesha 10, Bandung 40132, Indonesia*
[2]*Geomicrobiology-Biomining & Biocorrosion Laboratory, Microbial Culture Collection Laboratory, Biosciences and Biotechnology Research Center, Institut Teknologi Bandung, Ganesha 10, Bandung 40132, West Java, Indonesia*

Abstract. Utilization of indigenous bacteria should be considered to establish a successful biohydrometallurgical process. In this study, mixrotrophic iron-oxidizing bacterial consortia consisting of *Comamonas testosteroni*, *Alicyclobacillus ferrooxydans* and *Pantoea septic* which were isolated from Indonesian mineral ores were examined to determine their abilities to recover nickel from limonite and saprolite ores in the bioleaching experiments using stirred tank reactors. The nickel bioleaching experiments inoculated with the bacterial consortia were carried out using coarse limonite ores and weathered saprolite ores with pulp density of 10% w/v. Abiotic controls were also carried out replacing the inocula by the sterile medium. The bioleaching processes were monitored by measuring Ni and Fe contents and pH of the leaching solution as well as the total bacterial enzymatic activity measured as FDA hydrolytic activity. The effect of leaching on the mineralogy of laterite ores was investigated by the scanning electron microscope equipped with energy-dispersive spectroscopy (SEM-EDS) and X-ray powder diffraction (XRD). After 28 days of incubation, the FDA hydrolytic activity was observed in both bioleaching experiments containing limonite (17.2 μg fluorescein/mL) and saprolite ores (16.9 μg fluorescein/mL). The leached Ni and Fe in the bioleaching experiments containing limonite ores (30% Ni and 5.6% Fe) was greater than that in abiotic controls (1% Ni and 0.1% Fe) with the pH range of 2.5 to 3.5. However, the bacterial consortia were less capable of bioleaching of Ni (2.5%) with the similar leached Fe (6%) from the saprolite ores. In abiotic controls, the medium pH remained relatively constant (pH 6). It was concluded that these bacterial isolated as the consortium were capable of nickel bioleaching (precious metal) more effectively than iron (gangue metal), thus being applicable to the commercial processing of the difficult-to-process low-grade nickel laterite ores in Indonesia.

[a] Corresponding author: skchaerun@gmail.com

Influence of flotation cell volume and solids mass on kinetics of sulfide ore flotation

Michal Plawski[1,a] and Alicja Bakalarz[1]

[1]*Wroclaw University of Science and Technology, Faculty of Geoengineering, Mining and Geology, Wybrzeze Wyspianskiego 27, 50-370 Wroclaw, Poland*

Abstract. The paper presents studies on the influence of flotation cell capacity and mass of solids in the suspension on the flotation kinetics of sulfide copper ore. A sample of copper ore that was collected from the Polkowice Mine of KGHM Polska Miedz S.A. in Poland was used in the experiments. It was determined that neither the volume of flotation cell nor the mass of solids had influence on the type of kinetics equation of flotation. Copper-bearing minerals floated according to the second-order equation, while the remaining components according to the first-order equation. The kinetic rate constants and maximum recovery of the studied components decreased with increasing solids mass in the flotation cell, regardless of the capacity of the cell. The best results were obtained for tests using a 1.0 dm^3 cell, while the less favorable kinetics results were observed in the test with the smallest cell of 0.75 dm^3 volume. The obtained results can be helpful in choosing the most appropriate methodology of upgrading the sulfide copper ore from Poland in order to obtain the best kinetics results.

1 Introduction

Flotation is a process of selective separation of hydrophobic materials from hydrophilic ones. Three major groups of factors have influence on the flotation effectiveness, namely: chemistry, operation and equipment components. The most important factors of the aforementioned groups are: feed rate, particle size, mineralogy, flotation agents, pulp density, electrochemical properties of feed (pH, redox), temperature, air flow and cell design [1]. Therefore, it is important to take all of these factors into account in flotation operations. A change of one factor will automatically cause changes in the whole process [1-2].

Numerous papers have considered the effect of pulp density on flotation effectiveness and selectivity [3-6]. Mehrotra and Kapur [3] analyzed the influence of pulp density on quartz flotation. It was shown that for low pulp densities there was no influence on the flotation kinetics. For higher pulp densities, the rate constant of separation k decreased along with increasing the pulp density. According Mehrotra and Kapur [3], these observations could be different for various materials. The influence of pulp density (7.5% and 20% by weight) on flotation selectivity of a mixture of sphalerite and pyrite was investigated by Uçurum and Bayat [4]. Higher pulp density of flotation suspension caused more effective separation of these sulfides. Uçurum [5] studied the flotation kinetics of carbon for five different pulp densities, that is 5, 7.5, 10, 15, and 20% by weight. The author stated that the best results of flotation kinetics of carbon were observed under the conditions of medium values of pulp

[a] Corresponding author: michal.plawski@op.pl

density, i.e. 10%. According to Konieczny et al. [6], the yield of concentrate of Polish copper ore increased with the pulp density. Additionally, it was shown that the pulp density does not have significant effect on the recovery of copper in the concentrate.

Kinetics is one of the most important aspects of separation process, which determines the time needed for removal of valuable components of ores. The kinetics of separation is also a basic element of separation models relating to components recovery, products grade and properties of the feed [7-9]. Numerous kinetic models are available and applied in separation science and technology [10-11]. In this paper, only first-order and second-order kinetic equations are analyzed. The models are described in details elsewhere [11-12].

2. Experimental

2.1. Materials

For the purpose of conducting the experiments a sample of copper ore was collected from the Polkowice Mine of KGHM Polska Miedz S.A. in Poland. Before each flotation test, a sample of 900 g was wet-ground in a laboratory ball mill for 80 minutes, in the presence of stainless grinding media and tap water (0.5 dm^3). An industrial mixture of sodium ethyl and izobuthyl xanthates (MX) in the dose of 100 g/Mg was used as a collector. An aqueous solution of Corflot was utilized as a frother in the dose of 0.9 g per 1 dm^3 of water. Both reagents were prepared directly before flotation tests.

2.2. Procedures

Nine flotation tests were performed. Description of all experiments is presented in Tables 1-2. All experiments were conducted in the laboratory mechanical flotation machine with removable flotation cells and rotors. Three different flotation cells with various capacity were applied: 1.5, 1.0 and 0.75 dm^3. The conditioning time with collector and frother was 3 and 1 minute, respectively. During each flotation test the water solution of frother (0.9 g per 1 dm^3 of water) was added to retain the froth on the constant level. The air flow rate was regulated during each flotation test by using a rotameter. The speed of stirrer was 1900 rpm and the flow of air was 60 dm^3/h. During each flotation test seven concentrates were collected 45 sec. and 1.5, 4, 8, 15, 25, 40 minutes.

Table 1. Symbols of performed tests.

Mass of the solids/ capacity of flotation cell	400 g	300 g	200 g
1.5 dm^3	F1	F2	F3
1.0 dm^3	F4	F5	F6
0.75 dm^3	F7	F8	F9

Table 2. Pulp densities (*d*) in each flotation test.

Symbol of test	F1	F2	F3	F4	F5	F6	F7	F8	F9
d, kg/m^3	1 112	1 058	1 026	1 238	1 156	1 066	1 258	1 208	1 140

The flotation products were dried in a laboratory drier at 105 °C for 24 hours, and then weighted. The content of copper in each flotation product was determined by using an XRF analysis. The results of flotation were evaluated by analyzing and plotting kinetics. Two kinetics equations were taken into account in the description of the flotation kinetics of copper and useless components into the concentrate, namely: first-order and second-order equations. The first-order equation is:

$$\varepsilon = \varepsilon_{max}(1 - e^{-kt}), \quad (1)$$

while the second-order equation is:

$$\varepsilon = \frac{\varepsilon_{max}^2 kt}{1 + \varepsilon_{max} kt} . \quad (2)$$

In Equations (1-2) ε stands for recovery of a component in separation product, ε_{max} for maximum recovery of the same component in separation product, *k* for rate constant of separation, and *t* denotes separation time. All calculations were performed using the MATLAB software.

3. Results and discussion

Table 3 shows the results obtained by checking the adjustment of first and second order kinetic equations. The equation which the best describes the flotation kinetics of Cu and the rest of the components was chosen based on the values of determination coefficient (R^2) and standard estimation error (*SEE*). The equation which the best describes the kinetics flotation is bolded in Table 3.

Table 3. Analysis of the kinetics of copper and useless components for all conducted flotation tests.

Flotation test	Equation order	Copper			Useless components		
		k, min^{-1}	R^2	*SEE*	k, min^{-1}	R^2	*SEE*
F1	1	0.2843	0.9839	5.9404	**0.0612**	**0.9924**	**1.2672**
	2	**0.0059**	**0.9981**	**2.1403**	0.0031	0.9599	3.1504
F2	1	0.3559	0.9839	6.0128	**0.0641**	**0.9928**	**1.0550**
	2	**0.0073**	**0.9987**	**1.6651**	0.0035	0.9667	2.6895
F3	1	0.4374	0.9758	7.1567	**0.0672**	**0.9946**	**1.0651**
	2	**0.0087**	**0.9986**	**1.3414**	0.0032	0.9651	3.1159
F4	1	0.4416	0.9682	7.9609	**0.0782**	**0.9878**	**1.4839**
	2	**0.0088**	**0.9968**	**1.8208**	0.0042	0.9704	2.3449
F5	1	0.6333	0.9663	7.8581	**0.0905**	**0.9940**	**1.1401**
	2	**0.0126**	**0.9962**	**2.0188**	0.0049	0.9809	2.0238
F6	1	0.6724	0.9753	7.1557	**0.0885**	**0.9921**	**1.3285**
	2	**0.0130**	**0.9983**	**1.3687**	0.0040	0.9833	2.5842
F7	1	0.3055	0.9757	7.0836	**0.0843**	**0.9928**	**1.4312**
	2	**0.0064**	**0.9957**	**2.4593**	0.0040	0.9708	2.7109
F8	1	0.3813	0.9635	8.5353	**0.0743**	**0.9961**	**1.0987**
	2	**0.0077**	**0.9924**	**2.9903**	0.0030	0.9697	3.6516
F9	1	0.5257	0.9767	7.2930	**0.0692**	**0.9926**	**1.2836**
	2	**0.0100**	**0.9986**	**1.2582**	0.0030	0.9703	3.1829

Table 4 presents the maximum values of copper and useless components recoveries in the concentrate with their rate constants. As seen from Tables 3 and 4, flotation of copper-bearing minerals is the best described by the second-order kinetic equation, while the remaining (useless) components by the first-order kinetic equation. It is observed regardless of the capacity of flotation cell and content of solids.

The flotation results are also presented in Figs. 1-3 in the form of the recovery of either Cu or useless (remaining) components vs. flotation time. The fastest flotation of copper in all experiments is observed in the first 5–10 minutes. After 30 minutes of the flotation process the maximum recoveries of copper achieve a plateau level. The maximum recoveries of copper in the concentrate are in the range of 80–90% for each test. However, different results can be observed for useless (remaining) components. The maximum recoveries do not reach the plateau level even after 40 min of flotation. The maximum recoveries of useless (remaining) components in the concentrate range from 30 to 35%.

Table 4. Maximum values of recovery of copper and useless (remaining) components in all flotation tests.

Symbol of test	F1	F2	F3	F4	F5	F6	F7	F8	F9
Copper									
k, min^{-1}	0.0059	0.0073	0.0087	0.0088	0.0126	0.0130	0.0064	0.0077	0.0100
ε_{max}, %	81.02	82.50	85.04	84.08	84.98	87.33	83.71	87.23	87.41
Useless (remaining) components									
k, min^{-1}	0.0612	0.0641	0.0672	0.0782	0.0905	0.0885	0.0843	0.0743	0.0692
ε_{max}, %	27.73	26.32	30.69	29.90	30.01	35.22	34.42	36.98	34.86

Figures 4-6 show the relationships between the parameters of kinetics (maximum recovery and rate constant) estimated for copper and useless (remaining) components as well as mass of solids in the flotation cell. It can be seen that the rate constant for copper increases when the mass of solids in the flotation cell decreases (Fig. 4a). The highest values of constant k are observed in the tests with the 1.0 dm^3 cell. The lowest values of k are obtained in tests with the 1.5 dm^3 flotation cell. Similar results are observed for the maximum recovery of copper in the concentrate (Fig. 4b). The highest values of ε_{max} of copper are obtained in the tests with the smallest flotation cell used (0.75 dm^3), while the lowest maximum recoveries of Cu in the concentrate are observed for the biggest cell used (1.5 dm^3). The values of k and ε_{max} (Fig. 4c) show that the best results of copper kinetics are observed for the 1.0 dm^3 flotation cell in the presence of solids mass in the amount of 200 g. The worst results are obtained for the flotation cell with the biggest volume and capacity. Similar observation is obtained for the flotation kinetics of useless (remaining) components. However, different flotation results using the cell with the capacity of 0.75 dm^3 are obtained. The flotation rate constant (Fig. 5a) and maximum recovery (Fig. 5c) increase with the mass of solids in the feed.

As it can be seen in Fig. 6, the most beneficial results of kinetics (the highest values of k for copper and useless components) are received in the tests using the flotation cell in the capacity of 1.0 dm^3, regardless of the mass of solids in the feed. The less favorable results of kinetics are observed in tests with the smallest cell (0.75 dm^3).

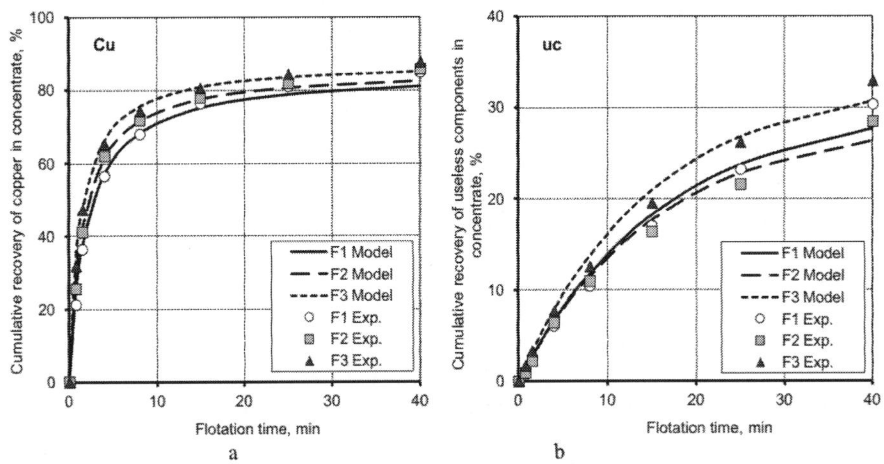

Figure 1. Flotation kinetics of copper (a) and useless (remaining) components (b) in 1.5 dm^3 flotation cell.

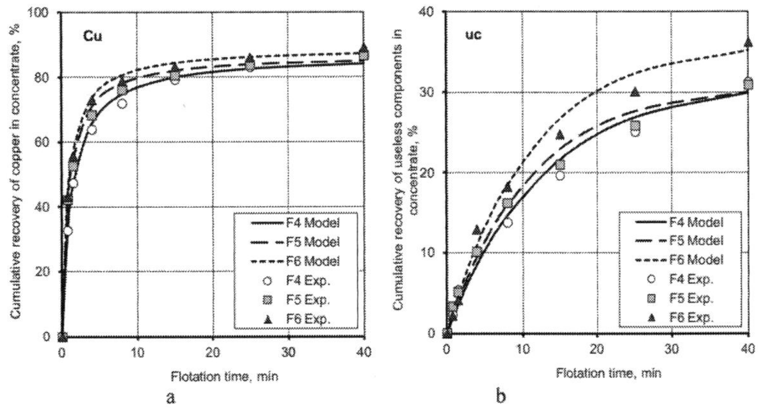

Figure 2. Flotation kinetics of copper (a) and useless (remaining) components (b) in 1.0 dm^3 flotation cell.

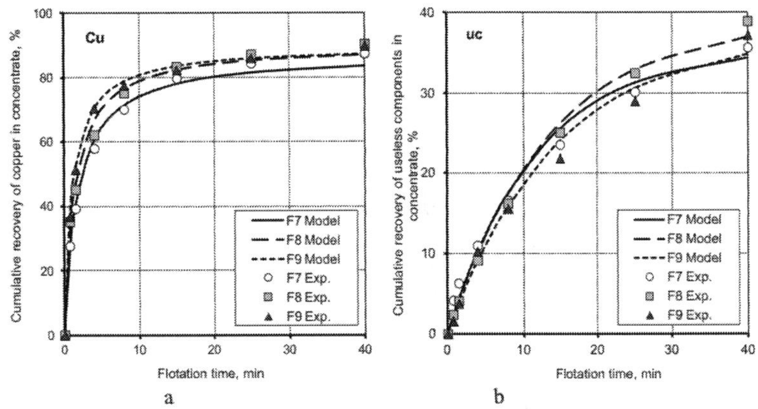

Figure 3. Flotation kinetics of copper (a) and useless (remaining) components (b) in the 0.75 dm^3 flotation cell.

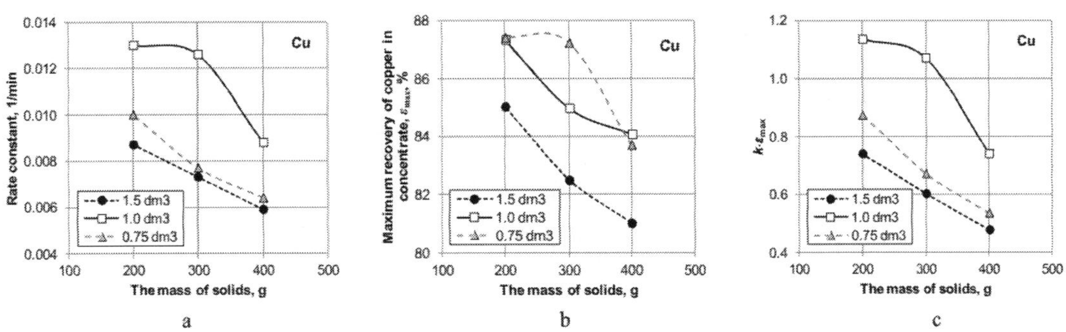

Figure 4. Relationship between kinetic parameters estimated for copper and mass of solids for all conducted flotation tests.

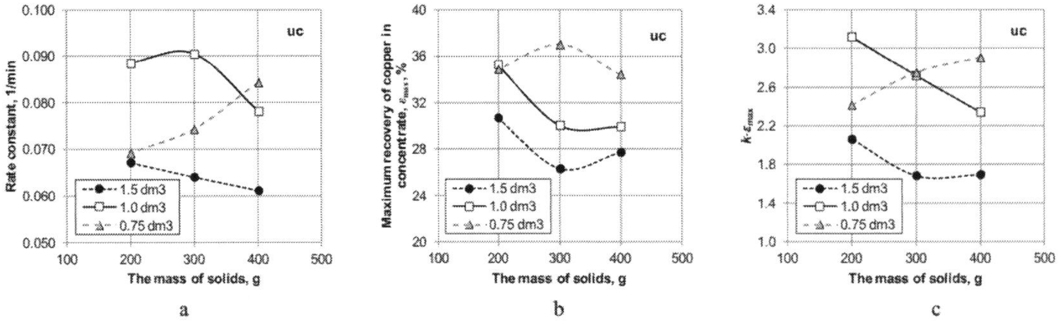

Figure 5. Relationship between kinetic parameters estimated for useless components and the mass of solids for all conducted flotation tests.

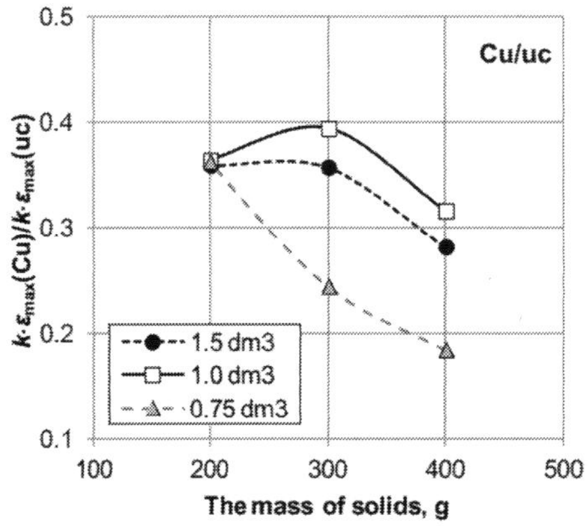

Figure 6. Relationship between kinetic parameters estimated for copper and useless components and mass of solids for all conducted flotation tests.

4. Conclusions

The study presents the influence of the flotation cell volume and mass of solids used on the flotation kinetics of the copper sulfide ore (Polkowice, KGHM Polska Miedz S.A.). Nine flotation tests were conducted. Three flotation cells with different cell volumes (0.75, 1.0. and 1.5 dm^3) and three various portions of solids (200, 300 and 400 grams) were used in the experiments. First-order and second-order kinetic equations were used to describe the process. It was shown that neither the capacity (volume) of the flotation cell nor the mass of solids had the influence on the type of kinetics equation for flotation of Cu and useless (remaining) components. The obtained results confirmed data presented in a previous paper [13]. In both papers, the flotation kinetics of copper components can be described by the second-order equation, while the useless (remaining) components by the first-order equation. The rate constants as well as maximum recoveries of both studied components decreased with increasing solids mass in the flotation cell, regardless of the capacity (volume) of the cell. The best kinetics results were obtained in the tests with the medium cell (1.0 dm^3), while the less favorable kinetics results were observed in tests performed in the smallest flotation cell (0.75 dm^3). The obtained results can be helpful in choosing the most appropriate methodology of upgrading of the sulfide copper ore from Poland in order to obtain the best kinetics results.

Acknowledgements

This work was partially financed by the Polish Statutory Research Grant B50182.

References

1. R.R. Klimpel, *The Influence of Frother Structure on Industrial Coal Flotation. High-Efficiency Coal Preparation* (Society for Mining, Metallurgy, and Exploration, Littleton, CO, 141, 1995)
2. S.K. Kawatra, *Flotation Fundamentals*, http://www.chem.mtu.edu/chem_eng/faculty/kawatra/Flotation_Fundamentals.pdf, (2016)
3. S.P. Mehrotra, P.C. Kapur, *Powder Technol.*, **9**, 213 (1974)
4. M. Uçurum, O. Bayat, Sep. *Purif. Technol.*, **55**, 173 (2007)
5. M. Uçurum, *Powder Technol.*, **191**, 240 (2009)
6. A. Konieczny, W. Pawlos, K. Ksiezniak, M. Krzeminska, E. Kasinska-Pilut, P. Piwowar, *CUPRUM J.*, **75**, 87 (2015)
7. R. Schuhmann Jr., *J. Phys. Chem.* **46(8)**, 891 (1942)
8. B.A. Wills, T. Napier-Munn, *Mineral processing technology. An introduction to the practical aspects of ore treatment and mineral recovery*, 7th edit (Elsevier, 2006)
9. J.S. Laskowski, *Thermodynamic and kinetic flotation criteria* (Gordon and Breach Science Publishers, New York, 1989)
10. P. Somasundaran, *I. J. Lin, Trans.* AIME **254**, 181 (1973)
11. M. Brozek, A. Mlynarczykowska, *Physicochem. Probl. Miner. Process.* **41**, 51 (2007)
12. F.H.B. de Castro, M.C. de Hoces, *Chem. Eng. Sci.*, **51(1)**, 119 (1996)
13. A. Bakalarz, M. Duchnowska, A. Konieczny, E. Kasinska-Pilut, W. Pawlos, R. Kaleta, P.B. Kowalczuk, J. Drzymala, A. Luszczkiewicz, *CUPRUM J.*,**2**, 55 (2015)

Preliminary study on collectorless flotation of chalcocite, bornite and copper-bearing shale in the presence of selected frothers

Jan Drzymala[1,a], Alicja Swebodzinska[1], Magdalena Duchnowska[1], Alicja Bakalarz[1], Andrzej Luszczkiewicz[1], and Przemyslaw B. Kowalczuk[1]

[1]*Wroclaw University of Science and Technology, Faculty of Geoengineering, Mining and Geology, Wybrzeze Wyspianskiego 27, 50-370 Wroclaw*

Abstract. Flotation of carbonaceous copper-bearing shale and copper sulfide minerals such as chalcocite and bornite in the presence of only frother was investigated. It was determined that all investigated solids did not float in pure water in the absence of flotation reagents. However, an addition of a frother rendered both shale and bornite floatable, while the recovery of chalcocite was negligible. These findings suggest that removal of carbonaceous matter from the ore by the so-called pre-flotation process can be a suitable procedure to reduce the amount of organic carbon in the concentrates provided that the feed does not contain bornite. Otherwise, simple pre-flotation must be replaced with a more sophisticated process in which flotation of bornite is suppressed.

1. Introduction

Many sedimentary type ores contain organic carbon in the form of kerogen, bitumen and graphite [1-4]. Also the Kupferschiefer sedimentary ore located in SW Poland contains organic carbon [5-7]. The presence of organic carbon in the ore creates difficulties during its processing by flotation as well as during flash smelting of copper concentrates [8]. Industrial trials conducted to remove organic carbon from the Kupferschiefer sedimentary ore were proved to be successful [9]. However, too much useful sulfide minerals floated together with carbonaceous shale in the so-called pre-flotation process, in which the ore is subjected to flotation in the presence of only frother (collectorless flotation) [10]. Such flotation is possible because shale, being a part of the ore and containing organic carbon, is naturally hydrophobic [11]. However, sulfide minerals can also be hydrophobic [12]. Therefore, it becomes essential to determine flotation properties of mineral components of the Kupferschiefer ore. For this purpose, laboratory flotation tests were performed in a monobubble Hallimond tube to determine flotation properties of selected components of the Kupferschiefer sedimentary ore, that is copper-bearing shale, chalcocite and bornite.

[a] Corresponding author: jan.drzymala@pwr.edu.pl

 DOI: 10.1051/e3sconf/20160801031

2. Experimental

2.1. Materials

Natural carbonaceous copper-bearing shale, chalcocite and bornite, all originated from the copper ore, called the Kupferschiefer sedimentary ore, mined by KGHM Polska Miedz S.A. located in Poland, were investigated. The identity and purity of the solids were checked by XRD and XRF analyses. The solids exhibited natural hydrophobicity with the contact angle greater than zero [11,13]. Distilled water, without pH regulation, was used in all experiments. Poly(ethylene glycol) alkyl ether ($C_nH_{2n+1}O(C_2H_4O)_mH$, $C_{16}E_{20}$) and methylisobutylcarbinol (MIBC) were used as the frothers. The reagents were obtained from Sigma-Aldrich (≥98% pure) and were used without further purification.

2.2. Procedures

Flotation tests were performed in a monobubble Hallimond tube with an increased length of the compartment containing the feed (Fig. 1). The purpose of using such a tall tube was to reduce the mechanical entrainment of particles during flotation. The volume of the Hallimond tube was 200 cm^3. The receiver of the Hallimond tube was calibrated to be able to determine the volumetric recovery of the floating particles. A solid sample of 3 grams and distilled water were mixed together and agitated for two minutes in a beaker, and next transferred to the Hallimond tube. In each flotation tests the air flow rate was $6.25 \cdot 10^{-7}$ m^3/s.

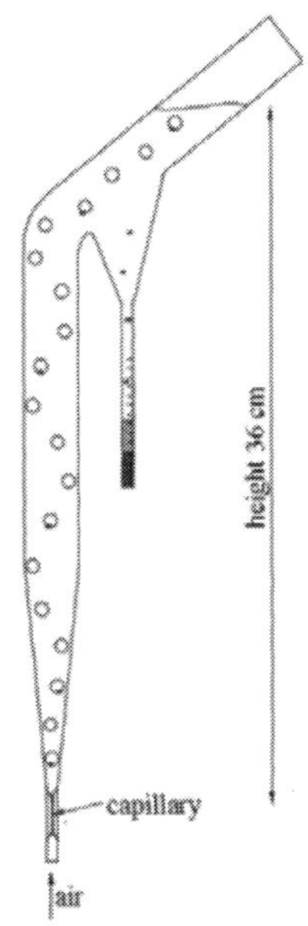

Figure 1. Monobubble Hallimond tube used in all experiments.

3. Results and discussion

The results of flotation of natural carbonaceous copper-bearing shale, chalcocite and bornite in pure water and in the presence of aqueous solution of frothers at different concentrations are shown in Figs. 2-4, respectively.

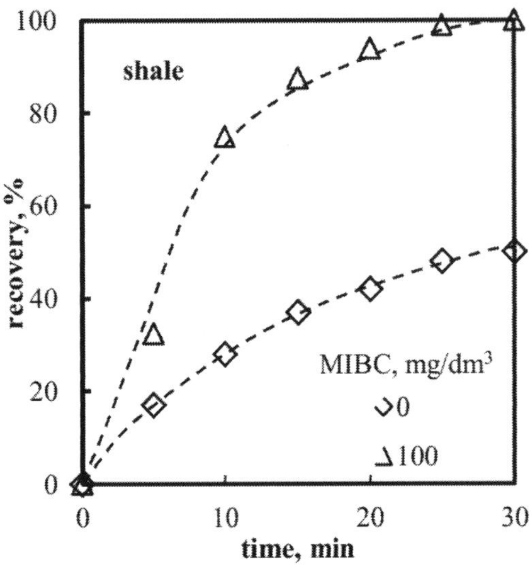

Figure. 2. Flotation kinetics of shale (0.075-0.040 mm) in water and the presence of MIBC.

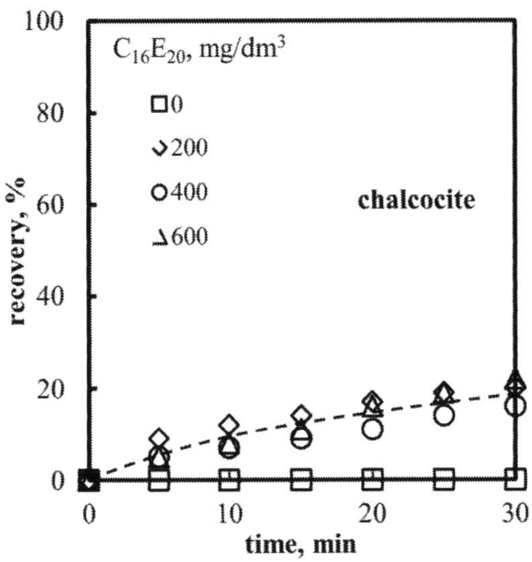

Figure. 3. Flotation kinetics of chalcocite (0.075-0.025 mm) in water and the presence of $C_{16}E_{20}$.

The data presented in Figs. 2-4 clearly show that there is no flotation of shale, chalcocite and bornite in pure water. Some particles recovery, after a long time of flotation, points to their entrainment, which still is accompanying the flotation process even though the Hallimond tube was modified to avoid excessive mechanical flotation. Figure 5 confirms that the recovery of particles after 30 minutes of flotation is partially due to entrainment and partially due to true flotation. The recovery in pure water is dependent of the density of the material subjected to flotation (Fig. 5).

The presence of frother in the flotation system renders shale and bornite floatable, while chalcocite is still non-floatable (Figs. 2-4). It indicates that separation of bornite from shale, when present together in a feed, is not possible, while separation of shale and chalcocite is feasible.

 DOI: 10.1051/e3sconf/20160801031

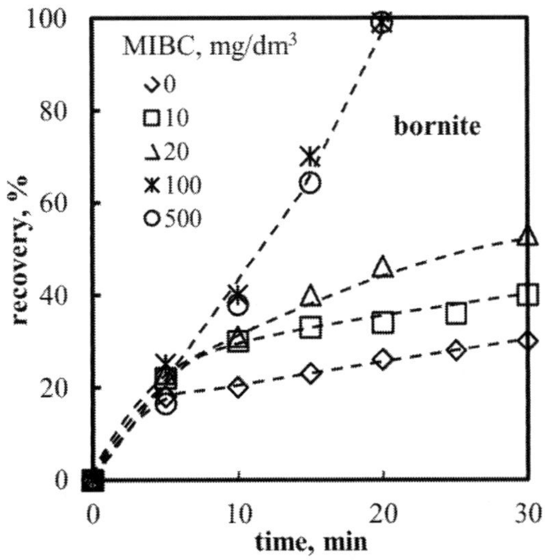

Figure 4. Flotation kinetics of bornite (0.100-0.040 mm) in water and the presence of MIBC.

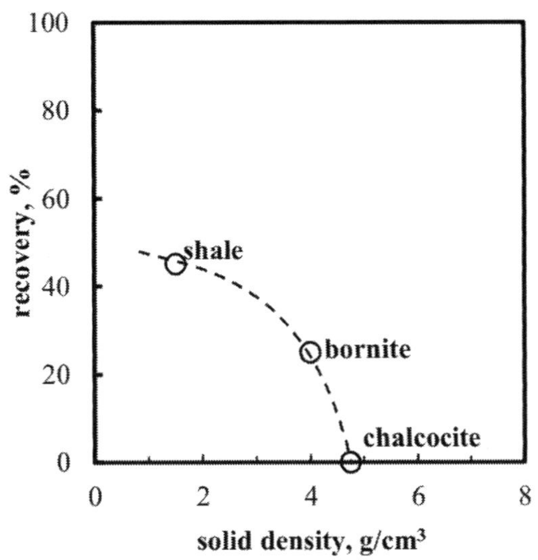

Figure 5. Influence of density of shale, bornite and chalcocite in pure water on entrainment in the investigated Hallimond tube. Flotation time 30 min. Density (in g/cm^3) of shale was 2.4-2.7 (in water about 1.5) [14], bornite 4.9-5.3 (in water about 4.0) and chalcocite 5.5-5.8 (in water about 4.6) [15].

4. Conclusions

Flotation data involving carbonaceous copper-bearing shale, bornite and chalcocite indicate that separation of shale from chalcocite by pre-flotation performed in the presence of only frother is possible, while separation of shale from bornite would require additional, more sophisticated operations in which flotation of bornite is suppressed. These findings can be useful in choosing the most appropriate procedure of processing of the Kupferschiefer sedimentary ore located in Poland, which is rich in organic carbon making its processing difficult.

Acknowledgements

This work was partially financed by the Polish Statutory Research Grant 0401/0186/16.

References

1. A. Hutton, S. Bharati, T. Robl, *Energy Fuels* **8**, 1478 (1994)
2. R.J. Pugh, *Miner. Eng.* **13(2)**, 161 (2000)
3. A. Konieczny, W. Pawlos, M. Krzeminska, R. Kaleta, P. Kurzydlo, *Physicochem. Probl. Miner. Process.* **49(1)**, 189 (2013)
4. P.B. Kowalczuk, B. Buluc, O. Sahbaz, J. Drzymala, *Physicochem. Probl. Miner. Process.* **50(2)**, 835 (2014)
5. Z. Sawlowicz, *Mineral. Pol.* **20**, 69 (1989)
6. Z. Sawlowicz, *Mineral. Pol.* **22**, 49 (1991)
7. Z. Sawlowicz, A.P. Gize, M. Rospondek, *Organic Matter and Mineralisation* **10**, 220 (2000)
8. P. Kijewski, R. Leszczynski, *Zesz. Nauk. Insty. Gospod. Suro. Mine. i Ener. PAN* **79**, 131 (2010)
9. A. Konieczny,M. Duchnowska, R. Kaleta, W. Pawlos, P.B. Kowalczuk, M. Krzeminska, A. Bakalarz, E. Kasinska-Pilut, J. Pr edki, A. Luszczkiewicz, J. Drzymala, in *Processing of XIth International Conference on Non-Ferrous Ore ICNOP'2015, (*Trzebieszowice, 147 2015)
10. M. Duchnowska, *Private communication* (2016)
11. P.B. Kowalczuk, E. Zaleska, O. Danczak, *Trans. Nonferrous Met. Soc. China* **25(1)**, 314 (2015)
12. J. Lekki, J. Drzymala, *Colloids and Surf.*, **44**, 179 (1990)
13. P.B. Kowalczuk, J. Drzymala, *Ind. Eng. Chem. Res.* **50(7)**, 4207 (2011)
14. K. Cependa, J. Drzymala, M.P. Lewicka, in *Lupek miedzionosny*, edited by J. Drzymala, P.B. Kowalczuk (WGGG PWr, 2014)
15. A. Manecki, *Encyklopedia mineralow* (Wydawnictwo AGH, 2004)

Purification of Dunino halloysite by H_2SO_4 leaching and magnetic separation

Piotr Sakiewicz[1], and Marcin A. Lutynski[2,a]

[1]*Division of Nanocrystalline and Functional Materials and Sustainable Proecological Technologies Institute of Engineering Materials and Biomaterials, Silesian University of Technology, Konarskiego 18a,C44-100 Gliwice, Poland*
[2]*Institute of Mining, Faculty of Mining and Geology, Silesian University of Technology, Akademicka 2,C44-100 Gliwice, Poland*

Abstract. The article presents results of purification and comminution of raw halloysite from the Dunino deposit. The aim of the study was to clean halloysite from Dunino deposit with acid (H_2SO_4) leaching and to remove iron impurities in order to obtain fine grained fractions smaller than 20 μm, which can be used in future for manufacturing polymer-aluminosilicate composites. The proposed method, where nanotubes (HNT) and nanoplates (HNP) were separated, consisted of several simple stages including crushing, milling magnetic separation and sedimentation. A multi-stage process resulted in three products: heavy iron oxide fraction, HNP and HNT mixture (<20 um) and remaining intermediate product (mostly iron, titanium oxides and aluminosilicates >20μm).

1 Introduction

Halloysite is an interesting example of phyllosilicates. As a member of clay mineral group together with other aluminosilicates (e.g. montmorillonite, kaolinite) is often used in many applications [1, 2] e.g as a filler in natural or modified forms in nanocomposites [3, 4]. Halloysite is a two-layer mineral with chemical formula $Al_2Si_2O_5(OH)_4$, consisting of one tetrahedral sheet of silica (SiO_4) linked through oxygen atoms to one octahedral sheet of alumina octahedral (AlO_6). As a mineral it belongs to the kaolinite subgroup, which is represented by the same chemical formula $Al_2Si_2O_5(OH)_4 \cdot nH_2O$, where n is the number of water molecules that occupy the interlayer spaces of the clay aggregates. The value of n is zero for kaolinite and up to 4 for halloysite. The raw halloysite mineral contains a layer of water in its interlayer space which results in an increase in layer thickness up to 1.01 nm (called 10 Å halloysite). The interlayer water molecules can be removed by heating at elevated temperature. This process turns down the interlayer distance ranging from 0.715 nm up to 0.75 nm thickness (called 7Å halloysite) [2, 5]. However, after a dehydration process, the halloysite plates stay separated.

There are several variations of the crystal structure of halloysite, most common include either halloysite tubules or plates. Halloysite plates are composed of either one or more plates loosely joined together with the so-called "gallery", where different atoms, ions and molecules can be absorbed. Platy halloysite is ready for delamination, intercalation and exfoliation. Halloysite from the Dunino

[a] Corresponding author: Marcin.Lutynski@polsl.pl

DOI: 10.1051/e3sconf/20160801032

deposit is characterized by the specific platy and tubular structure. The octahedral external surface is built up from the hydroxyl groups with outstanding hydrogen and has a positive surface charge when the oxygen surface of the tetrahedral sheet is negatively charged, like montmorillonite. This platy structure gives the possibility and facilitates ability to activate both the surface tetrahedral and octahedral sheet [1-5].

In the raw material from natural deposits in countries like New Zealand and America halloysite has usually a tubular structure. The external surface of the halloysite nanotube is siloxane surface and internal alumina surface, active groups Al-OH are situated on the edges of the tube. In plates, a higher share of the active Al-OH edges occur, besides due to their spatial distribution they are more readily available. It should be recognized that the individual layers of the slats forming the wall of the tubes can participate, only to a small extent, in a variety of processes (intercalation, sorption, formation of various bonds, etc.). Halloysite from the Dunino deposit has a unique structure of mixed nanoplates and nanotubes which can be in certain cases its advantage. Another important factor in the reactivity of halloysite is the substitution occurring in the crystal structure where the most common are mineral substitutions of Al and Fe in the Si tetrahedra and Fe, Mg, Ti for Al in octahedra. These substitutions cause local imbalances of electric charges and create a number of so-called active sites capable of forming bonds with many different substances [1-5].

Particular attention of scientist is paid to the unique structure of halloysite nanotubes (HNT) and nanoplates (HNP) [2,4,6,7,8]. The presence of iron and titanium oxides e.g.: magnetite, magnesioferrite, ilmenite, hematite and goethite in raw material is the major problem impeding the use of halloysite in nanotechnology and polymer industry [8,9]. Due to the high content of iron oxides the raw material from the Dunino deposit has a rusty-reddish colour. Even low quantity of iron oxides may cause strong pigmentation in aluminosilicates [8] and change the properties and colour of polymers containing them. Many physical and chemical processes such as magnetic separation and acid leaching are commonly used in purification of aluminosilicates [8-18].

Figure 1. Structure of raw material after first milling. Impurities in the form of iron and titanium oxides are clearly visible.

Figure 2. Structure of raw halloysite from the Dunino deposit after drying and crushing (SEM picture).

2 Experimental procedure and results

The previous work [9] allowed to purify raw material by HCl treatment and obtain a fraction consisting mainly of aluminosilicates with particle size smaller than 20 µm. The purification process developed earlier, however consisted of more than thirty stages and operations. To separate aluminosilicates from iron and titanium oxides by physicochemical separation the following operations were applied crushing, milling, filtration, drying, hydrochloric acid treatment, sedimentation (settling), polygradient magnetic separation in strong (2T) magnetic field. The use of polygradient magnetic separation in the magnetic field was sufficient to separate heavy magnetic minerals (e.g. Fe_3O_4), which were difficult to remove with the use of other methods. In this research however, authors focused on optimization of the following stages of the process: crushing, milling magnetic separation and sedimentation and what's more, leaching in hydrochloric acid was replaced by sulfuric acid leaching in order to enhance the purification process.

The first stage of preparation involved drying the raw material at the temperature of 100 °C for one hour. The ball mill was used for crushing the raw material (Fig. 1). The structure of dried and crushed raw material was observed under the scanning electron microscope (Fig. 2). In the next stage the material was sieved below 1 mm. The particle size analysis were carried out to check the particle size distribution of the unified material. Histogram showing the particle size distributions of raw material after drying, crushing and sieving is shown in Fig. 3. A dried, crushed and sieved material was homogenized and the mass of 400 g was designated as a feed for further processing. The feed was immersed in 550 cm^3 of water, stirred and washed. The resulting slurry was ground in a ball mill with steel balls of 5 to 10 mm in diameter.

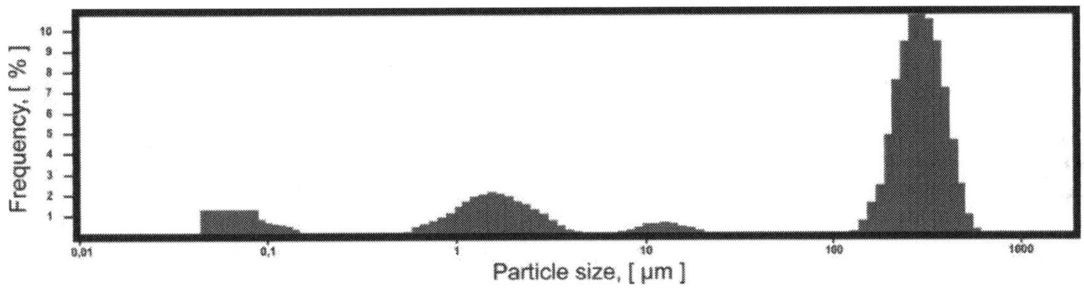

Figure 3. Histogram showing the particle size distributions of raw material after drying, crushing and sieving.

The material was washed and the density of the resulting slurry was $d = 1.35$ g/cm^3. The next stage of the experiment was chemical processing where 50 ml of H_2SO_4 was added to the slurry. After 90 min of leaching and mixing in reactor at 90 ^{0}C, the coarse material was washed out. The filter cake was washed with water and filtrated. The resulting material was subjected to magnetic separation in order to separate magnetite from nonmagnetic fraction. Detailed description of the experimental procedure is shown in Fig. 4.

Since the coarse fraction had still slightly reddish colour it was cleaned by slow magnetic separation one more time. A matrix magnetic separator was equipped with cylindrical neodymium magnets as a filtration medium. In all experiments the polygradient magnetic separation in magnetic field was used. The applied magnetic flux was 2 T. The magnetic fraction concentrate was marked as the sample no. 1.Consequently the slurry was settled out and the particle size fraction below 20 μm was separated in a sedimentation column. The obtained homogeneous fluid was separated when settled due to the density difference, in which ferruginous minerals of high specific density, are separated from the layered aluminosilicates. The resulting product consisting of light aluminosilicates fraction mostly HNT and HNP was marked as the sample no. 2. The particle size analysis of this fraction was done to check the particle size distribution of this material. Fig. 5 shows the histogram showing the particle size distributions of sample no. 2.

The resulting material was filtrated, dried and subjected to microscope analysis (Fig. 6). The remaining mixture of iron, titanium oxides and aluminosilicates over > 20 μm in particle size was marked as the sample no. 3. Consequently, the pre-thickened material was subjected to dehydration in a filter. The sample intended for the study was a mixture of magnetite, hematite and goethite crystals of size above 20 μm and other phases such as dispersed micro crystallites of halloysite with particle sizes ranging from a few to several μm and irregular aggregates of particle sizes 20 - 1000 μm (Fig.6). The results of physicochemical treatment of raw material from the Dunino deposit are shown in Table 1.

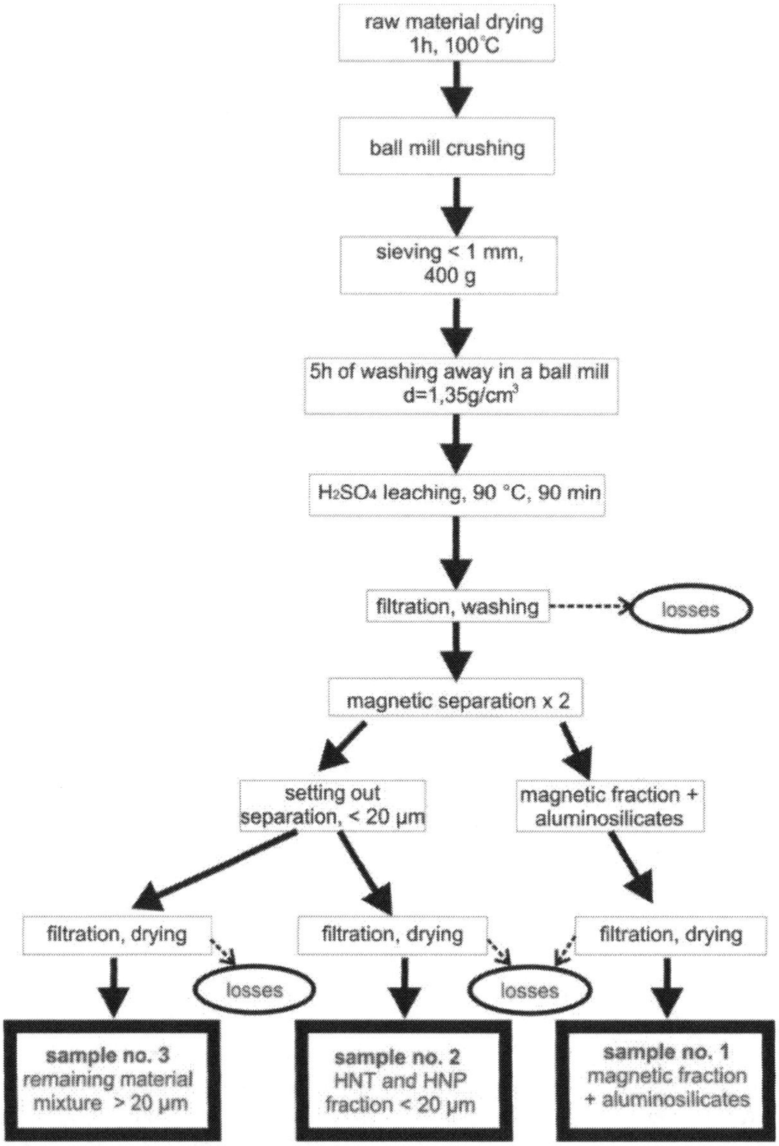

Figure 4. Detailed diagram of purification procedure of halloysite by H_2SO_4 leaching, magnetic separation and sedimentation.

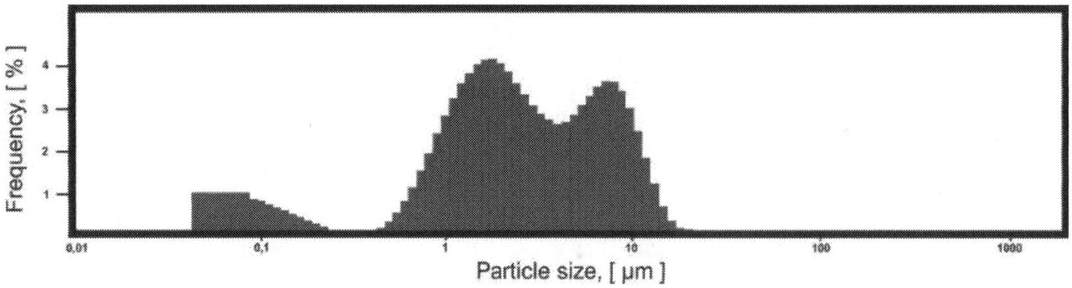

Figure 5. Histogram showing the particle size distributions in sample no. 2 (aluminosilicates HNT and HNP mixture < 20 µm).

Figure 6. Iron concentrate + aluminosilicates after H_2SO_4 treatment and double magnetic separation (mainly iron oxides).

Figure 7. The structure of „light" non-magnetic fraction of halloysite after H_2SO_4 treatment, settling out (float) (magnetic separation aluminosilicate, < 20 µm).

The procedure described above proved that iron oxides can be removed from halloysite by chemical leaching without either deformation of the crystalline structure or changes in the particle morphology. Scanning electron microscope observations (Fig. 7) confirm that light fraction sample consisted mainly of halloysite nanotubes (HNT) and nanoplates (HNP).

Table 1. Results of physicochemical treatment of raw material from "Dunino" deposit.

Sample no.	Processes applied	Yield [g]	Yield [%]
1	Iron concentrate + aluminosilicates after H_2SO_4 treatment and double magnetic separation	65	16.25
2	HNT and HNP mixture < 20 µm	101	25.25
3	Remaining material mixture > 20 µm	210	52.5
	Losses	24	6

3 Summary and conclusions

The aim of this work was to obtain fine-grained fraction of halloysite particles consisting of nanotubes (HNT) and nanoplates (HNP). Physicochemical processing consisting of acid leaching treatment using H_2SO_4, sedimentation and magnetic separation gave promising results. The proposed method where nanotubes (HNT) and nanoplates (HNP) were separated, consisted of several simple stages and can be easily scaled up to the industrial scale. The main advantage of this method was that the impurities in the form of iron and titanium oxides, which were present in the Dunino deposit, were removed efficiently and with small amount of loses (6% on average). The first and second stage of the process, drying and milling, helped to purify the light aluminosilicate fraction, and therefore the recovery in magnetic separation was much more efficient than in the previously developed method [9]. The most desirable product of this method was light fraction of aluminosilicate (< 20 µm) consisting mainly of halloysite nanotubes (HNT) and halloysite nanoplates (HNP) which can be used for the manufacturing of e.g. polymer - aluminosilicate composites. The average yield of this fraction was 25.25%.The yield of mixed fraction (> 20 µm) of aluminosilicate and the rest of the iron and titanium oxides was on average 52.5%. Although this fraction was not of high quality yet it can be used for manufacturing of high quality sorbents. The yield of heavy iron oxides fraction separated during magnetic separation was 16.25% on average and although it appears as a reject it can be utilized e.g. for manufacturing of pigments.

References

1. P. Yuan, P.D. Southon, Z. Liu, C.J. Kepert, *Nanotechnology*, **23**, 1 (2012)
2. P. Sakiewicz, R. Nowosielski, W. Pilarczyk, K. Golombek, M. Luty nski, *JAMME*, **48,** 2 (2011)
3. K. Hedicke-Hochstotter, G.T. Lim, V. Altstadt, *Compos. Sci. Technol.*, **69** (2009)
4. R. Kamble, M. Ghag, S. Gaikawad, B. Kumar Panda, *JASR*, **3**, 2 (2012)
5. E. Joussein, S. Petit, C. Fialips, P. Vieillard, D. Righi, *Clays Clay Miner.*, **54(4)** (2006)
6. M. Stechman, *Chemik*, **1**,1 (2001)
7. B. Makary, W. Pilch, A. Siwiec, Z. Tajchman, *Physicochem. Probl. Miner. Process.,* **22,** 127 (1990)
8. S. Saklar, A. Yorukoglu, *Physicochem. Probl. Miner. Process.*, **51(1)**, 83 (2015)
9. P. Sakiewicz, M, Luty nski J. Soltys, A. Pytli nski, *Physicochem. Probl. Miner. Process,C* **52(2)**, 991 (2016)
10. V.R. Ambikadevi, M. Lalithambika, *Appl, Clay Sci.,* **16**, 135 (2000)
11. K. Belkassa, F. Bessaha, K. Marouf, I. Batonneau, J. Comparot, A. Khelifa, *Colloids Surf., AC***421**, 26 (2013)
12. A. Ciesla, *Physicochem. Probl. Miner. Process.*, **37**, 169 (2003)
13. J. Drzymala, *Mineral Processing*, (Oficyna Wydawnicza PWR, Wroclaw, 2007)
14. A. Zhang, L. Pan, H. Zhang, S. Liu, Y. Ye, M. Xia, X. Chen, *Colloids Surf., A*, **396, 90** (2012)
15. G.D. Moeser, K.A Roach., W.H. Gren, T.A. Hatton, *AlChE J.*, **50**, 11 (2004)
16. J. Svoboda, T. Fujita., *Miner. Eng.*, **16**, 785 (2003)
17. E. Lewicka, *Prz. Gorn.,* **68,** 3 (2012)
18. M. Meireles, F. Bourgeois, M. Tourbin, P. Guiraud, C. Frances,CResearch *Report CNRS*, (2010)

Local and global assessment of flotation kinetics

Tomasz Ratajczak[1,a], Jan Drzymala[1], and Przemyslaw B. Kowalczuk[1]

[1]*Wroclaw University of Science and Technology, Faculty of Geoengineering, Mining and Geology, Wybrzeze Wyspianskiego 27, 50-370 Wroclaw*

Abstract. Most flotation data can be approximated with the first order kinetic equation. However, this equation frequently provides two, not one, parameters, that is the first order flotation rate constant k and maximum flotation recovery R_{max}. Currently, the most often way of evaluation of a set of flotation data is by using efficiency $e_{t0} = k \cdot R_{max}$, being the slope of flotation kinetic curve at zero flotation time (t=0). This parameter has a local character and is useful only in very limited cases. It was proposed in this work to use a global flotation kinetic efficiency (e) which can characterize the whole set of kinetic curves. For the considered in this work flotation data, a simple relation $R_{max} = e \cdot k$ was used. For other sets of flotation data, the global flotation kinetics efficiency very likely will be represented by another equation with the one-adjustable parameter.

1 Introduction

Flotation is a very complex process [1-3]. One of its aspects is kinetics [4-13]. The kinetics of process significantly influences the economics of flotation. An extensive review on kinetics of flotation was recently published by Gharai and Venugopal [13]. It shows that there are numerous equations which can be used to delineate the flotation kinetics.

According to many studies, most flotation processes follow the so-called first order kinetic equation [15]:

$$R = R_{max}\,[1\text{-}\exp(-kt)] \qquad (1)$$

where:
R – floating material mass recovery, %,
R_{max} – maximum flotation recovery, %,
k – first order rate constant, s^{-1},
t – flotation time, s.

Figure 1 shows the flotation data of Ni et al. [16]. As seen from Fig. 1a, the first order kinetics provides the best approximation of the investigated flotation data.

[a] Corresponding author: tomasz.ratajczak@pwr.edu.pl

 DOI: 10.1051/e3sconf/20160801033

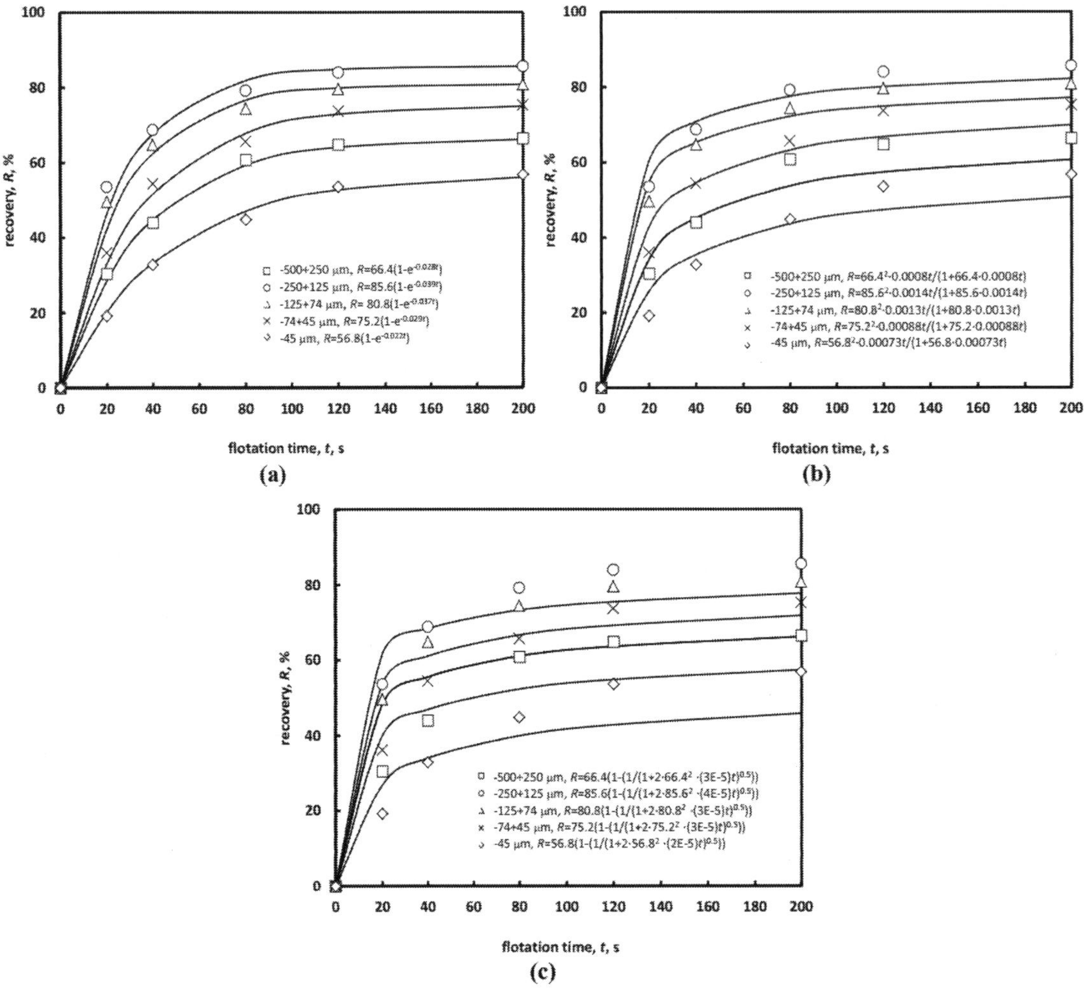

Figure 1. Kinetics of flotation of different size fractions of coal. Date points taken from Ni et al. [16]. The best approximation of flotation data is provided by the 1st order kinetic equation (a) in comparison to 2nd (b) and 3rd orders (c).

As predicted by Eq. 1, very often the final recovery of floating material is not 100% (Fig. 1). It complicates the analysis of flotation kinetics because one has to deal with not one but two variables, that is R_{max} and k simultaneously. As a result, there are many approaches to evaluate the flotation kinetics when R_{max} is less than 100%. Two of them seem to be the most useful and will be discussed in this paper.

2 Local efficiency of flotation kinetics

The first order kinetic equation can be differentiated to create new parameters which can be used to characterize the kinetics of flotation at the certain time. Obviously, such efficiency will be only local, that is valid for a specific flotation time. The well-known approach is the slope of the flotation kinetic plot R=f(t) at the start of the process, that is at zero time (t=0). This approach was proposed for the first time perhaps by Agar et al. [6], and next used in many investigations [12, 17-21]. The differential of Eq. 1

$$\frac{dR}{dt} = [R_{max} \cdot (1 - \exp(-kt)]' \tag{2}$$

provides:

$$\frac{dR}{dt} = (-R_{max}) \cdot (-k) \cdot \exp(-kt) \tag{3}$$

which at $t = 0$ assumes the form:

$$\frac{dR}{dt}|_{t=0} = R_{max} \cdot k. \tag{4}$$

Thus, the efficiency of flotation kinetics by this approach is given by the formula:

$$e_{to} = k \cdot R_{max} \tag{5}$$

where e_{to} is the efficiency of flotation, or in our nomenclature, the local flotation kinetics efficiency at t=0.

It is obvious that the local efficiency of flotation kinetics can be calculated for any other flotation time. When $t \to \infty$, then:

$$\frac{dR}{dt}|_{t\to\infty} = e_{t_\infty} = 0 \tag{6}$$

while for the case when one chooses to determine the local efficiency of flotation kinetics at the time when the flotation recovery reaches 50% of R_{max} the differential is:

$$\frac{dR}{dt}|_{t=1/2} = e_{t_{1/2}} = R_{max} \cdot k \cdot \exp(-0.5 \cdot k). \tag{7}$$

Figure 2 shows the graphical representation of the local efficiencies of flotation kinetics.

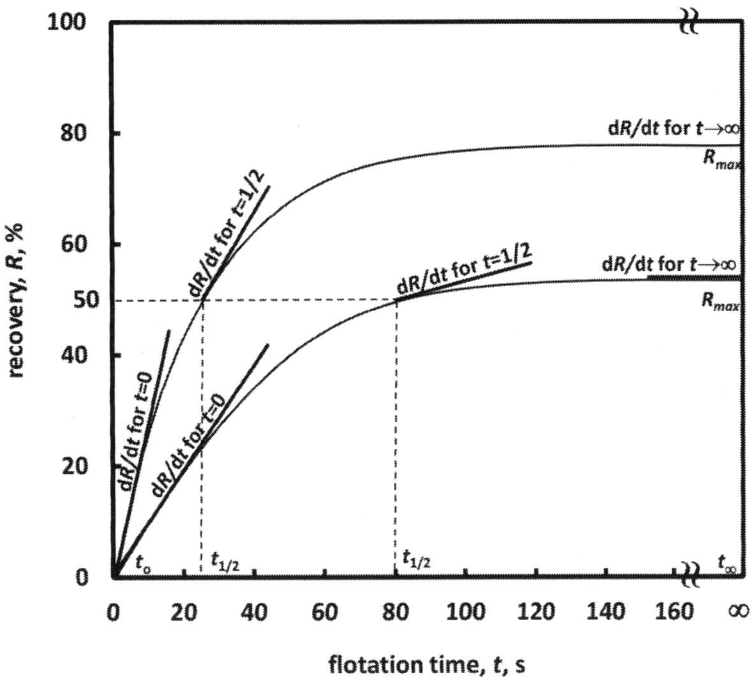

Figure 2. Graphical representation of local efficiencies of flotation kinetics in the form of a derivative dR/dt (slope) of the first order kinetic curve at t=0, t=1/2, and t=∞ (theoretical data).

Unfortunately, the local approach is not useful because it does not allow to compare two different flotation first order kinetic curves. The shortcomings of the local efficiency of flotation kinetics is given in Fig. 3. Figure 3 shows that when two curves intersect, then the efficiency e_{t0} is misleading because curve *a* delineates slower flotation kinetics which reaches higher maximum recovery R_{max}, while curve *b* represents faster flotation kinetics with lower R_{max}. Therefore, one does not know which flotation, in the light of the e_{t0}, is more efficient.

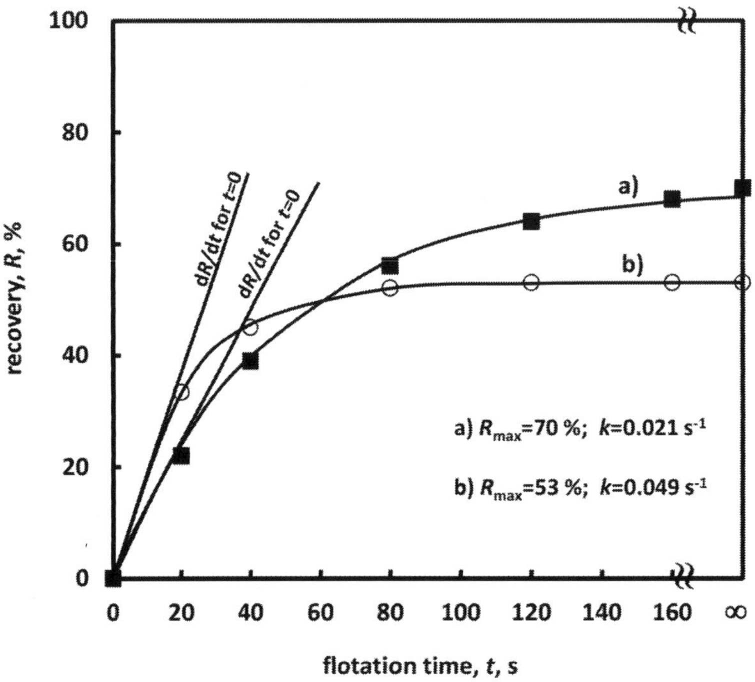

Figure 3. Kinetics of flotation and derivative e_{t0} = d*R*/d*t* at *t* = 0. Curve *a* delineates slower and higher flotation (k = 0.021 s^{-1}, R_{max} = 70%), while curve *b* represents faster and lower flotation (k = 0.049 s^{-1}, R_{max} = 53%).

3 Global efficiency of flotation kinetics

For the purpose of comparison of different flotation systems, when both follow the first order kinetic equation and have different R_{max} and k values, another approach is needed. The most obvious seems to be a relation between R_{max} and k, which has been used by many authors [13, 20]. However, a next step must be undertaken to find a relationship between R_{max} and k that can provide a parameter, which can be used to characterize the kinetics of the whole flotation system. Such a parameter can be called the global flotation kinetics efficiency. Figure 4 shows an attempt to determine the global flotation kinetics efficiency for a set of flotation experiments. The experimental data were adopted from Ni et al. [16]. As seen from Fig. 4, the relationship between k and R_{max} can be approximated with a linear equation having only one-adjustable parameter e, which is the global flotation kinetics efficiency of the considered system. For the data presented in Fig. 4 the equations is:

$$R_{max} = e \cdot k \, , \, R_{max} = 2315.2 \cdot k. \qquad (8)$$

It is expected that for other sets of flotation data the global flotation kinetics efficiency e will be different, and in some case, other than $e = k/R_{max}$. The one-adjustable parameter equation will be needed to delineate precisely such kinetic data.

DOI: 10.1051/e3sconf/20160801033

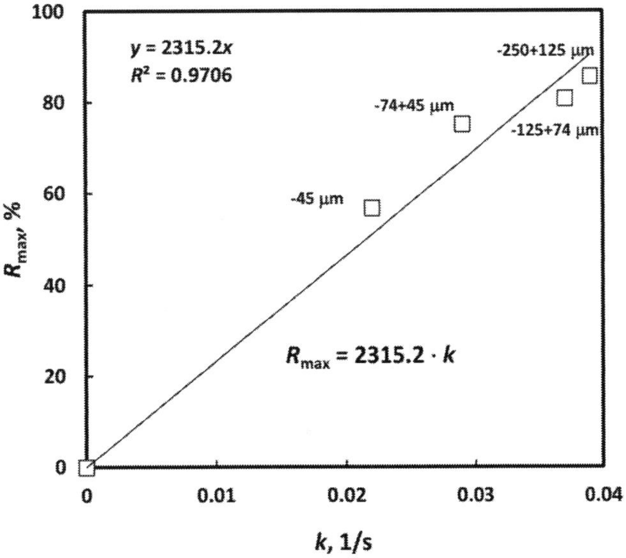

Figure 4. Relationship between k and R_{max} for kinetics of flotation of different size fractions of coal. Data points from Ni et al. [16]. The slope of the approximation line is the global flotation kinetics efficiency e.

4 Summary

The most useful and meaningful way of characterization and evaluation of flotation kinetics seems to be through the so-called global flotation kinetics efficiency parameter. It combines both the first order flotation rate constant k and maximum flotation recovery R_{max}. For the examined here set of flotation data the global flotation kinetics efficiency e was expressed by a simple relation $R_{max} = e \cdot k$. For other sets of flotation data, the global flotation kinetics efficiency very likely will be represented by another equation, but this requires a further study. The currently often used local flotation kinetics efficiency e_{to}, being a slope of the flotation kinetic curve at t=0, is useful only in very limited cases.

Acknowledgements

This work was partially financed by the Polish Statutory Research Grant 0401/0124/16.

References

1. A. M. Gaudin, *Flotacja* (Wydawnictwo Slask, Katowice, 1963)
2. V. I. Klassen, V. A. Mokrousov, *An introduction to the theory of flotation* (Butterworths, London, 1963)
3. B. Wills, T. Napier-Munn, *Wills' Mineral Processing Technology*. 7th ed. (Butterworth-Heinemann, London, 2006)
4. H. G. Zuniga. *Bol. Soc. Nac. Min.*, Santiago, **47**, 83 (1935)
5. T. M. Morris, *Tr. AIME*, **193**, 794 (1952)
6. G. E. Agar, R. Stratton-Crawley, T. J. Bruce, *Int. J. Miner. Process.*, **73**, 824 (1980)
7. N. Ahmed, G. J. Jameson, *Miner. Process. Extr. Metall. Rev.*, **5**, 77 (1989)
8. M. Brozek, A. Mlynarczykowska, *Physicochem. Probl. Miner. Process.*, **40**, 31 (2006)
9. M. Brozek, A. Mlynarczykowska, *Physicochem. Probl. Miner. Process.*, **41**, 51 (2007)

10. M. Brozek, A. Mlynarczykowska, *Wydaw. Insty. Gospo. Suro. Mine. i Ene. PAN* (Krakow, 2009)
11. O. Guven, M. S. Celik, W. Drelich, *Miner. Eng.*, **79**, 125 (2015)
12. A. Bakalarz, P. Nowak, M. Duchnowska, A. Konieczny, E. Kasinska-Pilut, W. Pawlos, R. Kaleta, P.B. Kowalczuk, J. Drzymala, A. Luszczkiewicz, *Czasopismo CUPRUM*, **2**, 75 (2015)
13. Y. Xing, X. Gui, J. Liu, Y. Cao, Y. Zhang, S. Li, *Physicochem. Probl. Miner. Process.*, **52(2),** 703 (2016)
14. M. Gharai, R. Venugopal, *An. Int. J.* **37**, 2, (2016)
15. N. Arbiter, C.C. Harris, In: D. W. Fuerstenau (Ed.), *Froth flotation* (AIME, Ann Arbor, MI, 215, 1962)
16. Ch. Ni, G. Xie, M. Jin, Y. Peng, W. Xia, *Powder Technol.*, **292**, 210 (2016)
17. M. Xu, *Miner. Eng.*, **11**, 33 (1998)
18. R. Sripriya, P. V. T. Rao, B. R. Choudhury, *Int. J. Miner. Process.*, **68**, 109 (2003)
19. S. Muganda, M. Zanin, S. R. Grano, *Int. J. Miner. Process.*, **98**, 182 (2011)
20. D. Chipfunhu, M. Zanin, S. Grano, *Miner. Eng.*, **24**, 50 (2011)
21. D. Chipfunhu, M. Zanin, S. Grano, *Chem. Eng. Res. Des.*, **90**, 26 (2012)

Monitoring of the pulsation cycle in a jig using a radiometric density meter

Stanislaw Cierpisz[1], and Jaroslaw Joostberens[2,a]

[1]*Institute of Innovative Technologies EMAG, Katowice, Poland*
[2]*Silesian University of Technology, Gliwice, Poland*

Abstract. Application of a radiometric density meter to monitor changes in coal bed density during a pulsation cycle in a jig has been presented. The signal from the radiation detector has been filtered in an adaptive filter to react properly to rapid changes in density. The radiometric density meter has been used to measure density of a coal layer 3 cm thick and 30 cm wide. The distribution of the material density along the bed height has been measured during field tests performed on the industrial jig. Changes in density have shown the process of material loosening and its compression during each cycle of pulsations.

1 Introduction

Raw coal is often beneficiated by gravity separation processes where coal grains are stratified according to their density in a pulsating coal/water medium in jigs (Fig. 1). These processes have been discussed in [1-3]. During subsequent water pulsations induced by opening and closing air valves, coal stratification occurs as grains of various density move upwards and downwards with different velocities. Grains of low density (low ash content) migrate to upper layers with grains of high density (high ash content) moving down to lower layers.

Stratification of grains due to their density is not ideal. Their upward and downward movement also depends on their diameters, shape and the way in which the material loosens within each pulsation cycle. Simulation models of fine coal segregation and stratification processes in jigs have been discussed elsewhere [4-5]. The material travels horizontally on a screen deck along the jig compartment with the flow of water. At the end of the jig compartment the stratified material is split to upper (concentrate) and bottom (refuse or middlings) products. This process is controlled by the refuse discharge system influencing the tonnage of the bottom product.

The most popular refuse discharge control systems in jigs use a metal float to monitor the thickness of the heavy fraction in a jig compartment. Although the float is the main sensor, its operation as a measuring unit has not been adequately studied so far. The control system stabilizes the position of the float at the chosen desired value adjusting the tonnage of the bottom product and in this way stabilizing, to some extent, the separation density. The separation density is the density of the coal layer reporting in half to the upper product and in half to the refuse. Monitoring and control

[a] Corresponding author: Jaroslaw.Joostberens@polsl.pl

systems applied commercially at present in jigs are limited to control operation of air valves in order to stabilize the shape of the bed pulsation and to control the tonnage of the discharged bottom product.

Laboratory and industrial tests show fluctuations of the position of a float in relation to the position of the measured layer. These fluctuations appear as a result of changes in the feed tonnage and the feed washability characteristics. The changes in the degree of material loosening produce changes in velocity of float during upward and downward movements in subsequent pulsation cycles. The investigations showed the fluctuations of the separation density of the order of 0.05 g/cm^3 and sometimes even more [6]. This can result both in fluctuations of tonnages and ash content in products, and in coal losses in refuse. Raw fine coal washing in jigs can be more efficient when a float in a refuse discharge control system is replaced by a radiometric density meter.

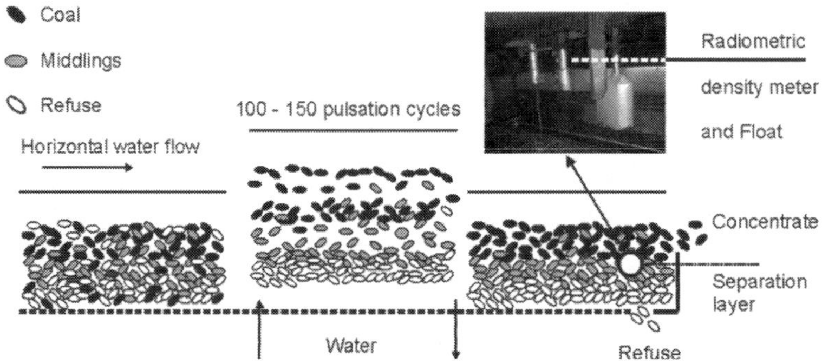

Figure 1. Illustration of a coal separation process in a jig.

First application of a radiometric density meter to control a jig operation (imperfection reduced from 0.15 to 0.08) was reported elsewhere [7]. Much more advanced system with a radiometric density meter, applied to control a jig operation (JigScan), was described elsewhere [8]. A radiometric density meter, applied to monitor the density of a jig's bed at different levels, appeared to be a good tool used for determination of both static and dynamic characteristics of a jig [9]. Results of these investigations have been presented in this paper.

2 Radiometric density meter

The radiometric density meter used in the experimental system consisted of a waterproof container for the radiation source (^{137}Cs) and the scintillation counter as the gamma radiation detector – Figure 2. The radiation beam was collimated (ca. 3 cm diameter) and absorbed by the 30 cm long layer of the pulsating media.

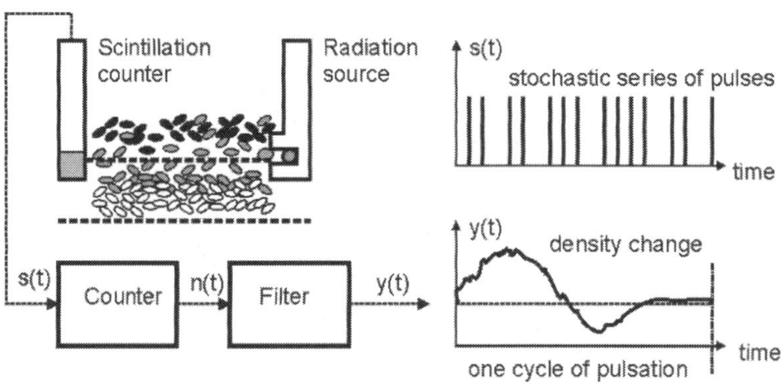

Figure 2. Radiation density meter.

The output signal from the radiation detector $s(t)$ is a nonstationary signal of series of pulses of Poisson distribution presented by equation (1):

$$p(n,k) = \frac{n(t)^k \cdot e^{-n(t)}}{k!} \tag{1}$$

where:
$p(n,k)$ – probability of counting k pulses in the time interval $(t, t + t_s)$,
$n(t)$ – mean intensity of pulses in time t.

The density of the bed was determined according to the absorption law (2):

$$\rho = 1 - \frac{\ln\left(\frac{I_x}{I_w}\right)}{\mu d} = 1 - \frac{\ln\left(\frac{Nx}{Nw}\right)}{\mu d} \tag{2}$$

where:
I_0 – incident radiation intensity ,
I_w – radiation intensity registered by the detector,
N_w – number of pulses counted by the detector for water,
I_x – radiation intensity detected by the detector for media (coal/water),
N_x – number of pulses counted by the detector for media (coal/water),
d – the thickness of the absorbent, cm,
μ – mass coefficient of gamma absorption, cm^2/g,
ρ – density of media, g/cm^3.

For small changes in the density, Equation (2) can be transformed and linearized:

$$n(t) = n_0 - k_n \cdot \rho(t) \tag{3}$$

where:
n_0 – intensity of pulses for the reference density (water),
k_n – coefficient.

The counter output signal is converted to the form u in discrete time moments i using the equation:

$$u[i] = \frac{1}{k_n} \cdot \left(n_0 - \frac{k[i]}{t_s}\right) \tag{4}$$

where:
$k[i]$ – number of pulses from the detector counted by counter during time $t+t_s$,
$i = t/t_s$.

The signal $u[i]$ consists of the measured value of the density and the stochastic error $e[i]$:

$$u[i] = \rho[i] + e[i] \tag{5}$$

The output signal from the filter $y[i]$ is given by equation:

$$y[i] = u[i] - \varepsilon[i] \ . \tag{6}$$

The output signal $u[i]$ is a stochastic signal which should be filtered as it consists of the useful signal $\rho(t)$ and a noise. One of the methods of predictive filtration of $u[i]$ is presented in details in [10]. The predictive filter is described by equation:

$$\rho[i] \approx y[i] = -b_1 \cdot u[i-1] - b_2 \cdot u[i-2] - ... - b_p \cdot u[i-p] \tag{7}$$

Coefficients b_j are calculated from the equation (8) for the minimum value of J:

$$J = \sum_{i=p+1}^{N} \left(u[i] + b_1 \cdot u[i-1] + ... + b_p \cdot u[i-p]\right)^2 \tag{8}$$

where
$t_s N$ – time of one cycle of the separation process in a jig.

The results of filtration of the output signal from the counter (using period t_s = 3 ms) are shown in Figure 3.

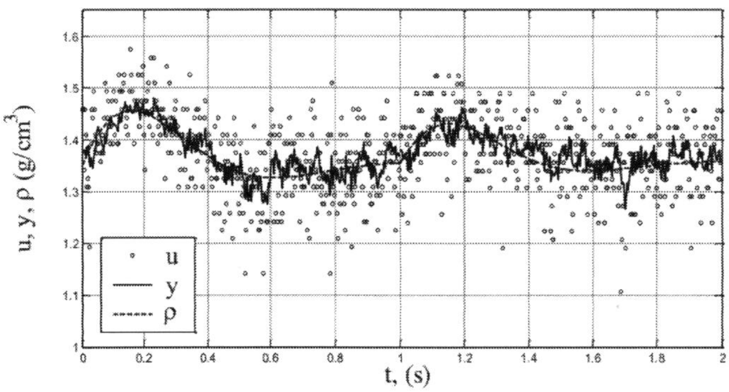

Figure 3. Output signal from the radiometric density meter after filtering in the predictive filter – two subsequent cycles of air/water pulsation.

3 Field tests

The radiometric density meter, presented in Fig. 2, was used to monitor changes in the jig's bed density during each cycle of air/water pulsations. The typical changes in the intensity of registered pulses from the detector and corresponding changes in the bed density, during one cycle of a jig operation and at the chosen bed level, are shown in Fig. 4. During the first 30 ms the bed is lifted by the water when the inlet air valve is opened, then follows the process of material loosening and finally the bed reaches its compressed state. The dynamic change in coal bed density depends on the point of density registration (position of the density measuring head), coal bed thickness, bed density distribution along the bed height and on parameters of the air pulsation cycle (pressure, time periods of air valves opening and closing).

The distribution of bed density was monitored for various positions (from -30 to +15 cm) of the measuring head, showed in Fig. 5a, above the bottom product discharge gate. The position "0" cm was accepted for the upper edge of the concentrate discharge gate. The thickness of the bed layer monitored in each point by the measuring head was ca. 3 cm and the length – 30 cm. Density of the bed was determined in compressed states of the material. The distribution of the bed density is presented in Fig. 5b.

The bed density in the compressed state varies from 1.0 g/cm^3 (water) at +15 cm above the edge of the upper gate to 1.35 g/cm^3 at -4 cm and to 1.45 g/cm^3 at -20 cm. The density at the position -4 cm can be accepted as the density of the separation layer which reports in half to concentrate and in half to the refuse as the amplitude of bed pulsations during the measurements was ca. 3 – 4 cm.

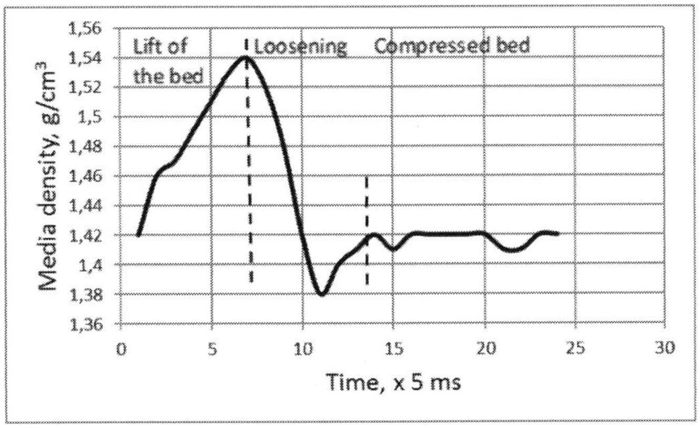

Figure 4. Typical change in the bed density during one cycle of pulsations (1.2 s) at the upper product split gate.

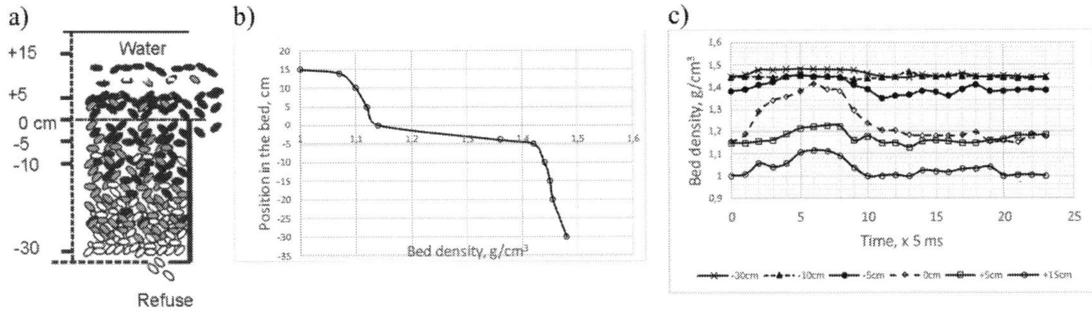

Figure 5. a) positions of measuring head during test; b) distribution of the bed density along its height; c) changes in the bed density over the one cycle of pulsations at various positions of measuring head.

The density of the dry material ρ_m can be estimated from the equation (9) for a given density of the medium ρ_o:

$$\rho_m = \rho_o + \frac{(\rho_o - 1) \cdot V_{H_2O}}{V_z} \tag{9}$$

where:
V_{H2O} – volume of water in a unit volume of medium,
V_z – volume of dry material.

The ratio V_{H2O}/V_z during the test was ca. 0.65, hence the separation density (dry material) was ca. 1.58 g/cm^3.

Changes in the bed density over the one cycle of pulsations at various positions of the measuring head are shown in Fig. 5c. The density of the bed in lower layers is fairly constant (from -30 cm to -5 cm) and for this reason changes in density when the bed is lifted are small. They become similar to Fig. 5 when loosening and compressing of the bed appears. Changes in the bed density over a cycle of pulsation cannot be a measure of the degree of material loosening but the period of time when material is in the loose state can be estimated. This information could be used for adjusting operation of air valves to stabilize the process of grains stratification.

Pulsations of the float (constant density of the float) during the increase of the feed tonnage are presented in Figure 6. The height of the float was 34 cm, its "0" position corresponded to the position of the float bottom at "– 17cm". Full range of the float position was 20 cm. The amplitude of pulsations varied from 3.5 cm up to 4.5 cm.

Changes in the bed density corresponding to the float pulsations are shown in Fig. 7. Initially, when the thickness of the bed is smaller, significant changes in density occur. The material at the height "-4 cm" is lifted, then loosens and does not reach the compressed state. An optimal shape of density changes, when material loosens and reaches the compressed state can be observed for the position of the bottom of the float from 10% to 20%. This information can be used for the control of the refuse discharge to stabilize the separation density in the jig.

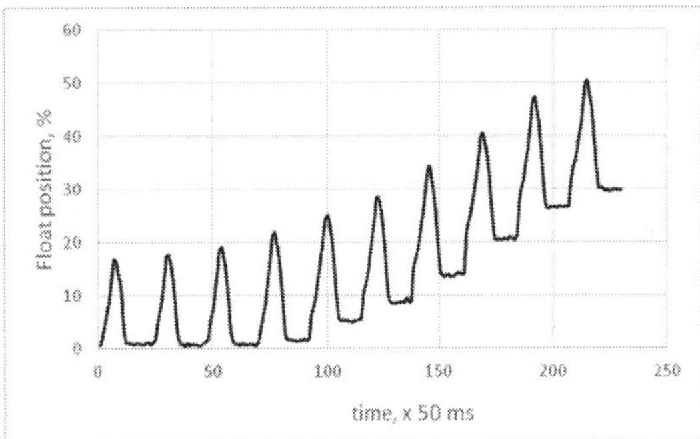

Figure 6. Pulsations of the float during the increase of the feed tonnage.

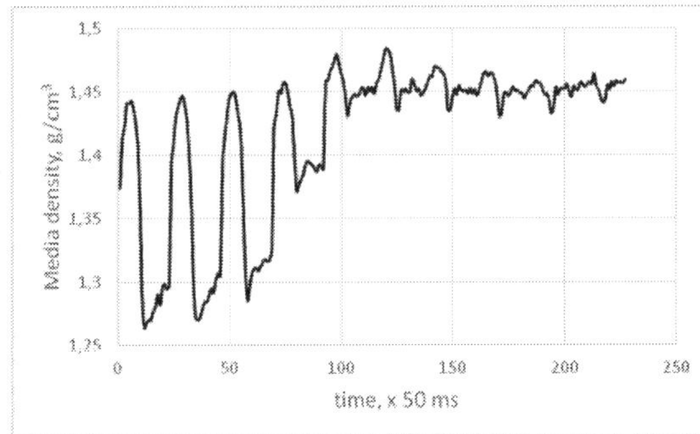

Figure 7. Changes in the bed density corresponding to changes in the float position (Fig. 6.) for 4 cm position of the measuring head.

4 Conclusions

The radiometric density meter can be effectively used for monitoring the distribution of density of the bed in a jig and for monitoring changes in density during a pulsation cycle. The signal from the meter contains significant stochastic noise due to random decay of the radiation source. For this reason it is necessary to apply a properly designed filter reducing the noise/signal ratio.

The change in the bed density during each cycle of pulsations, registered close to the edge of the upper product discharge gate, gives information on the bed lift, its loosening and compressing. The period of time of the bed loosening can be estimated from the shape of the signal and used for the control of operation of air valves.

Density of the bed in the compressed state indicates directly the density of the separation layer reporting in half to concentrate and in half to refuse. This signal can be used for better control of refuse discharge from the jig and better to stabilize the desired separation density.

References

1. J.W. Leonard, *Coal Preparation* (Society of Mining, Metallurgy, and Exploration, 1991)
2. R.P. King, *Modeling and simulation of mineral processing systems* (University of Utah, 2001).
3. G.J. Lyman, *Coal Preparation*, (1992).
4. R. Srinivasan, B.K. Mishra, S.P. Mehrotra, *Coal Preparation*, **20**, 145 (1999).
5. Y. Xia, F.F., E. Wolfe, *Int. J. Miner. Procces.*,**82(3)**, 164 (2007)
6. S. Cierpisz, *in Proccedings of XVI International Coal Preparation Congress*, (2013)
7. D. Bartelt, *in Proccedings of IVth International Coal Preparation Congress*, (1962)
8. G.J. Lyman, A. Jonkers, *Gluckauf – Forschungshefte*, **53**, 131 (1992)
9. S. Cierpisz, J. Joostberens, *Measurement*, **88**, 147 (2016)
10. S. Cierpisz, J. Joostberens, *in Proccedings of Conference EMTECH 2016*, Poland, (2016).

Application of the shrinking core model for dissolution of serpentinite in an acid solution

Agnieszka Didyk-Mucha[1], Agnieszka Pawlowska[1], and Zygmunt Sadowski[1,a]

[1]*Department of Chemical Engineering, Wroclaw University of Technology, Wybrzeze Wyspianskiego 27, 50-370 Wroclaw, Poland*

Abstract. Serpentinites have the potential to be used as carbo dioxide capture and storage materials. Acid dissolution is the first stage of this process. This study examined how a shrinking core model can be applied to the dissolution of serpentinite in sulphuric acid. The dissolution process was assumed to occur in two stages: an initial step involving the surface dissolution of serpentinite pieces, and a second step involving surface chemical reaction, described by the shrinking core model. The model formulation was completed by relating the dissolution rate of serpentinite to the surface changes.

1 Introduction

Mineral carbon dioxide sequestration is a promising method for reducing carbon dioxide emission. Mineral carbonation transform carbon dioxide into the form of geologically and thermodynamically stable magnesium carbonate. This process can be considered as a result of two steps: (1) dissolution of silicate minerals and (2) the reaction between either calcium or magnesium cations and CO_2 to create carbonates [1-6]. For aqueous carbonation, magnesium silicate minerals such as serpentinite and olivine are frequently used [7-9]. This process can be described by the following reactions

$$Mg_3Si_2O_5(OH)_4 + 3CO_2 = 3MgCO_3 + 2SiO_2 + H_2O \quad (1)$$

$$Mg_2SiO_4 + 2CO_2 = 2\ MgCO_3 + SiO_2. \quad (2)$$

It is evident that carbonation of silicate minerals involves two sub-processes: dissolution of silicate minerals and carbonate precipitation. In the aqueous carbonation process, dissolution of serpentinite was considered to be the rate-limiting step [10]. In a previous study, serpentinite was dissolved in HCl and HNO_3 acid solutions and hydromagnesite precipitation was obtained at pH level of 9 with addition of NaOH [11]. Generally, the carbonation process with the leached solution was promoted using a basic medium. Carbonation of a Mg-leaching solution was performed at pH 9.5 [12].

A significant problem has been previously observed with the diffusion-limiting SiO_2 layer in silicate minerals. Some effect had been undertaken to eliminate this problem by selective flotation of quartz [13]. However, selective separation of free quartz particles has not eliminated this problem. A shrinking core model has been considered for leaching of zinc ores containing silicates with

[a] Corresponding author: zygmunt.sadowski@pwr.edu.pl

sulphuric acid [14]. The presence of porous layer must be considered during the dissolution of serpentinite particles.

According to the shrinking particle model of the dissolution, the reaction is considered to take place first on the outer surface of serpentinite ore particles. The porous zone penetrates into the solid particles and the particles shrink during the leaching process. The kinetics of the leaching process involves three sub-processes that occur during leaching:

i. the external diffusion of acid to the solid surface,
ii. internal diffusion in the porous layer from the solution to the core surface and vice versa,
iii. leaching reaction on the core surface.

The rate of leaching (dissolution) is generally controlled diffusion through the porous layer or the dissolution reaction at the mineral surface of un-reacted particles [15].

One major factor that can affect interpretation of leaching data is the particle size distribution of a solid material. It was observed that the initial rate of leaching increased due to increase in the small particle fraction. In a previous study, the rate of leaching was observed to decrease during the latter part of the process due to the decrease in the large particle fraction [16]. The leaching process can also be described based on the residual volume of particles:

$$X(t) = 1 - [\frac{r(t)}{r_0}]^3 \ . \tag{3}$$

Furthermore, the rate of leaching was related to the surface area of solid particles [17]. The experimental results presented in a previous paper [18] indicated that the rate at with leaching was a function of surface area. The primary hypothesis made in this study was that the increase in the surface area of leached ore particles was directly proportional to decrease in the magnesium content of the ore particles during the initial dissolution period. In this study, we investigated dissolution of serpentinite rock particles from the Grochow mine. This serpentinite ore was the most interesting for carbon dioxide capture (CCS). The main aim of this study was to correlate the surface area of mineral particle to the corresponding dissolution kinetic.

2 Materials and methods

Mineral samples of serpentinite ore were delivered from Grochowa mine (Magnezyty "Grochow" S.A.), located in Lower Silesia (Poland). The ore material was dry-ground using a porcelain ball mill and wet sieved. Two size fractions were separated. A size fraction of +45 – 71 μm was used for the dissolution experiments. The mineral samples were analysed by an X-ray diffraction method (Fig. 1). The XRD pattern shows that the mineral samples consisted of 70 wt % serpentinite minerals (antigorite, lizardite) 10 wt % talc, 5 wt % clinochlore, 8 wt % actinolite, 3 wt % quartz , and 2 wt % magnesite. The results of the chemical analysis of mineral sample are presented in Table 1.

Table 1. The elemental analysis of serpentinite rock (XRF).

Element and compound	**Results of analysis [%]**
SiO_2	40.48
TiO_2	0.02
Al_2O_3	1.15
Fe_2O_3	8.49
MgO	35.35
CaO	1.3
Na_2O	0.17
K_2O	0.02

DOI: 10.1051/e3sconf/20160801035

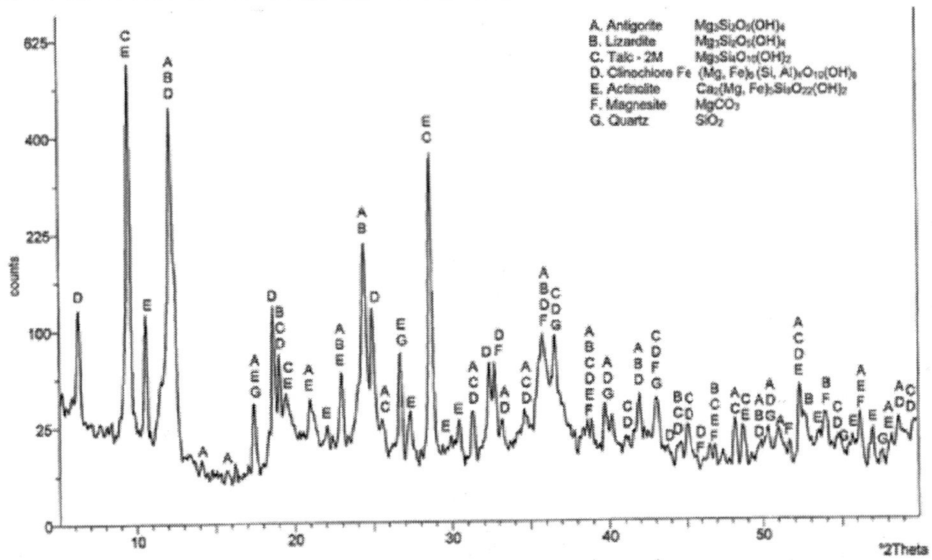

Figure 1. X-Ray diffractogram of serpentinite sample.

The particle size distributions of the mineral samples (Fig. 2) were determined using the Mastersizer 2000 (Malvern) particle analyser. The specific surface area of the mineral sample was measured using a FlowSorb II 2300 instrument. The BET method was employed to determine the surface area of investigated mineral fraction which was equal to 16.38 g m^{-2}.

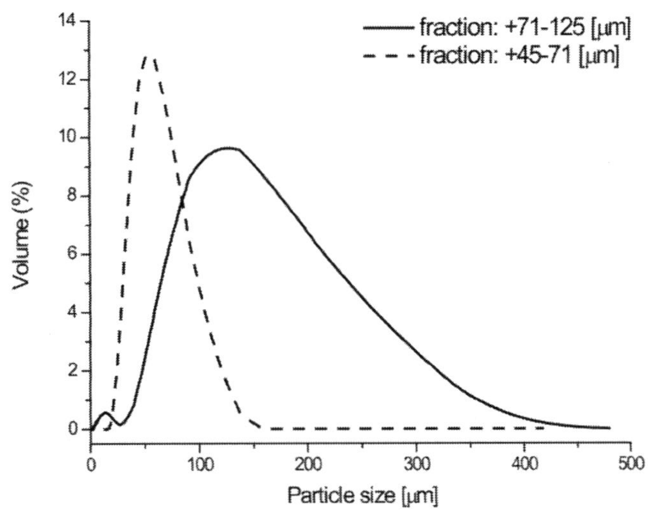

Figure 2. Particle size distributions of mineral samples (+45-71 and +71-125 μm fractions).

Dissolution experiments were carried out in closed Erlenmeyer flasks (200 ml) at room temperature (20 °C). Two grams of each mineral sample was dissolved in 100 ml of sulphuric acid solution. The concentrations of the acid solution were 1.0, 2.0, and 2.5 M, respectively. The concentrations of magnesium and calcium ions in the solution were periodically measured. The extent of magnesium and calcium leaching recovery (X) was calculated using formula

$$X_{Mg/Ca} = \frac{c_{Mg/Ca}V}{mY_{Mg/Ca}}100\ [\%] \tag{4}$$

where: $c_{Mg/Ca}$ the concentration of Mg^{2+} or Ca^{2+} ions in the solution [g/dm^3], v the solution volume [dm^3], m the weight of sample [g], $Y_{Mg/Ca}$ the content of Mg or Ca in the sample.

3 Results and discussion

Serpentine is the main component of serpentinite rock. Serpentine is a silicate mineral with tetrahedral silica-type and octahedral brucite-type layers [10]. The chemistry of serpentinite ore acid leaching process is well understood [11,12]. Partial decomposition of the silicate structure of serpentine minerals occurs during acid dissolution. Such behaviour is known as incongruent dissolution [7]. Dissolution of serpentinite using sulphuric acid could form the basis of a commercial technology used for carbonation of silicate minerals. In this work we tested dissolution of serpentinite using three sulphuric acid concentrations (1.0, 2.0, and 2.5 M). The kinetics of mineral dissolution were monitored in terms of concentration of calcium and magnesium ions in each solution. Figure 3 shows the changes in the concentration of calcium ions that were extracted.

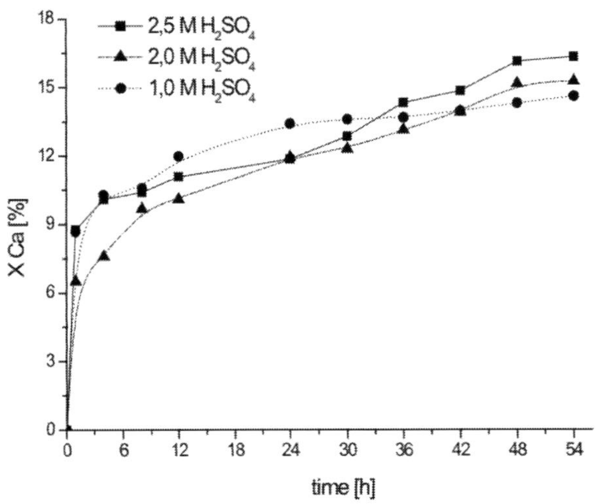

Figure 3. Kinetics of calcium extraction from serpentinite ore sample (+45-71 μm) at room temperature.

Data regarding magnesium extraction are presented in Fig. 4. The concentration of calcium and magnesium extracted indicate that the rate-limiting step of dissolution of serpentinite in H_2SO_4 is most likely due to formation of a passive silica layer.

Dissolution of serpentinite increases with increasing acid concentration. In a previous study, this concentration dependence was observed for dissolution of serpentinite ore in a trisodium citrate solution.

The effect of acid concentration on the particle diameter (d_{50}) of serpentinite ore particles (+45 - 71μm fraction) is illustrated in Fig. 5. As the serpentinite dissolution process proceeds, the particle radius decreases. The dissolution rate D is defined as the rate of change in the characteristic diameter d_{50} of the particle population.

$$\frac{dd_{50}}{dt} = -D(d_{50}). \tag{5}$$

The dissolution rate is also related to the particle surface area. The available surface area of serpentinite mineral particles was observed to vary throughout the dissolution process, as shown in

Fig. 6. The total surface area of the mineral particles (+45 –71 μm fraction) increased significantly during the second stage of the leaching process.

The surface area (*S*) represents the availability of dissolution reaction sites, and it is assumed that the concentration of ions varies directly with the surface area. The rate of dissolution can be expressed as follows.

$$\frac{dc}{dt} = kS. \tag{6}$$

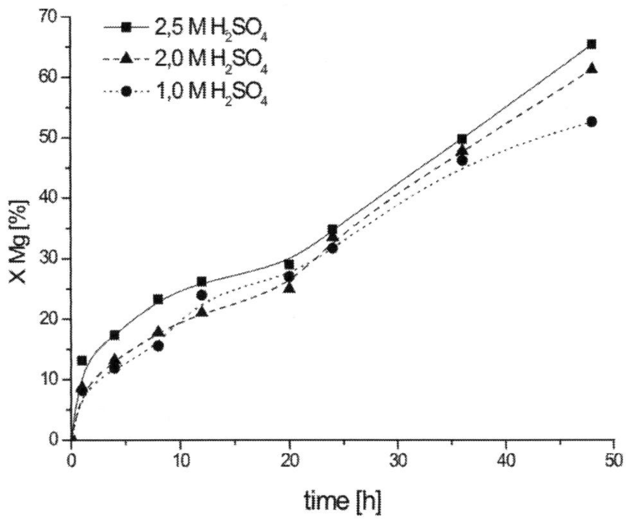

Figure 4. Kinetics of magnesium extraction from serpentinite ore sample (+45-71 μm) at room temperature.

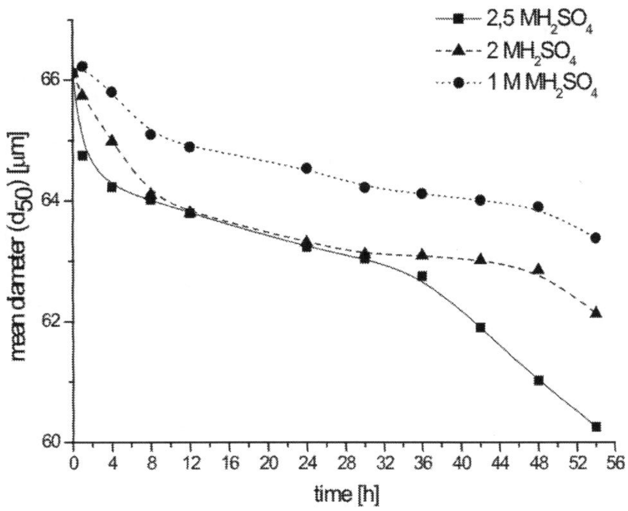

Figure 5. Mean diameter changes of serpentinite particles (+45 -71 μm) during an acid dissolution.

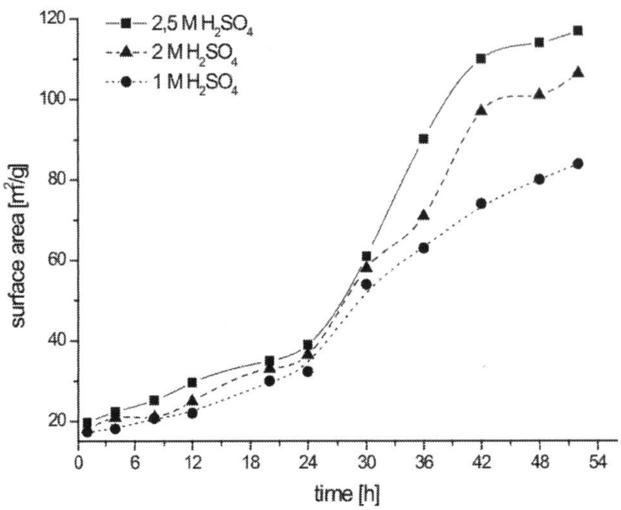

Figure 6. Surface area changes of serpentinite ore particles (+45 -71 μm) during acid dissolution.

According to [25], the particle surface area changed during mineral dissolution. The BET surface area of sandstones was observed to increase during dissolution [26]. Figure 6 indicates that after 24 h of dissolution the surface area of the serpentinite particles (+45 -71μm fraction) dramatically increased. This increase in the surface area corresponded to decrease in the particle radius (Fig. 5). Serpentine, $Mg_3Si_2O_5(OH)_4$, is a sheet silicate mineral that consists of layers of SiO_4 tetrahedra and $Mg(OH)_2$ octahedral. The silica-rich layer remains passive during dissolution. The observed increase in surface area can be correlated to formation of the silica porous layer during acid leaching.

To determine the mechanism of serpentine rock dissolution in sulphuric acid, the dissolution process was analysed according to shrinking particle and shrinking core models [27]. In this study, the kinetic analysis was performed based on experimental data regarding surface area changes, which indicate that the dissolution process may be explained by the formation of a porous layer. As shown in Fig. 7, dissolution of serpentinite particles can be divided into two stages. The ordinary shrinking core model does not apply in the initial stage. Modification of shrinking core model should be as follows.

The primary assumption made in constructing the model is that the increase in the surface area is directly proportional to the decrease in the metal (Ca, Mg) content in the solid particle at the surface:

$$\frac{dS}{dt} = kc_{Ca,Mg}S\ . \tag{7}$$

To simplify our calculation we introduced constant B as a substitute of $kc_{Ca,Mg}$. Thus, the new equation has the form

$$\frac{dS}{dt} = BS. \tag{8}$$

The solution of Equation (6) for the initial condition $S_{(t=0)} = S_0$ can be written as:

$$S(t) = S_0 e^{Bt}\ , \tag{9}$$

and in the logarithmic form as:

$$\ln(S) = Bt + \ln(S_0). \tag{10}$$

Logarithmic equation 10 properly describes the first stage of the serpentinite dissolution process (Fig. 8).

Now, we consider the second stage of dissolution. If one assumes that serpentinite particles are spherical and the chemical reaction is the rate-controlling step, then the shrinking particle model can be used. The kinetics of this process is described by the following equation:

$$1(1-\alpha)^{\frac{1}{3}} = K_{app}t \tag{11}$$

where $K_{app} = \frac{k[acid]^n}{3\rho_{solid}R_0}$ and R_0 is the particle radius.

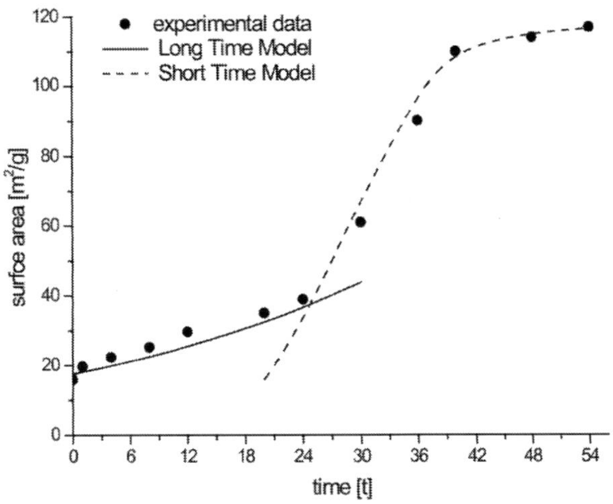

Figure 7. The "short time" and "long time" model data compared with the experimental data for +45-71 μm fraction (2.5 M H_2SO_4).

Formation of a porous (ash/inert) layer affects the surface area of serpentinite particles during dissolution. To describe the increase in the surface area during ash layer creation, we introduce the modified module $d(S_k\text{-}S)$. Thus, the surface changes can be described by a new Equation:

$$\frac{dS}{dt} = BSd(S_k - S). \tag{12}$$

For the initial conditions $S_{(t=0)} = S_0$, the solution of Equation (12) is

$$S(t) = \frac{S_0 e^{Bt}}{1 - \frac{S}{S_k}(1 - e^{Bt})}. \tag{13}$$

The above equation can be transformed into the following logistic equation:

$$\ln\left(\frac{S}{S_k - S}\right) = S_k Bt + \ln\left(\frac{S}{S_k - S_0}\right) = S_k Bt + D, \tag{14}$$

where: $D = \ln(\frac{S}{S_k - S}) = -2.0873.$

This logistic equation is a linear function that accurately describes only the final stage of the serpentinite ore dissolution process (Fig. 9).

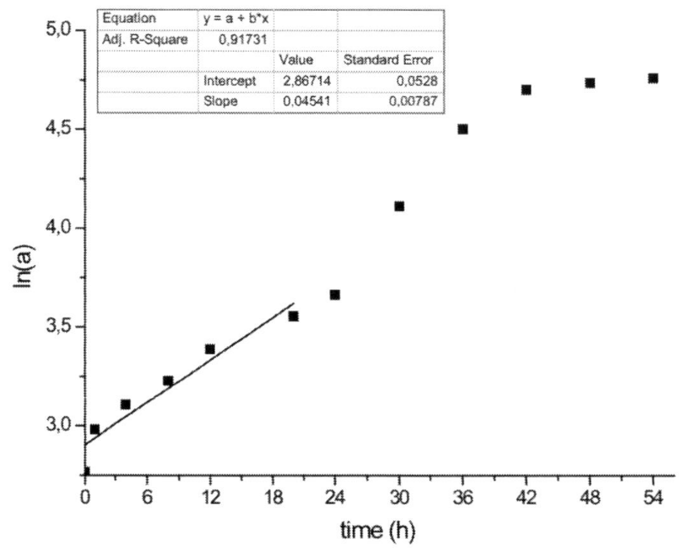

Figure 8. Linear dependence of surface changes at the initial period of leaching process.

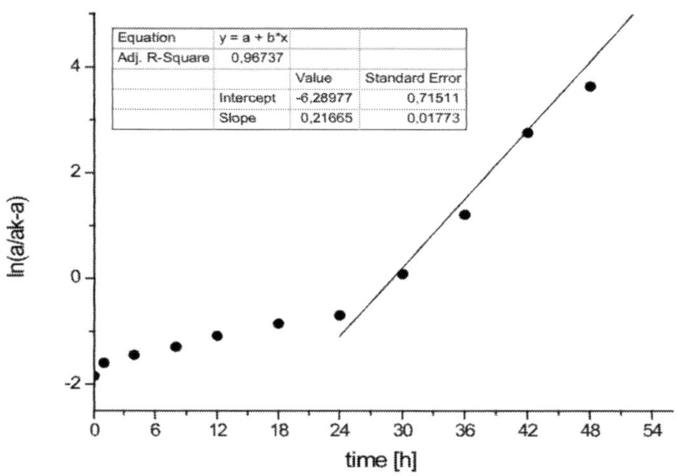

Figure 9. Linear dependence of surface changes at the final stage of leaching process.

The porous structure of surface layer after 56 hours of leaching is presented in Fig. 10 (SEM picture), while Fig. 11. presents the picture of surface before leaching.

 DOI: 10.1051/e3sconf/20160801035

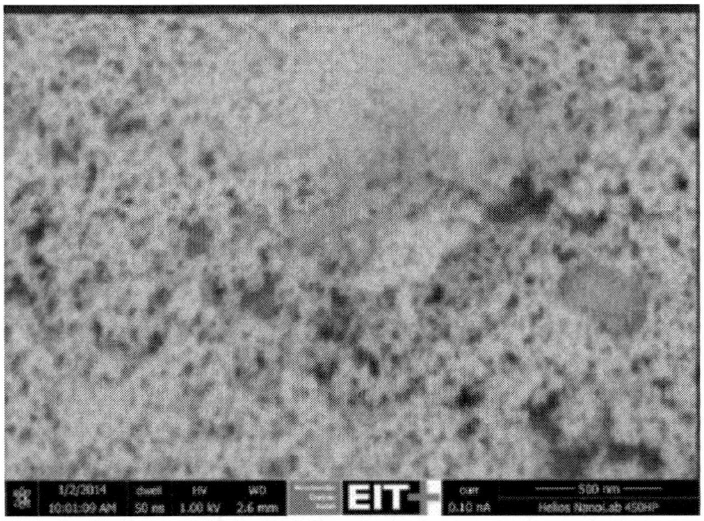

Figure 10. Surface of serpentine ore after 56 hours of leaching.

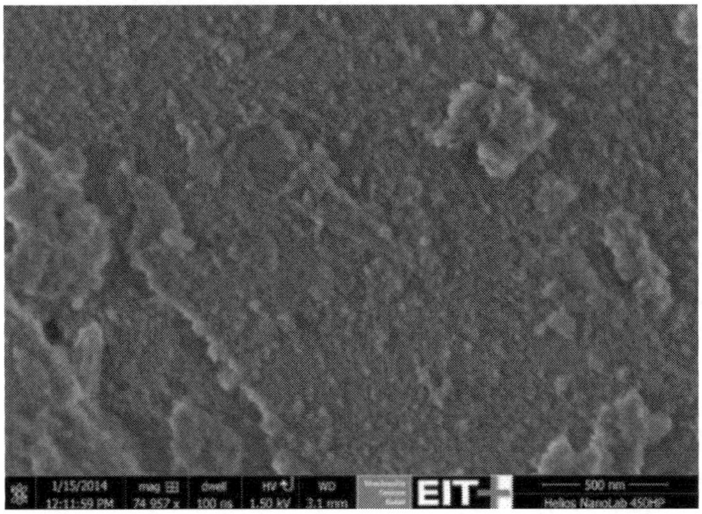

Figure 11. Surface of serpentinite ore befor leaching.

4 Conclusions

This paper discusses how dissolution of serpentinite in acid media can be modelled. Based on the experimental data, the following conclusions can be drawn.

1. The kinetics of serpentinite dissolution in H_2SO_4 solutions can be described by equations based on changes in the surface area.
2. There is a correlation between the extent of calcium and magnesium recovery and the increase in the particle surface area.
3. The dissolution process should be divided into two separated stages.
4. The shrinking core model is able to describe the second stage of the serpentinite rock dissolution process.

Acknowledgements

The work was financed by a statutory activity subsidy from the Polish Ministry of Science and Higher Education for the Faculty of Chemistry of Wroclaw University of Technology.

References

1. A.Orlando, D.Borrini, L.Marini, *Appl. Geochem.*, **26**, 1569 (2011)
2. T.A.Haung, A.I.Munz, R.A.Kleiv, *Energy Procedia*, **4**, 5029 (2011)
3. R.G.McDonald, B.I.Whittington, *Hydrometallurgy*, **91**, 35 (2008)
4. S.Teir, H.Revitzer, S.Eloneva, C-J.Fogelholm, R.Zevenhoven, *Int. J. Miner. Process.*, **83**, 36 (2007)
5. M.M.Maroto-Valer, D.J.Fauth, M.E.Kuchta, Y.Zhang, J.M.Andresen, *Fuel Process. Technol..*, **86**, 1627 (2005)
6. J.Baldyga, M.Henczka, K.Sokolnicka, *Chem. Eng. Res. Des.*, **89**, 1841 (2011)
7. S.C.M.Krevor, S.K.Lackner, *Int. J. Greenh. Gas Con.*, **5**, 1073 (2011)
8. M.Yoshimoto T.Koyama, *Chem. Eng. Technol.*, **36(8)**, 1430 (2013)
9. I.S.Romao, LM.Gando-Ferreira, M.Manuela, V.G. da Silva, R.Zevenhoven, *Miner. Eng.,* **94**, 104 (2016)
10. A.Park, A.He. L-S.Fan, *Chem. Eng. Sci.*, **59**, 5241 (2004)
11. S.Teir, R.Kuusik, *Int. J. Miner. Process.*, **85(1-3)**, 1 (2007)
12. X.Wang, M.M.Maroto-Valer, *Fuel*, **90**, 1229 (2011)
13. A.M.Didyk, Z.Sadowski, *Physicochem. Probl. Miner. Process.*, **48(2)**, 607 (2012)
14. V.Safari, G.Arzpeyma, F.Rashchi, N.Mostoufi, *Int. J. Miner. Process.*, **93**, 79 (2009)
15. H.Rutto, C.Enweremadu, *Korean J. Chem. Eng.*, 29 (1), 1 (2012)
16. P.K.Gbor, C.QJia, *Chem. Eng. Sci.*, **59**, 1979 (2004)
17. H.M.Lizama, M.J.Fairweather, Z.Dai, T.D.Allegretto, *Hydrometallurgy*, **69**, 109 (2003)
18. A.Szubert, R.Lupinski, Z.Sadowski, *Physicochem. Probl. Miner. Process.*, **40**, 211 (2006)
19. C.Fischer, R.Gaupp, *Geochim. Cosmochim. Acta,* **69(5)**, 1213 (2005)
20. B.Kieffe, F.C.Jove, H.E.Oelkers, J. Schott, *Geochim. Cosmochim. Acta*, **63(21)**, 3525 (1999)
21. H.M.Lizama, *Miner. Eng.*, **18**, 23 (2004)

E3S Web of Conferences 8, 01036 (2016) DOI: 10.1051/e3sconf/20160801036
MEC2016

Multidimensional statistical and visualization methods in description of grained materials

Tomasz Niedoba[1,a], Paulina Pieta[1], Agnieszka Surowiak[1], and Dariusz Jamroz[2]

[1]*AGH - University of Science and Technology, Faculty of Mining and Geoengineering, Department of Environmental Engineering and Mineral Processing Cracow, Poland*
[2]*AGH – University of Science and Technology, Faculty of Electrical Engineering, Automatics, Computer Science and Biomedical Engineering, Department of Computer Science*

Abstract. As far as coal is concerned, the data are usually considered to be independent variables, and such an approach is not always appropriate. Therefore, the paper focuses on the multidimensional analysis, which allows to conduct the comparisons of the coal types and to determine the relationship between their specific characteristics. The paper presents an analysis of variance and an observational tunnels method, which enabled to examine the differences between three types of coals: 31, 34.2 and 35. In order to achieve the experimental aim, a number of laboratory analyses of coal were carried out. Such parameters as combustion heat, ash contents, sulfur contents, volatile parts contents and analytical moisture were determined for samples, including the mass and density of these fractions, so seven various features were specified for each coal. The results of the observational tunnels method and the ANOVA application confirmed that the accepted parameters were sufficient for the proper identification of the coal sample origin and their size fractions. As a result of ANOVA, the volatile matter content was determined as the feature on which identification of the coal types can be based, regardless of size fraction. However, the remaining parameters affect the study groups randomly. While applying the methods of multidimensional analysis, some limitations were encountered that indicated the complex structure of the grained material.

1. Introduction

Mineral resources are characterized by different physical, chemical and mechanical properties. This fact determines their complex nature which can be described by a multidimensional vector of grain property set. The grained material is a set of random variables, and this property makes it possible to carry out a multidimensional analysis, which allows a more complete examination of objects and material characteristics. There is a tendency for a comprehensive study of various phenomena and materials to be carried out in order to explore thoroughly a number of factors affecting them. As it is shown in the paper, the multidimensional statistical and visualization methods are appropriate to consider the problems associated with minerals engineering. The examples described in this paper confirmed the high potential that they carry. Consideration of grained materials classification and

[a] Corresponding author: tniedoba@agh.edu.pl

identification constitutes only one of many issues which can be faced in the field of mineral processing. A number of works serve as an evidence of the widespread use of the analysis of variance and the observational tunnels method in different areas of science [4, 10, 11, 14, 16-18]. However, the analysis carried out for different types of coal confirms that these methods can be successfully connected to the mineral processing. The methods of multidimensional data visualization have their origins in statistics and computer science as well as they allow for graphical interpretation of large amounts of data. The multidimensional data analyses are complicated, but they are conducted more frequently in the era of computerization [15, 21].

2. Experiment

In order to introduce the topic of the paper, a number of laboratory analyses of coal were carried out. Coals classified, according to the Polish classification, as types 31, 34.2 and 35 came from some mines located within the Upper Silesian Industrial Region. At the beginning, an analysis of particle size distribution of each coal type was screened on a set of sieves of following sizes: –1.00, –3.15, –6.30, -8.00, –10.00, –12,50, –14.00, –16.00 and –20.00 mm. After that the density analysis was performed in zinc chloride suspensions of various densities (1.3, 1.4, 1.5, 1.6, 1.7, 1.8 and 1.9 g/cm^3). The density fractions were determined by the following characteristics: particle size, density, ash content, volatile matter content, sulfur content, heat of combustion and analytical moisture. It was assumed that the resulting measured values were continuous random variables [17]. The results of laboratory analyses allowed to carry out the one-dimensional analysis of variance and the visualization of multidimensional data by the observational tunnels method. Energy and coking coals can be distinguished in classification of coals types. The coal properties are usually considered as independent variables, and such an approach is not always appropriate. Therefore, the paper focuses on the multidimensional analysis, which allows to conduct the comparisons of the coal types and to determine the relationship between specific characteristics of this energy source.

3. Results and discussion

3.1. Visualization of 7-dimensional data by observational tunnels method

A large amount of experimental data concerning an object can bring particular difficulties in their interpretation and reasoning. On the other hand, separate examination of variables affecting the phenomenon will limit a spectrum of relationship occurring among them in reality. A qualitative multidimensional analysis can therefore be applied using methods of data visualization, which change multidimensional features of either phenomena or material vectors into a form in which it is possible to transfer data onto either a two- or three-dimensional plane to examine certain areas of homogeneity. The results of these analyses can be helpful in characterizing grained materials, as well as in constructing mineral processing models [18]. Up until now, the methods applied to provide a visual image of multidimensional data were used to examine different problems. Those methods are: grand-tour [6], principal component analysis [12], neural networks [7, 13], parallel coordinates [3], scatter-plot matrices method [2] and mosaic plots [5]. In the paper, the results of another multidimensional visualization of data are presented [8, 9]. The observational tunnels method makes it possible to achieve an external view of the observed multidimensional sets of points using the tunnel radius. [18].

Methodology of the observational tunnels method was described elsewhere [8]. It can be stated that the method used the parallel and local orthogonal projections of an extent limited by the maximal radius of the tunnel. This connection made it possible to observe selected parts of a space in order to obtain important information on an object. Applying only one of these methods does not provide such opportunities. The method of projection used in this paper is presented in a demonstrative manner in Fig. 1 [8, 10, 11, 16, 17].

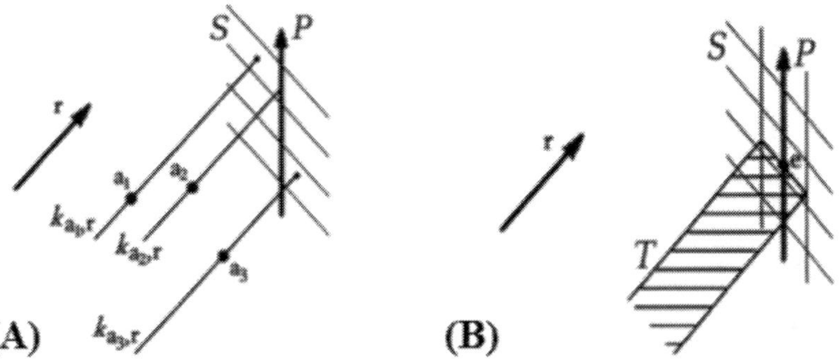

Figure 1. Presentation of projection rules on plane P in observational tunnels method (A) and the way of choosing observational tunnel T (B). Source: [18].

Mathematical tools used to describe the space, in which the objects will be placed, are: observation point a, observed space X, observational plane $P=\delta(w,\{p_1,p_2\})$ (1), hypersurface $S_{(s,d)}$ (2), line parallel to the r and passing through a called the set $k_{a,r}$ (3), tunnel radius b_a (4), projection of an observation point $e\in P\cap k_{a,r}$. The tools are defined as follows:

$$\delta(w,\{p_1,p_2\}) = \{x\epsilon X: \exists\beta_1,\beta_2\epsilon F, such\ that\ x = w + \beta_1\cdot p_1 + \beta_2\cdot p_2\} \quad (1)$$

$$S_{(s,d)} = \{x\epsilon X: (x-s,d) = 0\} \quad (2)$$

$$k_{a,r} = \{x\epsilon X: \exists\psi \in F\ , such\ that\ x = \psi\cdot r + a\} \quad (3)$$

$$b_a = \psi\cdot r + a - w - \beta_1\cdot p_1 - \beta_2\cdot p_2 \quad (4)$$

where:

$$\psi = \frac{(w-a,r)}{(r,r)} \quad (5)$$

$$\beta_1 = \frac{(\psi r+a-w,p_1)}{(p_1,p_1)} \quad (6)$$

$$\beta_2 = \frac{(\psi r+a-w,p_2)}{(p_2,p_2)} \quad (7)$$

$r\in X$ - a proper direction of protection onto observational plane P,
$\{\beta_1,\beta_2\}$ - a position of the projection of observed point a,
$\beta_1,\beta_2\in F$, such that $e=w+\beta_1 p_1+\beta_2 p_2$,
$\psi\in F$, such that $e=\psi r+a$ - the distance projection of the observed point a.

The methodology of the observational tunnels method and the drawing procedure are accurately described elsewhere [8-11, 18]. The observational tunnels method was used to analyze the data describing three types of hard coal. This method served to visualize the multidimensional data, in which each of seven parameters was treated as a separate dimension, where the 7-dimensional space was created. While analyzing simultaneously all the features of coals, such views were received which did not permit the indication of a clear differentiation between different types (Fig. 2). It means that the analyzed data are complicated. Moreover, the observation tunnels method does not allow to provide constructive conclusions on the identification of size and density fractions regarding the data set [9-11]. Only individual analysis of the size fractions in pairs allowed their identification. The views obtained through the analysis confirmed that the assumed characteristics are sufficient to identify the correct size fractions, for which the coal type originates. It was concluded on the basis of separate areas in the space occupied by the considered size fractions, which was not obtained in examination of all 9 size fractions at the same time. A similar analysis was performed for all possible pairs, only confirming the aforementioned conclusions [9-11, 18]. Therefore, the following charts are presented for the separated size fraction (1-3,15 mm).

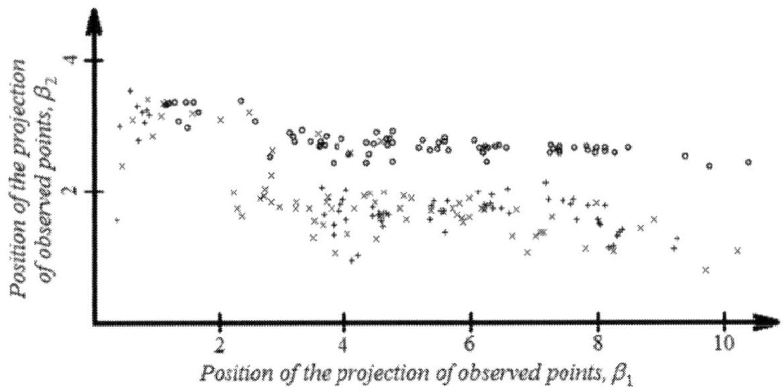

Figure 2. View of 7-dimensional data for coal 35 ("o") gather in other part of the space than coal 34.2 ("+"). From this view it is not possible to conclude about possible separation of coal of type 31 ("×"). Source: [10].

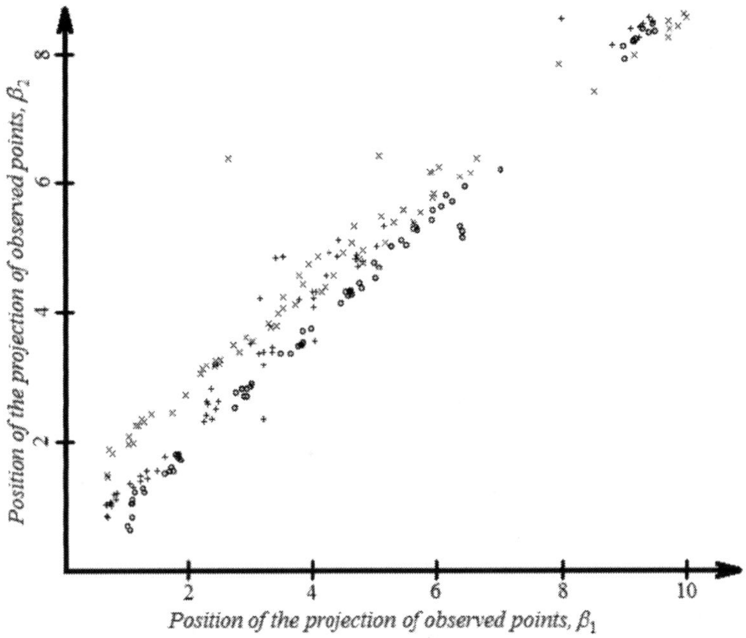

Figure 3. View of 6-dimensional set of data created after removal of parameter analytical moisture. It is visible that the data representing coal of type 35 ("o") is located in other part of the space than data representing coal type 31 ("×"). Source: [10].

DOI: 10.1051/e3sconf/20160801036

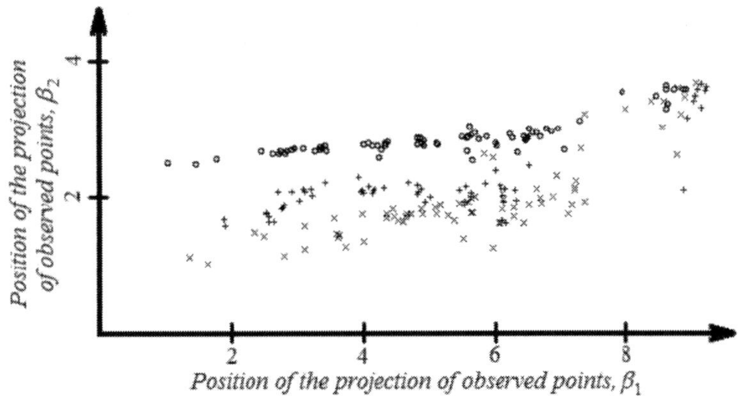

Figure 4. View of 6-dimensional data created after the removal of parameter analytical moisture. It is visible that the data representing coal type 35 ("°") is located in other part of the space than data representing coal type 34.2 ("+"). Source: [10].

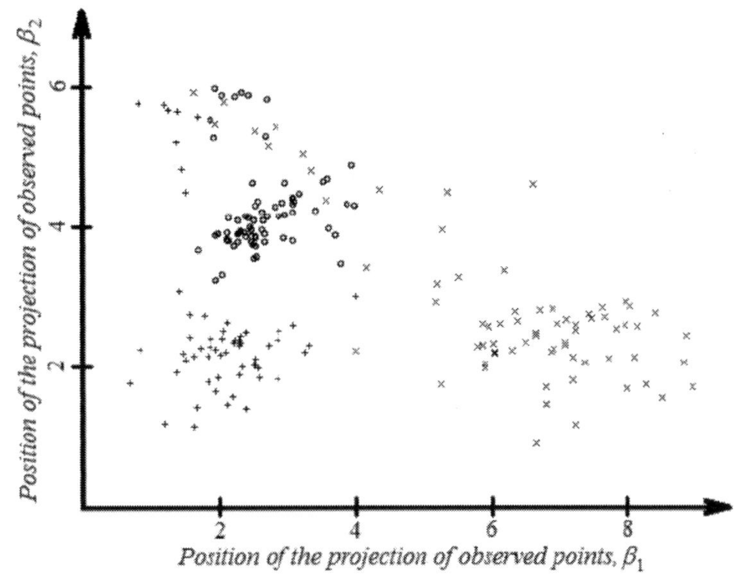

Figure 5. View of 3-dimensional data created from features: analytical moisture, sulfur contents and volatile parts contents. It is visible that the data representing coal type 34.2 ("+") is located in other part of the space than data representing coal type 31 ("×") as well data representing coal type 34.2 ("+") is located in other part of the space than data representing coal type 35 ("°"). Source: [10].

The accepted properties of coal were considered by the correlation analysis. Based on this analysis, it was stated that the influence of analytical moisture on the differentiation of coal types was small. Therefore, it was examined how its omission affected the results of the observational tunnels method (Figs. 2-4). As it was noticed before, the seven-dimensional data created from the seven coal features described above were sufficient to the proper identification of coal type, but the same data were not sufficient for this purpose after the removal of the parameter analytical moisture [10-11, 18].

The results of the correlation analysis showed that the most correlated coal features were density, combustion heat, ash contents and volatile parts contents. It can be presumed that these parameters were the most significant for coal type identification [10]. Based on correlation matrix results, the three-dimensional space was created for the mentioned features. On the basis of Figs. 6-7 it can be concluded that the accepted characteristics were sufficient for the proper identification if the certain sample originated from coal type 31 and 35, 31 and 34.5 or 34.2 and 35.

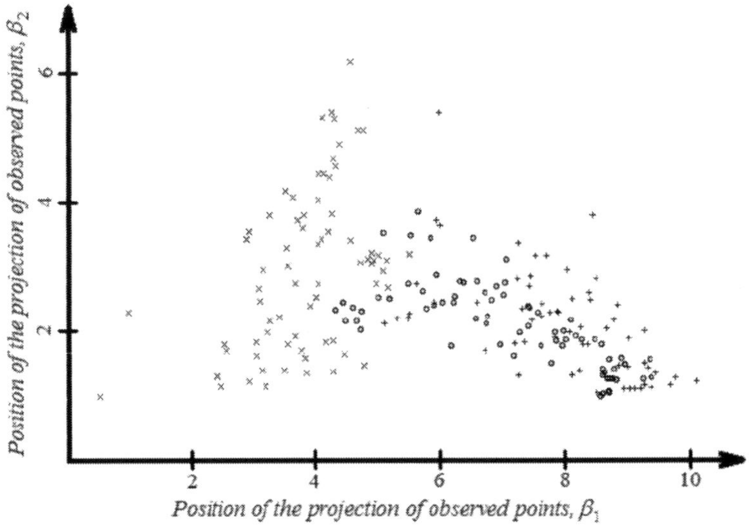

Figure 6. View of 3-dimensional data created from features: analytical moisture, sulfur contents and combustion heat. It is visible that data representing type coal 31 ("×") is located in other part of the space than data representing coal type 34.2 ("+") and other part of the space than data representing coal type 35 ("o").

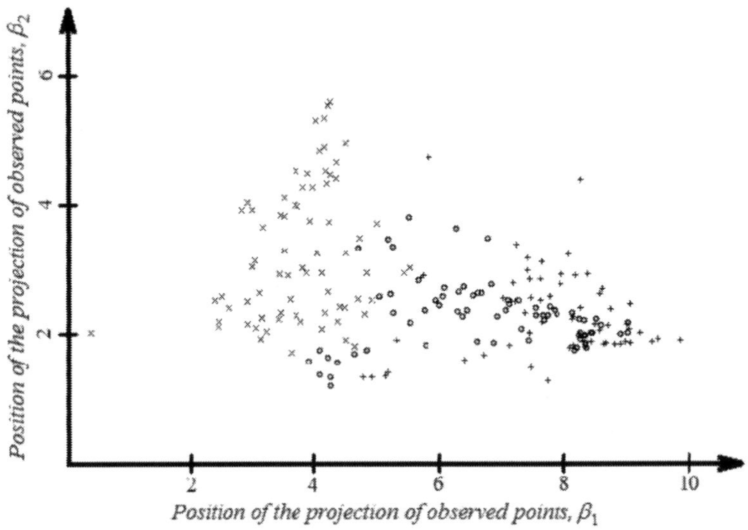

Figure 7. View of 3-dimensional data created from features: analytical moisture, sulfur contents and ash contents. It is visible that data representing coal type 31 ("×") is located in other part of the space than data concerning coal type 34.2 ("+") and other part of the space than data representing coal type 35 ("o").

At the same time it was impossible to get one view from which such conclusion could be drawn. It can be the proof that the structure of analyzed data is complicated [10-11, 18]. The next attempt was aimed to verify the influence of potential replacement of one of the coal features with another one highly correlating with it. They were characteristics highly related to each other. That is why in a set of three coal properties (analytical moisture, sulfur contents and volatile parts contents), which were sufficient for the correct identification of coal type as shown above, the parameter volatile parts contents was replaced with the combustion heat and ash contents (Figs. 5-6). This proved that the replacement of the volatile parts contents with a mentioned features also resulted in a loss of information allowing the proper identification of all three coal types despite the high value of correlation index between the replaced coal properties [10]. It can be deduced that the volatile matter content is an essential characteristic of identifying coal types.

3.2. Examination of hard coals diversity by applying the analysis of variance

Experimental studies of certain phenomena, which are influenced by many factors, can successfully be subjected to a multidimensional parametric analysis. Variability of measurements and observations of the same objects is a widespread phenomenon which results from a number of random factors occurring during an experiment, which are simultaneously able to disturb the value of parameters. The problem of diversity was elaborated by Fisher, who developed the analysis of variance. The analysis of variance is a parametric statistical analysis, so the basis to carry out the ANOVA is to verify the assumptions concerning the homogeneity of variance and the normal distribution of these variables. For the one-dimensional analysis, the null-hypothesis assumes that the mean values μ_i of the observed characteristics in each group are equal ($H_0{:}\mu_1 = \mu_2 = \mu_3 = \cdots = \mu_n$). The alternative hypothesis H_1 ($H_1{:}\ \mu_1 \neq \mu_2 \neq \mu_3 \neq \cdots \neq \mu_n$) is opposite to the assumption of H_0. In order to check the impact of parameters on a phenomenon it is necessary to calculate a value of F according to formula (8) and to determine the critical value F_α read from the F-distribution tables [19]:

$$F = \frac{\frac{\sum_{i=1}^{a} n_i(\bar{x_i}-\bar{x})^2}{a-1}}{\frac{\sum_{i=1}^{a}\sum_{j=1}^{n_i}(x_{ij}-\bar{x_i})^2}{n-a}} = \frac{n-a}{a-1}\cdot\frac{\sum_{i=1}^{a} n_i(\bar{x_i}-\bar{x})^2}{\sum_{i=1}^{a}\sum_{j=1}^{n_i}(x_{ij}-\bar{x_i})^2} \quad (8)$$

where: n_i - the number of random variables of the studied properties,
$\bar{x}$ - the general average for all observations,
$\bar{x_i}$ - the arithmetic average of a characteristic in the studied group,
x_{ij} - the value of a single property measurement.
n - the number of independent observations x_{ij} equal $n = \sum_{i=1}^{a} n_i$
a - the number of independent variables.

The methodology of verifying the assumption and description of different ANOVA schemes is accurately described elsewhere [20]. In the next part of this article the experiment is presented, in which ANOVA is successfully applied to study the characteristics of grained material.

A grained material has many properties that determine its classification. Such features are: content and a liberation efficiency index of valuable minerals, particle size, particle density, physicochemical properties and other physical and geometrical characteristics [1, 17, 22, 23]. In this case the ANOVA made it possible to determine main properties of coal, which differentiated the aforementioned coal types, and a comprehensive study enabled a thorough characterization of variability properties of the tested grained material. According to the scheme of the ANOVA application the fundamental assumptions were verified by Shapiro-Wilk and Brown-Forsythe tests. It allowed to exclude two properties (sulfur content, moisture analysis) from further analysis, which breached the assumption of homogeneity of variance. The F test was conducted for four features that is density, heat of combustion, ash content and volatile matter content. The results are presented in Table 1.

The F test allowed to classify the characteristics influencing the differentiation of coal types taking into account size fractions. Looking at the overall results it turned out that only the volatile matter content was a feature differentiating all tested coals in every size fraction (Fig. 8D). It can be therefore concluded that the value of this parameter was the major factor by which coal was classified according to its properties and use. This assumption indicated that a completed identification of coal types, taking into account size fractions date, was complicated and possible only for the volatile matter content. A more detailed analysis was possible thanks to the drawn graphs showing the distribution of features values in the size fractions. The 31, 34.2 and 35 coal types were significantly different from each other in size fractions: 3.15-6.3 8-10, 10-12.5, 16-20 mm regarding four parameters. Identification of the coal types in the aforementioned size fraction taken separately could theoretically be based on four characteristics (density, heat of combustion, ash content and volatiles matter content), as indicated by *F* test results and its graphic interpretation. However, the complex structure of the analyzed data did not allow identification of hard coal types with an overall analysis of the observed characteristics in all size fraction. This was caused by the occurrence of some interactions between the values of features, which were depicted in the graphs (Fig. 8).

Table 1. Results of *F* test (significant results have been highlighted).

Size fraction [mm]	The values of empirical *F* statistics for the studied characteristics			
	Density	**Heat of combustion**	**Ash content**	**Volatile matter content**
0.5-1	1.466594	**5.487025**	2.467011	**44.13694**
1-3.15	1.134825	0.073532	1.815244	**3.441177**
3.15-6.3	**4.583223**	**6.009332**	**6.946015**	**73.70378**
6.3-8	2.399904	**13.15684**	**4.973166**	**20.0778**
8-10	**6.414866**	**10.15947**	**7.465963**	**15.53309**
10-12.5	**6.549542**	**8.846621**	**7.088464**	**14.30923**
12.5-14	0.335797	2.175174	2.395407	**6.058163**
14-16	**7.623756**	**24.66782**	0.098286	**14.31082**
16-20	**12.08655**	**12.44888**	**10.54814**	**15.00584**
The verification of assumptions	$F_{\alpha}(2,97)$=3,0903			

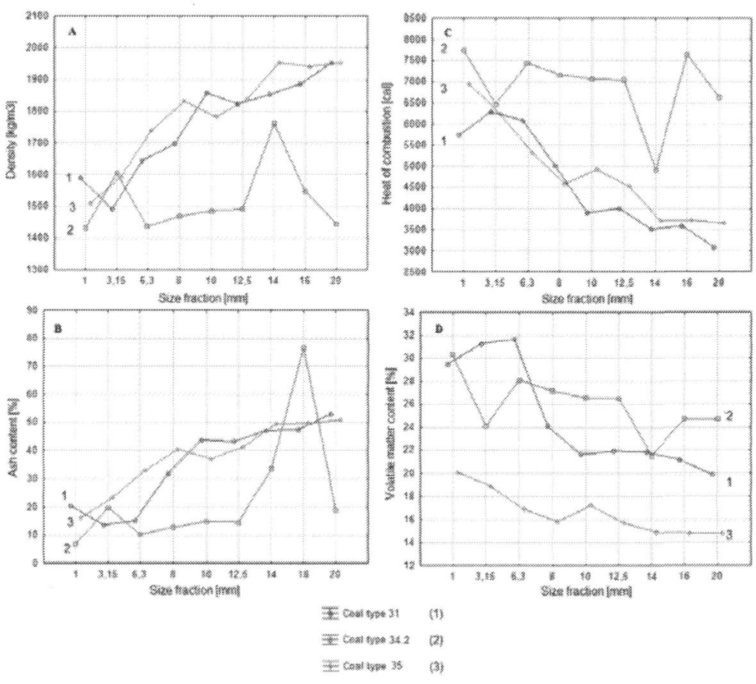

Figure 8. Graphs of weighted mean parameters (A - density, B - ash content, C - heat of combustion, D - volatile matter content) of 31, 34.2, 35 coal types.

4. Conclusions

A multidimensional description of grained material characteristics was used for evaluation of the differentiation and classification of three types of coal. This choice was dictated by the adjustment of methodology to the nature of the analyzed raw materials, which had different characteristics.

The results of the observational tunnels method application confirmed that the accepted parameters were sufficient for the proper identification of the coal sample origin and size fractions. A complicated structure of the analyzed data determined the fact that the recognition of coals or size fraction was only possible while comparing them in pairs. The visualizations of six-dimensional data made it impossible to identify the coal type. They only allowed to recognize coal types 35 and 31 as well 35 and 34.2 in pairs. Despite of the low value of the correlation coefficient for the analytical moisture, this property turned out the crucial significance for the classification of coal types. However, three-dimensional data, taking into account only the most significant characteristics (according to the correlation analysis) were enough for the proper identification of coals and the volatile matter content was necessary for appropriate recognition of coal origin.

As a result of one-dimensional analysis of variance, the volatile matter content was determined as the feature on which the identification of the coal types can be based, regardless of the size fraction. However, the remaining parameters (density, heat of combustion, ash content) affected the study groups randomly. Nevertheless, analyzing the types of coal separately in size fraction: 3.15-6.3, 8-10, 10-12.5, 16-20 mm, it can be stated that their identification was possible on the basis of all the analyzed features. In case of applying the ANOVA and the multidimensional data visualization, some limitations were encountered that indicated the complex structure of the grained material.

The results of the ANOVA and the observational tunnels method allow to conclude that the data structure is complicated. In both cases it is possible to recognize the coals in the selected size fraction on the basis of the four most important parameters: density, heat of combustion, ash content and volatiles matter content. The volatile matter content is a crucial feature that allows to classify coals according to their origin.

Acknowledgement

The paper is the effect of the scientific project no. 15.11.100.021.

References

1. Brozek M., Surowiak A., *Arch. Min. Sci.*, **55** (1), 39 (2010)
2. Eick S.G., Wills G.J., *EJOR*, **81** (3), 445 (1995)
3. Gennings C., Dawson K.S., Carter W.H., Jr. Myers R.H. *Biom.*, **46**, 719 (1990)
4. Güler T. Akdemir U., *Trans. Nonferrous Met. Soc.*, China, **22**, 199 (2012)
5. Heike H., *Metrika* **51**, 11 (2000)
6. Hurley C., Buja A., *SIAM*,**11** (6), 1193 (1990)
7. Jain A.K., Mao J., in: *Proc. IEEE Internat. Joint Conf. On Neural Networks*, Baltimore, MD, **3**, 35 (1992)
8. Jamroz D., Ph.D. Thesis, AGH, University of Science and Technology, Cracow, Poland (2001)
9. Jamroz D., *Arch. Min. Sci.*, **59** (2), 413 (2014)
10. Jamroz D., Niedoba T., *Physicochem. Probl. Miner. Process.*, **50** (1), 185 (2014)
11. Jamroz D., Niedoba T., *Physicochem. Probl. Miner. Process.*, **51** (2), 769 (2015)
12. Li W., Yue H.H., Valle-Cervantes S., Qin S.J., *J Process Contr*, **10** (5), 471 (2000)
13. Mao J., Jain A.K., *IEEE Trans. Neural Networks*, **6** (2), 296 (1995)
14. Nakhaei F. Irannajad M., Sam A., Jamalzadeh A., *Physicochem. Probl. Miner. Process.*, **52** (1), 252 (2015)

15. Niedoba T., in *proceedings of the XIV Balkan Mineral Processing Congress*, Tuzla, Bosnia and Herzegovina, **1**, 57 (2011)
16. Niedoba T., *Physicochem. Probl. Miner. Process.*, **49** (1), 175 (2013)
17. Niedoba T., *Mineral Resources Management*, Cracow, (2013)
18. Niedoba T., Jamroz D., *Arch. Min. Sci.*, **58** (4), 1317 (2013)
19. Pieta P., *5th EYEC*, Warsaw, Poland, 5, 198 (2015)
20. Stanisz A., *StatSoft*, Cracow (2006)
21. Tumidajski T., Saramak D., AGH, Cracow (2009)
22. Tumidajski T., *MRM*, **23** (6), 111 (2010)
23. Tumidajski T. Niedoba T., Saramak D, *AGH J. of Mining and Geoengi.*, **36** (4), 167 (2012)

Optimisation Platform for copper ore processing at the Division of Concentrator of KGHM Polska Miedz S.A.

Bogdan Kuzba[1], Witold Pawlos[1,a], Andrzej Konieczny[1], and Malgorzata Krzeminska[1]

[1]*KGHM Polska Miedz S.A., Oddzial Zaklady Wzbogacania Rud, ul. Kopalniana 1, 59-101 Polkowice, Poland*

Abstract. The idea of Optimisation Platform is to create an innovative system. It is dedicated to technology and cost efficiency improvement of process realized at the Division of Concentrators of KGHM Polska Miedz SA. This highly sophisticated tool is based on visual, acoustic and vibrating detection systems. The range of its functionality was described in this work. Three main utility modules were described: froth flotation image processing (FloVis), grinding and classification monitoring (MillVis) and belt conveyors control unit (ConVis). The effects of implementation of the system under KGHM conditions were described. It is concluded that the Optimisation Platform is one of the most promising solution for improvement of technology and economy performance at the Division of Concentrators of KGHM Polska Miedz S.A.

1 Introduction

Nowadays the optimisation systems of process control is one of the most dynamic developing area at the Division of Concentrators of KGHM Polska Miedz S.A. Their importance has grown since acquisition of a great volume of variable data. Managing such a number of parameters is out of control for a human operator of the industrial machinery. An optimisation platform is the answer for the current demand regarding data processing. Implementation of platform utilities was possible thanks to realisation of research and development projects. The first introduced solution was flotation cells control unit based on recording the visual data system called FloVis. The successful in implementation of FloVis determined the direction of new research. As a result, a grinding control system (MillVis) and a belt conveyors control system (ConVis) were developed. Finally, the whole utility was created, based on the advanced knowledge, including data analysis, processing visual, acoustic and vibrating signals as well as process optimisation technology.

2 Structure of Optimisation Platform

The Optimisation Platform has a module structure. Its specific modules with their functionality are presented on Fig. 1.

[a] Corresponding author: witold.pawlos@kghm.com

 DOI: 10.1051/e3sconf/20160801037

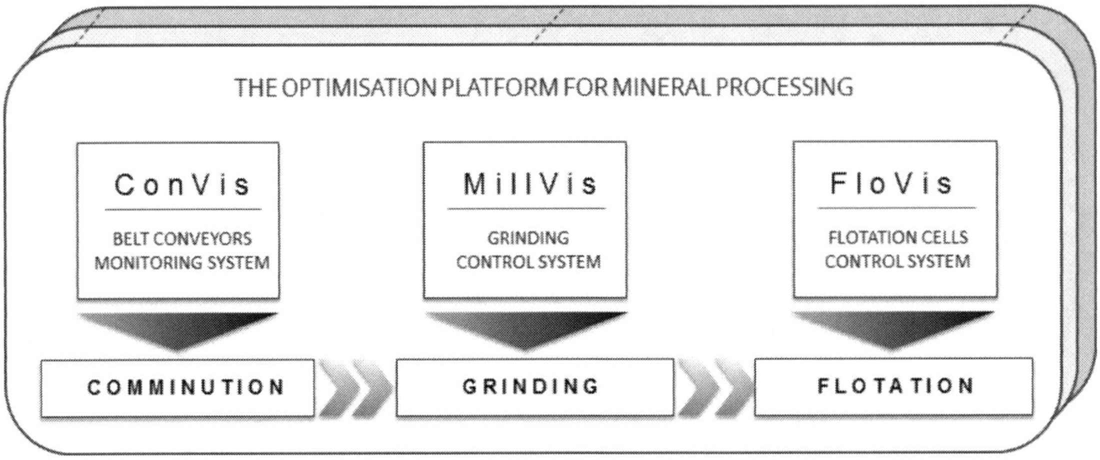

Figure 1. Modules of Optimisation Platform.

At the first stage of production, the ConVis system will be used. It is dedicated to control working parameters of belt conveyors operating in comminution (crushing) circuits. The physical state of belt and its operating condition is measured by the linear scanning laser and visual camera. The working parameters are: cross-section geometry of the belt, its velocity and load of material. Simultaneously, recording of vision pictures enables estimation of mass and volume flow of the transported material. The important functionality of the ConVis system is detection of abnormal position of the belt caused by potential belt damage. The system is currently under implementation. The economical evaluation of results will be possible after project commissioning.

Grinding and classification processes are supported by a MillVis system based on recording visual, acoustic and vibrating signals. MillVis is acquiring visual pictures of ore material conveying to the first stage of grinding. The system allows for detection of increased number of coarse grains (over 15 mm) and approximated estimation of lithology constitution of rocks [5]. Those parameters influence directly the performance of the comminution process conducted in mills. Even roughly estimated, they are useful for maintaining the optimised throughput. Hence, the information supplied by the system is a base for making decision on operational regime of the grinding sections. The system delivers data on real-in-time state of grinding mill and ground material. After implementation of the system, a broad range of process data is available due to visual pictures and also acoustic and vibrating signals recorded during mill's run. Three application dedicated for input data processing are presented in Table 1.

Table 1. MillVis system applications [3].

Application	Measured parameter	Task
Vision	Visible picture of mill's feed	Assessment of variability of mill's feed granulation and lithology composition
Vibration	Analysis of vibrating signal of mill's shell	Diagnostic of mechanical state of liners and lifters
Acoustic	Analysis of acoustic signal in mills space	Assessment of mill's effectiveness due to grinding media load and rotation speed

The MillVis system is responsible for optimising the grain size of the feed in grinding and classification circuits while being prepared for flotation process. Thanks to MillVis implementation, the grinding and classification bottleneck can be eliminated. The MillVis system influences positively the electric energy effectiveness, creating 1.3 million PLN annual cost decrease. MillVis allows for either reduction of a number of running mills or their more effective use.

The major system dedicated to the control the flotation process is the visual system for optimisation of the flotation cells operation (FloVis). Pictures of flotation froth are digitally processed after being recorded by cameras placed above the surface of the flotation cells. The stages of images processing are presented in Fig. 2.

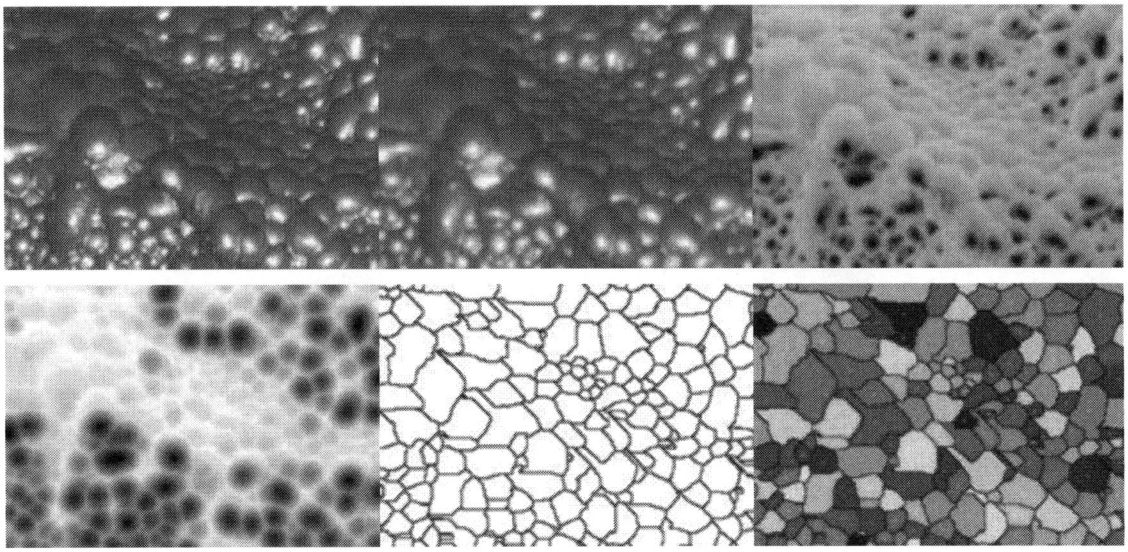

Figure 2. Stages of foam image processing [2].

The developed image processing algorithms determine the values of visual parameters, characteristic for actual froth quality. Visual parameters include the froth structure, bubbles variability, their velocity, froth transparency and RGB colour components of the froth. The flotation process operational parameters are monitored and corrected on the basis of froth visual parameters [1]. Implementation of this system has a great importance due to a crucial role of flotation as "the heart" of the ore processing technology developed at KGHM. Moreover, the optimised operating regimes of flotation circuits influence positively technology effectiveness and cost reducing [2].

FloVis is a master visual control system equipped with an important functionality of flotation process optimisation. The FloVis application results in better process stability. Moreover, it positively influences the main process operational indexes [5] leading to:

- decreasing Cu content as mass% in final tailings by 0.038%,
- increasing concentrate Cu grade by 1.34%,
- decreasing of concentrate yield circulated in cleaning circuits by 12%,
- increasing operational copper recovery in flotation cells by at least 0.7%.

The already designed and employed application modules of Optimisation Platform are dedicated to local monitoring and control of production sections. The development of fully sophisticated platform assumes a strict connection between the FloVis system, measured parameters of coarse comminution (ConVis) and grinding (MillVis) as well as maximisation of metal recovery from the ore as a result. Such a solution as the Optimisation Platform is an example of synergy, where cooperation and communication between several modules give a better technical and economy result than summarized results of each of them working separately.

It needs to be added that currently the Optimisation Platform does not include the last production step realized at KGHM, that is dewatering of the copper concentrate. However, a feasibility study of this project has been launched. The component, being designed, is named DryVis being a system of monitoring thermal dewatering process, including the control of fuel (natural gas) consumption.

3 Summary

The Optimisation Platform is a global, breakthrough solution being implemented in the ore processing industry. It is a sophisticated, fully automatic, still developing tool for changing conditions of working at the Division of Concentrators of KGHM. Thanks to its implementation, a reliable and modern system of control and production optimisation is available. It ensures support for production engineers and operators in the 24/7 access regime. The Optimisation Platform forms a perfect tool to a further development of production processes. It improves decision making processes, eliminates errors and helps to maximize economy benefits. The Optimisation Platform is a base of Intelligent Production Chain. It provides security, technology effectiveness as well as possibility of successful and flexible production process management from ore delivery (ConVis), its preparation for flotation process (MillVis), and finally copper concentrate production (FloVis). In the ore processing industry, the innovativeness and usefulness of such solutions were used for creation similar products by global research centres. However, their functionality is much smaller than the Optimisation Platform.

References

1. W. Pawlos, M. Wrobel, T. Legierski, A. Zamora, M. Rdest, W. Legierski, M. Wrobel, J. Moszumanski, S. Garlowski, Report, Project no: 108/2006, (2006)
2. A. Konieczny, W. Pawlos, M. Jach, R. Pepkowski, M. Krzeminska, R. Kaleta, *Gor. i Geol.*, **6(2)**, 49 (2011)
3. T. Legierski, Sz. Ogonowski, M. Kurzydlo, A. Zamora, S. Kazior, D. Foszcz, Report no 12012, (2013)
4. T. Legierski, Sz. Ogonowski, M. Kurzydlo, A. Zamora, S. Kazior, D. Foszcz, Report (2013)
5. A. Konieczny, M. Krzeminska, W. Pawlos, K. Ksiezniak, *in Proceedings Mineral Engineering Conference* (Istebna, Poland, 2014)

E3S Web of Conferences **8**, 01038 (2016) DOI: 10.1051/e3sconf/20160801038
MEC2016

Research on choice of xanthates blend dedicated to processing technology at KGHM Polska Miedz S.A. concentrators

Witold Pawlos[1,a], Andrzej Konieczny[1], and Malgorzata Krzeminska[1]

[1]*KGHM Polska Miedz S.A., Oddzial Zaklady Wzbogacania Rud, ul. Kopalniana 1, 59-100 Polkowice, Polska*

Abstract. A proper choice of reagents quality and quantity has a great influence on the flotation process effectiveness. Attempts to develop a method to create sodium xanthates blends used as a collector for copper sulphides flotation was undertaken at the Division of Concentrators of KGHM Polska Miedz S.A. Laboratory tests involving different blends of sodium xanthates were performed followed by the data analysis of the results. The obtained results proved that it was profitable to prepare a blend consisted of the so called "weak" and "strong" xanthates with a minimum difference in the alkyl chain length of two carbon atoms. The results of research will be helpful for development of specific xanthates blends dedicated to technology applied at KGHM under industrial conditions.

1 Introduction

One of the assumptions of effective flotation process is to properly developed strategy in flotation reagents management. It is a giant challenge to technical staff to compose an accurate blend of xanthates and consumption rate at each feeding point. Currently, the implemented practice is based on empiric confirmation of reagent effectiveness on a laboratory scale tests followed by verification of the obtained data under industrial conditions. However, the ore beneficiation process is a dynamic and multistage flotation system with variations specific for a given ore processing plant. Therefore, some of suggested by other investigators reagents were found unsuccessful under specific industrial conditions.

The level of complexity of flotation and a crucial role of flotation reagent in achieving high production indexes such as metal recovery, concentrate grade and production costs were the reason of taking serious attempts by scientists to improve the sulphide ore flotation regime. One of the proposals, based on theoretical assumptions, was to establish a methodology of preparing xanthates blends for flotation.

Majority of research results led to the only known conclusion that a specific type of xanthate was responsible for good flotation results, conducted under specific reagent conditions, including its type, length and structure of hydrocarbon radical. However, this information is helpful for preparing the

[a] Corresponding author: witold.pawlos@kghm.com

reagents blend, but is not sufficient, and must be supported by considerable experience and intuition. This work describes new possibilities revealed after a thorough analysis of subject.

2 Blends of xanthates

Xanthates are the most important sulphide minerals flotation collectors. Since implementation of froth flotation for the ore processing technology, xanthates has been the basic reagents dedicated to sulphide ores. Because they have a very broad range of activity, the choice of a specific xanthate with suitable chain length and type is not easy. In practice, it is assumed that increasing the chain length will result in increasing the surface activity of collector and its consumption rate can be reduced. However, increasing carbon chain length causes decreasing reagent selectivity.

Introducing into a flotation system only one type of xanthate is not sufficient to ensure the optimised chemical conditions of flotation slurry. This is necessary for proper metal content in the concentrate and high recovery, in addition to process economy. There is no solitude collector which is able to activate the surface all sulphide types present in a polymetallic ore. Better process results can be achieved by introducing a blend of few collectors.

Dosing xanthate blends to a flotation process is widely known since fifties of the last century [1]. However, this subject has been investigated only partly since then. The good results of using the collectors blends in a flotation process are recognized and they are confirmed by laboratory experiments as well as on the full industrial scale. The most important benefits include increased recovery of the valuable components, improved process kinetics, increased process selectivity, improved floatability of the extreme grains: fine and coarse, controlled froth production and reduced process costs due to decreasing dose of the main collector, often the most expensive one. The proper performed blend preparation makes room for improvement of ore processing parameters as well as for cost reduction.

The components of collectors blend are mainly based on experience and knowledge of specialists and technology engineers. Usually, the mixtures are tested in the laboratory scale, and then potentially positive results are verified under industrial conditions. In literature some general phenomena are described, which are helpful in preparing the xanthate blends. The rules of selection of flotation reagents blends was established by Glembotskii [1]. According to him, larger differences in the structure of a collectors blend causes more visible synergic. Hence, the basic composition of xanthates blend includes so called "weak" xanthate, e.g. ethyl or isopropyl and "strong" xanthate with either isobutyl or amyl radical.

At Division of Concentrators of KGHM Polska Miedz S.A. the results of laboratory tests of using blends of xanthates with different carbon chains were investigated.

3 Methods and test results

The Polkowice concentrator feed was investigated. The laboratory flotation experiments were conducted for various xanthate types: ethyl, isopropyl, isobutyl and isoamyl. The reagents with short alkyl chains (C2-C3) were defined as weak, while others with longer chains (C4-C5) were described as strong.

The results of flotation experiments were analysed using the Fuerstenau upgrade curve and selectivity factor a. The ideal upgrading occurs for a value equal to 100 and for its increasing values, process selectivity decreases [2]. The obtained upgrading curves are illustrated in Fig. 1 and a values for specific xanthate blends are compared in Fig. 2. The obtained results proved that the assumptions were reflected under KGHM industrial conditions. The minimal values of selectivity index (e.g. the highest selectiveness) are obtained for mixtures consisting of only weak or strong xanthates. If the difference in the length of the carbon chain of reagents present in a blend becomes more significant, the flotation performance achieves better effects. The best results were observed for blending C2 and

DOI: 10.1051/e3sconf/20160801038

C5, while the worst were noticed for the C2 and C3 combination, when there was only one carbon atom difference between alkyl chains' length. Similar phenomenon was observed in the blends consisting of two weak or strong xanthates. These blends contain reagents having difference between alkyl chain too small to observe visible positive effect.

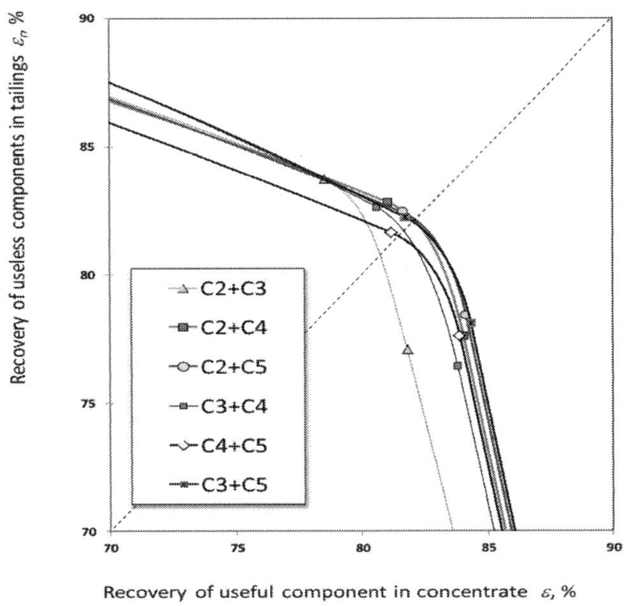

Figure 1. A part of the Fuerstenau upgrading curves for different xanthate blends (based on [3]).

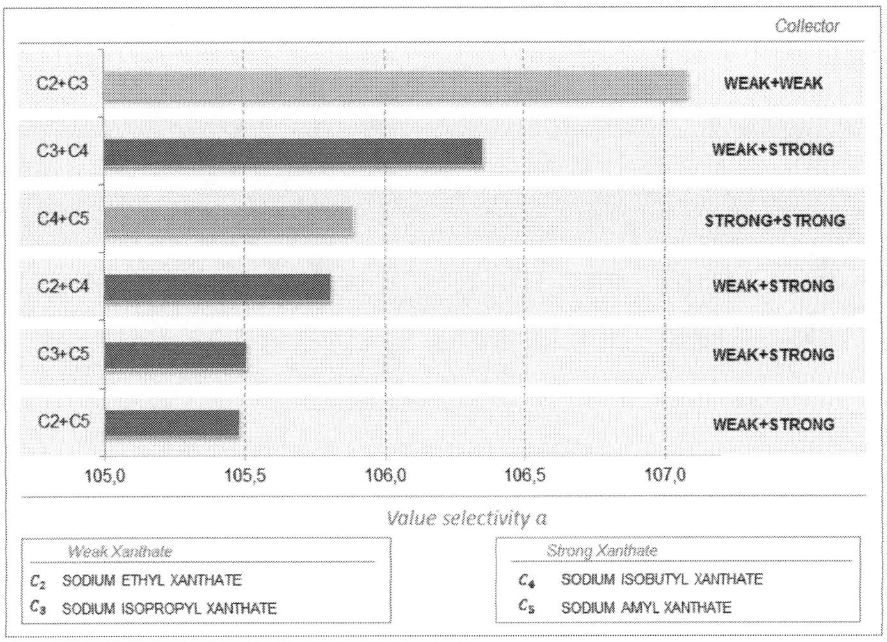

Figure 2. Results of implementation of different xanthate blends (based on [3]).

It can be concluded that the difference of only one atom of carbon in the alkyl chain is not sufficient to ensure the optimum flotation conditions. If xanthates are blended, the rule of combining weak and strong xanthates must be respected, providing difference of minimum of two carbon atoms between their alkyl chains.

4 Summary

One of the most important factor of flotation performance in copper ore processing is the proper selection of flotation reagents combination. It influences achieving such parameters as recovery, concentrate grade, process costs reduction as well as environment protection. The research and development activity of Division of Concentrators of KGHM Polska Miedz S.A. is focused on selecting new, more efficient than currently used, blend of flotation reagents. It was assumed that preparing new combination of reagents blends should be performed after validation of a confirmed method of preparing xanthate blends. There is a broadly known relation between the type of carbon radical, its structure and length as well as flotation performance. However, the efforts to establish rules of design xanthate blends have not been undertaken yet. Instead, the technologist activity was driven by intuition and experience only. The analysis of previously obtained results of flotation experiments was assumption of achieving the method of combining xanthate blends. It was proven that developing the xanthates blend must respect combining the so-called weak and strong xanthates together. The minimum difference of two atoms of carbon between alkyl chains length should be ensured.

In addition, it should be underlined, that the kind of functional group of reagent are not less important than structure and length of alkyl chain, while selecting reagents blend for sulphide flotation. Since the xanthates blend should be supported by additional reagents, this question will be investigated in future research.

References

1. A. Glembotskii, *Tsvetnye Metally*, vol. **4** (edited by S. Vynogradova, Trans., 6, 1958)
2. J. Drzymala J., H. A. M. Ahmed, *Int. J. Miner. Process.*, **76**, 55 (2005)
3. M. Wrobel, E. Kasinska-Pilut E., A. Krzyzosiak, Report (2010)

E3S Web of Conferences **8**, 01039 (2016) DOI: 10.1051/e3sconf/20160801039
MEC2016

Evaluation of bioleaching column test of sulphide copper ore and copper concentrate using preconditioned surface

Alvaro Videla Leiva[1,a] , and Mario Vera Vélizand[2]

[1]*Mining Engineering Department, Pontificia Universidad Catolica de Chile, Chile*
[2]*Environmental Engineering Deparment, Pontificia Universidad Católica de Chile, Chile*

Abstract. Bacterial activity can be related to the presence of Fe^{+3} in the solution, which drives copper oxidation during the secondary copper low grade sulfide ore leaching process. The present work evaluates improvements in kinetics of leaching when *ferroxidans* are preconditioned in an inert surface helping to build a biofilm which improves metabolism of the colony. The present work shows evaluation using laboratory columns to perform bioleaching during a 30 days period under three conditions: a base column with no inert surface, a column with loofa available for bacterial growth and conditioning, and finally a column with the loofa surface ground and distributed in the column among the particles.

[a] Corresponding author: avidela@ing.puc.cl

DOI: 10.1051/e3sconf/20160801040

Hard coal preparation in a single stage crushing circuit for gasification purposes

Bartlomiej Ryszka[1,a], and Daniel Saramak[1]

[1]*AGH, University of Science and Technology, Faculty of Mining and Geoengineering, Department of Mineral Processing and Environmental Engineering*

Abstract. The article concerned the problem of coal preparation for gasification purposes. The main aim was to design the coal crushing circuit, which would produce the material within the particle size range between 0.5 and 5 mm. The one-stage crushing circuit based on hammer crusher was proposed. There laboratory tests of coal crushing in two circuits: one-stage circuit based on a hammer crusher and two-stage circuit based on jaw and roller crushers were also performed. The results showed that application of the one-stage circuit appeared to be more favorable both in terms of energy consumption and, if considered, the product particle size.

1 Introduction

In a gasification process, coal treated with oxidizing agent is being transformed from the fossil fuel into the gas fuel, and it takes place in an installation with a dedicated gasification reactor as a main device. The technology is consistent with current world energy policy, which focuses on development of clean coal technologies with low CO_2 emission and low waste production. Petrographic diversity of coal generates a justified need of application of suitable technological processes that would precede the main gasification operation [1]. The primary objective of these preparatory processes is to achieve for coal certain parameters, which are needed in order to meet requirements of a applied gasification reactor type. Proper coal preparation is one of the main factors that determine possibility of conducting stable and consistent oxidation reaction. This is the reason of mineral engineering being first and important part of gasification technology. Depending on the reactor type, parameters required for coal processing are different for each technology and this paper focuses on coal feeded for gasification in circulating fluidized bed (CFB) reactors [2]. In this work the required specifications of coal and possible methods of their achievement through application of mineral processing methods with the lowest possible energy consumption and simplicity of proposed technological solutions will be presented.

Conversion of coal from fossil fuel into a gas, during gasification, takes part due to the influence of the oxidizing agent, which can be a steam, O_2 or CO_2. The main idea consists of injecting coal into a reactor and treating it with hot stream of reaction gas at the same time, allowing coal particles to be lifted by that stream and creating dynamic suspension in reactor chamber. This phenomenon is called fluidization. The coal residence time in the reactor chamber is between 10 to 100 seconds and

[a] Corresponding author: dsaramak@agh.edu.pl

temperature oscillates around 800-1050 °C. Such temperature does not guarantee full coal oxidization, however, conducting the process in higher temperature could cause unwanted coal slagging. In order to achieve full coal conversion into gas, coal has to undergo a regasification process. As a final result, the gas obtained in that process consists of 17% H_2 , 53% CO, 15% CO_2, 10% H_2O, 2% CH_4.

The results of oxidization process are connected with the physical and chemical properties of coal feed, such as the reactivity, particle size, ash content, calorific value coking properties as well as chlorine and sulfur content [3]. The typical values of these parameters are presented in Table 1.

Table 1. Required coal parameters for CFB reactor.

Type of parameter	Value required
Reactivity	High
Particle size, mm	0.5-5
Ash content	Unlimited
Ash melting temperature, °C	> 1100
Moisture content, %	< 5

The other issue is a technological process of coal preparation for gasification purposes. It is necessary to optimize the technological circuit in terms of energy and effectiveness, considering the fact that the gasification process itself generates the costs, which in overall assessment of this technology might indicate that gasification is the process with rather low possible benefits.

2 Methodology and results

The main aim of this paper is to present an approach to design the crushing circuit, which may result in lower processing costs, and as a result it may increase possible benefits of application of the gasification process, especially in terms of energy consumption [4-5]. The suitable circuit of coal processing was designed, which the primary purpose was to achieve the coal particle size from 0.5 to 5 mm. Application of one crushing stage demands an application of a hammer impact crusher – the device capable to achieve a high comminution degree (Sx) up to 30. Application of this crushing device type has also many other significant advantages. It has ability of adjusting to variable feed parameters with replaceable grid (installed grid defines the output mesh size), hammers and impact plates and also possibility of changing the rotor speed. All these features guarantee the device responsiveness and a fixed particles size distribution, even if feed parameters change in time. The hammer crusher, especially the one working in the closed circuit, also provides high cubicity of its products, which results in greater surface area, and thus higher coal reactivity.

To confirm the accuracy of hammer crusher application into the coal preparation circuit for gasification purposes, suitable tests in the laboratory scale were performed. Two types of crushing circuits were tested: first one as a two-stage with the jaw crusher operated in the closed circuit, followed by the roller crusher, and the second one, as a one-stage, with the hammer crusher. The feed material particle size was the same for both circuits and ranged from 5 to 31.5 mm. The jaw crusher in the first circuit has operated with the gap value 5 mm, while the gap in the downstream roller crusher was 4 mm. The jaw crusher product below 5 mm, was the final product. The hammer crusher operated with a grid 10 mm. The test results were analyzed in terms of productivity the product below 5 mm, and from the energetic point of view. The results are presented in Table 2.

The results of laboratory scale investigations show that application of the single crushing stage results in greater yield of particle size fraction 0-5 mm, which was 95%, while the two-stage crushing circuit results in only 65%. Furthermore, the one-stage crushing circuit is more favorable also in terms of energy consumption with minor energy demand per 1 Mg of crushed material.

DOI: 10.1051/e3sconf/20160801040

Table 2. Results of laboratory tests.

Parameter under investigation	One-stage circuit	Two-stage circuit
Energy consumption (E_{sp}), MJ/Mg	11.16	13.86
Yield of product below 5 mm (γ_{-5}), %	95	65

3 Summary

Coal gasification processes are becoming more important, especially from the scope of increasing energy demands together with technological development of the world. On the other hand, it is necessary to find a proper selection of coal preparation circuit, in order to achieve the required effectiveness of its work.

The results of investigations presented in this paper show that application of a hammer crusher gives more favourable results both in production of the required size of product and in terms of energy consumption. These two factors are the most significant in the presented case and the obtained results prove that application of one-stage crushing is justified both from technological and economic point of view. Moreover, it guarantees process stability, simplicity of its course, and also lower investments cost, resulting from smaller number of the required devices.

References

1. D. Krawczykowski, J. Marciniak-Kowalska, A. Surowiak, *Pol. J. Environ. Stud.*, **22**, 6 (2013)
2. T. Chmielniak, M. Sciazko, A. Sobolewsk, *Karbo*, **1** (2012)
3. J. Marciniak-Kowalska, T. Niedoba, A. Surowiak, T. Tumidajski, *Arch. of Min. Sci.*, **59**, 3 (2014)
4. T. Gawenda, *AGH J. of Min. Geoeng.*, **4** (2010)
5. T. Gawenda, *AGH J. of Min. Geoeng.*, **2** (2011)

Improving gravel and crushed aggregates washing technology efficiency through application of high pressure washer

Jaroslaw Lagowski[1], and Daniel Saramak[2,a]

[1] *HAVER TRADING Sp. z o.o., Poland*
[2]*AGH University of Science and Technology, Faculty of Mining and Geoengineering, Department of Mineral Processing and Environmental Engineering, Poland*

Abstract. Removal of fine clay impurities is the most common method of mineral aggregates processing. Various washing devices operate with different effectiveness and capacities. The paper concerns the investigations of aggregate washing operations by using high-pressure washers. A comparative analysis of two types of washers, that is log washer and high pressure washer, shows that the second device operates at much higher effectiveness measured through the percentage of removal of clay impurities. The energy consumption is also more favorable for the high pressure washer.

1 Introduction

In the year 2011 in total over 300 million Mg of gravel and crushed aggregates were produced. In subsequent years the trend has been downward and currently reaches a stable level of about 200 million Mg of rock minerals annually. The market value of mineral aggregates mainly depends on the quality of raw material. Therefore, production is increased by application of various mechanical processing operations, like crushing, screening, liberation and separation of undesirable fractions such as clay and organic impurities [1, 2].

One of the most commonly used method to improve the aggregates quality is removal of fine clay impurities through application of washing operations. In this process various devices such as vibrating screens equipped in water spraying systems (wet screening), drum washers - vibratory and rotary, log washers (trough, single and multi-shafts), turbo-washers and high pressure washers can be utilized [3]. All of these devices operate with different technological efficiencies, and hence it is not always possible to produce aggregates with required qualitative characteristics. The article presents issues related to operations of high pressure aggregates washing and their influence on the efficiency improvement of aggregates refining process, by removing clay components.

[a] Corresponding author: dsaramak@agh.edu.pl

2 Methodology and results

Principle of operation is based on high pressure water stream (jet), which energy is assumed to be greater than clay fractions adhesive forces. Water penetrating the feed layer, breaks adhesive forces and causes liberating the minerals from impurities. In addition, by rotating of washing rotor equipped with nozzles, the feed particles are put into the motion, and therefore the mechanical friction effect is obtained. It assists liberation of raw material and increases the process efficiency [4-5]. Distribution of working layers has been presented in Fig. 1.

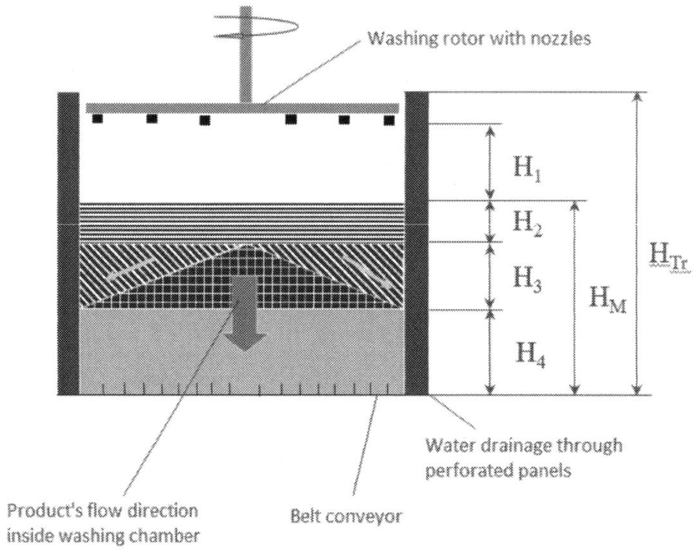

Figure 1. Functional diagram of high pressure washer and layers distribution inside the device, where: H_{Tr} – height of washing chamber, H_M – height of washed-bed material, H_1 – distance between washing nozzles and material, H_2 – washing area, water stream penetration depth, H_3 – dewatering area, H_4 – height of transported material's layer.

Washing tests of aggregates in the semi-industrial washer, for three size fractions that is 0–16, 0–31.5 and 0–40 mm, were carried out. The working capacity of the washer corresponds to about 80 Mg/h of capacity for the industrial-scale device and the working pressure at nozzles was 140 bar. Table 1 presents content of clay fractions for each material before and after washing. The washing results for the log washer device, are shown in Tab. 1.

Table 1. Clay impurities contents obtained for feed and washing products.

Particle size fraction	Content of clay in feed, %	Content of clay in product, %
0 – 16	24.6	0.5
0 – 31.5	14.5	0.4
0 – 40	12	0.4
Log washer 0 – 31.5	19	4.1

Comparing the results obtained by using both technologies, it can be seen that the aggregates washed in high pressure washer contained ten times less dust at the water consumption level between 1–1.2 m^3/Mg/h. Energy consumption for the high pressure washer was 1.08 MJ/Mg. At the same time, log washer circuit consumed 1.7 m^3/Mg/h of water and 1.80 MJ/Mg of energy. The log washer circuit demands an application of additional screening operation in order to separate the size fractions 0-2 mm and above 80 mm. The high pressure washer capacity was also 30% higher, comparing to the log washer circuit.

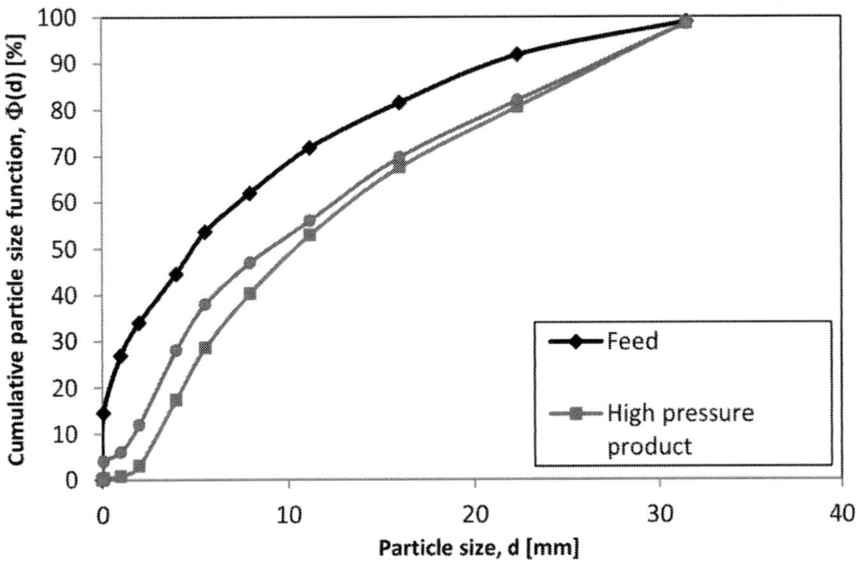

Figure 2. Particle size distribution of feed and washing products in particle size fraction 0 – 31.5 mm.

3 Conclusions

The results of the paper show that application of the high pressure washing technology into aggregate processing circuits significantly influenced effectiveness and capacity of the washing technology for natural and broken aggregates. The detailed analysis of results show that regardless of particle size composition of the feed, the effectiveness of washing process was high both for narrow and wide particle size ranges. However, it should be noted that the type of impurities significantly affected the course and effectiveness of the washing process. Clays, which mainly consist of the dusty fractions, due to their total surface area were more difficult to dissolution and, as a result, required a higher water consumption during the process, comparing to clays containing coarser particles. Nevertheless, the high-pressure washer was capable to remove these impurities in a very efficient way.

Despite the above, the high pressure washer, on average, generated lower processing costs through lower water and energy consumption. From technical point of view, the high pressure washer installation required much less building area and lower number of associated facilities. Combining the high effectiveness of operation of high pressure washers with a potentials of control of their work [3] it should be highlighted that high pressure washing technology was of a major significance in improvement of crushed aggregates quality. Further investigations, especially in terms of more effective control and modelling of work of washing devices, may be helpful within this issue.

References

1. A. Battaglia, *Mining and Metallurgy Publ. House*, Katowice, (1963)
2. R. Kobialka, Z Naziemiec, *Mining Institute Works*, Wroclaw Unive. of Tech., 91, **28** (2000)
3. J. Lagowski, T. Lubieniecki, D Saramak, *Inzynieria Mineralna*, **1**, 14 (2013)
4. Z. Naziemiec, T Gawenda, *Mining Institute Works*, 105, **41** (2005)
5. Z. Naziemiec, *Institute of Glass, Ceramics and Building Materials*, **5(3)** (2010)

Flotation of zinc and lead oxide minerals from Olkusz region calamine ores

Krystian Cichy[1,a], Cezary Regula[2] and Grazyna Pajor[2]

[1]*Institute of Non-Ferrous Metals, ul. Sowinskiego 5, 44-100 Gliwice, Poland*
[2]*ZGH "Boleslaw" S.A., Poland*

Abstract. The paper presents chemical and mineralogical characteristics of calamine ore from the Pomorzany mine. A flowsheet for recovery of sulphide minerals of zinc and lead in the form of the Zn-Pb bulk concentrate was presented. In the following part, preparation of the feed for flotation of Zn-Pb oxide minerals and optimal conditions for separation from it iron sulphide minerals, represented by marcasite, were determined. In the final section the results of flotation of Zn-Pb oxide minerals with anionic collector AM2 belonging to the hydroxyamide group of collectors and a cationic collector in the form of a coconut amine, being a mixture of primary aliphatic amines, were presented. Basing on the obtained results, a technological flowsheet for the recovery of Zn-Pb sulphide and oxide minerals from the calamine ore of the Pomorzany mine was presented.

1 Introduction

Lead and zinc was produced in Poland from calamine ores by using the Waelz process. The extracted ore was subjected to roasting in the presence of coke as a reducing agent. As a result the so-called raw zinc oxide, which was the raw material for production of metallic zinc and lead was achieved. The high costs of the process were the reason for stopping, in the early 90-ies of the twentieth century, production of lead and zinc from the calamine ore using this method.

The calamine ores are the final results of transformation of sulphide deposits of Zn-Pb under the influence of surface water enriched in oxygen. The zinc and lead minerals are usually in the form of carbonate minerals such as smithsonite and monheimite (zinc) and cerussite (lead). Their formation is a result of reaction of acidic sulphate solutions arising from oxidation of metal sulphides contained in the ore with the surrounding carbonate rocks such as calcite and dolomite. As a result formation of secondary carbonate minerals of zinc and lead takes place.

Due to the similarity of surface properties of carbonate minerals of zinc and lead to the surrounding waste rock, which is mostly dolomite, their flotation has not yet been sufficiently resolved. So far, the studies have shown that the only practical flotation method of oxide ore of Zn-Pb beneficiation is the so-called Rey method involving sulphidizing the surfaces of minerals by sodium sulphide and their flotation with primary aliphatic amines having the hydrocarbon chain length of C_{12}-C_{16}. [1, 2]

---

[a] Corresponding author: krystianc@imn.gliwice.pl

Due to the fact that the surface of carbonate minerals has in aqueous medium a negative charge and amine is a cationic collector, it is necessary to separate slimes from the feed prior to flotation because the cationic collector can be adsorbed by the electrostatic forces and it will finally result in transfer the slimes to the froth product.

The necessity of the initial desliming of the feed significantly reduces in practice the range of application of this method due to high losses of metals in the removed slime fraction. Another disadvantage of this method is a good collecting ability of aliphatic amines with respect to the sulphide minerals of iron, which have, as natural sulphides, under these conditions, much better floatability than sulphidized carbonate minerals of Zn-Pb.

The necessity of the initial desliming of the feed and flotation of iron sulphide minerals considerably reduces the use of this method due to the high losses of metals in the removed products. Its use can be economically justified only in case of enriching the very rich Zn-Pb oxide ores.

The aim of this study is to investigate whether flotation recovery of Zn-Pb oxide minerals is possible with other than amines collectors, which do not require costly and troublesome sulphidation.

2 Characteristics of studied ore

The subject of research was the calamine ore from the Pomorzany mine which is currently the only active mine in Poland that extracts the Zn-Pb ore. The chemical analysis of studied ore is shown in Table 1.

Table 1. Chemical analysis of studied calamine ore from the Pomorzany mine.

Content, %										
Zn	**Zn(O)**	**Zn(S)**	**Pb**	**Pb(O)**	**Pb(S)**	**Fe**	**Fe(O)**	**Fe(S)**	**CaO**	**MgO**
5.58	3.61	1.97	0.95	0.73	0.23	43.70	29.93	13.77	2.44	0.93

The chemical analysis indicates that the tested ore is almost an iron ore with significant additions of zinc and lead, in which there is practically no waste rock. A basic mineralogical composition of the ore is as follows:

Mineral	Distribution, %
galena	0.3
cerussite	1.0
sphalerite	2.9
marcasite	29.6
monheimite	13.3
goethite	46.8
dolomite	6.2
Sum	100.0%

The main mineral in the studied ore is goethite constituting nearly 47% of the composition, while the second the most abundant component is marcasite, which is sulphide mineral of iron in the amount of approx. 29.6%. In total iron minerals constitute more than 76% of the ore. The main zinc minerals are monheimite and sphalerite occurring in amounts of approx. 13.3% and 2.9%, while lead minerals are cerussite and galena in the amounts of approx. 1.0% and 0.3%. Such a composition of the ore shows that the basic difficulty in the process of enrichment is the presence of iron oxide minerals represented by goethite forming the smallest particles of slime.

3 Determination of the optimal conditions for the recovery of zinc and lead sulphide minerals

Flotation concentrate containing Zn-Pb oxide minerals can be directed for processing by the Waelz process provided that firstly the Zn-Pb sulphide minerals are removed, because their presence in the process is undesirable due to the sulphur content. Accordingly, the first step is to determine the optimal conditions for flotation of galena and blende.

Due to a significant amount of goethite, which is a potential source of slime fraction generated in grinding, the ore grinding is limited to the standard particle size <0.2 mm carried out in a laboratory rod mill. The analysis of particle size distribution and chemical content of feed for flotation is given in Table 2.

Table 2. Analysis of particle size distribution of feed for flotation.

Grain class mm	Yield %	Content, %						
		Zn	Zn(O)	%Zn(O)/Zn	Pb	Pb(O)	% Pb(O)/Pb	Fe
> 0.2	4.32	5.30	3.31	62.45	0.91	0.73	80.22	43.70
0.2 – 0.1	10.28	6.80	4.03	59.26	0.76	0.62	81.60	39.40
0.1– 0.071	10.32	6.20	3.84	61.94	0.95	0.70	74.01	46.17
0.071-0.035	14.00	5.47	3.41	62.30	1.00	0.76	76.14	43.15
< 0.035	61.08	5.34	3.62	67.79	0.96	0.69	71.70	42.25
Sum	100.00	5.60	3.64	65.09	0.94	0.70	73.77	42.55

The presented particle size distribution shows that in the feed for flotation there is a significant amount of ultrafine grains <0.035 mm of approx. 60%, which is the result of overgrinding of goethite. A small amount of galena in the studied ore was the reason for separation of the Zn-Pb sulphide minerals in the collective enrichment flowsheet. The study of bulk flotation process was conducted in the laboratory pneumatic-mechanical flotation machine of IZ type with a chamber volume of 2.5 l. In the first stage, the tests of the bulk flotation of galena and blende were carried out with the feed without desliming, and in the next stage with the deslimed feed. Desliming was performed in a laboratory hydrocyclone ϕ 50 mm with particle cutsize of approx. 10 μm. The flotation conditions were as follows: density 400 g/l, $CuSO_4$ 350 g/t, NaX iso-amyl 25 g/t, Mileflot X-23 15 g/t, 20 g/t and 25 g/t, Corflot 5 g/t, flotation time 15 min. Table 3 presents the results of the preformed tests of Zn-Pb bulk flotation.

Table 3. Results of bulk flotation of Zn-Pb sulphide minerals from the feed not deslimed and after desliming.

Flotation conditions	Product	Yield %	Content, %			Recovery, %		
			Zn	Pb	Fe	Zn	Pb	Fe
Not deslimed feed	Feed	100.00	5.52	0.94	43.36	100.00	100.00	100.00
	C ZnS + PbS	13.26	8.42	0.90	37.40	20.23	12.70	11.44
	T ZnS + PbS	86.74	5.08	0.95	44.27	79.77	87.30	88.56
Deslimed feed	Feed	100.00	5.54	0.96	43.92	100.00	100.00	100.00
	Slime	15.76	5.30	1.03	42.00	15.08	16.91	15.07
	HC underflow	84.24	5.58	0.95	44.28	84.92	83.09	84.93
	C ZnS + PbS	7.52	23.37	1.62	28.95	31.73	12.69	4.96
	T ZnS + PbS	76.72	3.84	0.88	45.78	53.19	70.40	79.97

The introduction of desliming operation resulted in a significant improvement of the enrichment indices, both in terms of improving concentrate quality and metals recovery.For determination of the final indices of recovery of Zn-Pb sulphide minerals corresponding to the industrial conditions, the flotation tests were carried out using the flowsheet with the cleaning flotations to obtain Zn-Pb concentrate as in commercial operation. This flowsheet is shown in Fig. 1, while the balance of the enrichment process in Table 4.

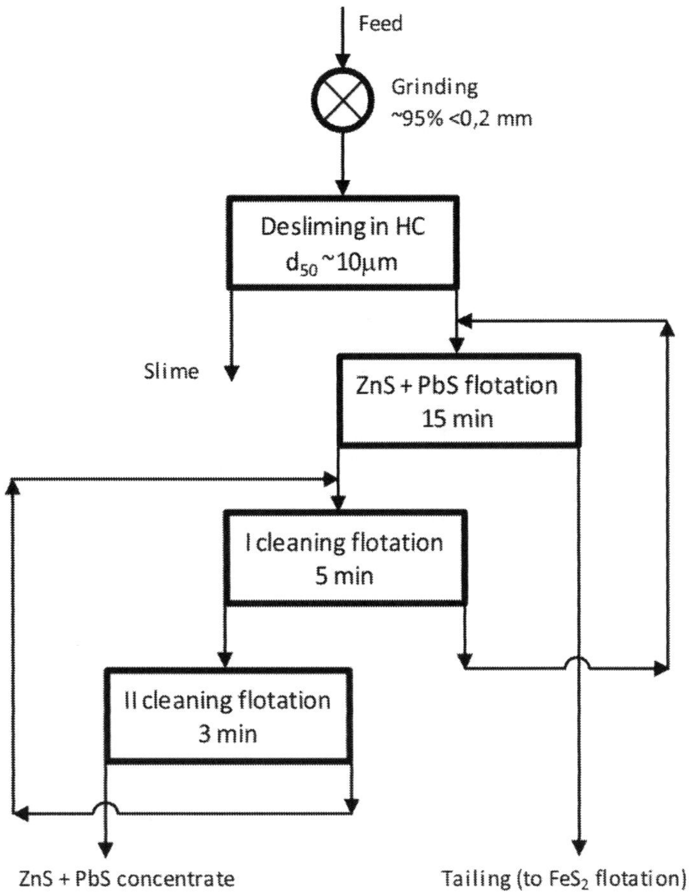

Figure1. Flowsheet for Zn-Pb sulphide minerals recovery from calamine ore of the Pomorzany mine.

Table 4. Balance of the recovery process of Zn-Pb sulphide minerals from calamine ore of the Pomorzany mine.

Product	Yield	Content, %			Recovery, %		
	%	Zn	Pb	Fe	Zn	Pb	Fe
Feed	100.00	5.61	0.97	43.51	100.00	100.00	100.00
Slime	15.53	5.27	1.06	42.38	14.59	16.97	15.13
HC underflow	84.47	5.67	0.95	43.72	85.41	83.03	84.87
C ZnS + PbS	2.59	53.24	3.54	5.32	24.62	9.47	0.32
T ZnS + PbS	81.88	4.17	0.87	44.93	60.79	73.56	84.56
M1 ZnS + PbS	6.99	12.66	1.58	30.18	15.78	11.37	4.85
M2 ZnS + PbS	2.08	25.40	2.66	18.81	9.40	5.70	0.90

4 Determination of the optimal conditions for flotation of iron sulphide minerals

The next stage of preparation of ore for flotation of Zn-Pb oxide minerals was to determine the conditions of flotation of iron sulphide minerals.

The mineralogical analysis shows that about 30% of the ore is iron sulphide represented by marcasite. In practice, it is known that almost all of the collecting reagents for flotation of Zn-Pb oxide minerals are also good collectors of sulphide minerals including iron minerals. In the case of avoiding flotation of marcasite prior to flotation of oxides, the oxide concentrate would contain significant quantities of marcasite, reducing its quality both in terms of zinc content and increased sulphur content. It is therefore apparent that there is a need for this operation, before flotation of oxides.

The material used in this study was the tailing of bulk flotation of galena and blende. The examined parameters of marcasite flotation was the dose of sodium ethyl xanthate and the addition of sodium sulphide to refresh its surface.

Flotation conditions were as follows: density approx. 350 g/l, NaX ethyl 75 g/t, 100 g/t, 100 g/t + Na_2S 500 g/t, Corflot 5 g/t, time of flotation 15 min. The results are shown in Table 5.

Table 5. Results of FeS_2 flotation from Zn-Pb bulk flotation tailings from calamine ore of the Pomorzany mine.

Flotation conditions	Product	Yield %	Content, %			Recovery, %		
			Zn	Pb	Fe	Zn	Pb	Fe
NaX ethyl - 75 g/t	Feed	100.00	4.14	0.87	44.91	100.00	100.00	100.00
	C FeS_2	19.78	0.65	0.77	41.60	3.11	17.50	18.32
	T FeS_2	80.22	5.00	0.89	45.73	96.89	82.50	81.68
NaX ethyl - 100 g/t	Feed	100.00	4.16	0.83	44.72	100.00	100.00	100.00
	C FeS_2	23.87	0.62	0.75	40.25	3.56	21.57	21.48
	T FeS_2	76.13	5.27	0.86	46.12	96.44	78.43	78.52
Na_2S - 500 g/t NaX ethyl - 100 g/t	Feed	100.00	4.18	0.85	44.80	100.00	100.00	100.00
	C FeS_2	26.44	0.71	0.80	44.28	4.49	24.88	26.13
	T FeS_2	73.56	5.43	0.87	44.99	95.51	75.12	73.87

The results indicate that, for the maximum removal of iron sulphide minerals before flotation of oxide minerals of zinc and lead, it is necessary to add to the flotation suspension sodium sulphide in the amount of approx. 500 g/t. Under these conditions, the maximum degree of marcasite removal with the iron content of the froth product close to the theoretical Fe content of the mineral is noted. Tailing from the flotation of marcasite constituted the feed for the flotation of oxide minerals of zinc and lead.

5 Study of collecting properties of reagents from the hydroxyamide group and primary aliphatic amines for Zn-Pb oxide minerals

Literature survey shows that for flotation of oxide minerals of zinc and lead, besides the aliphatic primary amines, hydroxyamides may also be used. In terms of chemistry, this reagent is a potassium salt of hydroxyamide, while the physical form is paste which is highly soluble in water. The tested reagent was AM2 hydroxyamide (AXIS HOUSE, Republic of South Africa). According to the manufacturer, this reagent shows favourable results for flotation of oxide minerals of non-ferrous metals, especially copper.

For the purpose of flotation a 2% aqueous solution was used. First, flotation of oxide minerals of zinc and lead from calamine ore (without sulphide zinc, lead and iron minerals) was carried out without feed sulphidation. The parameter of the flotation process was the dose of this reagent. The flotation conditions were as follows: density approx. 300 g/l, AM2 100 g/t, 150 g/t and 200 g/t, Corflot 5 g/t, flotation time 12 min. The flotation test results with the addition of this reagent are shown in Table 6.

Table 6. Flotation results of Zn-Pb oxide minerals from calamine ore of the Pomorzany mine using AM2 hydroxyamide without sulphidation process.

Flotation conditions	Product	Yield	Content, %			Recovery, %		
		%	Zn	Pb	Fe	Zn	Pb	Fe
AM2 100 g/t	Feed ($T\ FeS_2$)	100.00	5.32	0.88	44.83	100.00	100.00	100.00
	C ZnO + PbO	11.26	7.32	0.93	37.62	15.49	11.90	9.45
	T ZnO + PbO	88.74	5.07	0.87	45.74	84.51	88.10	90.55
AM2 150 g/t	Feed ($T\ FeS_2$)	100.00	5.27	0.85	44.67	100.00	100.00	100.00
	C ZnO + PbO	14.32	7.64	0.95	38.94	20.76	16.00	12.48
	T ZnO + PbO	85.68	4.87	0.83	45.63	79.24	84.00	87.52
AM2 200 g/t	Feed ($T\ FeS_2$)	100.00	5.24	0.87	44.32	100.00	100.00	100.00
	C ZnO + PbO	17.67	8.16	0.91	39.73	27.52	18.48	15.84
	T ZnO + PbO	82.33	4.61	0.86	45.31	72.48	81.52	84.16

Table 7. The results of the flotation of Zn-Pb oxide minerals from calamine ore of the Pomorzany mine with AM2 hydroxyamide, with the addition of sodium sulphide in an amount of 8.0 kg/t.

Flotation conditions	Product	Yield	Content, %			Recovery, %		
		%	Zn	Pb	Fe	Zn	Pb	Fe
AM2 100 g/t	Feed ($T\ FeS_2$)	100.00	5.30	0.85	44.37	100.00	100.00	100.00
	C ZnO + PbO	15.37	9.43	1.03	35.12	27.35	18.62	12.17
	T ZnO + PbO	84.63	4.55	0.82	46.05	72.65	81.38	87.83
AM2 150 g/t	Feed ($T\ FeS_2$)	100.00	5.34	0.89	44.53	100.00	100.00	100.00
	C ZnO + PbO	17.46	9.15	1.00	36.12	29.92	19.62	14.16
	T ZnO + PbO	82.54	4.53	0.87	46.31	70.08	80.38	85.84
AM2 200 g/t	Feed ($T\ FeS_2$)	100.00	5.24	0.87	44.32	100.00	100.00	100.00
	C ZnO + PbO	21.12	8.84	0.97	37.48	35.63	23.55	17.86
	T ZnO + PbO	78.88	4.28	0.84	46.15	64.37	76.45	82.14

The structural formula of AM2 reagent is given below.

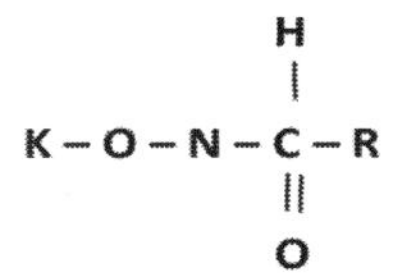

$R = C_6$

As appears from the presented data, this reagent under conditions without sulphiding operation practically does not exhibit any collecting properties with respect to the oxide minerals of lead and zinc. The Zn content in the froth product is slightly higher than that of the feed and the recovery increase with increasing dose of the reagent. It is due to the increased entrainment of fine particles to the froth product due to increased foaming. In the second step, the tests were conducted after sulphidation of Zn-Pb oxide minerals in the feed using sodium sulphide in an amount of 8.0 kg/t. Other flotation conditions remained unchanged. The results are summarized in Table 7.

The obtained results clearly show that despite sulphidation of flotation feed, this reagent does not show good collecting properties for the Pb-Zn oxide minerals. An increase in the Zn content is noticeable in the froth product (approx. 9% Zn) and increased operating recovery of zinc to about 35%. However, from a practical point of view, the results are unsatisfactory.

The second examined reagent was coconut amine of the Clariant company. In the chemical sense it is a mixture of amines of various lengths of hydrocarbon radical from C_8 to C_{18}. Its fractional composition is given in Table 8.

Table 8. Fractional composition of coconut amine from Clariant company.

Trade name	Chemical name	State of matter (20° C)	Minimal concentration of active substance, %	Fractional composition					
				C_8	C_{10}	C_{12}	C_{14}	C_{16}	C_{18}
Genamin CC 100	coconut fatty amine	liquid	96	6	6	52	18	9	8

The experimental materials, were as previously, the tailings from marcasite flotation. Two process parameters were tested, namely the dose of sodium sulphide and dose of coconut amine. Flotation results, for varying amount of sodium sulphide, are shown in Table 9 and for varying amounts of coconut amine in Table 10.

Table 9. Results of the flotation of Zn-Pb oxide minerals from calamine ore of the Pomorzany mine with varying amounts of sodium sulphide.

Flotation conditions	Product	Yield %	Content, %			Recovery, %		
			Zn	Pb	Fe	Zn	Pb	Fe
Na_2S 4.0 kg/t	Feed (T FeS_2)	100.00	5.12	0.91	45.65	100.00	100.00	100.00
C. amine	C ZnO + PbO	11.66	9.63	1.13	37.11	21.94	14.48	9.48
200 g/t	T ZnO + PbO	88.34	4.53	0.88	46.78	78.16	85.42	90.52
Na_2S 6.0 kg/t	Feed (T FeS_2)	100.00	5.12	0.91	45.65	100.00	100.00	100.00
C. amine	C ZnO + PbO	23.20	14.2	0.86	31.9	64.36	21.93	16.21
200 g/t	T ZnO + PbO	76.80	2.38	0.93	49.8	35.70	78.48	83.78
Na_2S 8.0 kg/t	Feed (T FeS_2)	100.00	5.12	0.91	45.65	100.00	100.00	100.00
C. amine	C ZnO + PbO	22.40	16.3	1.02	28.95	71.35	25.07	14.21
200 g/t	T ZnO + PbO	77.60	1.89	0.88	50.47	28.65	74.93	85.79

The tests carried out indicate that the optimum dose of sodium sulphide for the studied ore is approx. 8.0 kg/t, and the amine use at the level of approx. 200-220 g/t. Increasing the dose does not affect significantly the improvement of zinc recovery, while noticeable is the reduction of its content in the froth product. Significantly better results of zinc and lead oxide minerals flotation, obtained by the use of coconut amine, gave rise to a further tests performed in order to obtain a higher zinc content in the froth product by introducing a cleaning flotation of raw Zn (O) + Pb (O) concentrate step.

As a result of introduction of cleaning flotation, an oxide concentrate containing approx. 19.8% Zn, and approx. 1.2% Pb with a recovery of zinc of about 38% and lead approx. 13.3% was finally obtained. The full balance of the enrichment process of studied calamine ore is shown in Table 11, while the flowsheet in Fig. 2.

Table 10. Results of the flotation of Zn-Pb oxide minerals from calamine ore of the Pomorzany mine with varying amounts of coconut amine.

Flotation conditions	Product	Yield %	Content, %			Recovery, %		
			Zn	Pb	Fe	Zn	Pb	Fe
Na_2S 8.0 kg/t	Feed ($T FeS_2$)	100.00	5.23	0.95	46.34	100.00	100.00	100.00
C. amine 150 g/t	C ZnO + PbO	19.18	15.3	0.91	29.37	56.12	18.37	12.16
	T ZnO + PbO	80.82	2.84	0.96	50.37	43.88	81.63	87.84
Na_2S 8.0 kg/t	Feed ($T FeS_2$)	100.00	5.10	0.91	45.52	100.00	100.00	100.00
C. amine 200 g/t	C ZnO + PbO	22.34	16.3	1.02	28.95	71.35	25.07	14.21
	T ZnO + PbO	77.38	1.89	0.88	50.47	28.65	74.93	85.79
Na_2S 8.0 kg/t	Feed ($T FeS_2$)	100.00	5.23	0.95	46.33	100.00	100.00	100.00
C. amine 260 g/t	C ZnO + PbO	28.83	13.71	1.1	28.72	75.52	33.36	17.87
	T ZnO + PbO	71.17	1.8	0.89	53.47	24.48	66.64	82.13

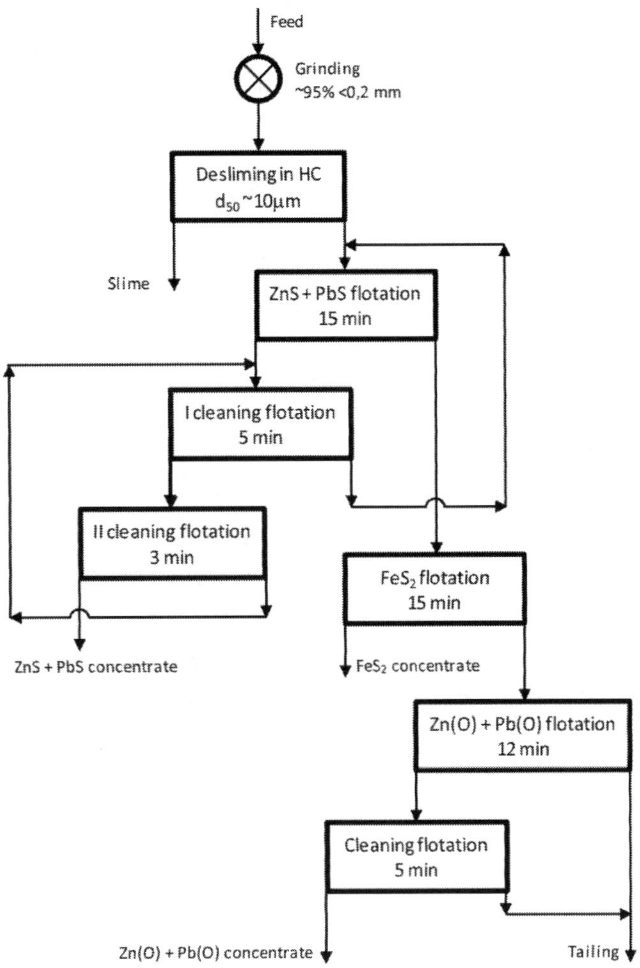

Figure 2. Flowsheet for recovery of Zn-Pb sulphide and oxide minerals from calamine ore of the Pomorzany mine.

Table 11. Balance of the recovery process of Zn-Pb sulphide and oxide minerals from the calamine ore of the Pomorzany mine.

Product	Feed %	Content, %			Recovery, %		
		Zn	Pb	Fe	Zn	Pb	Fe
Feed	100.00	5.61	0.97	43.51	100.00	100.00	100.00
Slime	15.53	5.27	1.06	42.38	14.59	16.97	15.13
HC underflow	84.47	5.67	0.95	43.72	85.41	83.03	84.87
C ZnS + PbS	2.59	53.24	3.54	5.32	24.62	9.47	0.32
C FeS_2	21.64	0.71	0.80	44.28	2.74	17.85	22.02
C Zn(O) + Pb(O)	10.75	19.82	1.20	26.14	37.98	13.30	6.46
T Zn(O) + Pb(O)	49.49	2.28	0.83	49.30	20.07	42.41	56.07

6 Summary

The studied calamine ore from the Pomorzany mine was characterized by an unusual mineral composition. With an average zinc content approx. 5.6% Zn, including 3.6% Zn (O), and lead approx. 0.95%, with the approx. 0.73% Pb(O), the iron content was approx. 43.7% Fe of which approx. 13.8% Fe was present as an iron sulphide and approx. 29.9% Fe as an iron oxide. The mineralogical analysis showed that the main component of the ore was goethite, which part was of the approx. 46.8%, and marcasite in an amount of approx. 29.6%. Oxide minerals of zinc in an amount of approx. 13.3% mainly occured as monheimite being ferruginous variety of smithsonite.

The high share of goethite in ore makes it necessary to introduce, prior to flotation, the desliming process in order to remove from the flotation feed the finest colloidal particles composed mainly of iron hydroxides passing to the slurry in the grinding process. The introduction of preliminary desliming process allows for separation from the ore the sulphide Zn-Pb bulk concentrate containing approx. 53.2% Zn, and approx. 3.5% Pb with a recovery of zinc of approx. 24.6% and lead of approx. 9.5%.

The high share of marcasite in the ore requires its separation before flotation of Zn-Pb oxide minerals. The most preferred conditions of this process are after initial sulphidizing its surface with sodium sulphide in an amount of approx. 500 g/t, and then flotation using sodium ethyl xanthate in an amount of approx. 100 g/t for 15 min. Under these conditions, approx. 22% of the ore in the form of iron sulphide concentrate containing approx. 44.6% Fe is extracted to the froth product.

Among the studied collecting reagents for flotation of Zn-Pb oxide minerals, coconut amine, which is a mixture of primary aliphatic amines with a carbon number in radicals of C_8-$C_{18.}$, shows the best flotation properties. The second tested reagent from the hydroxyamide group does not show any significant collecting properties for the Zn-Pb oxide minerals, both under conditions without and with addition of sodium sulphide.

The most beneficial flotation results of Zn-Pb oxide minerals are obtained under condition of the addition of sodium sulphide to the suspension in the amount of approx. 8.0 kg/t and coconut amine in the amount of approx. 200-220 g/t. In the studied flowsheet, an Zn-Pb oxide concentrate containing approx. 19.8% Zn, and approx. 1.2% of Pb with a recovery of zinc of approx. 38% and lead of approx. 13.3% was achieved. A relatively low content of zinc in concentrate is due to its occurrence in monheimite, which contains, next to Zn atoms also Fe atoms in its crystal lattice.

References

1. K. Cichy, R. Luczak, A. Manka, *IMN report No. 6997* (2011)
2. K. Cichy, R. Luczak, A. Manka, *IMN report No .7029* (2012)

Solvent extraction of scandium from lateritic nickel- cobalt ores using different organic reagents

Ece Ferizoğlu[1,2,a], Şerif Kaya[1], Yavuz A. Topkaya[2]

[1]*Metallurgical and Materials Engineering, METU,06800 Çankaya Ankara, Turkey*
[2]*Meta Nikel Kobalt A.Ş. (Zorlu* Group), 06520 Balgat Ankara, *Turkey*

Abstract. Scandium is the most important and strategic metal that can be recovered as a by-product from lateritic nickel-cobalt ores. In this research, different extractants were investigated in order to extract scandium from a sulfate medium by a using a solvent extraction method. Generally, the organic extractants are classified as acidic, neutral and basic organophosphorus compounds. However, in solvent extraction of scandium, the acidic and neutral organophosphorus compounds are preferred due to their higher extraction efficiencies. Thus, the aim of the present study was to compare the scandium extraction efficiencies of some acidic and neutral organic reagents. For this reason, Ionquest 290 (Bis(2,4,4-trimethylpenthyl) phosphonic acid), DEHPA (Di(2-ethylhexyl) phosphoric acid), Cyanex 272 ((Bis(2,4,4-trimethylpentyl) phosphinic acid) which are acidic organophosphorus compounds, and Cyanex 923 (Trialkylphosphine oxide), which is a neutral organophosphorus compound, were used. The extraction capacities of these organics were studied with respect to the extractant concentration at same pH and phase ratio. As a result of the study, DEHPA was found to have higher scandium extraction efficiency with lower iron extraction at pH=0.55 at a phase ratio of 10:1=A:O.

[a] Corresponding author: eceferi@gmail.com

Leaching of pyrite in cupric chloride solution

Heini Elomaa[1,a], Mari Lundström[1]

[1]*Aalto University, School of Chemical Technology, Department of Materials Science and Engineering, P.O. Box 16200, FI-00076 AALTO, Finland*

Abstract: Pyrite (FeS_2) is the most abundant sulphide mineral in the world and may contain significant amounts of gold locked in its structure. Traditionally, pyrite has to be oxidized in order to break down the mineral matrix for better gold liberation by e.g. roasting, pressure leaching and bioleaching [1]. Due to environmental and health concerns, as well as public opinion, alternative methods are developed to replace gold cyanide leaching, which has been the preferred gold leaching process. Gold is known to form a stable complex with chloride as $[AuCl_4]^-$ or $[AuCl_2]^-$. In chloride leaching of pyrite in the presence of $[Cu^{2+}]$ which acts as a catalyst/oxidant, the pyrite structure can be simultaneously broken down and the gold dissolves into a solution from which it can be later recovered [2, 3]. In this study, dissolution of the pyrite concentrate in cupric chloride solutions was investigated. The effect of cupric ion concentration (16, 32 g/L), temperature, (60, 75, 90 °C), chloride ion concentration (50, 150 g/L), pH (1, 1.5) and mixed chloride-bromide ion concentration (75+75 g/L) on pyrite dissolution were studied during 6 hour batch leaching experiments. Redox potentials between 0.495-0.511 V vs. Ag/AgCl resulted in pyrite dissolution of 31-34%, whereas redox potentials between 0.523-0.573 V vs. Ag/AgCl resulted in 45-68% pyrite dissolution.

References

1. O. Marsden and C. I. House, *The Chemistry of gold extraction*, 2nd edit. (Colorado, 2006)
2. V. Miettinen, M. Haapalainen, R. Ahtiainen, J. Karonen, in *Proceedings of ALTA Metallurgical Seirvices*, 187 (2013)
3. M. Lundström, R. Ahtiainen, T. Haakana and J. O'Callaghan, in *Proceedings of ALTA Metallurgical Services*, 16 (2014)

[a] Corresponding author: heini.elomaa@aalto.fi

 DOI: 10.1051/e3sconf/20160801045

The effect of feed overload at secondary coal separation with product recirculation

Sergiusz Boron[1,a], Adam Heyduk[1], Joachim Pielot[1]

[1] *Silesian University of Technology, Department of Electrical Engineering and Control in Mining, 44-100 Gliwice, ul. Akademicka 2, Poland*

Abstract. For coal separation with product recirculation, there is a risk of an excessive increase in the amount of feed. Protection against this effect can be achieved by an adequate control of the recirculated material amount. The paper presents a control analysis and economic results of separation in either one or two jigs. The advantage of a slight jig performance oversizing and application of video monitoring in their control systems were justified.

1 Secondary separation with product recirculation

Secondary separation means repeated separation of intermediate product (aggregates of useful mineral and gangue). This product is usually crushed in order to separate the grains of minerals and gangue [2], but sometimes crushing may not be advisable [13]. There can be also applied recirculation (Fig. 1) of the part of the concentrate (system *1 jig°rC*) or of an intermediate product when using three-product jigs (system *2 jig rCW*). Product recirculation is applied to metal ore flotation [16]. Also for the coal flotation it is possible to obtain better beneficiation results by appropriate configuration and recirculation of the flotation products [6]. However, there are very few studies on coal gravity separation with product recirculation [9, 12, 15, 17, 18]. Improved washing efficiency can be expressed by less erroneous grains in the concentrate and the waste [10].

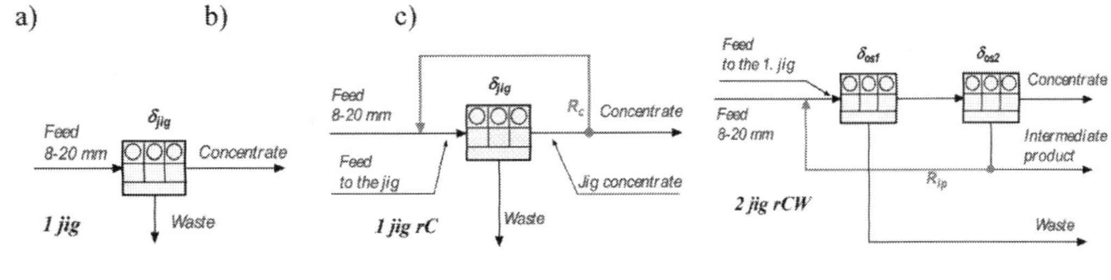

Figure 1. Diagrams of considered technological systems with jigs. Symbol *rCW* means recirculation of intermediate fractions from the first jig and (at the same time) sinking fractions from the second jig.

All of the following results of simulation predictions have been carried out with the use of array models of jig separation curves [5]. Secondary enrichment is particularly important when coal is

[a] Corresponding author: Sergiusz.Boron@polsl.pl

resistant to washing. So, the analysis has been limited only to the hardly washable feed, which density-quality characteristics are given in Table 1 [12].

Table 1. Density-quality characteristics of the raw feed stream (8-20 mm).

No. of dens. fraction	Fraction density g/cm³	Yield of fraction %	Ash content %	Total sulphur content %	Calorific value kJ/kg
1	<1.30	12.15	4.67	0.84	30 680
2	1.30–1.35	17.96	7.40	0.86	29 630
3	1.35–1.40	10.95	10.99	0.97	28 300
4	1.40–1.50	8.47	17.92	1.10	25 750
5	1.50–1.60	7.43	26.61	1.24	22 550
6	1.60–1.70	7.02	35.81	1.25	19 160
7	1.70–1.80	3.95	43.81	1.13	16 220
8	1.80–1.90	4.04	51.03	1.12	13 560
9	1.90–2.00	2.57	57.08	1.39	11 330
10	>2.00	25.45	75.84	2.75	4 420
	Total	100.00	33.67	1.46	19 960

In the *1 jig rC* system (Fig. 1b) ratio R_c indicates how much of the concentrate stream from the jig goes back to repeated separation. Similarly, a recirculation factor R_{ip} in the *2 jig rCW* system (Fig. 1c) determines what part of the of intermediate product stream returns again to the input.

2 The effect of feed amount overload

Repeated separation of the part of the concentrate from the jig (Fig. 1b) means that with an increase in recirculation ratio R_c there increases significantly the relative amount of feed to the jig (Fig. 2a). The largest increase takes place at the highest separation density in a jig, because in this case there is a greater participation of the concentrate separated again. When changing raw coal beneficiation or density separation there can occur an uncontrolled increase in the feed amount. A better solution to this problem is to recirculate the intermediate product stream because in this case recirculate mainly erroneous particles (Fig. 2b, results obtained for a constant separation density in the first jig – 1.950 g/cm^3 and three separation densities in the second jig: 1.450, 1.700 or 1.950 g/cm^3). Also in this case the feed amount may be dangerously increased, but it is possible to tune the control in order to avoid this effect. Feed overload can be avoided when the separation densities in both jigs (or more generally in both separators) do not differ much. For example, if in the first jig separation density is equal to 1.700 g/cm^3, and in a second jeig is equal to 1.950 g/cm^3, then the increase in the feed amount, with the full recirculation of intermediate product (R_{ip}=100%) is less than 39% and at equal separation densities (1.950 g/cm^3), this increase is 3%.

The dangerous effect of an excessive increase in the feed amount can be controlled using a system of visual analysis of particle size distribution [3], so that the reaction to the excessive increase in the feed amount can done be rapid changes in the recirculation ratio or by the relevant changes in separation density values.

DOI: 10.1051/e3sconf/20160801045

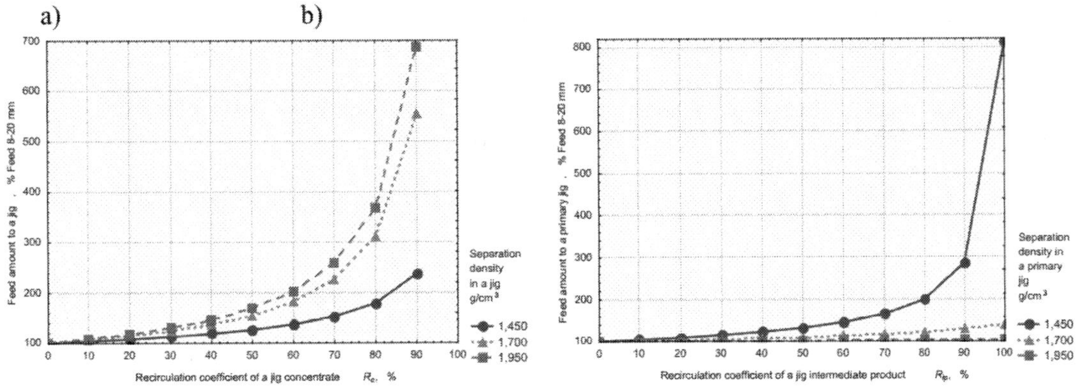

Figure 2. Relative feed flow (quantity) to a jig (in system *1 jig rC* – Fig. a) or primary jig (in system *2 jig rCW* – Fig. b) at changes of recirculation coefficient of a concentrate from jig R_c or intermediate product R_{ip}.

In order to assess whether the use of recirculation systems gives better results than in the case of a single jig, there have been calculated maximum production values at various concentrate quality levels in all the systems presented in Fig. 1.

3 Optimum control with product recirculation

Algorithm of desired quality production maximization [12] has been used again for optimization calculations. The production value (*PV*) in a *1 jig rC* system (Fig. 1b) is defined by the equation:

$$PV_i(\delta_{jig}, R_c) = M_{Ci}(\delta_{jig}, R_c) \cdot P_{Ci}(\delta_{jig}, R_c) - M_{Wi}(\delta_{jig}, R_c) \cdot C_W, \tag{1}$$

and is an objective function of the maximization algorithm, so there is a need to find its maximum:

$$\max_{(\delta_{jig}, R_c) \in R^n} PV_{Ci}(\delta_{jig}, R_c) \tag{2}$$

with a strict limitation of the concentrate ash content value:

$$A_C = A_{Ci}, \tag{3}$$

for subsequent values of A_{Ci} 9, 12, 15 and 18%

where: δ_{jig} – jig separation density, PLN/cm^3
R_c – partial recirculation coefficient of the jig concentrate, %
M_C, M_W – concentrate and waste masses, Mg
P_{Ci} – concentrate unit price, dependent on the coal quality, PLN/Mg [8]
C_W – the unit cost of waste disposal, PLN/Mg.

Setting the relevant jig concentrate ash content values, the production maximization algorithm finds optimum separation densities in a jig, at different recirculation ratio values. It has been assumed that the density distribution in the jig can be changed in the range of 1.300–1.950 g/cm^3.

In the discussed system with partial jig concentrate recirculation *1 jig rC*, adjusted jig separation densities increase with increasing R_c value (Fig. 3), because the jig feed ash content decreases (as there is an increase in share *n* of the recirculated concentrate) [7, 12]. The production value increases with good concentrate quality with a distinct maximum (Fig. 4). It is assumed that the maximum production value possible to achieve in a single jig without recirculation *1 os* (Fig. 1a) is reference value and is regarded as 100%. The problem is a very large increase in the jig feed amount, at high values of R_c. When R_c value does not exceed e.g. 40-50%, then, in accordance with Fig. 2a, the increase in jig feed amount is small. Separation in jig, slightly oversized in terms of performance, allows for even a few percent increase in the production value (Fig. 4) in the case of a high quality concentrate, without installing additional processing machines. Larger jig also makes possible to respond flexibly to increased raw coal amount because at increasing raw coal stream it is simple to reduce the ratio of the recycled product.

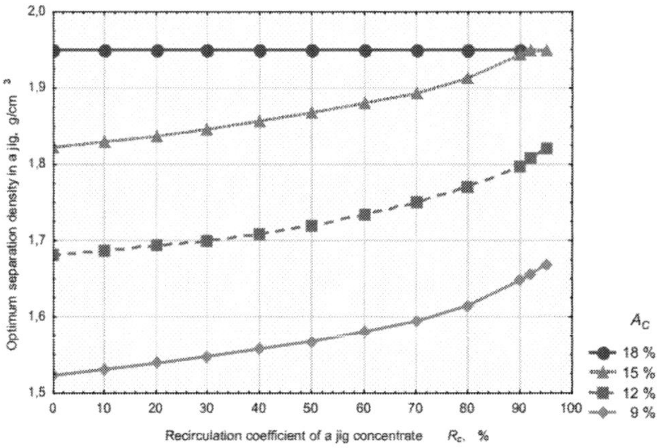

Figure 3. Optimum separation densities in a jig at changes of recirculation coefficient of concentrate from jig R_c in *1 jig rC* system.

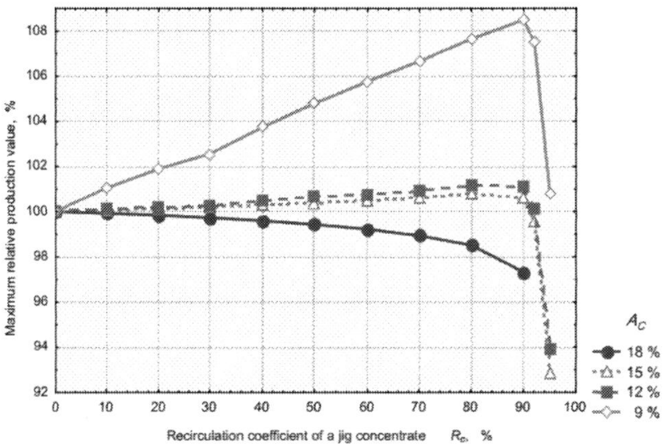

Figure 4. Maximum relative production value from a jig at changes of recirculation coefficient of concentrate from jig R_c in *1 jig rC* system.

Similarly as in Fig. 1b, either a portion or all of the intermediate product can be re-entered to the input of the three-product separator or to the input of the system of two two-product separators. This layout is illustrated in Fig. 1c. A recirculation ratio defines what portion of the intermediate product flow returns back to the input, and is mixed with the raw input (feed stream of 8-20 mm).

The concentrate production value has been established as an objective function for the maximization algorithm. For a system with a single jig without recirculation *1 jig* (Fig. 1a):

$$PV_i(\delta_{jig}) = M_{Ci}(\delta_{jig}) \cdot P_{Ci}(\delta_{jig}) \ . \tag{4}$$

In the *2 jig rCW* system (Fig. 1c) with intermediate product recirculation it is possible to obtain a desired concentrate quality in various combinations of control signals. It has been assumed that all the intermediate product recirculates (which is acceptable at small differences between separation densities in both jigs), and thus the recirculation ratio assumes the value of R_{ip}=100%. The production value is determined by the equation:

$$PV_i(\delta_{jig1},\delta_{jig2}) = M_{Ci}(\delta_{jig1},\delta_{jig2}) \cdot P_{Ci}(\delta_{jig1},\delta_{jig2}). \quad (5)$$

There are maximized functions (4) and (5) (in the whole range of possible to obtain values of the concentrate ash content) with en equivalence constraint of subsequently asked 1% increments of the concentrate ash content:

$$A_C = A_{Ci}. \quad (6)$$

Entering to the production value maximization algorithm consecutive values of the concentrate ash content A_{Ki} there can be obtained optimum separation densities for jigs (within a predetermined range of 1.300-2.200 g/cm^3). These calculated optimum separation densities are shown in Fig. 5.

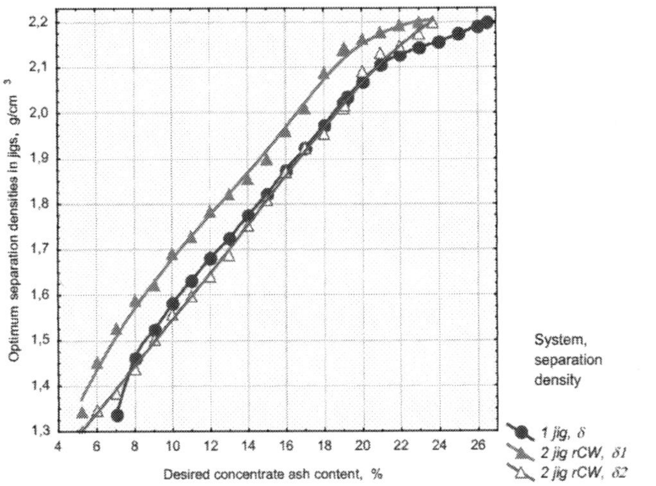

Figure 5. Optimum separation densities in one jig (*1 jig*) and in system (*2 jig rCW*) with recirculation of an intermediate product for R_{ip}=100%.

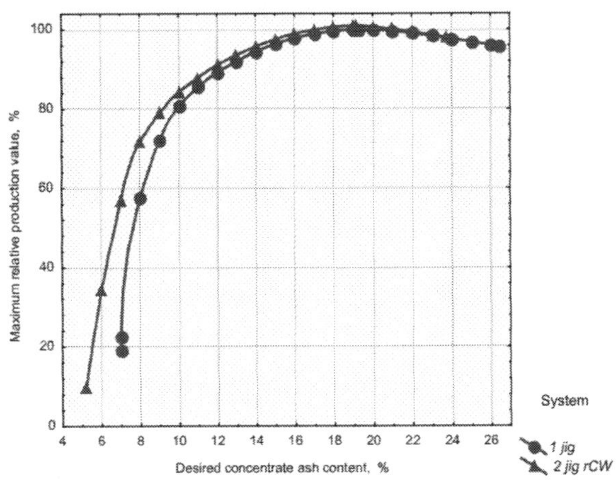

Figure 6. Maximum relative production value of a concentrate in one jig (*1 jig*) and in *2 jig rCW* system with recirculation of an intermediate product for R_{ip}=100%.

 DOI: 10.1051/e3sconf/20160801045

Changes in the maximum production value of the desired quality are illustrated in Fig. 6. In each case in a system with recirculation there can be achieved a higher production value than in the case of a single-jig layout. This is particularly important for small desired values of the concentrate ash contents. Furthermore, as shown elsewhere [4], the use of product recirculation at the desired high quality of the concentrate is economically profitable. Figure 7 shows the relative amounts of feed input to the jigs.

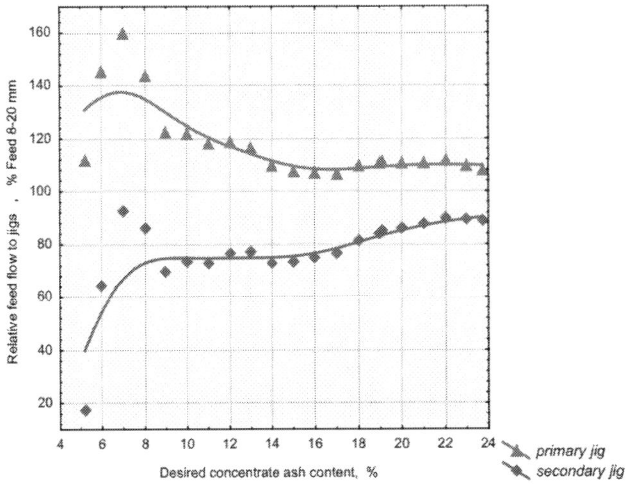

Figure 7. Relative feed flow to jigs in *2 jig rCW* system with recirculation of an intermediate product for R_{ip}=100%.

4 Summary

In the case of the concentrate recirculation in a roughly exponential manner increases the amount of feed to the separators. This unfavorable effect is entirely possible to be omitted in the case of the intermediate product recirculation when the separation densities distribution in both separators do not differ significantly from each other.

A dangerous effect of an excessive feed amount increase can be controlled using a system of machine vision analysis of particle size distribution [3]. Therefore, a reaction to this excessive feed amount increase may be performed by rapid changes in a recirculation ratio or relevant changes in density distribution.

The use of product recirculation – in some cases – enables an increase in the production value. For coal washing in a single jig, the increase in output value is obtained by using the secondary washing of the small part of the concentrate stream. At low values of the desired concentrate ash content (Fig. 4). The increase in production value may even be in a several percent range.

Partial concentrate recirculation is the easiest (practically without investments) way to increase the production value of good quality concentrate. However this is connected with a considerable increase in the feed amount. It seems that the useful value of the concentrate recirculation ratio is restricted to about 40-50% (Fig. 2a).

In systems with the intermediate product recirculation there can be always obtained higher production value than the output value of the single-separator system. This is particularly important at low desired concentrate ash contents values.

The use of product recirculation systems requires a slightly oversized jigs, but it is more cost effective than installing additional processing machines. Improvement in the operational functioning of the intermediate product recirculation circuit can be obtained by the use of video monitoring systems allowing to react flexibly to the changes in feed amount by dynamic changes in recirculation coefficient values.

References

1. W. Blaschke, *Przeróbka węgla kamiennego – wzbogacanie grawitacyjne* (Wydawnictwo IGSMiE PAN, Kraków, 2009)
2. J. Blaschke (i in.), *IM*, Kraków, **4 (17)** (2006)
3. S. Boron, A. Heyduk, J. Pielot, *in Proceedings of XIX Mineral Engineering Conference 2014*, (Istebna, 2014)
4. S. Boron, J. Pielot, A. Wojaczek, *Miner. Resou. Manag.*, **30(2)**, 67 (2014)
5. F. Goodman, J. McCreery, *Coal Prep. Comp. Model,* **1** (1980)
6. K. Kalinowski, *Network Integrators Associates* (Parkland, Florida, 2007)
7. K. Kalinowski, J. Pielot, *Mechanizacja i Automatyzacja Górnictwa*, **12(431),** 41 (2006)
8. U. Lorenz, W. Blaschke, Z. Grudziński, *Monography* (Wydawnictwo IGSMiE PAN, Kraków, **112**, 2002)
9. G.H.Luttrell, *Coal Prep.* **24**, 35 (2004)
10. J. Martyniak, *Wiadomości Górnicze*, **5**, 235 (1996)
11. J. Pielot, *Mechanizacja i Automatyzacja Górnictwa*, **12 (443)**, 23 (2007)
12. J. Pielot, *Monograph* (Wydawnictwo Politechniki Śląskiej, Gliwice, **306**, 2011)
13. J. Pielot, *in Proceedings of Mineral Engineering Conference MEC 2015* (Szczawnica, 20, 2015)
14. J. Pielot, *in Proceedings of Mineral Engineering Conference MEC 2015* (Szczawnica, 24, 2015)
15. A.I.A. Salama, M.W. Mikhail, *in Proceedings of XII International Coal Preparation Congress*, (Krakow, **4**, 43 1994)
16. K. Trybalski, *Kontrola, modelowanie i optymalizacja procesów technologicznych przeróbki rud*, (Wydawnictwo AGH, Kraków, 2013)
17. R.B. Venkoba, P.C. Kapur, *Int.l J. Miner. Process.* **89**, 23 (2008)
18. H. Yagun, L. Shan, L. Maixi, K. Yali, L. Huaiyu, *Coal Prep.*, **22**, 93 (2002)

DOI: 10.1051/e3sconf/20160801046

Production value optimization with intermediate jig product recirculation

Joachim Pielot[1,a]

[1] *Silesian University of Technology, Department of Electrical Engineering and Control in Mining, 44-100 Gliwice, ul. Akademicka 2, Poland*

Abstract. The paper describes the result of the production value increase in systems with two and three jigs and recirculation of the interemediate product compared to the output value obtained from a single jig. The increase is greater when the reference ash content in the concentrate is lower.

1 Secondary separation with a recirculation of the intermediate product

Either a part or the whole intermediate product stream can again enter the input of the three-product separator or the system of two two-product separators. This layout is shown in Fig. 1. The rate of recirculation R_{ip} determines which part of the intermediate product flow returns back to the input and is mixed with the input raw feed stream.

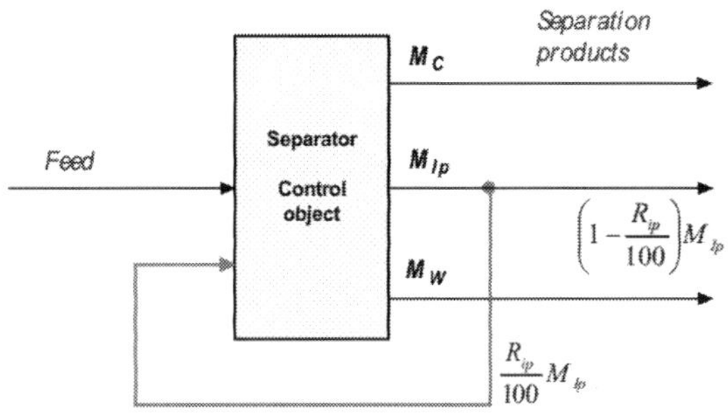

Figure 1. Three-product gravitational separator as a control object with either partial or complete recirculation of the intermediate product; M_C,M_{Ip}, M_W – concentrate, intermediate product and waste masses.

According to the concept shown in Fig. 1 it is possible to create two versions of separation with intermediate product recirculation, as illustrated in Fig. 2. Symbol *rWC* means recirculation of intermediate fractions obtained from the portion of the waste fractions from the first separator and the lighter fractions from the second separator. The *rCW* symbol means recirculation of intermediate fractions obtained from the part of the concentrate fraction part from the first separator and sinking

[a] Corresponding author: Joachim.Pielot@polsl.pl

fractions from the second separator. Recirculation coefficients R_{ip1} and R_{ip2} in both systems determine what part of the intermediate product stream goes back again to the separation process input.

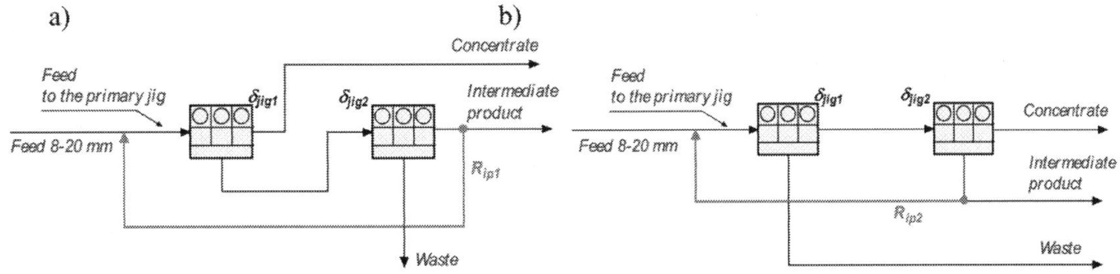

Figure 2. Systems of two two-product jigs with recirculation of an intermediate product: a) *2jig rWC*, b) *2jig rCW*

2 Optimum control with product recirculation

As the value of production there has been adopted a product of the concentrate mass and its quality-dependent unit price, as determined from the 4th version of the sales formula of the year 2002 [6].

As a goal function of the maximization algorithm it has been assumed a value of the concentrate output:

$$PV_i(\delta_{jig1},\delta_{jig2},R_{ip}) = M_{Ci}(\delta_{jig1},\delta_{jig2},R_{ip})\cdot P_{Ci}(\delta_{jig1},\delta_{jig2},R_{ip}), \tag{1}$$

which is maximized:

$$\max_{(\delta_{jig1},\delta_{jig2},R_{ip})\in R^n} PV_i(\delta_{jig1},\delta_{jig2},R_{ip}) \tag{2}$$

with an equivalence restriction of the concentrate ash content:

$$A_C = A_{Ci}, \tag{3}$$

at subsequent A_{ci} values of 6, 9, 12, 15 and 18%:
where: PV – production value, PLN,
δ_{jig} – jig separation density (decision variables of a production maximization algorithm), g/cm^3,
R_{ip} – partial recirculation coefficient of the of the intermediate product, %,
M_C – concentrate mass, Mg,
P_{Ci} – concentrate unit price, dependent on the coal quality, PLN/Mg [8].

All simulation forecasts were carried out using array models of jig separation curves [5]. There was included a hardly-washable feed with characteristics as in the paper [3].

Figure 3 illustrates obtained dependence of the relative production value in a *2 jig rWC* layout. The reference level is different in each case (with different requested ash content values) maximum output value of production achieved with the desired concentrate ash content of the concentrate in a system without intermediate product recirculation. With the increase recirculation ratio of intermediate and taken into account the criterion of maximizing the objective function (2) the value of production increases. This is roughly equivalent to an increase in the production value with jig concentrate recirculation (Fig. 4) [3]. The advantage of the intermediate product recycling, however, is that the feed amount in this case grows very slightly (recirculate only grains with densities similar to the separation density). Increase in the feed amount of the order of a few percent is several times lower than in the case of concentrate recirculation.

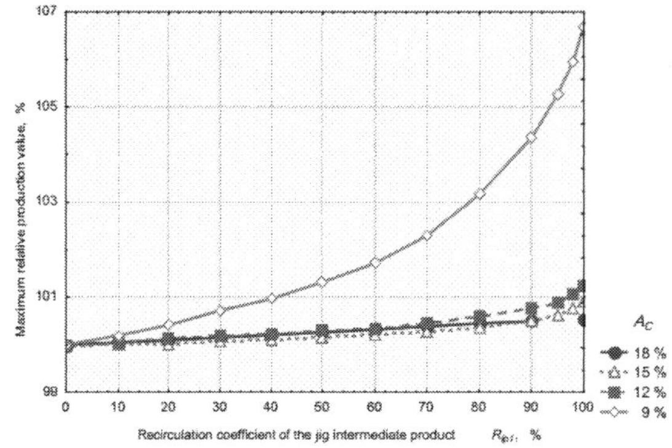

Figure 3. Maximum relative production value of a concentrate from jigs in system *2 jig rWC* at changes of recirculation coefficient of intermediate product R_{ip1}.

Figure 4 shows an example of optimally selected separation densities in a *2 jig rCW* system, at the lowest desired concentrate ash content value – 6%, and at 12%. As the value of the recirculation coefficient increases, there is a noticeable increasing difference between the chosen separation densities in both jigs.

a) b)

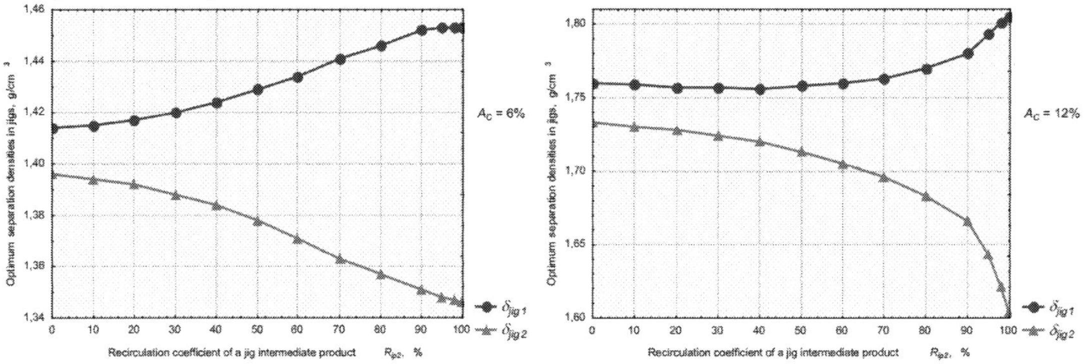

Figure 4. Optimum separation densities in both jigs in system *2jig rCW* at changes of recirculation coefficient of intermediate product R_{ip2}: A_k= 6% (a), A_k= 12% (b).

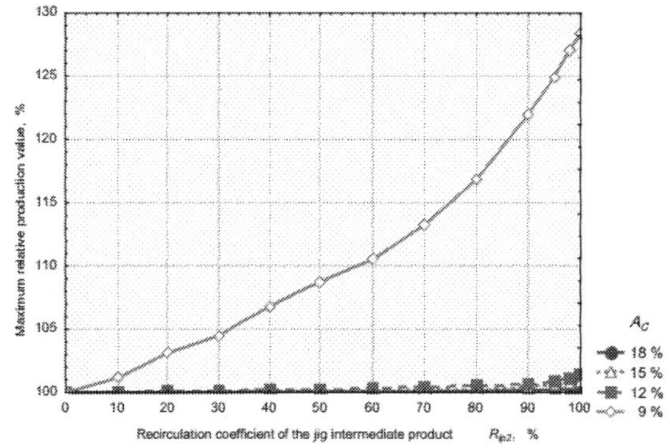

Figure 5. Maximum relative production value of a concentrate from jigs in system *2 jig rCW* at changes of recirculation coefficient of intermediate product R_{ip2}.

3 Comparison of the maximum output value in jig systems with intermediate product recirculation

Output production values for two jig systems with intermediate product recirculation presented in Section 2 are given in relative terms, so it is not visible, where the absolute value of the system output is higher. In order to compare the value of production, the calculations were performed with desired concentrate ash content growing of 1%.

Since the maximum value of output production can be obtained by the full recirculation of intermediate product (Figs. 3 and 5) further analysis was therefore limited to those cases (R_{ip1}=100% or R_{ip2}=100%). There was made a comparison of production values achieved in the four systems. The first system was a single jig (without recirculation), two more systems are presented in Fig. 2, and the fourth system was a group of three jigs with two completely recirculating intermediates, as shown in Fig. 6. The maximum value of a single jig output production was a reference level (100%).

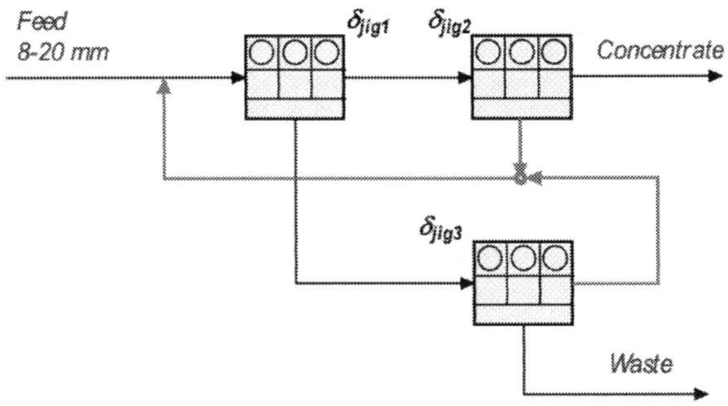

Figure 6. System of three two-product jigs *3os 2r* with recirculation of two intermediate products.

The objective function of the maximization algorithm is a concentrate production value. For one jig:

$$VP_i(\delta_{jig}) = M_{Ci}(\delta_{jig}) \cdot P_{Ci}(\delta_{jig}) \ . \tag{4}$$

In the case of the two-jig systems shown in Fig. 2 (R_{ip1} =100% or R_{ip2} = 100%):

$$VP_i(\delta_{jig1}, \delta_{jig2}) = M_{Ci}(\delta_{jig1}, \delta_{jig2}) \cdot P_{Ci}(\delta_{jig1}, \delta_{jig2}) \, . \tag{5}$$

In the case of the three-jig circuit shown in Fig. 6:

$$VP_i(\delta_{jig1}, \delta_{jig2}, \delta_{jig3}) = M_{Ci}(\delta_{jig1}, \delta_{jig2}, \delta_{jig3}) \cdot P_{Ci}(\delta_{jig1}, \delta_{jig2}, \delta_{jig3}) \, . \tag{6}$$

There are maximized functions (4), (5) and (6) at equivalence restrictions and at subsequently asked 1% concentrate ash content increments:

$$A_C = A_{Ci}. \tag{7}$$

The calculation results are presented in Fig. 7. In each system with intermediate product recirculation there can be achieved higher production value than the production of the single-jig system. This is particularly important for small values of desired concentrate ash content. The highest production value is achieved in a three jig system *3 jig 2r,* slightly lower production value can be achieved in a two-jig system *2 jig rCW*. Comparing both two-jig systems it can be said that at a better desired concentrate quality higher production value can be provided by a *2 jig rCW* system p (Fig. 2b), and at worse quality – *2 jig rWC* system (Fig. 2a).

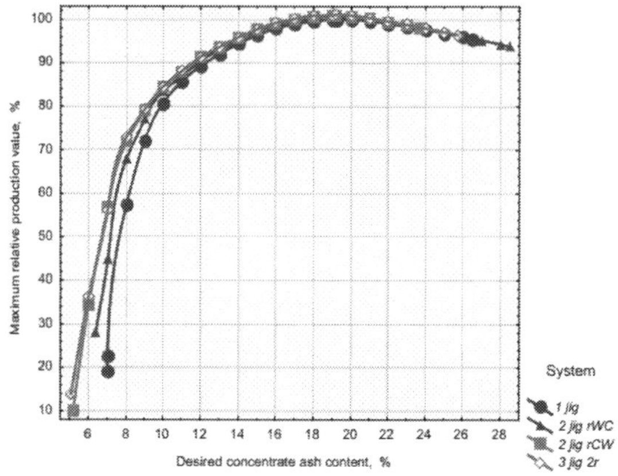

Figure 7. Maximum relative production value of a concentrate from one jig and in three systems at complete recirculation of an intermediate product.

4 Summary

In any system with intermediate product recirculation the output value is higher than the output value of a single separator. The increase in a production value is more significant when the reference concentrate ash content value is lower. Therefore, the application of intermediate product recirculation is profitable for investment and operation [4].

A comparison of the three jig separation systems shows that the highest output value is achieved in a *3 jig 2r* system with three jigs, slightly lower production value can be achieved by a *2 jig rCW* layout. Comparing the two systems with two jigs it can be said that with a better set point of concentrate quality higher output value (Fig. 7) can be achieved in a *2 jig rCW* system (Fig. 2b), while at the lower concentrate quality the*2 jig rWC* system (Fig. 2a) provides higher output value.

Gravity separators are characterized by a certain separation inaccuracy enrichment. It is a result of the fact that the separation curves have a shape deviating from the ideal curve. The use of secondary separation with a recirculation of the intermediate product allows for the effect equivalent to the separation curve shape improvement of a single separator [8].

References

1. W. Blaschke, *Wydawnictwo IGSMiE PAN* (Krakow, 2009)
2. J. Blaschke (i in.), *IM,* **4(17)** (2006)
3. S. Boron, A. Heyduk, J. Pielot, *in Proceedings of Mineral Engineering Conference*, MEC (2016)
4. S. Boron, J. Pielot, A. Wojaczek, *Miner. Resour. Manag.*, **30(2),** 67-82 (2014)
5. F. Goodman, J. McCreery, *Coal Prep. Compu. Model,* **112** (1980)
6. U. Lorenz, W. Blaschke, Z. Grudzinski, *Monograph* (Wydawnictwo IGSMiE PAN, Krakow, **112**, 2002)
7. J. Pielot, *Monograph*, (Wydawnictwo Politechniki Slaskiej, **306**, 2011)
8. J. Pielot, *in Proceedings of Mineral Engineering Conference MEC 2015,* (Szczawnica, 24, 2015)

Sensitivity problems of control system of coal preparation processes

Roman Kaula[1,a]

[1] *Silesian University of Technology, Department of Electrical Engineering and Control in Mining, 44-100 Gliwice, Akademicka 2, Poland*

Abstract. Control of technological processes of coal preparation takes place in the presence of wide disturbances. An important problem is the choice of the controller which is robust for a variety of disturbances. No less important problem in the control process is the tuning of the controller parameters. In the paper the analysis of influence of changes in object model parameters on the course of the controlled value was carried out. For the controller settings, calculated according to model parameters research was carried out on object with other values of parameters. In the studies a sensitivity analysis method was used. The sensitivity analysis for the three methods of tuning PI controller for control systems of coal preparation processes characterized by dynamic properties of the inertial element with time delay was presented. Considerations were performed at various parameters of the object on the basis of the response of the control system for a constant value of the set point. The assessment of the considered tuning methods based on selected indices of control quality was realized.

One of the primary tasks of the control system is to maintain the value of the controlled variable (process output) at the desired level. An important problem from the point of view of the control quality is also getting the appropriate course of the controlled variable and reduce the influence of disturbances running on the system. Compliance with these requirements may be accomplished by an appropriate selection of the structure of the controller and its settings. Many methods of tuning the controller use a dynamic model of a control object (determined on the basis of dynamic characteristics of the controlled process). The description of the properties of the dynamic process by using a mathematical model is helpful in the design of the control system. Knowing the dynamic model and its parameters we can analytically determine the controller settings for the assumed criterion of the control quality. Different processes with the same dynamic properties can be described by a dynamic model of the same structure which differs in only parameter values. Such an approach makes it possible to apply the same type of controller in control systems of various industrial processes. This also applies to the processes of coal preparation.

The dynamic properties of many processes of coal preparation can be characterized by the inertial model of the first order (described time constant T) with time delay τ [1-4].

The PI controllers have two tuning parameters to adjust: the proportional gain (k_r) and integral time (T_i). One of the basic methods of tuning of the PI controller is the method of Ziegler-Nichols of the identification test [5]. The Z-N method may be the initial method of tuning for many processes of the coal preparation where the dynamics has the properties of the

[a] Corresponding author: rkaula@polsl.pl

inertial system of the first order with time delay. Many authors dealing with the tuning of PI controllers parameters for processes with dynamics of the first order inertial model proposes settings based on the reduction of the time constant [6]. In these methods (R) it is assumed that the parameter T_i of the integral term is equal to the time constant $T_i=T$ of the system. Differences in the settings are the result of the adopted tuning criteria of the parameter k_r. In the paper [7] a new approach for tuning PI controller for inertial object with time delay has been proposed. In the direct method (D) of controller tuning an additional criterion was used. The condition on the phase margin from the Nyquist criterion was applied.

It should be noted that the identification of the object parameters (including the time constant) under industrial conditions is usually performed during normal operation (with the influence of disturbances). Thus, determined parameters of the dynamic model may be different from the actual values of the process. The control system with controller which is tuned on the basis of such the model may not satisfied assumed requirements of the control quality. In the paper we analyse the impact of changes of the object model parameters on the course of the controlled variable. In the study the sensitivity analysis has been used.

One of the important indicators to assess the control quality is a parameter describing the sensitivity of the controlled variable on change in the object parameters. Changes of two parameters $\alpha=\{T,\tau\}$ for the object were analyzed. The controlled variable in the Laplace form Y(s) can be written:

$$Y(s)=K_{cl}(s)Y_{ref}(s) \tag{1}$$

where: $Y_{ref}(s)$ – reference value of the controlled object,
$K_{cl}(s)$ – closed loop transfer function.

Thus, the changes of the controlled variable and the changes of the model parameter $\alpha=\{T,\tau\}$ are as follows:

$$Y(s,\alpha)+\Delta Y(s,\alpha)=K_{cl}(s,\alpha)Y_{ref}(s)+\Delta K_{cl}(s,\alpha)Y_{ref}(s). \tag{2}$$

Relative sensitivity of the transfer function of the closed system according to parameter $\alpha=\{T,\tau\}$ can be expressed by the formula:

$$S_{Kcl(\alpha)}=\frac{\dfrac{\Delta K_{cl}(s,\alpha)}{K_{cl}(s,\alpha)}}{\dfrac{\Delta\alpha}{\alpha}}\cong\frac{\alpha}{K_{cl}(s,\alpha)}\frac{\partial K_{cl}(s,\alpha)}{\partial\alpha}. \tag{3}$$

After transformations:

$$\Delta K_{cl}(s,\alpha)=\frac{K_R(s)}{(1+K(s,\alpha)K_R(s))^2}\Delta K(s,\alpha) \tag{4}$$

where: $K(s,\alpha)$, $K_R(s)$ – the transfer function of object and controller,

$$\Delta K(s,\alpha_T)=\frac{k}{T_2s+1}e^{-\tau s}-\frac{k}{T_1s+1}e^{-\tau s},$$

$$\Delta K(s,\alpha_\tau)=\frac{k}{Ts+1}e^{-\tau_2 s}-\frac{k}{Ts+1}e^{-\tau_1 s},$$

T_1, τ_1 – parameter values for which the controller parameters have been determined,

T_2, τ_2 – actual values of the system parameters.

The paper presents the sensitivity analysis for the three methods of tuning PI controller. Based on the dependences presents in works [5-7] the values of the PI controller parameter settings (depending on the parameters *k*, *T*, τ of the dynamics model) have been calculated.

Data of the parameters of the dynamic model from work [3] were taken. In this work the dynamic model parameters of the coal flotation process using various methods of identification were determined. Depending on the chosen method of identification and the set point of the system, different values of system parameters (time constant *T* and the delay τ) were obtained. The time constant *T* in the range of values (200-800) s, while the time delay τ in the range of values (200-400) s have been identified. It is therefore advisable to analyze the control system for the process with parameters that differ even tens of percent from parameters used to controller settings.

Table 1. Indices of the control quality.

no	ΔT	Δτ	$\frac{T \pm \Delta T}{\tau \pm \Delta \tau}$	**Integral criterion in the relative terms** $\frac{I_2}{I_{2\,min}}100$ **,%**			**Rise time t_n, s (Settling time t_u, s)**		
				ZN	R	D	ZN	R	D
1	+0.5T	0	3	190	295	237	343 (1200)	1015 (2861)	557 (2400)
2	-0.5T	0	1	267	208	170	136 (5200)	1324 (2100)	667 (1378)
3	0	+0.5τ	4/3	318	303	278	214 (4800)	686 (2640)	400 (2520)
4	0	-0.5τ	4	117	211	151	786 (1757)	1324 (1862)	760 (1120)
5	+0.5T	+0.5τ	2	278	352	316	331 (3062)	920 (3280)	480 (2400)
6	+0.5T	-0.5τ	6	139	250	184	643 (1114)	1311 (1714)	827 (2193)
7	-0.5T	+0.5τ	2/3	unstable	257	271	-	386 (2143)	240 (3680)
8	-0.5T	-0.5τ	2	100	171	119	124 (1945)	1600 (2400)	1000 (1680)

The simulation studies of the response of the control system on step change of the set point were carried out. In each considered case a constant value of gain $k = 1$ of the object was assumed. The controller settings for T=500 s and τ=250 s were calculated. The study on the system with parameters $T \pm 0{,}5T$ and $\tau \pm 0{,}5\tau$ was carried out. Changes in one and both parameters simultaneously were assumed.

The results are shown in Table 1. To assess the control quality, the criterion of the integral of the squared control error (in relative terms) and parameters: rise time t_n and settling time t_u were used.

The sensitivity analysis for three methods of the tuning PI controller in the control systems of coal preparation processes characterized by the dynamic properties of the inertial system with time delay was carried out. The studies of the control system at the object parameters which differed to a certain range from *T* and τ (Table 1) were performed.

Responses of the control system for a step change in the set point were obtained. Based on the courses it can be noted that for all considered cases (except one) regardless of the choice of the method of the parameter settings of the PI controller, the controlled variable assumes a steady state value. This shows a large robustness of the PI controller on disturbances connected with the change of the dynamic properties of the object (or inaccurate identification of the parameters). Only at controller settings according to the ZN criterion for case 7 the system is unstable. In this case the ratio $((T\text{-}\Delta T)/(\tau\text{-}\Delta\tau))=2/3$ is less than the ratio $T/\tau=2$ for which the settings of the PI controller were tuned. The situation in which the ratio of $((T\text{-}\Delta T)/(\tau\text{-}\Delta\tau))$ is less than the T/τ ratio refers also to cases 2 and 3. In these cases we may notice the presence of the larger oscillations of the controlled variable.

The best results of the control (the control quality) from the point of view of the integral index for cases 4 and 8 were obtained. This occurs when the parameter of the time delay τ (accepted for the settings) is larger than the actual time delay of the object. Analyzing the transient nature of the controlled variable and the quality parameters it should be noted that in these cases were achieved better control quality than at the control system in which the controller settings for the actual object parameters were determined. Comparing indices of the control quality for the analyzed methods of the tuning of the controller settings you will notice the largest variability for the ZN method and the smallest for the method of reduction (R). The transients of the controlled variable for the tuning method of ZN in most cases are oscillatory. The oscillations are not present in the transients of the controlled variable for the method of the reduction (R). This is achieved at the expense of significant deterioration of other quality parameters. The transients have the longest settling time and rise time for the settings according to the criterion of the reduction (R). The value of the integral index is the smallest in five cases for the method of ZN and in two cases for the direct method (D). In most cases the shortest settling time of the transient in the direct method (D) was obtained.

For the assumption of larger time delay in the controller settings, relative to the real delay of the system (identified parameter larger than actual), the quality of control was improved. The obtained results in identifying the parameters of the system model and a new approach to the tuning of the PI controller can be very helpful.

References

1. S. Cierpisz, *in Proceedings of IX Konferenca Automatyzacji Procesów Przeróbki Kopalin*, (Szczyrk, 49, 2003)
2. S. Cierpisz, *Monograph,* Wydawnictwo Politechniki Śląskiej, Gliwice, **403** (2012)
3. J. Joostberens, *MiAG,* **12,** 34 (2011)
4. K. Kalinowski K., R. Kaula, *MiAG,* **1,** 35 (2000)
5. J. Ziegler, N. Nichols, *Trans. ASME,* **64,** 759 (1942)
6. A. O'Dwyer, *Handbook of PI and PID controller tuning rules*, Imperial College Press, London (2003)
7. R. Kaula, *GSM-MRM,* **31(1),** 141 (2015)

Stability of green tea nanoscale zero-valent iron

Tomasz Suponik[1,a], Marcin Lemanowicz[2], and Pawel Wrona[1]

[1]*Silesian University of Technology, Faculty of Mining and Geology,44-100 Gliwice ul. Akademicka 2, Poland*
[2]*Silesian University of Technology, Faculty of Chemistry, 44-100 Gliwice ul. Strzody 7, Poland*

Abstract. Colloidal suspension of nano zero-valent iron (nZVI), obtained as a result of mixing green tea (GT) solution with iron solutions ($FeCl_2$ and $FeSO_4$ separately to form two suspensions: GT-nZVI$_{FeCl2}$ and GT-nZVI$_{FeSO4}$) was assessed in terms of stability. The particles of suspensions had a tendency rapidly aggregation to form larger aggregates, and then settled. The absolute value of zeta potential (ζ) of suspensions did not exceed 17 mV. To prevent the aggregation of nZVI particles, the pHs were changed by the addition of Na_2CO_3 to obtain higher values of ζ. For both suspensions, the zeta potential at a satisfactory level was obtained at pH values higher than 7.5. Then, the potential was lower than - 40 mV for GT-nZVI$_{FeCl2}$ and lower than - 25mV for GT-nZVI$_{FeSO4}$. This means that the first suspension had a better stability than second. The isoelectric point (IEP) of GT-nZVI$_{FeCl2}$ was in the range from 4.5 to 5, while in the case of GT-nZVI$_{FeSO4}$ all measured ζ were negative. To investigate the effect of increased ionic strength (IS) on the stability of suspensions at different pH values, NaCl and $MgSO_4$ were added to the irons solutions. The final IS in the GT-nZVI was 12 mM. In both of the suspensions the absolute values of ζ were lower than previous and were about 25 and 20 mV, for GT-nZVI$_{FeCl2}$ and GT-nZVI$_{FeSO4}$, respectively.

1 Introduction

The application of zero-valent iron (nZVI) for the removal of heavy metals and halogenated organic compounds, as well as pesticides and herbicides is well established. Wastes resulting from the mining of non-ferrous metal ores and coal, as well as from the manufacturing process of non-ferrous metals accumulated at industrial waste disposal sites contain metal minerals and, as a result, may affect the quality of ground and surface waters in the southern provinces of Poland. Technical limitations [1] and the high cost of excavation in the case of deep construction have caused, that Wang and Zhang [2] proposed, to use Fe(0) as nano-sized particles. The suspension of nZVI is used for in situ remediation due to its large surface area, rapid reactivity and better injectability into aquifers [3]. An overview of techniques for nZVI production were presented in the article by Tosco et al. [4]. An effective, relatively cheap and environmentally friendly way of producing nZVI is the method of using green tea (GT) [5, 6]. A key advantage of this method is its simplicity. GT-nZVI may be safely produced even outside the laboratory. Polyphenolic compound, tannic acid, determines the reducing properties of the tea [7].

[a] Corresponding author: tomasz.suponik@polsl.pl

Besides the control of the reactivity of GT-nZVI, the important issues, that need to be analyzed for successful full-scale applications, are stability against aggregation. This study was carried out to obtain a high stability of GT-nZVI, in order to pump the suspension into correct places of the aquifer and remove contaminants.

2 Materials and methods

2.1 Synthesis of GT-nZVI

Nanoscale zero-valent iron particles was prepared in aqueous solutions by way of the reduction of ferrous iron Fe(II) using a solution of green tea (GT). The method was inspired by article written by Mystrioti et al. [6]. The 1:2 volume ratio of the GT solution-to-the Fe(II) solution were vigorously mixed in a flask reactor (see Fig. 1) at room temperature for an additional 15 minutes after the titration of GT. The solutions in the flask were mixed with a mechanical stirrer at 120 rpm. During titration at a rate of 1.0 cm^3/s, the solution of GT was introduced to reduce ferrous ions to zero-valent iron Fe(0).

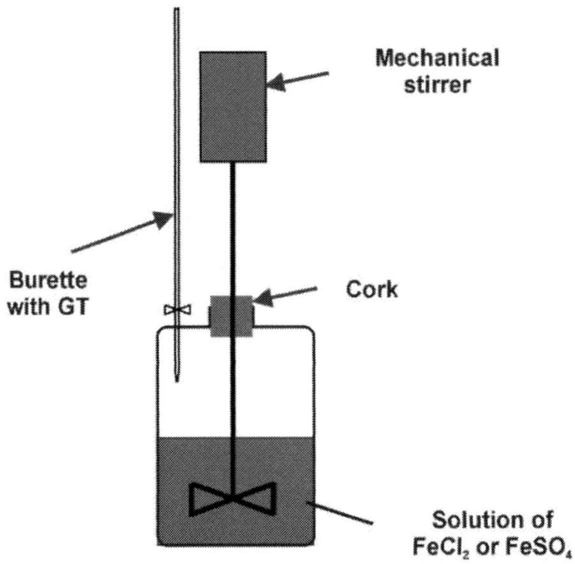

Figure 1. Laboratory set-up for GT-nZVI synthesis.

The Loyd brand of green tea was the source of polyphenolic compounds. An immersion of the dry green tea in deionized water at a temperature of 80-90 °C for 30 minutes was the way of preparing the solution. It was filtered by the vacuum filtration, using a filter paper of 0.45 μm pore size. The ratio of green tea leaves to deionized water amounted to 1.35 g/50 ml, while the concentration of iron in the solutions varied depending on the experiment.

As a source of Fe(II), two chemicals were assessed at various concentrations. These were: $FeCl_2 \cdot 4H_2O$ and $FeSO_4 \cdot H_2O$. Thus, the two nano-irons were finally tested: GT-$nZVI_{FeCl2}$ and GT-$nZVI_{FeSO4}$. These chemicals were dissolved in deionized water for a short time before introducing the GT solution.

The pH, conductivity and oxidation-reduction potential (ORP) in $FeCl_2$ and $FeSO_4$ solutions, as well as in the GT solution are presented in Table 1. The values shown in Table 1 were measured in appropriate solutions, which then formed the GT-nZVI suspensions, i.e. before mixing the solution of Fe(II) with the solution of GT.

Table 1. pH, conductivity, ORP in $FeCl_2$, $FeSO_4$ and GT solutions that later created GT- nZVI suspensions.

Parameter, unit	Solutions of $FeCl_2$ for the preparation of GT-nZVI$_{FeCl2}$ containing:				Solutions of $FeSO_4$ for the preparation of GT-nZVI$_{FeSO4}$ containing:				Solution of GT
	0.04 wt %Fe	0.10 wt %Fe	0.10 wt %Fe IS	0.15 wt %Fe	0.04 wt %Fe	0.10 wt %Fe	0.10 wt %Fe IS	0.15 wt %Fe	
pH	2.41	2.22	2.18	2.09	2.98	2.66	2.63	2.46	6.19
Temperature, °C	20.1	21.0	20.3	20.8	19.1	18.7	19.9	19.4	19.5
ORP, mV	421	448	460	500	369	390	402	423	156
Cond., mS/cm	2.20	5.07	5.93	6.38	1.31	2.23	2.73	2.91	0.76

2.2 Stability of GT-nZVI

After formation of colloidal suspensions, it turned out that the GT-nZVI particles had a tendency to rapidly agglomerate forming larger aggregates due to the Van der Waals and magnetic forces, rendering them undeliverable to the targeted contaminant locations. The zeta potential is the major factor determining the stability of nanoparticles and mobility in aquifers. The absolute value of zeta potential of suspensions did not exceed 20 mV (see Fig. 2). To prevent aggregation of nZVI particles the pHs were changed by the addition of Na_2CO_3 to obtain higher values of zeta potential, as shown in the paper by Suponik [8]. The chemical was added to the flask reactor directly after the preparation of the nZVI. The suspension was then mixed with a mechanical stirrer at 120 rpm for 10 minutes.

To investigate the effect of increased ionic strength (IS) on the stability of suspensions, NaCl and $MgSO_4$ were added in some experiments to the Fe solution, i.e. just before titration of the green tea. The final concentration of these chemical substances in the GT-nZVI, i.e. after the addition of GT, were 8 mM and 1 mM, for NaCl and $MgSO_4$, respectively. The final IS in the GT-nZVI was 12 mM.

The following parameters were analyzed in the study: zeta potential, pH, ORP, temperature, conductivity, and finally a grain-size distribution of the particles in one chosen suspension. The measurements (except the measurement of particle size) began 10 minutes after the preparation of the samples. The pH, conductivity, ORP and temperature were measured with the Knick PORTAMESS meters, while zeta potential and grain-size distribution were measured with a zeta-meter (Laser Zee PenKem Model 501) and a laser granulometer (Malvern Nano S-90), respectively.

3 Results and discussion

The zeta potential measurements as a function of pH are shown in Fig. 2. Results indicated that in both cases the pH above 7.5 corresponds to the maximum suspension stability of GT-nZVI. For measurements with a higher ionic strength of GT-nZVI the zeta potential was ca. – 25 mV, while in other cases, this was lower than – 40 mV for GT-nZVI$_{FeCl2}$ and lower than – 25 mV for GT-nZVI$_{FeSO4}$, which means that the suspensions had good stability. Thus, the addition of Na_2CO_3 caused a significant increase in the stability of the GT-nZVI.

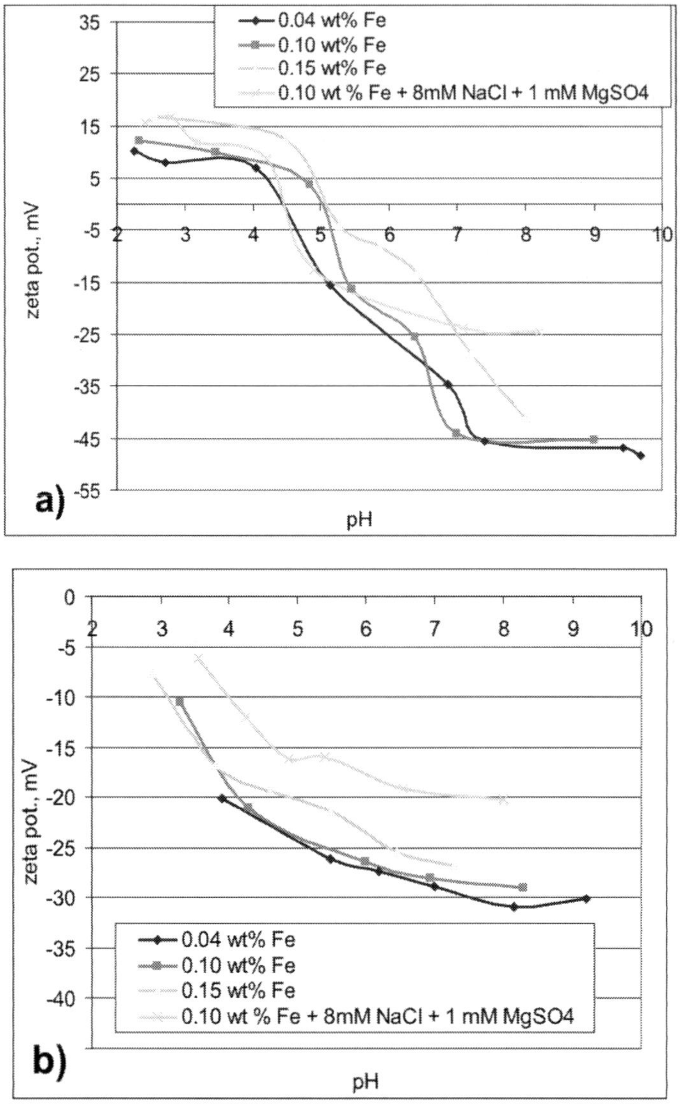

Figure 2. Zeta potential of GT-nZVI as a function of pH for: a) GT- $nZVI_{FeCl2}$, b) GT- $nZVI_{FeSO4}$.

When particles have low absolute zeta potential values, then flocculation process occurs. This was noticeable macroscopically for all the samples having a pH lower than 5.5. The isoelectric point (IEP) in the presented study for GT-$nZVI_{FeCl2}$ was in the range from 4.5 to 5. In the case of the second suspension, all measured zeta potentials were negative. The trends indicate that the IEP was then 2.5.

As provided by Ji [9] and Sun et al. [10] the IEP was found to be between 8.0 and 8.5. In both cases, the sodium borohydride was applied as a reducer. Comparatively, the values of IEP for typical products formed on the surface of the Fe(0), such as the α-FeO(OH), γ-FeO(OH), $Fe^{2+}Fe^{3+}{}_2O_4$, Fe_2O_3, $Fe(OH)_3$ are as follows: 7.3 to 9.4; 5.6 to 7.4; 5.5 to 8; 6.0 to 9.5; 6 - 9, respectively. All this data can be found in the review written by Kosmulski [11].

The polyphenol molecules were probably adsorbed onto iron particles and made the IEPs of GT-nZVI much lower. The polyphenol molecule contains hydroxyl groups. These groups seem to strongly interact with nano-particles of iron.

In a similar study, Yang [12] concluded that carboxymethylcelluloses (these molecules were used to stabilize iron nanoparticles) were adsorbed onto the nano-irons. These molecules resulted in the IEP of iron nano-particles being also lower.

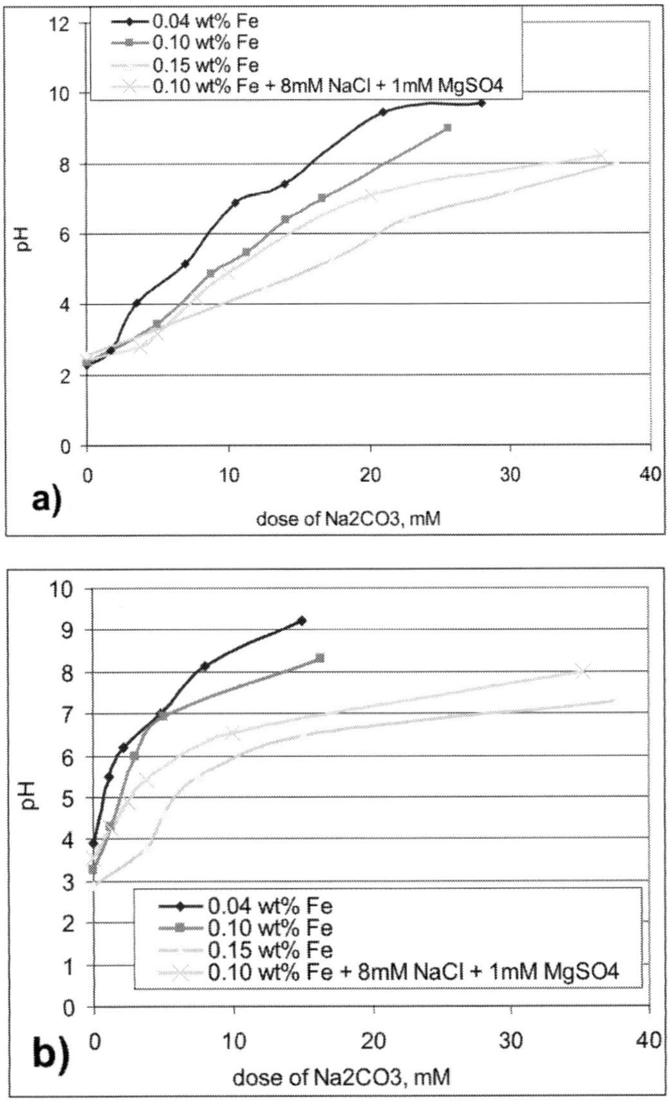

Figure 3. pH as a function of dose of Na_2CO_3 for a) GT- $nZVI_{FeCl2}$ and b) GT- $nZVI_{FeSO4}$.

As a result of pH increasing (see Fig. 3), the iron ions (not reduced earlier) have probably been precipitated out in the form of iron hydroxides, oxides or, in proper conditions, also in Fe(0) form. This, in turn, meant a decrease in the concentration of ions, thereby reducing conductivity in the initial stage of the addition of Na_2CO_3 - this was particularly evident for GT-$nZVI_{FeCl2}$ (Fig. 4a). When more Na_2CO_3 was added, the loss of ions was compensated, and conductivity began to rise again (see Fig. 4). These phenomena have already been observed in the previous paper of the author [8].

In all tests, the ORP decreased significantly with an increase in the dose of Na_2CO_3, and thus, an increase pHs (Fig. 5). This can be associated with the precipitated iron compounds (iron hydroxides,

oxides and Fe(0)) that have a reducing property. In addition, higher values of pH also cause a lower ORP for the same reactions.

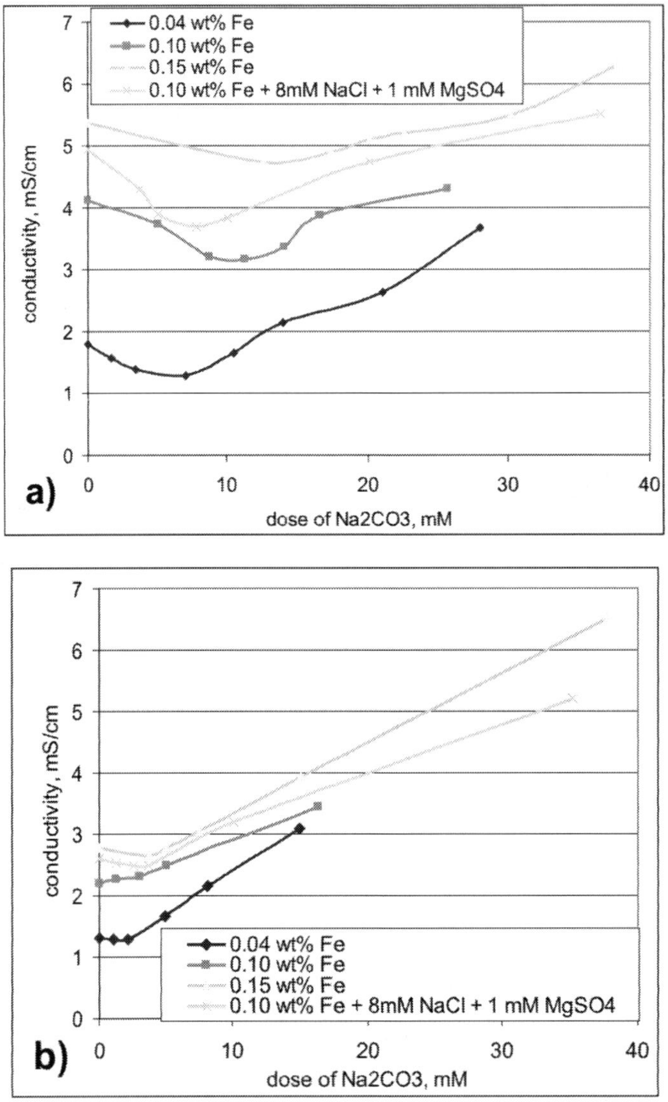

Figure 4. Conductivity as a function of dose of Na_2CO_3 for: a) GT- $nZVI_{FeCl2}$, b) GT- $nZVI_{FeSO4}$.

In both cases of nano-irons, other relationships were noted. The lower iron concentration in solutions for the preparation of GT-nZVIs, the lower the ORP, for a given dose of Na_2CO_3. Colloidal suspensions of GT-nZVI with a lower concentration of iron are characterized by higher values of pHs (see Fig. 3).

Based on the study, comparing the two colloids (GT-$nZVI_{FeCl2}$ and GT-$nZVI_{FeSO4}$), the GT-$nZVI_{FeCl2}$ was assessed as being more stable. For this suspension, the zeta potentials at pH above 7.5 were much lower than for GT-$nZVI_{FeSO4}$. In addition, GT-$nZVI_{FeCl2}$ was characterized by lower values of ORP, which could be very important in assessing the reactivity of the material. However, in order to obtain pH above 7.5, a larger dose had to be added to get GT-$nZVI_{FeCl2}$ than in the case of GT-$nZVI_{FeSO4}$.

 DOI: 10.1051/e3sconf/20160801048

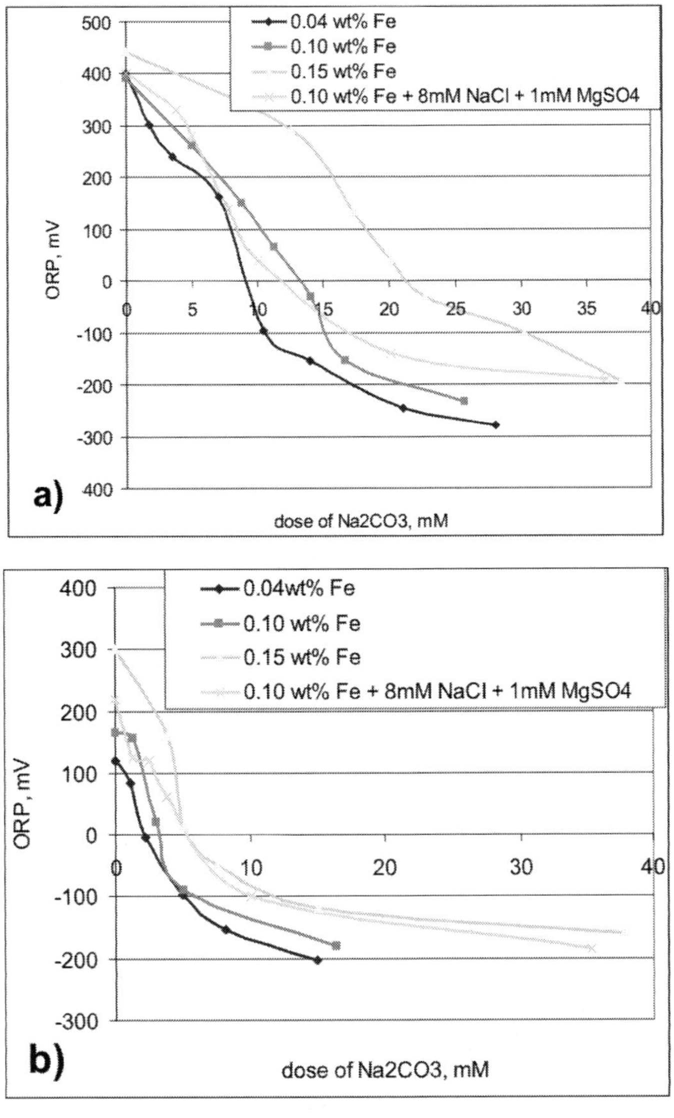

Figure 5. ORP as a function of dose Na_2CO_3 for: a) GT- nZVI$_{FeCl2}$, b) GT- nZVI$_{FeSO4}$.

Figure 6 presents the grain-size distribution of the particles in GT-nZVI$_{FeCl2}$ at pH 7.5. The results indicate that the iron particles were in the range from 80 to 900 nm, what means that they might create a colloidal suspension under the appropriate conditions. In the paper by Sun [10], the nano-irons, obtained by way of reduction of Fe(III) solution using a borohydride, had a size in the range from several to 900 nm. The lack of particles with a size of several nm in the studies using GT, may be due to the fact that measurements were made after more than an hour after synthesis. At this time, the finest particles could be aggregated.

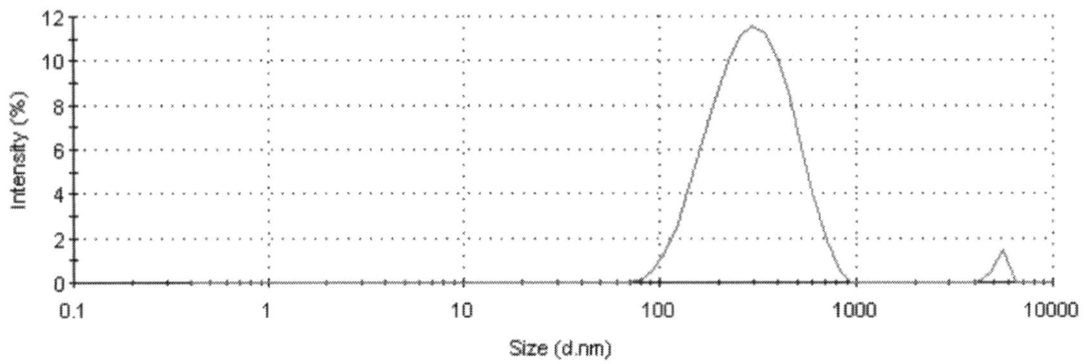

Figure 6. The grain-size distribution of the particles in GT-nZVI$_{FeCl2}$ at pH 7.5.

4 Conclusions

The suspensions of GT-nZVI, obtained as a result of mixing solutions of Fe(II) with a solution of green tea, were not stable. The particles of suspensions had a tendency to rapidly aggregate to form larger aggregates and then settled. The pHs of the suspensions were ca. 2.5 for GT-nZVI$_{FeCl2}$ and between 2.9 and 3.9 for GT-nZVI$_{FeSO4}$, while the absolute values of zeta potentials were between 10 and 17 mV, and between 6 and 20 mV, respectively, for these suspensions. By adding Na_2CO_3, the pHs increased and consequently the absolute values of zeta potential increased and reached 45 mV for GT-nZVI$_{FeCl2}$ and ca. 30 mV for GT-nZVI$_{FeSO4}$. The grain-size distribution of the particles for the first suspension, at a pH of 7.5, showed the particles were in the range from 80 to 900 nm.

The isoelectric point of GT-nZVI$_{FeCl2}$ was in the range from 4.5 to 5.0, while in the case of GT-nZVI$_{FeSO4}$ all the measured zeta potentials were negative. The addition of NaCl and $MgSO_4$ to increase ionic strength, meant that in both of the suspensions the values of zeta potential were about - 25mV.

References

1. V. Zolla, F.S. Freyria, R. Sethi, A. Di Molfetta, *J. Environ. Qual.*, **38(3)**, 11 (2009)
2. W.X. Zhang, C.B. Wang, H.L. Lien, *Catal. Today*, **40(4)**, 8 (1998)
3. L. N. Shi, Y. Zhou, Z. Chen, M. Megharaj, R. Naidu, *Environ. Sci. Pollut. Res.*, **20(6)**, 9 (2013)
4. T. Tosco, M. P. Papini, C. C. Viggi, R. Sethi, *J. Clean. Prod.*, **77**, 11 (2014)
5. M. N. Nadagouda, A. B. Castle, R. C. Murdock, S. M. Hussain, R. S. Varma, *Green. Chem.*, **12(1)**, 8 (2010)
6. C. Mystrioti, N. Papassiopi, A. Xenidis, D. Dermatas. M. Chrysochoou, *J. Hazard. Mater.*, **281**, 6 (2015)
7. Y. Joshi, S. Khandekar, *Nanoscale and microscale phenomena* (Springer, 2015)
8. T. Suponik, *Silesian University of Technology Monograph,* **8**, 11 (2016),
9. Y. Ji, *Colloids Surf., A*, **444**, 8 (2014)
10. Y. P. Sun, X. Q. Li, J. Cao, W. X. Zhang, H. P. Wang, *Adv. Colloid. Interfac.*, **120(1)**, 9 (2006)
11. M. Kosmulski, J. *Colloid Interface Sci.*, **426**, 4 (2014)
12. J. Yang, H. Sun, *Water Air Soil Pollut.*, **226(9)**, 15 (2015)

Challenges related to flotation cleaning of oil shales. Issues due to compositional and surface features and post-grinding surface behavior

N. Emre Altun[1,a]

[1]*Midlle East Technical University, Mining Engineering Department, Ankara, Turkey*

Abstract. Oil shale is an important energy resource alternative. Despite its recognition as an unconventional oil source, oil shale is also considered as an important solid fossil fuel alternative to coal and lignites due to the solid form and remarkable extent of organic content. Utilization possibilites, similar to coal and lignites, have been considered in the past decades and direct use of oil shales in thermal power production has been possible in countries like Estonia and China. In the perspective of utilization of oil shales in a similar manner to coal and lignites, problems and restrictions related to the inorganic ash-making and potentially pollutant constituents are applied. In this respect, cleaning of this important energy source through mineral processing methods, particularly by flotation, is an outstanding option. However, on the basis of unique features and distinctive characteristics, treatment of oil shales like a type of coal is a big perception and may be highly misleading. This paper discusses specific challenges regarding flotation behavior of oil shales with reference to the surface characteristics and behavior of oil shale entities – probably the most important aspect that determines the efficiency and success of the flotation based cleaning process.

1 Introduction

Energy supply is still a major issue for human civilization. Increasing demand for energy in the globe necessitates evaluation and usage of various alternative sources. Despite efforts towards utilization of renewables, non-renewable solid fossil fuels are still an outstanding energy resource. On accounts of factors such as secure supply, lower costs of utilization and extensive and well-dispersed reserves in many countries, not only coal, but other solid fossil fuel reserves, become even more important. Among alternatives to coal and lignite, oil shale is probably the most important solid fossil fuel option, with huge reserves, a variety of utilization possibilities and beneficiation routes similar to those applied to coal and lignites.

The type of organic matter in oil shales consists mainly of kerogen and with lower amounts bitumen. In this respect, oil shales have been treated as an unconventional oil source [1, 2]. However, being solid and having a combustible potential, using oil shales as a solid fuel substitute is also a known practice. Estonia is particularly an outstanding example for such a practice and in use of oil shales in thermal power production, similar to coal and lignite usage [3]. Considering oil shales as an

[a] Corresponding author: ealtun@metu.edu.tr

alternative solid fossil fuel rather than an unconventional oil source introduces envrionmental restrictions, which apply to utilization of coal such as ash, post-combustion gases, etc.

Despite the possibility of application of cleaning methods such as flotation and/or gravity concentration techniques, oil shales possess extremely specific characteristics based on their origin, formation and occurrence. These critical differences, related to the surface behavior and/or type of organic matter, require a comprehensive evaluation of the physical, chemical and physicochemical characteristics of oil shales prior to establishing a utilization route.

This paper specifically discusses distinctive surface characteristics of oil shales, both in their original forms as well as after grinding, as a primary step before further physical/physicochemical processes that could be applied. The objective is also to assess these distinctive features as an explanation to some of the critical challenges during flotation cleaning of oil shales to reduce the amount of ash making and potentially pollutant constituents, related to use of oil shales as a substitute to coal and lignites.

2 Issues related to composition and surface characteristics of oil shales

Figure 1 and 2 show XRD spectra of a high and low quality oil shale samples from Turkey, respectively [4, 5]. The spectra show that oil shales may consist of variety of inorganic minerals, i.e. ash making components as well as those resulting in harmful post-combustion emissions. All these inorganic minerals, typical for almost any oil shale deposit, show the potential and necessity for application of mineral processing methods for cleaning of oil shales, i.e. rejection of inorganic constituents to achieve a cleaner and environmentally sound enery resource alternative. In addition, remarkable extent of sulfur involving constituents such as pyrite in oil shales point to the requirement of cleaning procedures to avoid/decrease post-combustion SO_2 emissions. It should be noted that, despite the presence of a variety of minerals, the inorganic matter in oil shales is dominated by carbonate minerals, i.e. calcite and dolomite, with calcite being the major carbonate. As opposed to many types and particularly low-ranking coals, the organic matter is abundant in oil shales, especially in those with lower quality.

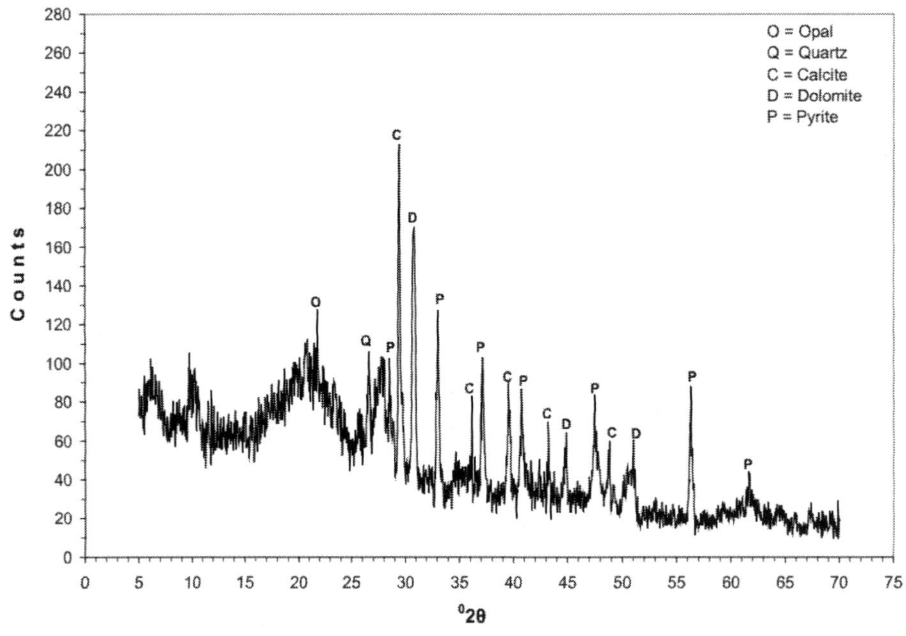

Figure 1. XRD spectrum of a high-quality oil shale [4].

 DOI: 10.1051/e3sconf/20160801049

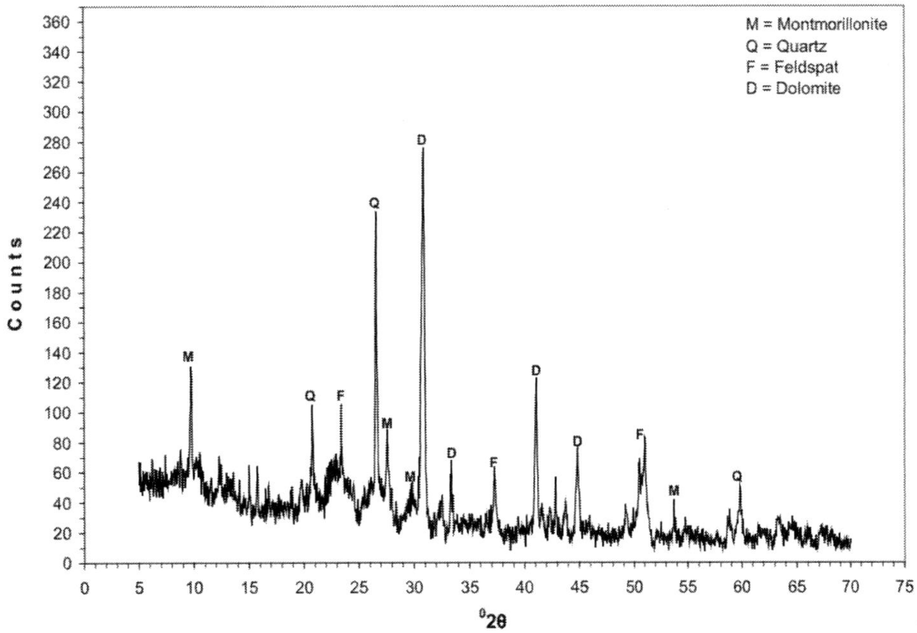

Figure 2. XRD spectrum of a low-quality oil shale [5].

The relative abundance of inorganic matter and carbonates is a remarkable challenge in application of flotation as oil shales cleaning [4, 5]. In direct flotation, high extent of the inorganic matter in the pulp presents a challenge against the bubble-organic matter interaction. The smears of organics on the surfaces increases the possibility of entrapment and eventually causies dilution of the organic-rich concentrate. All these factors reduce the efficiency and effectiveness of oil shale cleaning using the direct flotation process. The results of such challenges and the insufficient cleaning of oil in some details in the previous studies shales using direct flotation were well expressed in some details in previous studies [4, 5].

The surface features of oil shale particles is another major challenge in flotation cleaning of oil shales. Figure 3 shows the FTIR spectra of a high-quality oil shale [6]. The spectra were retrieved using as-received oil shale sample. In addition to characteristic bands and peaks corresponding to the organic matter, apparent bands and strong, high-intensity peaks for functional groups were recorded (Fig. 3), such as those for phenolic and carboxylic hydroxyls, carboxyl and carbonyl groups. Despite evidence of the rich organic character, significant functional bands corresponding to hydroxyl, carboxyl and carbonyl groups in this typical spectrum point to a highly humic nature. The high -OH band at 3422 cm^{-1} as well as strong forms of carboxyl and carbonyl bands correspond to a humic character, even higher than low-rank, immature lignites [7, 8]. The oxygen-rich hydroxyl, carboxyl and carbonyl functionals are polar in nature. The remarkable presence of these polar functionals imposes a hydrophillic nature to oil shale particles during flotation [9], and this is similar to the poor floatability of low-rank lignites. These findings suggest that anticipating a hydrophillic behavior, even for high quality oil shales with good organic content, is a perception and high wettability of oil shale particles due to the polar surface behavior.

 DOI: 10.1051/e3sconf/20160801049

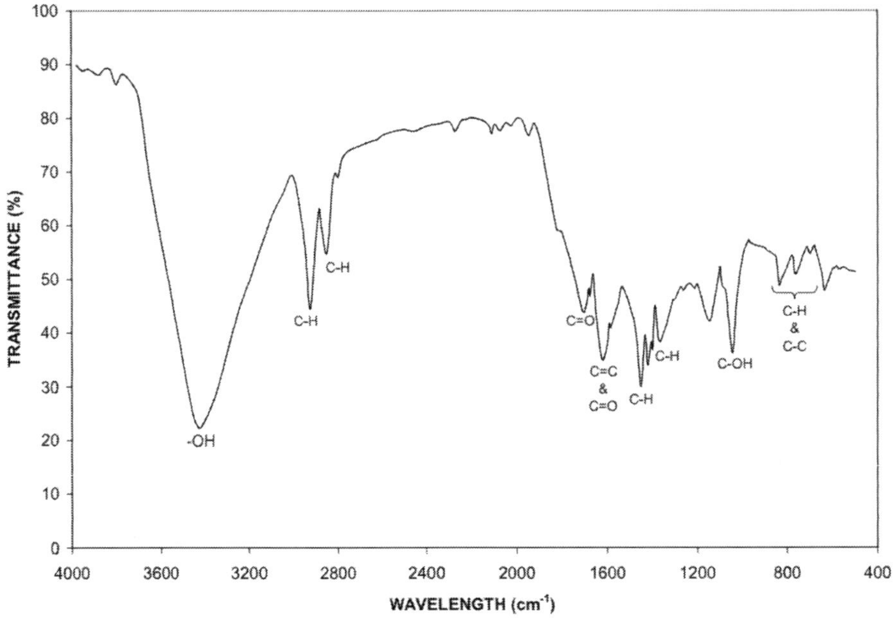

Figure 3. FTIR spectrum of a high-quality oil shale [6].

3 Issues related to post-grinding surface characteristics

Besides the hydrophillic behavior of oil shale particles in the flotation pulp, the release of organic matter in relatively liquid form, is an important difficulty in flotation based oil shale cleaning. This difficulty corresponds to the incidental, i.e. undesired and uncontrolled release of bitumen as the grind size becomes finer and subsequent coating of particles, regardless of either inorganic or organic, by this bitumen film [6]. The bitumen film is organic. The thin film around inorganic/organic entities caused by the smearing of released bitumen may significantly interfere with the selectivity of flotation, since this film randomly renders inorganic/organic particles hydrophobic. This non-selective hydrophobic behavior will eventually result in flotation of inorganic entities along with organic ones, imparting separation between the ash making and organic-rich constituents.

Fine grinding is inevitable and acts as a preparatory step for downstream processes, aiming at the disintegration of the organic and inorganic constituents prior to flotation cleaning of solid fossil fuels. Figure 4 demonstrates the impact of incidental release of bitumen during grinding of oil shales on the surface characteristics of oil shale particles [6]. With increasing fineness from the raw sample down to -25 μm, the characteristic bands and peaks corresponding to the organic content of the oil shale became more intense and apparent. The aliphatic C-H (due to CH_3 and CH_2) and aromatic C-C and C=C bands gradually increased intensities, reaching their maximum at -25 μm grind size (Fig. 4). The C-H bands at 2926 and 2850 cm^{-1} took a sharper form with increasing fineness. The C-H peak at 2926 cm^{-1} occurred with a broader form at -50 and -25 μm, as compared to coarser sizes. The intensity increase was also recorded for the characteristic C-H peak at 1452 cm^{-1} and the C-H and C-C bands at 850 and 675 cm^{-1}, respectively, became stronger as the grind size was made finer.

DOI: 10.1051/e3sconf/20160801049

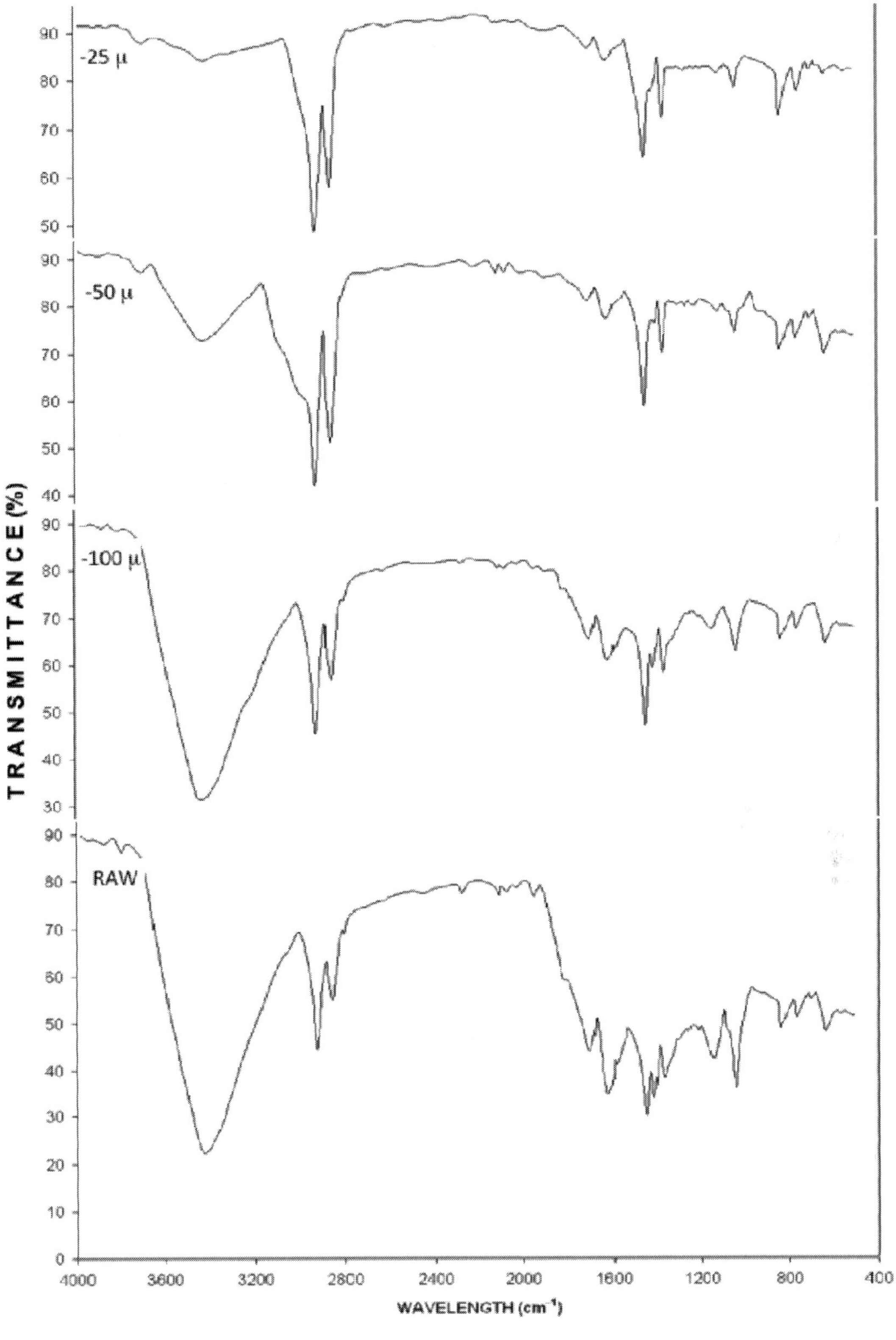

Figure 4. Comparative FTIR spectra of oil shale at different grind sizes [6].

Besides significant changes in the forms and intensities of peaks and bands that correspond to the organic matter in oil shales, the peak intensities of the polar functional groups also changed with respect to grind size. The extent of the functional groups relatively decreased with increasing fineness as revealed by the reducing intensities of the -OH bands of phenolic hydroxyls and C=O and C-OH peaks of carbonyl and carboxyl groups. The distinctive high-intensity of -OH band at 3422 cm^{-1} in the spectrum of the raw oil shale took a very slight form in the -25 μm sample (Fig. 4). The decreasing intensity of the polar functional groups and the increasing intensities of the peaks of organic groups showed that the oil shale particle surfaces became dominated by the organic matter with increasing fineness. The release of bitumen with fine-grinding of oil shale and coating of oil shale particles non-selectively by an organic film was also demonstrated by the XPS analysis of the surface of oil shale particles, as presented in Table 1 [6].

The presented data suggest and confirm that the humic character of oil shale particle surfaces changed to an organic-dominant form in prolonged grinding periods. This change corresponds to the shift of the natural hydrophilicity of oil shale particles to the hydrophobic behavior. Yet, this change does not present an advantage with respect to rejection of ash making and polluting constituents through flotation, because this hydrophobic behavior, as imposed by the uncontrolled release of bitumen, is non-selective, and randomly applies to any particle during the grinding period, as mentioned earlier.

Table 1. Surface characteristics of oil shale particles with respect to grind size [6].

Grind size (μm)	Elemental composition (%)						Carbon functionalities (%)				
	C	O	Ca	Mg	Si	Fe	C-H	C-C	C=C	C=O	C-OH
-200	30.60	23.88	21.78	13.56	3.76	5.74	18.22	13.94	10.52	44.32	12.92
-100	34.36	22.45	20.98	12.89	3.45	5.17	19.56	14.52	11.00	42.8	12.02
-50	54.98	16.88	14.78	8.46	1.52	2.92	30.98	21.65	15.28	23.89	8.14
-25	64.90	14.58	11.42	5.68	1.02	1.68	34.07	23.98	16.37	18.94	6.56

4 Conclusions

Previous work show that cleaning of oil shales through flotation is not a straight forward process. Behavior of oil shale particles in the flotation pulp is determined by oxygen-rich polar surface functionals, that impose natural hydrophillicity to oil shale particles. Remarkable extent of inorganic minerals, such as carbonates, is another negative aspect, that may keep the efficiency and effectiveness of flotation cleaning limited. This is due to interference and smearing of inorganic particles with/on organic entities, dilution of low-ash concentrate due to in-bubble entrapment/inocculation of fine inorganics. Another major issue regarding the surface behavior is the uncontrolled release of organic matter during grinding of oil shale. It is likely that release of organic matter results in randomly coating of oil shale particles by an organic film, particularly as the grind size is made finer. This smearing effect eventually results in the loss of selectivity, which in turn, may reduce the separation possibility between organic and inorganic entities. It is therefore suggested that the grind size should be carefully decided, grinding should be strictly controlled and over-grinding should be avoided in preparing oil shales for flotation cleaning.

References

1. T.F. Yen, G.V. Chilingarian, *Oil Shale* (Elsevier, Amsterdam, 1976)
2. B.P. Tissot, D.H. Welte, *Petroleum Formation and Occurrence* (Springer Verlag, Germany, 1984)
3. E. Vali, I. Valgma, E. Reinsalu, Oil Shale, **25(2S)**, 101 (2008)

4. N.E. Altun, C. Hicyilmaz, J.-Y. Hwang, A.S Bagci, *Energ. Fuel.*, **20** (2006)
5. N.E. Altun, C. Hicyilmaz, J.-Y. Hwang, A.S. Bagci, *Fuel Process. Technol.*, **87**, 783 (2006)
6. N.E. Altun, *Oil Shale* **26 (3)**, 382 (2009)
7. A.F. Gaines, *The Infrared Spectra of Coals* (NATO ASI Series C, Kluwer Academic Publishers, Dordrecht, 1988)
8. D.W. Van Krevelen, *Coal,* 3rd edn. (Elsevier, Amsterdam, 1993)
9. G. Atesok, M.S. Celik, *Fuel*, **79**, 1509 (2000)

Viscosity and density models for copper electrorefining electrolytes

Taina Kalliomäki[1,a], Arif T. Aji[1], Jari Aromaa[1], and Mari Lundström[1]

[1]*Department of Materials Science and Engineering Aalto University, Vuorimiehentie 2, 02150 Espoo, Finland*

Abstract. Viscosity and density are highly important physicochemical properties of copper electrolyte since they affect the purity of cathode copper and energy consumption [1, 2] affecting the mass and heat transfer conditions in the cell [3]. Increasing viscosity and density decreases the rate in which the anode slime falls to the bottom of the cell [4, 5] and lowers the diffusion coefficient of cupric ion (D_{Cu2+}) [6]. Decreasing the falling rate of anode slime increases movement of the slime to other directions than downward [4, 5]. If the anode slime ends up on the cathode, the impurities could entrap into coating [4]. Due to that the aim is to keep the viscosity and density sufficiently low [4]. According to the studies of Price and Davenport [3], Subbaiah and Das [2], Devochkin et al. [7] as well as Jarjoura et al. [8] increasing the concentration of copper, nickel and sulfuric acid increases both viscosity and density, while temperature decreases these properties. In addition, Price and Davenport [1] researched the effect of impurities arsenic and iron as well as Subbaiah and Das [2] the effect of iron, manganese and cobalt. All of these researchers found that those impurities increased both viscosity and density. The density and kinematic viscosity of copper electrorefining electrolytes have not been extensively researched under electrorefining conditions. The kinematic viscosity is also an important factor in the equation defining D_{Cu2+} using a rotating disc electrode technique [6]. The errors in the viscosity values cause significant error to D_{Cu2+}. Thus, this work introduces mathematical models for the density and kinematic viscosity. The kinematic viscosity of the test electrolytes was measured with a Ubbelohde capillary viscometer from Schott-Geräte GmbH and the density with a glass tube oscillator DMA 40 Digital Density Meter from Anton Paar K. G. The temperature (50, 60, 70 °C) and electrolyte composition were used as variables. The composition variables investigated were copper (40, 50, 60 g/dm^3), nickel (0, 10, 20 g/dm^3) and sulfuric acid (130, 145, 160 g/dm^3) in all models, and additionally the effect of arsenic acid for the viscosity was studied (0, 15, 30 g/dm^3). The electrolytes used in these tests were prepared from $CuSO_4 \cdot 5H_2O$ (99–100 %), $NiSO_4 \cdot 7H_2O$ (99–100 %), H_2SO_4 (95–97 %) and arsenic acid (containing As 151700 mg/dm^3, Bi 6.2 mg/dm^3, Se 0.07 mg/dm^3, Te 18.6 mg/dm^3, Ag 0.2 mg/dm^3, Cu 4794 mg/dm^3, Ni 1688 mg/dm^3, Pb 28.62 mg/dm^3 and Sb 3954 mg/dm^3). The results were normalized using known water values for viscosity as well as water and air values for the density. Based on these results the models for density (ρ, g/cm^3) and kinematic viscosity (ν, mm^2/s) were designed, refined and analyzed using modeling and design tool MODDE, Equation 1 for density and 2 for kinematic viscosity:

[a] Corresponding author: taina.kalliomaki@aalto.fi

$$\rho = 1.03473 + 0.00216707\,[Cu] + 0.000535362\,[H_2SO_4] + 0.00234682\,[Ni] - 0.000713461\,T \quad (1)$$

$$\nu = 0.07989 + 0.002868\,[Cu] + 0.0005293\,[H_2SO_4] + 0.003349\,[Ni] + 0.004768\,[As] - 0.007038\,T - 2.175 \cdot 10^{-05}\,T[As] \quad (2)$$

The designed model of density (Equation 1) and viscosity (Equation 2) were in a good agreement with the results of Price and Davenport [1] and Devochkin et al [7] when the effect of As was not included (Figs. 1 and 2). However, the density values of Subbaiah and Das [2] were slightly lower than the obtained this work, and it should be noted that the model of Subbaiah and Das [2] was constructed of results mainly measured at 30 °C. That is outside the range of the temperatures used in the electrorefining process. Moreover, the viscosity values defined with the current model were higher than the equivalent values of Price and Davenport [1], when the effect of As was taken into account (Fig. 2).

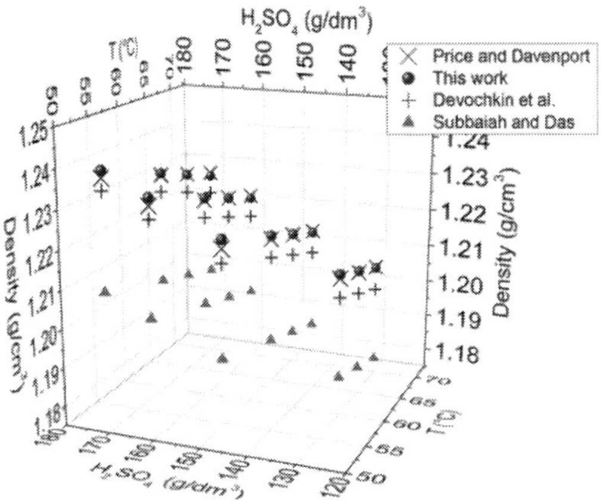

Figure 1. Copper electrorefining electrolyte density as a function of H_2SO_4 and temperature, plotted using models from this work, and models of Price and Davenport[1], Subbaiah and Das [2] and Devochkin et al. [7] with [Cu(II)] = 50 g/dm^3 and [Ni(II)] = 18 g/dm^3.

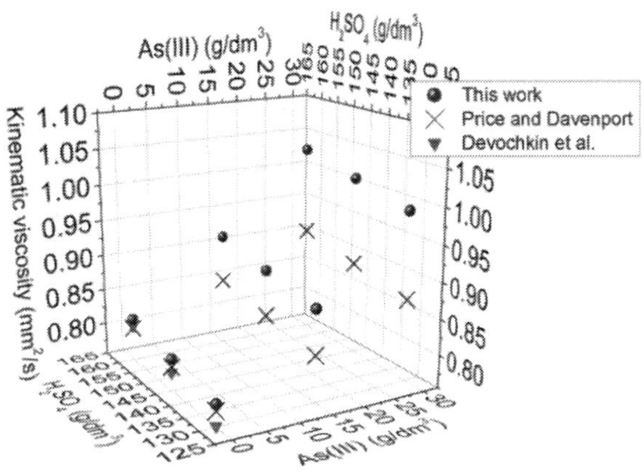

Figure 2. Copper electrorefining electrolyte kinematic viscosity as a function of As(III) and H_2SO_4 plotted using models from this work, and models of Price and Davenport [1] and Devochkin et al. [7] and with [Cu(II)] = 50 g/ dm^3 and [Ni(II)] = 18 g/ dm^3.

References

1. D.C. Price, W.G. Davenport, *Metall. Trans. B*, **12(B)**, 639 (1981)
2. T. Subbaiah, S.C. Das, *Metall. Trans. B*, **20(B)**, 375 (1989)
3. D.C. Price, W.G. Davenport, *Metall. Trans. B*, **11(B)**, 159 (1980)
4. W.G. Davenport, M. King, M. Schlesinger, A.K. Biswas, T. Robinson, *Extractive Metallurgy of Copper*, 4th edn.(Elsevier Science Ltd., 2002)
5. Y. Shi, Z. Ye, An, *in Proceedings of 4th International Symposium on High Temperature Metallurgical Processing* (2013)
6. M.S. Moats, J.B. Hiskey, D.W. Collins, *Hydrometallurgy*, **56**, 255 (2000)
7. A.I. Devochkin, I.S. Kuzmina, E.V. Salimzhanova, L.I. Petukhova, Tsvetnye Metally, **2015(6)**, 67 (2015)
8. G. Jarjoura, M. Muinonen, G.J. Kipouros, Canadian Metall. Q., **42(3)**, 281(2003)

Separation of ultrafine particles from class F fly ashes

Ilker Acar[1,a], Thomas L. Robl[2], and M. Umit Atalay[3]

[1]*Ataturk University, Faculty of Earth Sciences, Department of Petroleum and Natural Gas Engineering, Erzurum, Turkey*
[2]*University of Kentucky, Center for Applied Energy Research, Lexington, USA*
[3]*Middle East Technical University, Department of Mining Engineering, Ankara, Turkey*

Abstract. In this study, ultrafine particles were recovered from Çatalağzı (CFA) and Sugözü (SFA) thermal power plant fly ashes using a specific hydraulic classification technology. Since fly ashes have a high tendency to be flocculated in water, settling experiments were first designed to determine the more effective dispersant and the optimum dosage. Two different types of the superplasticizers (SP) polymers based on sulphonate (NSF, Disal) and carboxylate (Glenium 7500) were used as the dispersing agents in these settling experiments. Hydraulic classification experiments were then conducted to separate ultrafine fractions from the fly ash samples on the basis of the settling experiments. According to the settling experiments, better results were achieved with the use of Disal for both CFA and SFA. The classification experiments showed that the overflow products with average particle sizes of 5.2 μm for CFA and 4.4 μm for SFA were separated from the respective as-received samples with acceptable yields and high enough recoveries of -5 μm (ultrafine) particles. Overall results pointed out that the hydraulic classification technology used provided promising results in the ultrafine particle separations from the fly ash samples.

1 Introduction

Fly ash can be defined as the main waste residue resulted from the combustion of either ground or powdered coal and occurred as a fine powder of mainly spherical and glassy particles [1-3]. Coal properties, combustion conditions and storage methods determine the mineralogy, chemistry and physical properties of fly ash. Therefore, fly ash derived from burning of different types of coal (e.g., anthracite, bituminous and lignite coal) has different chemical compositions and physical properties [4-5].

Fly ash has a wide-range of particle size varying from sub-micron to 500 μm. Specific gravity of fly ash varies in the range 2.1-2.6, while bulk density ranges from 1 to 1.8 g/cm^3 [5]. Chemically, fly ash includes major (>1%), minor (1-0.1%) and trace (<0.1%) elements. Since it is a combusted material, the major and minor elements are present as oxides. In order of decreasing amounts, these elements are commonly O, Si, Al, Ca, Fe, C, K, Mg, H, Na, Ti, N, P, and Ba. The other elements are mostly present in trace amounts in fly ash [6].

[a] Corresponding author: ilker.acar@atauni.edu.tr

Fly ash utilization has increased in recent years due to its various environmental, economic and technical benefits. Firstly, less environmental pollution occurs in fly ash disposal area. Secondly, lower disposal costs and less area requirement for disposal are also provided with the use of this combustion waste. Further environmental and economic benefits take place when fly ash is used as a partial replacement material for some important natural resources. In addition, fly ash can be used as an essential additive in some sector to improve the properties of final products, such as special pozzolanic additive in cement and concrete industries [7].

According to the American Coal Ash Association (ACAA) and European Coal Combustion Products Association (ECOBA), utilization of total produced fly ash in 2013 in US was 43.7%, while in 2010 this utilization was 43.0% in Europe (EU-15 Countries). The majority of these utilizations were in the cement and concrete industries for both US and Europe. Thus, for the remaining material, disposal practices are used involving ash ponds, lagoons and landfills. However, all of them can be considered as environmentally undesirable use of land resources [8]. In order to increase the utilization potentials, heterogeneous as-received fly ashes should be subjected to a treatment for removal of contaminants and separation of particles into size ranges providing fly ash utilization in a wide-range of added value applications. One of the highest-value fractions in coal fly ashes is ultrafine particles which have a great potential as a suitable filler material for plastic polymers, concrete and cement [9-10].

Today, hydrocyclone and hydraulic classifiers are the most widely used classifiers in mineral processing. For separation of fine particles, hydrocyclones are commonly used in most applications down to 75 μm and smaller size cyclones can be used to separate fine particles down to a few microns. However, in addition to the problem of plugging due to small openings of the cyclone, the capacity is also reduced under this condition. On the other hand, conventional hydraulic classifiers are usually designed to obtain clean coarse fractions without slimes. The hydraulic classifiers have a major disadvantage known as "blanket settling" which can be defined as the hindering of smaller particles' upward movements due to coarser particles. As expected, the hydraulic classifiers generally provide very low yield and recovery for separation of very fine particles [11].

Hydraulic classification is mainly based on the size and density of a material. In hydraulic classification systems, fly ash generally tends to flocculate and these flocs are considerably stable even for high pH values. In literature, some known chemicals such as sodium silicate and potassium hydroxide have been used to disperse the flocs. However, these attempts have not given satisfying results for classification of fly ash. Some good results have been obtained with potassium metaphosphate and pyrophosphate [9]. On the other hand, these results have been erratic and highly substrate dependent. Therefore, it is required to find another dispersants producing high yield.

Some chemicals known as "superplasticizers" (SP), which are a group of organic water soluble polymers with long molecules of high molecular mass, are largely used in the cement and concrete industry because of their contributions to dispersion and water reduction properties. According to some projects, these chemicals can also be used as effective dispersing agents for separation of ultrafine particles [9, 11].

In this study, it was aimed to separate the ultrafine fractions (-5 μm) from the as-received Çatalağzı (CFA) and Sugözü (SFA) fly ashes with high enough recoveries and acceptable yields using a specific hydraulic classification technology. For this purpose, settling experiments were first designed to determine more effective dispersant and the optimum dosage. The hydraulic classification experiments were then conducted to separate ultrafine particles from the as-received class F fly ashes on the basis of the settling experiments.

2 Experimental

2.1 Materials

The coal fly ash samples used in this study were obtained from Çatalağzı and Sugözü thermal power plants in Turkey. Both of the fly ashes were resulted from burning of bituminous coals. A Spectro IQ X-ray fluorescence (XRF) spectrometer was used to determine major chemical constituents of the fly ashes. Loss-on-ignition (LOI) tests were performed in accordance with ASTM C 311. A Malvern Hydro 2000S particle size analyzer was used to determine particle size distributions of the samples. Porosity measurements were conducted using a mercury porosimeter. Density measurements were carried out using a helium pycnometer. Specific surface area values were determined with BET nitrogen adsorption method. Table 1 shows the major chemical constituents and basic physical properties of the as-received fly ash samples.

Table 1. Major chemical constituents and basic physical properties of the samples.

Chemical composition (%)	**Çatalağzı (CFA)**	**Sugözü (SFA)**
SiO_2	57.09	63.35
Al_2O_3	27.46	22.01
Fe_2O_3	6.56	7.90
CaO	1.97	1.45
MgO	2.38	2.31
Na_2O	0.31	0.79
K_2O	3.89	1.95
P_2O_5	0.11	0.16
TiO_2	1.13	0.91
SO_3	<0.01	<0.01
LOI (%)	1.68	2.15
Physical properties		
Density (g/cm^3)	2.08	2.36
Average particle size (μm)	39.29	21.42
Total porosity (%)	19.99	6.97
Surface area (m^2/g)	0.8284	1.1715

According to Table 1, the major chemical constituents of the fly ash samples were SiO_2, Al_2O_3 and Fe_2O_3. This result indicated that both of the fly ashes met the chemical requirements to be a class F fly ash consistent with ASTM C 618 due to their high amounts of silica, aluminum oxide and ferric oxide and low amounts of sulfur trioxide and LOI values. Since the fly ash samples were bituminous coal combustion residues, both of them had low lime contents. The differences in average particle sizes between the fly ashes can also clearly be seen from Table 1. Çatalağzı fly ash has an average particle size of 39.29 μm, nearly two times greater in comparison to Sugözü fly ash. Table 1 also shows that Sugözü fly ash had the higher density and surface area and lower porosity values than Çatalağzı fly ash.

Mineralogical compositions of the samples were determined using X-ray powder diffraction (XRD). The XRD results indicated that quartz, mullite and hematite were the main crystalline phases for both samples. Scanning electron microscopy (SEM) was used to examine microstructural properties of the fly ashes. According to the SEM observations, the fly ash samples mainly contained small and glass like particles with spherical shape and also larger irregularly shaped particles.

2.2 Methods

In this study, ultrafine fractions of Çatalağzı (CFA) and Sugözü (SFA) thermal power plant fly ashes were separated using a specific hydraulic classification technology. In order to accomplish this, fly ash particles had to be dispersed prior to classification. Therefore, settling experiments were first conducted to determine more effective dispersant and its optimum dosage. Based on these settling experiments, the ultrafine particles were recovered from the as-received fly ashes using a specific hydraulic classifier.

2.2.1 Settling experiments

Two different types of superplasticizers (SP) polymers based on sulphonate (NSF, Disal) and carboxylate (Glenium 7500) were used as the dispersing agents in the settling experiments. Unlike NSF dispersant (Disal), Glenium 7500, which is a high range water reducing admixture, is based on the next generation of polycarboxylate technology. Dispersant dosages of 0.5, 1.0, 1.5, 2.0, 2.5 and 3.0 mg/g of fly ash were tested in these experiments. Also, control experiments were done without dispersants to observe the differences. In these experiments, pulps with 1.5% solid content were prepared in a 250 ml graduated cylinder. After the required amount of dispersant added to the fly ash pulp, the graduated cylinder was well mixed for a minute. The pulp was taken down to 100 ml level using a vacuum pump after 5 minutes of settling time. After filtration and drying, the sample was weighed. Weight percentages of solid in the floating products were used as the measure of dispersant efficiency.

2.2.2 Hydraulic classification

The hydraulic classification process developed at the CAER (University of Kentucky, Center for Applied Energy Research) laboratory is based on a method for cross-flow of particles in slurry for selective separation of fine particles. Particle separation in this process was accomplished with the use of particle velocity differences in water (influenced by density, size and shape of particles). The particular system was a modified version of lamella hydraulic classifier (LHC), in which interior dividers were placed for effective removing of coarse particles from the system by increasing the settling area. The bottom of the classifier was separated into four underflow discharge areas for selective removal of particles with sequentially decreasing size. Ultrafine particles were obtained from overflow discharge area [11-12]. The lamella hydraulic classifier (LHC) used in this study is schematically illustrated by Figure 1.

In this hydraulic classification process, the important process parameters were the feed pulp density, feed rate, underflow discharge rates, type and dosage of dispersant and divider spacing. Prior to each experiment, slurry feed rates and the dispersant dosage were carefully adjusted to the desired values. Then, the classifier was filled with water up to the level of discharge pipe, and discharge rates of underflows were adjusted to remove the coarse particles from the system. After that, the classification process started. The slurry and mass flow rates of feed, overflow and underflow products were checked two times by taking the samples from each for a definite period of time during the process. After the experiment, percentage of solids and loss-on-ignition (LOI) values of feed, overflow and underflow were measured. Particle sizes of the all fractions were measured using a Malvern Hydro 2000S particle size analyzer. Figure 2 shows the lamella hydraulic classifier (LHC) during the experiment.

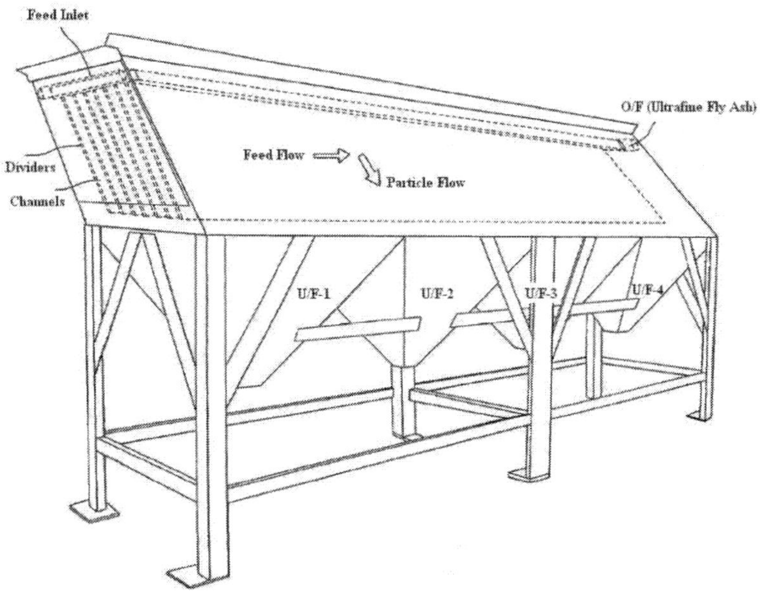

Figure 1. Schematic representation of the lamella hydraulic classifier (LHC).

Figure 2. The lamella hydraulic classifier (LHC) during the experiment.

In this process, the product grade was defined as the mean particle size of the overflow on volumetric basis. In addition to overflow mean size, yield and recovery were also calculated as measures of processing efficiency according to equations:

$$Product\ yield = \frac{Mass\ of\ product\ solids}{Mass\ of\ feed\ solids} \quad (1)$$

$$Product\ recovery = \frac{Mass\ of\ product\ solids\ finer\ than\ 5\mu m}{Mass\ of\ feed\ solids\ finer\ than\ 5\mu m}\ . \quad (2)$$

Some codes were given to the as-received fly ashes (FA) and their ultrafine fractions (UFA) through the experiments. These codes were CFA (Çatalağzı FA), CUFA (Çatalağzı UFA), SFA (Sugözü FA) and SUFA (Sugözü UFA).

 DOI: 10.1051/e3sconf/20160801051

3 Results and discussion

3.1 The results of settling experiments

Two different types of superplasticizer polymers based on sulphonate (NSF, Disal) and carboxylate (Glenium 7500) were used in these experiments to determine their dispersive effects on CFA and SFA samples. Figures 3 and 4 illustrate the results of settling experiments for CFA and SFA, respectively.

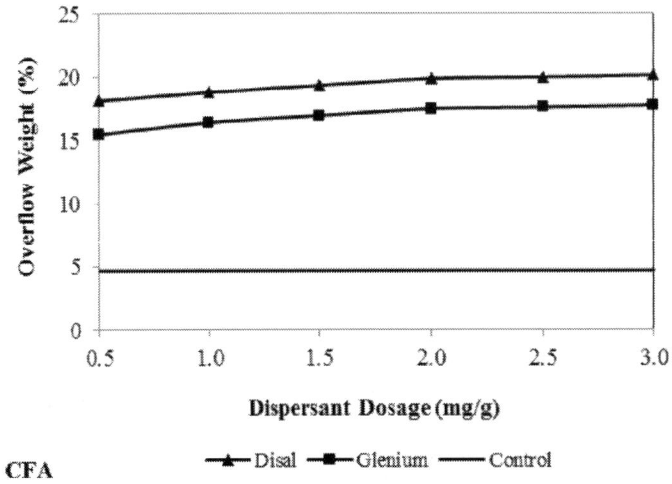

Figure 3. The results of settling experiments for Çatalağzı fly ash (CFA).

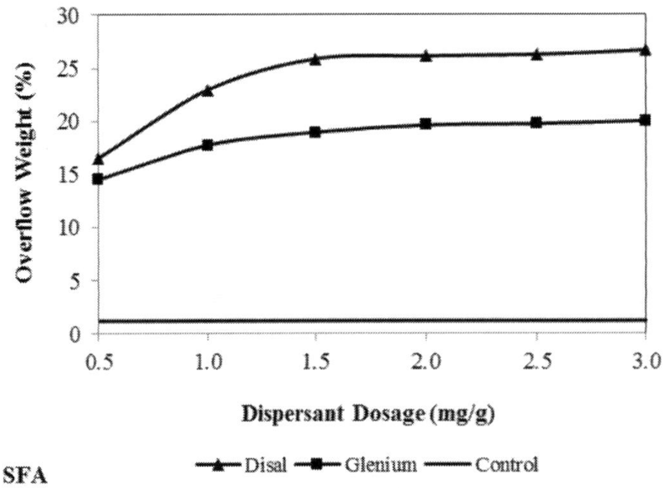

Figure 4. The results of settling experiments for Sugözü fly ash (SFA).

Figures 3 and 4 show that overflow weight percentages of the fly ash samples increased with increasing dispersant dosage. However, for both dispersants, very small increases in the overflow weight percentages were observed beyond the dispersant dosages of 2.0 mg/g for CFA and 1.5 mg/g for SFA. Although both dispersants showed the same patterns, better results were achieved with the use of Disal for both CFA and SFA. Figures 3 and 4 also indicate that very low overflow weight percentages were obtained without the addition of dispersants.

3.2 Hydraulic classification

3.2.1 Çatalağzı fly ash

The classification conditions and obtained results for Çatalağzı fly ash are shown in Tables 2 and 3, respectively. According to Table 3, from CFA with average particle size (d_{50}) of 39.29 μm, the ultrafine fraction (CUFA) with average particle size of 5.25 μm was separated with 10.75% yield and 53.28% recovery of -5 μm particles in the overflow (O/F) product.

Table 2. Classification conditions for CFA.

Operating conditions	
Feed pulp density	1.5% solids by wt. (0.015 kg/l)
Feed rate	11 l/min
Underflow discharge rates	U/F-1 = 300 cc/min, U/F-2 = 200 cc/min U/F-3 = 200 cc/min, U/F-4 = 200 cc/min
Dispersant	Disal
Dispersant dosage	2.0 mg Disal/g of fly ash

Table 3. Classification results for CFA.

Products	**Weight (%)**	**Solid (%)**	d_{50} **(μm)**	**Vol. % of -5 μm**	**-5 μm recovery (%)**
U/F-1	38.80	16.90	63.26	3.84	15.41
U/F-2	27.98	18.98	47.25	4.08	11.81
U/F-3	13.90	8.60	36.62	4.94	7.10
U/F-4	8.56	5.31	21.46	8.10	7.17
O/F (Ultrafine)	10.75	0.16	5.25	47.91	53.28
Total (feed)	**100.00**	**1.49**	**39.29**	**9.67**	

3.2.2 Sugözü fly ash

Classification conditions and the obtained results for Sugözü fly ash (SFA) are shown in Tables 4 and 5, respectively. Smaller underflow rates were chosen this time because of the smaller particle size of SFA compared to that of CFA. According to Table 5, the average size of 4.41 μm particles was obtained with 17.51% yield and 63.13% recovery of -5 μm particles in the overflow product (SUFA). In other words, finer particle size with higher yield and recovery was obtained in this experiment compared to the hydraulic classification test of CFA.

Overall results obtained from the classification tests of Çatalağzı and Sugözü fly ashes indicated that the hydraulic classification technology developed at CAER gave good results for recovery of very fine particles with acceptable yields and high enough recoveries.

Table 4. Classification conditions for SFA.

Operating conditions	
Feed pulp density	1.5% solids by wt. (0.015 kg/l)
Feed rate	11 l/min
Underflow discharge rates	U/F-1 = 200 cc/min, U/F-2 = 150 cc/min U/F-3 = 150 cc/min, U/F-4 = 150 cc/min
Dispersant	Disal
Dispersant dosage	1.5 mg Disal/g of fly ash

Table 5. Classification results for SFA.

Products	**Weight (%)**	**Solid (%)**	**d_{50} (μm)**	**Vol. % of -5 μm**	**-5 μm recovery (%)**
U/F-1	32.18	21.45	34.13	5.43	11.27
U/F-2	25.69	23.08	32.63	5.52	9.15
U/F-3	13.63	12.35	26.90	6.40	5.63
U/F-4	11.00	9.96	16.31	9.76	6.93
O/F (Ultrafine)	17.51	0.23	4.41	55.90	63.13
Total (feed)	100.00	1.46	21.42	15.50	

4 Conclusion

The hydraulic classification technology used to separate ultrafine fractions from the as-received samples was resulted in promising results. The overflow products with average particle sizes of 5.2 μm for CFA and 4.4 μm for SFA were separated from the respective as-received samples with average particle sizes of 39 and 21 μm. The ultrafine fraction of CFA (CUFA) was separated with 10.75% yield and 53.28% recovery of -5 μm particles in the overflow product while 17.51% yield and 63.13% recovery were obtained for the ultrafine fraction of SFA (SUFA). Finer particle size could be obtained with this hydraulic classification technology. However, in this case, yield and recovery values would decrease. The results obtained from all the experiments conducted within the scope of classification indicate that the hydraulic classification technology developed at CAER gave good results for the recovery of ultrafine particles.

References

1. S. V. Vassilev, R. Menendez, D. Alvarez, M. Diaz-Somoano, M. R. Martinez-Tarazona, *Fuel* **82**, 1793 (2003).
2. ASTM C 618, *Standard specification for coal fly ash and raw or calcined natural pozzolan for use in concrete*, ASTM International, West Conshohocken, PA, USA (2012).
3. BS EN 450-1, *Fly ash for concrete: Definition, specifications and conformity criteria*, BSI, London, UK (2005).
4. I. Acar, M. U. Atalay, *Fuel* **106**, 195 (2013).
5. S. Jala, D. Goyal, *Bioresour. Technol.* **97**, 1136 (2006).
6. V. Vassilev, C. G. Vassileva, *Fuel* **86**, 1490 (2007).
7. M. Ahmaruzzaman, *Prog Energy Combust* **36**, 327 (2010).
8. I. Acar, M. U. Atalay, *Fuel* **180**, 97 (2016).
9. T. L. Robl, J. G. Groppo, *Technology and methodology for the production of high quality polymer filler and super-pozzolan from fly ash,* United States Patent (Patent No: US 6,533,848 B1) (2003).
10. M. R. Jones, M. J. McCarthy, L. Zheng, T. L. Robl, J. Groppo, *in Proceeding of World of Coal Ash (WOCA) Conference* (2009).
11. T. L. Robl, J. G. Groppo, *Method for hydraulically separating carbon and classifying coal combustion ash*, United States Patent (Patent No: US 7,963,398 B2) (2011).
12. M. J. McCarthy, M. R. Jones, L. Zheng, T. L. Robl, J. G. Groppo, *Fuel* **111**, 430 (2013).

Coal recovery from a coal waste dump

Zenon Rozanski[1,a], Tomasz Suponik[1], Piotr Matusiak[2], Daniel Kowol[2], Jan Szpyrka[1], Michal Mazurek[1], and Pawel Wrona[1]

[1]*Silesian University of Technology, Faculty of Mining of Geology, 44-100 Gliwice, Poland*
[2]*Institut of Mining Technology KOMAG, 44-100 Gliwice, Poland*

Abstract. The possibilities and efficiency of coal recovery from the waste material located at the Central Coal Waste Dump in Poland were presented in this paper. The waste material includes significant amount of fly ash. Research conducted into determination of energetic properties of such wastes showed that the average ash content was 75.75% and the average gross calorific value was 7.81 MJ/kg. Coal was gravitationally separated from the waste material in a pulsatory jig and in a spiral washer including size fractions: 30-5 and 8-0 mm (this was crushed to a size <3.2 mm), respectively. The application of the pulsatory jig (pulse classifier) allowed to obtain a high-quality energetic concentrate with the ash content lower than 12% and the gross calorific value higher than 26 MJ/kg (with average yield 7.8%). The spiral separator gave much worse results. The average gross calorific value for the concentrate was 11.6 MJ/kg, with the high ash content 56.5% and yield approximately 26%.

1 Introduction

The Central Coal Waste Dump Przezchlebie (CCWD) is located in the Upper Silesia region of Poland. It was built in '60 of previous century. Previously it was the location of former sand mine which was flooded in 1955. The CCWD served for following hard coal mines: Makoszowy, Pstrowski, Bielszowice, Krupiński, Zabrze, Bobrek and Gliwice. In addition CCDW was used as fly ash landfill for Rybnik power plant. Fly ash storage was stopped in 1990 and coal waste was ended in 1999.

The CCWD is the one of the largest sites of this type within the region. The area is approximately 150 ha. There is a lack of detailed description of storage process and amount of waste material which is left at the dump, although it is estimated. It is assumed that currently there is about 25 mln m^3 of waste material and probably 65% of total value can be classified as mining wastes and 35% as power plant wastes.

Waste material located in the CCWD is not neutral for the environment. Considering chemical composition of wastes, there is a possibility of eluviations of dissolved components and their transport to water. This process intensifies when degree of weathering process increases and pH of water infiltrating through the dump decreases. The content of: SiO_2, Al_2O_3, TiO_2, Fe_2O_3, CaO, MgO, K_2O+Na_2O in the material from the CCWD was as follows: 45.22%, 18.48%, 0.99%, 4.06%, 1.40%, 1.41%, 1.21%. The content of the eluted elements was: Ba = 576 mg/kg, Cr = 76 mg/kg, Zn = 167 mg/kg, Co = 49 mg/kg, Mn = 1573 mg/kg, Cu = 130 mg/kg and Ni = 85 mg/kg [1]. The mentioned

[a] Corresponding author: Zenon.Rozanski@polsl.pl

above content is similar to typical values obtained for coal waste dumps within the Upper Silesia region [2, 3].

Some amount of coal and sulphur in coal wastes creates fire hazard which is caused by natural processes of coal and pyrite oxidation. It is connected with gas and heat emissions to the atmosphere [4, 5].

In 2009 local authority started to do activities for the development of the site. Now, there are several directions of further works: recovery of coal and aggregates, application of ashes to construction of dump solid and technical reclamation of the site. These activities can give rational economic profits and can decrease negative impact of the dump on the environment.

To find optimal solution, the waste material was put into the analysis and research. The one of the aims was to determine possibility of coal recycling from the dump. The research into this matter was conducted with two methods based on gravitational separation - spiral separators and pulsation jigs. After separation of the size fraction of > 30 mm as the waste material, the tests were carried out for two different sizes:
- 30 – 5 mm – separation at pulsatory jig,
- 8 – 0 mm – separation at spiral washer (after crushing the particles to the size of <3.2 mm).

2 Selected properties of waste material from the dump

2.1 Preparation of the samples

The samples of waste material from the CCWD were taken according to Polish standard PN-ISO 18283:2008 [6]. The samples were taken from three excavations. The depth of sampling was 3-6 meters (measured from the surface of the dump). Mass of each sample was approximately 20 kg.

Preparation of laboratory samples was based on mixing up of raw samples to prepare averaged and general samples. Then, they were quartered and lowered to obtain samples no. I, II and III adequately for each excavation. To achieve proper analytical grain class the samples were preliminary and secondary crushed. The samples were taken into sieve analysis. Then the samples were creamed in a mortar to obtain grain size lower than 0.2 mm. It allowed to separate analytical samples. In addition, the samples were dried (except samples used to moisture determination). Among other parameters, hygroscopic moisture, ash content, net calorific value etc. were determined.

According to Polish standard PN-EN 1097-7:2008 [7] the density of the samples was as follows: 2.29872, 2.28103 and 2.00579 g/cm^3. Thus average value was 2.19518 g/cm^3.

2.2 Determination of energetic features

Average free moisture content (W_{ex}), hygroscopic moisture content (W_h) and ash content (A^a) in the analytical samples were determined by the method given in Polish standards PN-80/G-04511 [8], PN-80/G-04512 [9]. Net calorific value and gross calorific value Q_j^a were determined according to Polish standard PN-ISO 1928:2002 [10]. Self-ignition of the samples as a result of oxygen supply was not detected during the tests conducted in calorimeter bomb. Therefore, benzoic acid was added to the samples as an activator. The acid was included during determination of gross calorific value. The total sulphur content S_t^a and organic coal C^a were determined with application of SC132 device (from LECO company) and by the Tiurin method. The results are given in Table 1.

Considering standards referred to aggregates, determined value of total sulphur content S_t^a in wastes (0.263-0.669%) does not disqualify tested material as an aggregate, although the content of organic coal is too high (above 1 %). Considering this result it is impossible to use the material as an aggregate according to requirements of the standard PN-S-06102:1997 [11]. When organic coal content is higher than 2% it is impossible to use the material for soil works purposes (according to PN-S-02205:1998 [12]) .

Table 1. Energetic features of waste material from CCWD.

Parameter	Sample I	Sample II	Sample III	Average
A^{a},%	77.49	79.05	70.70	75.75
C^{a},%	12.07	10.08	12.91	11.69
S_t^a,%	0.263	0.669	0.395	0.442
W_{ex},%	2.1	2.6	3.8	2.8
W_h,%	1.67	2.08	3.08	2.28
Q_j^a, MJ/kg	6.99	7.75	8.69	7.81
Benzoic acid content, %	13.8	14.3	11.7	13.3

Research into energetic parameters of the samples gave following values: coal content C = 10,08-12.91%, ash content A^a = 70.7 – 79.05%, gross calorific value Q_j^a = 6.99 – 8.69 MJ/kg. Obtained parameters (the range of ash content and coal substance content which was proved by significant gross calorific value) indicated that examinations were advisable and it is possible to recover coal from the waste material. Considering the experience of different companies involved in coal wastes processing it can be stated that there are economically proved methods of coal recovery from the dumps. The process can be profitable when coal content is higher than 8%.

Although coal recovery is justified not only because of economical purposes it also allows to reduce fire hazard at a dump and improve geochemical and geotechnical parameters of spin-off. It is very important in the case of application of this material for dump reclamation as well as outside the dump (e.g. for road constructions and other engineering applications). Although this direction of waste material application requires technological tests and selection of efficient recycling method.

3 Methodology of recovery process and results

Washing in heavy liquids is the one of the most efficient and economic method of separation of coal grains and deeds. This method is also applied by Haldex S.A., Zower S.A. during coal recovery from coal waste dumps [13, 14]. However, due to high content of fly ash in the waste material this method was rejected at the first stage of examinations. The grains of fly ash affected negatively on parameters of heavy liquid what reduced significantly the efficiency of the process.

3.1 Coal recovery in spiral washers

The research into possibility of coal grains recycling from CCWD was conducted at technical scale with application of spiral washer Reichert LD4. Wastes in raw state were not sufficient for direct tests because of too large size of the grains. Thus, they were crushed in a jaw crusher to obtain proper granulation of the material.

Coal wastes preparation process was preceded by the analysis of feed. Table 2 shows grain classes of the waste material *d*, their yield γ and ash content A^a.

DOI: 10.1051/e3sconf/20160801052

Table 2. Grain classes and their output, total ash content.

Grain class *d*, **mm**	**Yield** γ, **%**	**The sum of yields** $\Sigma\gamma$, **%**	**Total ash content** A^a, **%**
>3.2	20.96	20.96	86.37
3.2-2.5	11.36	32.33	86.45
2.5-2.0	9.78	42.11	85.62
2.0-1.5	9.54	51.65	65.80
1.5-1.0	16.26	67.91	60.07
1.0-0.75	7.04	74.95	69.74
0.75-0.5	8.16	83.11	71.46
0.5-0.3	6.70	89.81	80.31
0.3 - 0.2	4.83	94.64	81.66
0.2-0.1	3.82	98.46	82.44
< 0.1	1.54	100.00	81.75

The granulometric analysis was conducted by separation of the material into many limited grain classes to obtain the optimum range of grain size which was the least burdened by the ballast (ash content). The highest mass fraction was detected for grains in classes >3.2 mm and 1.5-1.0 mm. A high level of ash content was detected in all measured classes. The significant silt content in the material could be noticed even macroscopically, as well as other mineral admixtures. In fact, they cause such high ash content.

The testing stand included the feed bin and LD 4 washer (built of two beds having six reels and dewatering screen). The feed was delivered gravitationally from the bin with high-speed agitator and then it was mixed-up with the air from a compressor. The movement of the agitator and air stream caused that feed did not embedded in the bin and had constant density.

The preparation process was conducted with the following technological parameters:

- density of feed β= 300 g/dm^3,
- the time of feed mixing in the bin with the agitator (included additional aerodynamic stirrers for improvement of averaging of feed before washing) was 3 minutes,
- the amount of additional water delivery to the washer was 500 dm^3/h.

Table 3. Results of qualitative and quantitative analysis of separation products, n/m - not measured, n/a - not added.

Parameter, unit	**Concentrate**	**Middlings**	**Rejects**
S_t^a ,%	0.45	0.31	0.40
W_h, %	1.59	1.34	1.33
A^a ,%	56.5	74.8	83.1
Q^a , MJ/kg	11.58	3.42	n/m
Benzoic acid content, %	n/a	n/a	n/a

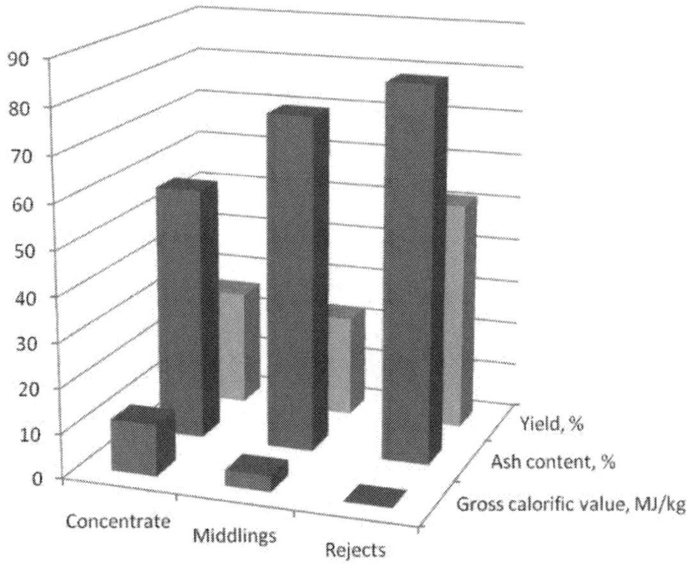

Figure 1. Parameters of products after spiral washers.

Table 4. Results of qualitative and quantitative analysis of separation products.

Particle size mm	Concentrate		Middlings		Rejects	
	Yield, γ_c,%	Ash content A_c^a, %	Yield, γ_m,%	Ash content A_m^a, %	Yield, γ_r,%	Ash content A_r^a, %
>3.2	35.17	66.80	15.58	85.95	12.14	86.37
3.2-2.5	19.37	60.56	8.84	84.69	5.88	86.45
2.5-2.0	11.71	56.37	11.23	81.94	6.41	85.62
2.0-1.5	8.89	45.10	11.61	77.16	8.11	85.80
1.5-1.0	9.61	27.14	20.35	61.50	18.83	84.07
1.0-0.75	2.43	18.14	8.46	57.60	10.25	81.74
0.75-0.5	3.27	21.49	7.55	47.14	13.64	81.46
0.5-0.3	2.98	15.96	5.67	44.81	11.46	80.31
0.3-0.2	1.95	35.80	4.34	56.79	8.18	81.66
0.2-0.1	3.03	63.82	4.56	72.57	3.88	82.44
<0.1	1.60	76.91	1.82	79.75	1.20	81.75
Sum	100.00	-	100.00	-	100.00	-
Average	-	53.92	-	69.74	-	83.42

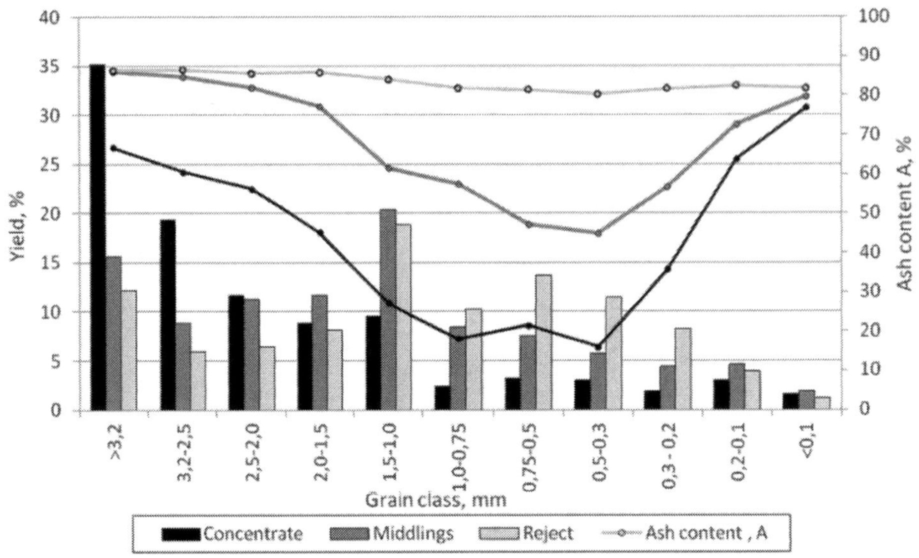

Figure 2. Qualitative and quantitative analysis of separation products.

Three products were obtained as the results of washing process: concentrate, middlings and rejects. Their yields were γ_c = 26.0%, γ_m = 22.67% and γ_r = 51.33%. The products were taken into the qualitative and quantitative analysis. The results of the analysis of obtained products are presented in Table 3 and Figure 1. The results including grain class classification of these products are presented in Table 4 and Figure 2.

The analysis of density separation gave following observations. Middlings do not have value in use as well as rejects (with high ash content in tested classes). It indicates on a lack of coal grains in the products. It should be mentioned that the feed and obtained products of separation (tested in particular grain classes) are given as average values. They were obtained from several series of tests. In the case of concentrate, the material can be divided into three main groups of grains. Size fractions 3.2-1.5 mm has relatively high ash content what can be caused by former washing processes when grains of clean coal are separated from feed and this class was storied at the dump as rejects. The second grain class is 0.2 – 0 mm. Considering this class it can be noticed that there is dependence between increase of ash content in the material and decrease of grain size. The average ash content in these fractions was α_{avg} = 58.70% (the lowest grains of stone with silt admixture mostly made this class).

The size fraction of 1.5 – 0.3 mm was the most valuable. The average ash content was α_{avg} = 23.12 %, and mass fraction was γ = 18.28 %. This size fraction had some energetic potential, which can be used, e.g. as an admixture to fine coal being used in power plants.

3.2 Coal recovery in a pulse classifier

The research into the recovery of coal grains in the size fraction 30-5 (0) mm was undertaken "in-situ" by industrial tests. Komag pulsatory classifier K-102 (a pulsatory jig) was in use. The functioning of the jig is based on a typical gravitational washing of the minerals performed by stratification of feed at an open grid plate (sieve deck) in pulsated water medium [15]. Stratification is conducted according to grain fractions and density of the components of feed. As the result of washing process three products were achieved: concentrate, rejects and spin-off (as a result of loss of fine grains through the sieve deck), however, spin-off was not under the tests during this research.

Inter-screen product from two-deck open grid classifier with a mesh size 30 mm (the upper deck) and 5 mm (the lower deck) created feed which was then transported to washing. Loading of the pulsatory jig during the tests was ~ 80 Mg/h. Due to density content, the determination of ash content and gross calorific value in obtained fractions was started, in the laboratory analysis, just after de-slurrying of the sample (the removal of grains <0.5 mm). Feed which was transported to washing had high ash content 77.91% and low gross calorific value 4.44 MJ/kg. The fraction with density >1.8 g/cm^3 was the main part of feed (90.53%).

The results of undertaken research showed that in washed post-coal wastes the content of grains of flammable substance can be significant. In this case the yield of concentrate was 7.66%. The average ash content was 19.96% and gross calorific value was 26.16 MJ/kg. The rejects from washing (with the yield 92.34%) had trace coal grains content (<1.5 g/cm^3) 0.15%, high ash content 83.61% and very low gross calorific value 2.45 MJ/kg [16,17].

The results of physicochemical analysis of feed and of the products of washing are presented in Table 5 and Figure 3.

Table 5. Results of physicochemical analysis of feed and the products of washing – size fraction 30-5 (0.5) mm.

Density of fraction g/cm^3	**Feed**			**Concentrate**			**Rejects**		
	Yield, γ_f,%	**Ash content, A_f^a,%**	**Gross calorific value, Q_f^a,MJ/kg**	**Yield, γ_c,%**	**Ash content, A_c^a,%**	**Gross calorific value, Q_c^a,MJ/kg**	**Yield, γ_r,%**	**Ash content, A_r^a,%**	**Gross calorific value, Q_r^a,MJ/kg**
< 1.3	3.05	3.35	31.75	39.44	3.18	32.63	0.03	4.73	31.35
1.3-1.4	1.13	11.35	28.79	14.20	10.72	29.80	0.04	12.79	27.95
1.4-1.5	1.07	22.50	24.48	13.05	21.88	25.25	0.08	22.73	24.25
1.5-1.6	0.89	31.52	21.63	9.85	31.39	21.89	0.15	32.25	21.56
1.6-1.7	1.73	38.60	18.87	11.33	38.18	19.08	0.93	39.75	18.61
1.7-1.8	1.60	47.35	15.35	4.53	44.99	16.45	1.36	49.05	15.15
1.8-2.0	3.69	55.36	12.31	4.50	54.58	12.65	3.62	55.72	12.42
> 2.0	86.84	84.86	1.92	3.10	77.79	4.28	93.79	85.81	1.65
Sum	100.0	-	-	100.0	-	-	100.0	-	-
Average	-	77.91	4.44	-	19.96	26.16	-	83.61	2.45
Total yield. γ, %	100.0			7.66			92.34		

The parameters describing separation process and its efficiency were determined on the base of the results of the analysis. The parameters were: density of separation (d_{50}), probable dissipation (E_p) and imperfection (I).

The results of washing of mine wastes showed that pulsatory jig worked with a high efficiency of separation: imperfection coefficient was I = 0.124, probable dissipation was E_p=0.081 g/cm^3, and density of separation was d_{50}=1.651 g/cm^3.

Additional laboratory tests of the concentrate samples were carried out during exploitation of the pulsatory jig. It turned out that the product has very advantageous qualitative parameters [18]. The results showed in Table 6 are given under working conditions.

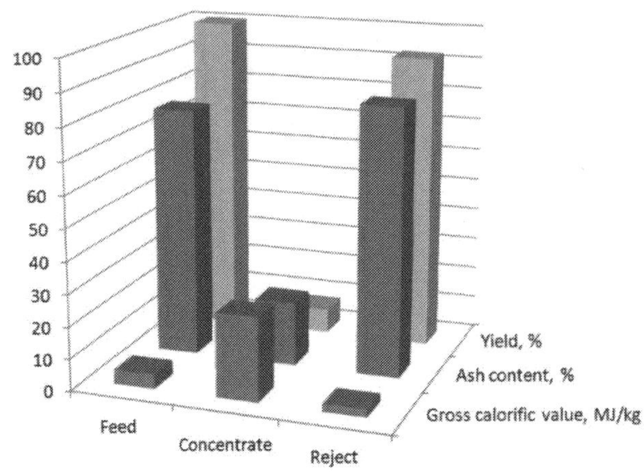

Figure 3. Parameters of feed and products of washing - industrial tests.

Table 6. Parameters of concentrate after jigging.

No. sample	Total moisture W_t, %	Ash content A_c^a, %	Gross calorific value, Q_c^a ,MJ/kg
1	8.1	12.0	26.14
2	7.8	11.3	26.79
3	8.3	11.7	26.65
4	9.2	10.8	26.46
average	8.35	11.45	26.51

4 Conclusions

The application of pulsatory jig for separation of wastes in 30-5 mm size fraction allowed to obtain a high-quality energetic concentrate with the ash content lower than 12% and gross calorific value higher than 26 MJ/kg. A spiral separator (applied in 8-0 mm size fraction) gave much worse results. The average gross calorific value for the concentrate was 11.6 MJ/kg, with the high ash content 56.5%. It was caused by the presence of fly ashes in the feed. Perhaps, removal of this component during wet screening could increase the energetic features of concentrate. To increase the parameters of concentrate as well as its yield the type of separator and/or the settings of the device should be optimized.

Efficiency of coal recovery from CCWD Przezchlebie was different and depended on applied device and enriched grain classes during washing tests. Properties of the material are significant reason of such a state (including presence of fly ashes and presence of coal in particular size fraction).

To sum up, the recovery of coal is necessary because it can cause not only financial profits but also can improve safety by reduction of fire hazard and negative impact on the environment. The material without coal could be applied for reclamation of the site or outside the dump as an aggregate. Furthermore, the reclaimed site can be an interesting place for further investments.

References

1. Z. Różański, Final project Report, Ekostaż/24/2014 in frame The Human Capital Operational Programme (2014) (unpublished)
2. Cz. Rosik-Dulewska, *Podstawy gospodarki odpadami* (1999)
3. Z. Bzowski, *Wiadomości Górnicze*, **7(8)** (2008)

4. Z. Różański, *Górnictwo i Geologia*, **4(3)**, 7 (2009)
5. J. Sułkowski, J. Drenda, Z. Różański, P. Wrona, *Gospod Surowcami Min* **24(3/1)**, 15 (2008)
6. PN-ISO 18283:2008 *Pobieranie i przygotowanie próbek węgla do badań* (2008)
7. PN-EN 1097-7:2008 *Badania mechanicznych i fizycznych właściwości kruszyw* (2008)
8. PN-80/G-04511 *Paliwa stałe. Oznaczanie zawartości wilgoci*
9. PN-80/G-04512 *Paliwa stale - Oznaczanie zawartości popiołu metodą wagową*
10. PN-ISO 1928:2002 *Paliwa stałe - Oznaczanie ciepła spalania metoda spalania w bombie kalorymetrycznej i obliczanie wartości opałowej* (2002)
11. PN-S-06102:1997 *Drogi samochodowe. Podbudowy z kruszyw stabilizowanych mechanicznie* (1997)
12. PN-S-02205:1998 *Drogi samochodowe - Roboty ziemne - Wymagania i badania* (1998)
13. T. Górski, B. Słowiński, *Energetyka i Ekologia*, **10(11)**, 739 (2003)
14. P. Kucharzyk, *IM,* **5(2)**, 5 (2005)
15. P. Matusiak, D. Kowol, R. Nieckarz, *Maszyny Górnicze*, **4**, 5 (2012)
16. P. Matusiak, D. Kowol, *Mining Science – Mineral Aggregates*, **23(1)** (2016)
17. P. Matusiak, D. Kowol, M. Łagódka, *KOMEKO 2016,* 15 (2016)
18. D. Kowol et al. *ITG KOMAG*, (2016) (unpublished)

E3S Web of Conferences **8**, 01053 (2016) DOI: 10.1051/e3sconf/20160801053
MEC2016

Total gas in-place calculations for the Baltic-Podlasie-Lublin basin shales in Poland

Marcin A. Lutynski[1,a], and Miguel A. Gonzalez Gonzalez[1]

[1]*Silesian University of Technology, Faculty of Mining and Geology, 44-100 Gliwice, Poland*

Abstract. The total gas in-place calculations reflecting Polish reservoir conditions were performed using a corrected version for the free gas in-place, where not only the pore conditions were taken under consideration but also the volume that was already occupied by the sorbed gas. Shales show very different capacities between samples mainly due to in-situ conditions.

1 Introduction

During the first decade of XXI century and, to the lesser degree, also Canada experienced "energetic revolution" related to production of unconventional resources of gas and oil. A significant impact on gas and oil prices as well as in energy security was achieved due to the fast increase of production of these resources. Technologic and economic success of this sector of petroleum industry in North America has inspired attempts to export this concept to other continents, including Europe, at the beginning of the second decade of XXI century. Most intensive exploration of shale gas and oil, apart from US and Canade, took place in the Lower Paleozoic Baltic-Podlasie-Lublin Basin in Poland [1,2].

According to the International Energy Agency, Poland has become one of the countries with potentially the largest shale gas deposits. An accurate assessment of these deposits is a key factor to allow Poland to reduce the foreign dependency of this gas from Russia and other former Soviet republics [3, 4].

2 Methodology

In order to estimate the total gas in-place, equation 1 is used.

$$G_{st} = G_a + G_f \tag{1}$$

This equation is divided in volumetric and surface components. In this equation G_{st} is the total gas in-place, G_f is the volumetric component involving hydrocarbons stored in the pore space as free gas, and G_a is the surface component, with the gas physically adsorbed on the large surface area of the micropores [5].

[a] Corresponding author: Marcin.Lutynski@polsl.pl

The surface component is calculated using equation:

$$G_a = G_L \cdot \frac{p}{p+p_L} \tag{2}$$

where G_L is the Langmuir volume, p is the pressure and p_L is the Langmuir pressure. Equation 3 is used to calculate the volumetric component,

$$G_f = \frac{32{,}0368}{B_g}\left[\frac{\phi(1-S_w)}{\rho_b} - \frac{1{,}318x10^{-6}M}{\rho_s}\left(G_L\frac{p}{p+p_L}\right)\right] \tag{3}$$

where B_g is the gas formation volume factor, ϕ is the porosity of the rock, S_w the water saturation, M the molar mass, ρ_b the bulk rock density and ρ_s the sorbed phase density.

This model takes under consideration, not only the adsorbed gas, but also the free gas stored inside the pores to give a more accurate value of the total gas in-place. It is particularly important that, unlike in coals, the ratio of adsorbed gas-to-free gas is much lower. For calculation of the total free gas in-place, the adsorbed gas must be calculated and subtracted from this calculation and corrected for the water saturation [5]. This methodology may give a new overview on the comprehensive assessment of CH_4 potential of Polish shales as a strategic reservoir for CH_4.

3 Calculations and results

Two samples were tested with completely different parameters as shown in Table 1. Sorption measurements were performed using the high precision manometric sorption setup, located in the Institute of Mining at the Silesian University of Technology (Figure 1.) Calculations of the sorption capacities and Langmuir models are shown in Fig. 2.

Figure 1. High precision manometric setup at the Silesian University of Technology, Poland.

Parameters related to the in-situ conditions of the basin were supplied by the drilling company and some of them are shown in Table 1.

Taking under consideration the in-situ characteristics of the basin and the tests performed in the laboratory both components of Equation (1) were calculated. The surface component, dependent on the sorption capacities, and thus on the rock composition, showed proportional values to the sorption capacities calculated with 2.06 sm^3/t for sample no. 1 and 3.37 sm^3/t for sample no. 2. The volumetric component mostly depended on the porosity and water saturation of the samples. It greatly impacted the results obtained for sample no. 1 with only 0.12 sm^3/t of free gas, while sample no. 2 had 3.92 sm^3/t of free gas. These results are shown in Figure 3 for comparison.

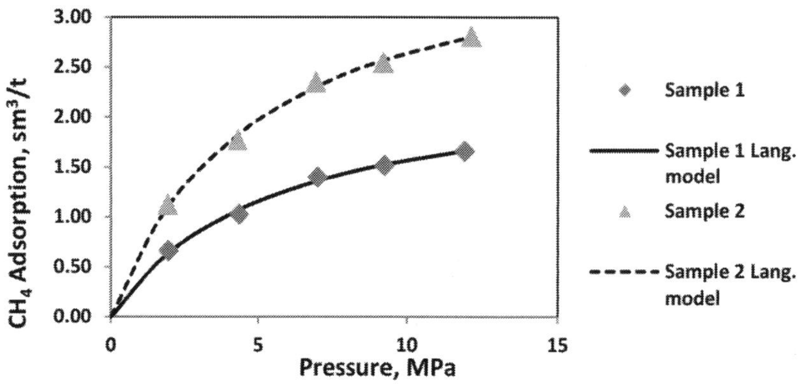

Figure 2. Methane sorption measurements and Langmuir model for Polish shales.

Table 1. Some of the in-situ parameters of the samples.

	Sample 1	Sample 2
Porosity, %	5.16	9.50
Water saturation, %	70.61	35.00
Bulk rock density, g/cm^3	2.67	2.40
Total Organic Carbon, %	1.25	4.19

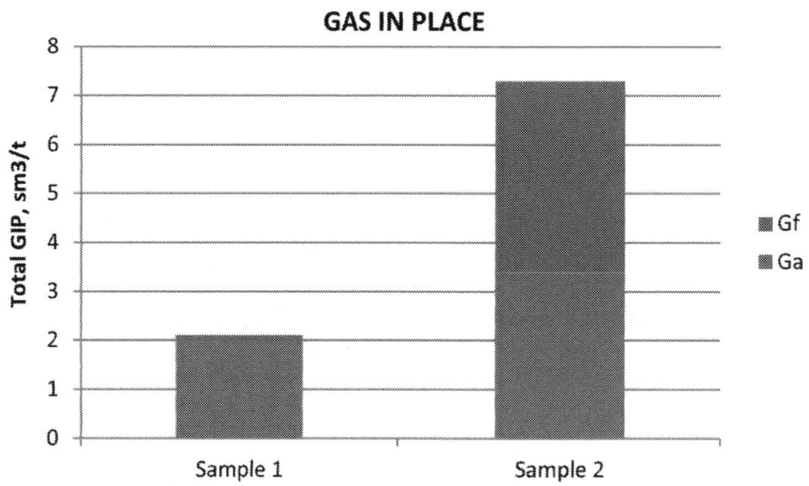

Figure 3. Results obtained in the total gas-in-place calculations.

4 Results and conclusions

Results obtained in the study showed that the biggest impact while calculation of the total GIP, was not produced by the rock composition, but by the basin conditions. Thus, a sample that during a preliminary analysis may seem to be the worse choice due to its composition, could be as good as higher organic matter sample under the adequate conditions. In the data exposed, it can be observed that sample no. 2 is better than sample no. 1, related to the surface component, by a 70%. However when taking under consideration the free gas, this value increases up to 250%,

To get a better idea of this, we decided to apply the basin data from sample no. 2 to the sample no. 1, with a hypothetical result of gas-in-place of 6.7 sm^3/t. These new values are shown in Figure 4. With these suppositions, the total gas in place difference, due to the rock composition, is reduced.

For the correct understanding of the data, the subtracting factor from Equation 3 has to be taken under consideration. The higher the sorption capacity, the lower total free gas calculation. It is due to the volume of pore that is taken by the sorbed gas on the surface of pores.

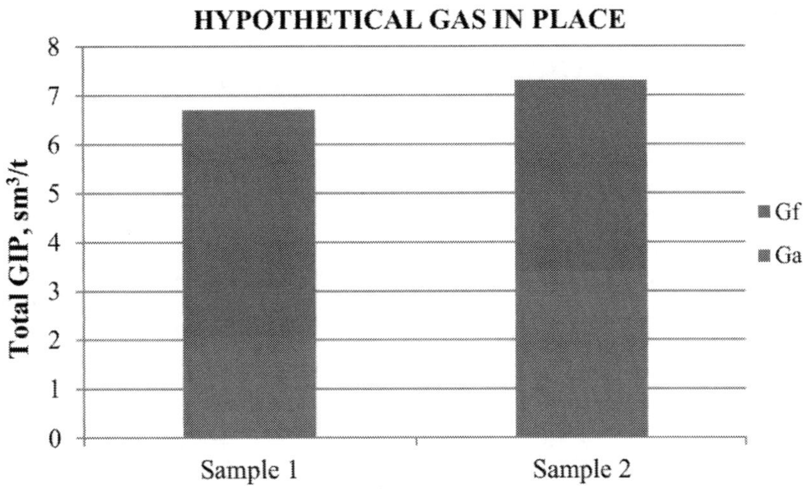

Figure 4. Hypothetical results obtained using the same main parameters for both samples.

Acknowledgments

The research leading to these results has received funding from the Polish-Norwegian Research Programme operated by the National Centre for Research and Development under the Norwegian Financial Mechanism 2009-2014 and Project Contract No. PL12-0109.

References

1. *Assessment of shale gas and shale oil resources of the Lower Paleozoic Baltic-Podlasie-Lublin basin in Poland* (Polish Geological Institute, 2012)
2. *Oil and Gas in Poland, new opportunities,* (Polish Geological Institute, 2015)
3. *World Shale Gas Resources: An Initial Assessment of 14 Regions Outside the United States*, (EIA, 2011)
4. *Natural Gas: Information.* (EIA, 2015)
5. R.J. Ambrose, R.C. Hartman, M. Diaz-Campos, I. Yucel Akkutlu and C.H. Sondergeld, *in proccedings of SPE Unconvencional Gas Conference (*SPE 131722, 2010).

DOI: 10.1051/e3sconf/20160801054

Carbon dioxide sorption on EDTA modified halloysite

Patrycja Waszczuk[1] , Marcin Lutynski[2,a], Miguel A. Gonzalez Gonzalez[2], Adam Smolinski[1], and Natalia Howaniec[1]

[1]*Głowny Instytut Gornictwa, Plac Gwarkow 40 – 166 Katowice, Poland*
[2]*Politechnika Sl aska, ul Akademicka 2, 41-100 Gliwice, Poland*

Abstract. In this paper the sorption study of CO_2 on EDTA surface modified halloysite was conducted. In the paper chemical modification of halloysite from the Dunino deposit (Poland) and its influence on sorption of CO_2 are presented. A halloysite samples were washed with water-EDTA 1% solution, centrifuged to separate liquid and impurities and dried. The samples were tested for the sorption capacity using a manometric method with pressure up to 3 MPa. A Langmuir adsorption model was fitted to the data. The results showed that EDTA had a limited effect on the increase of sorption potential at low pressure and the samples exhibited similar results to that ones treated solely with the water solution.

1 Introduction

Emission reduction and improving carbon capture technologies are currently the important issues in a carbon constrained world. It is especially desirable in countries where the main power source are fossil fuels, in particular coal. It is necessary to clean exhaust gases, which contain also SO_2, NO_X and other, mainly acidic, oxides.

A sorption study was conducted on a modified halloysite to verify modification of surface properties. Halloysite is an aluminosilicate clay mineral, consisting of aluminium, silica and hydrogen, with formula $Al_2Si_2O_5(OH)_4$. It naturally occurs with montmorillonite and kaolinite with small additions of other clay minerals, often contaminated with ilmenite. Other impurities in the raw halloysite material are iron and titanium oxides. In the past, halloysite was used as a petroleum cracking catalyst, also known for its sorption properties. It also can be used in nanocomposites as a filler [1-5].

It is suggested that modification of halloysite can improve sorption selectivity and/or increase sorption capacity of CO_2. The purpose of this research was to find modification of halloysite, which helps to improve selectivity of the sorbent. Typical sorbent reactivation can be divided into chemical and thermal treatment methods. In the case of chemical treatment the surface of sorbent is modified by chemical impregnation which can significantly increase the sorption capacity. As a method of modification surface impregnation with EDTA was chosen. EDTA is a ethylenediaminetetraacetic acid, widely known for its properties as a chelating agent and limescale dissolvent. It has the ability to complex, among others, heavy metal ions, but the most important is its ability to sequest Fe^{3+} ion.

[a] Corresponding author: Marcin.Lutynski@polsl.pl

A halloysite is strongly polluted with an ilmenite, which decreases its sorption properties. It was assumed that the EDTA supported the purification process and modified the surface at the same time. There is only one paper which shows EDTA action on halloysite [6]. The paper contains description of the devastating effects of acid on structure of halloysite immediately after mixing both substances. But after a longer time (90 days) a recrystallisation and expansion in diameter of nanotubes was observed.

2. Materials and methods

2.1 Materials

For the purpose of the study, raw halloysite from the Dunino deposit in western Poland was chosen. The halloysite Dunino deposit is located in the Lower Silesia and contains over 10 million tons of a homogeneous raw material mined using an open pit method. Halloysite extracted from this deposit is a product of basalt weathering.

The method of activated sample preparation involved immersing the sample in impregnating solution of 1% EDTA for an hour, shaking for 30 minutes and centrifuging aqueous at 2500 RPM for 10 minutes. The other raw sample of halloysite was prepared in the same manner with the exception that instead of EDTA and water mixture, pure demineralized water was used.

Consequently, the samples were dried in the oven at 105 °C for 2 to 24 hours, depending on the collected volume of mixtures. Before placing the samples in the cell of the sorption apparatus the samples were vacuumed.

The halloysite sample modified with EDTA was marked as a hall-EDTA and the sample prepared with demineralized water was marked as a hall-water. For the purpose of comparison we have also performed tests on raw halloysite, neither water immersion nor centrifuging, acquired straight from the deposit. The raw halloysite was crushed, sieved to the fraction below 0.1 mm and dried for 1 hour. This sample was marked as a hall-raw.

2.2 Methods

Sorption experiments were performed using the manometric technique with the custom made sample cells and setup. The process of adsorption removes sorbate gas molecules from the free gas phase, and thus results in a decrease of the gas pressure within the experimental system. The most fundamental operational procedure to quantify gas adsorption is the Gibbs approach, where the amount of adsorbed gas ($n_{adsorbed}$) is defined as the difference between the total amount of gas (n_{total}) present in the system and the amount occupying the void volume (V_{void}), i.e. the volume not occupied by the solid sample. This latter quantity is calculated from the molar concentration (c_{gas}) in the gas phase which is obtained from an equation of the gas state under corresponding pressure and temperature conditions:

$$n_{sorbed} = n_{total} - c_{gas} \cdot V_{void} = n_{total} - c_{gas}(V_{sample\ cell} - V_{sample}). \quad (1)$$

Prior to the start the helium experiment the void volume of the sample cell (V_{void}) is measured. For this purpose an inert gas (helium) is used. The density of gas is calculated with the McCarthy equation of state for helium. Multiplication of the void volume by the density of the gas ($V_{void} \cdot \rho^{CO2}(T,P)$) gives the "non-sorption" reference mass, i.e. the amount of gas that would be accommodated in the measuring cell for no sorption. The excess sorption mass is the difference between the gas mass that is transferred into the measuring cell up to a given pressure step and the non-sorption reference mass:

$$m_{excess}^{CO_2} = m_{transferred}^{CO_2} - V_{void} \cdot \rho^{CO_2}(T,P). \quad (2)$$

Sorption measurement is done in consecutive steps and the gas mass transferred from the reference cell into the sample cell at each step (N) can be calculated with the following formula:

$$m_{transferred}^{CO_2} = \sum_{i=1}^{N} V_{ref} \cdot \left(\rho_i^{f,CO_2} - \rho_i^{e,CO_2}\right). \quad (3)$$

Superscripts in the equation i.e. *f*, CO_2 and *e*,CO_2 refer to the density of gas in the filling phase (when the reference cell is not connected with the sample cell) and equilibrium phase (when the reference cell is connected with the sample cell and the pressure is equilibrated), respectively. Densities are calculated with the use of high-accurate equations [7]. The schematic diagram of the manometric sorption methodology is presented in Figure 1.

The manometric setup used for the experiments consists of the sample and reference cells, valves, tubing, pressure transducers and temperature sensors (Fig. 2). The gas is injected into the setup by a pump and prior to start of the experiment the system is evacuated with the use of the vacuum pump. The pressure is monitored with the use of a pressure transducers. The pressure transducers are connected to the data acquisition system. The temperature is monitored with PT100 sensors which are connected to the multi-channel temperature data-logger, which communicates with the data acquisition system. The sample cells were designed and constructed at the Silesian University of Technology. Fittings and valves used in the setup are high pressure fittings made of stainless steel with ¼' tubing. Setup is immersed in pure water in custom made bath with electronic temperature stabilization of 0.1 °C [8].

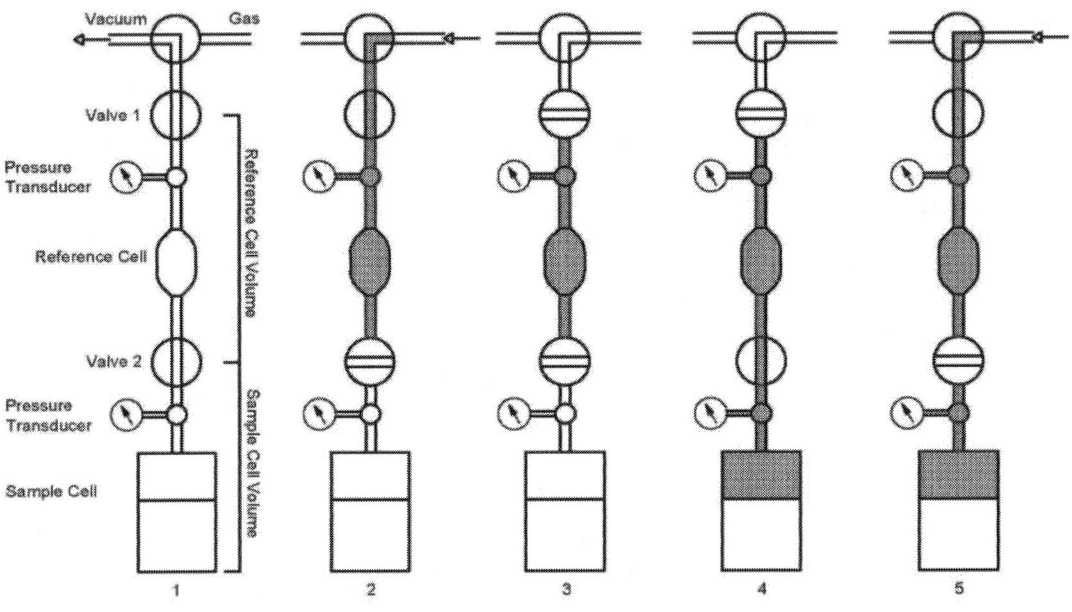

Figure 1. Schematic diagram of manometric sorption methodology, where each figure indicates each step: 1 - evacuation of the system, 2 - filling of the reference cell, 3 - closure of the valve 1 and thermal equilibration, 4 - valve 2 is opened and gas is transferred into the sample cell (start of the sorption process), 5 - when pressure equilibrium is reached, valve 2 is closed and refilling of the reference cell starts.

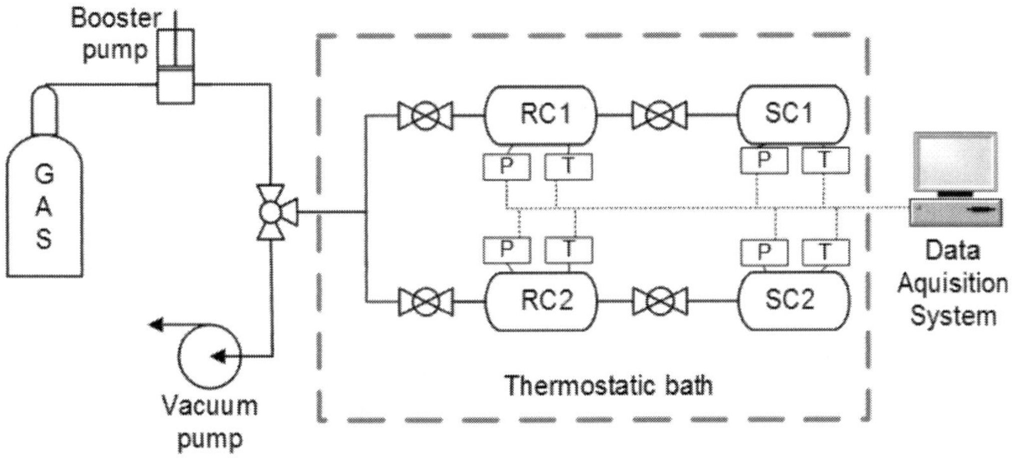

Figure 2. Scheme of high-pressure sorption manometric setup.

3. Results and discussion

An interesting phenomena occurred during sample preparation. After centrifuging the hall-EDTA and hall-water samples stratification occurred and the samples had different colours of separated liquid. Also sharper boundaries between layers of different densities could be seen (Fig. 3).

Figure 3. View of samples after separation in centrifuge: (left) with EDTA solution and (right) with demineralized water.

In order to acquire the hall-EDTA and hall-water samples for the experiments a "light" fraction was removed from the tube. The "dark" part of the mixture was probably iron and other metal oxides impurities. The results of sorption tests are presented in Fig. 4. Experiments were conducted at pressure of 3 MPa. The equilibrium time for each step was 1-2 hours. The experimental results were fitted with the Langmuir model and the calculated Langmuir constants are presented in Table 1. The results of experiments are also presented in Fig. 4. It was noticed that maximum adsorption was for the raw halloysite sample, next EDTA modified halloysite, and then halloysite washed with water.

The Langmuir model was calculated with the following equation:

$$V = \frac{V_L \cdot P}{P_L \cdot P}, \tag{4}$$

where P is pressure (MPa), V_L is the Langmuir volume (mmol/g) and P_L is the Langmuir pressure (MPa).

DOI: 10.1051/e3sconf/20160801054

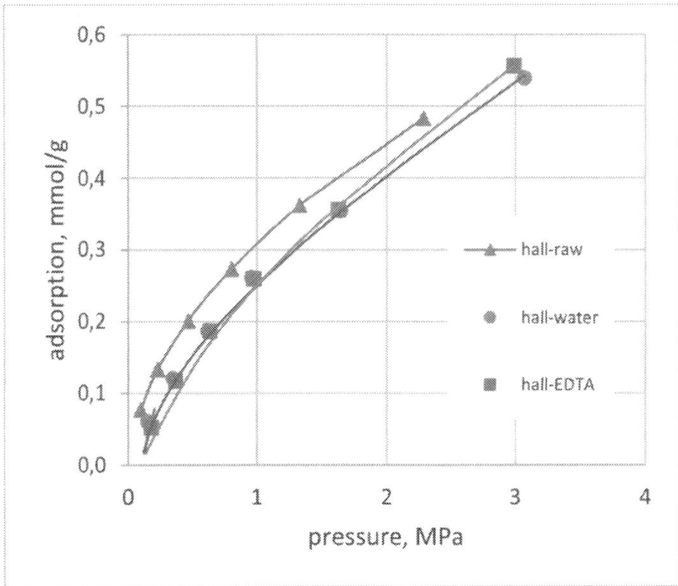

Figure 4. CO_2 excess sorption isotherms on the hall-EDTA, hall-water and hall-raw samples at 45 °C.

Table 1. Calculated Langmuir constants for samples.

Sample	V_L, mmol/g	P_L, MPa
Hall-EDTA	1.256	3.841
Hall-water	1.024	2.854
Hall-raw	0.713	1.192

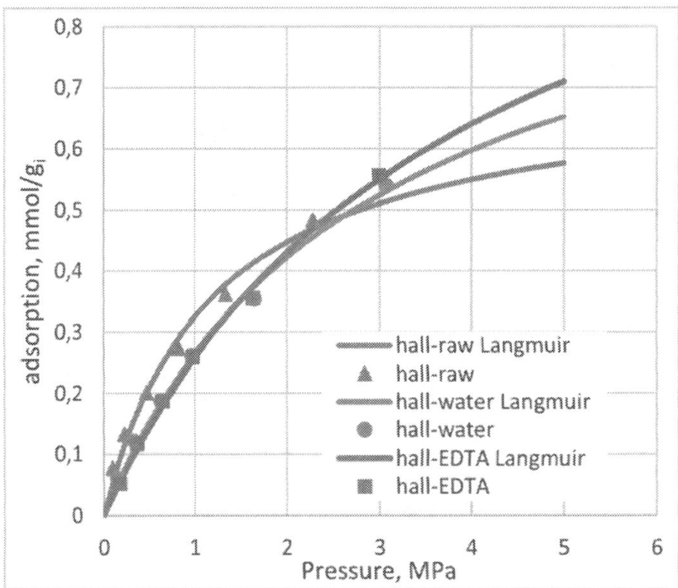

Figure 5. CO_2 excess sorption isotherms on the hall-EDTA, hall-water and hall-raw samples at 45 °C fitted and extrapolated with the Langmuir model.

Although the highest sorption capacity is visible in the hall-raw sample up to the pressure of 1 MPa it reaches the plateau level faster than other samples. After applying the model it could be noticed that for higher pressure (above 3 MPa) sorption is the highest for the hall-EDTA sample. This is reflected in the calculated Langmuir constants, where V_L has the highest value for the hall-EDTA sample (meaning the highest sorption capacity for infinite pressure). Lower V_L values are for the hall-water and hall-raw samples, respectively.

4. Conclusions

It can be noticed, that addition of EDTA slightly improves halloysite sorption capacity. Probably it is not due to surface modification but rather enhances separation between halloysite and irony contamination in the presence of EDTA. EDTA rather large and branched molecules, which could occupy the pores being the centers of adsorption for gas. The gaseous phase probably is reluctant to sorption on the EDTA molecule.

A tendency to increase the sorption capacity is observed at pressure above 3 MPa since the Langmuir volume for the hall-EDTA and hall-water samples is higher than for the hall-raw sample. It is not clear whether it is due to either the model equation or at higher pressure an unidentified phenomenon occurs impossible to explain based on the tests performed. A further research of sorption at 3 MPa as well as the surface analysis of modified material is necessary to check if there is any difference between the modified and unmodified materials.

References

1. P. Yuan, P.D. Southon, Z. Liu, C.J. Kepert, *Nanotechnology*, **23**, (2012)
2. P. Sakiewicz, R. Nowosielski, W. Pilarczyk, K. Gołombek, M. Lutynski, *JAMME* (2011)
3. K. Hedicke-Hochstotter, G.T. Lim, V. Altstadt, Compos. Sci. Technol., **69** (2009)
4. R. Kamble, M. Ghag, S. Gaikawad, B. Kumar Panda, *J. Adv. Sci.*, **3**, 2 (2012)
5. E. Joussein, S. Petit, C. Fialips, P. Vieillard, D. Righi, *Clay Clays Miner.*, **54(4)** (2006)
6. A. La Inglesia, E. Galan, *Clay Clays Miner.*, (1975)
7. W. Wagner, R. Span, *Int. J. Thermophysics*, (1993)
8. M. Lutynski, *Bulletin of the Polish Academy of Science*, **62(1)** (2014)

Current methods and possibilities to determine the variability of Cu content in the copper ore on a conveyor belt in one of KGHM Polska Miedz S.A. mines

Leszek Jurdziak[1], Witold Kawalec[1], and Robert Krol[1,a]

[1]*Wroclaw University of Science and Technology, Faculty of Geoengineering, Mining and Geology, ul. Na Grobli 15, 50-421 Wrocław*

Abstract. In this paper the methodology of ore sampling on a belt conveyor in one of KGHM Polska Miedz S.A. mines was presented together with the results of analysis of the variation of ore tonnage and quality. The ore was transferred from one mine to another one through the analysed transfer conveyor on the route from the mine Division of Mining to copper plants. The transported ore was sampled in person from Division of Concentrators on a regular basis for metal accounting purposes between both mines. The ore quality control became a significant problem since it is impossible to predict with satisfactory advance both Cu content in the feed as well as its lithology composition which is required to improve and optimise the enrichment efficiency.

1 Introduction

In SW Poland near Lubin town, KGHM Polska Miedz S.A. operates 3 underground mines that is Lubin, Rudna and Polkowice-Sieroszowice, 3 Ore Enrichment Divisions of Concentrators (OEDs) and two copper smelter/refinery complexes.

Copper is extracted by using a room-and pillar method by drilling and blasting, cyclically loaded and hauled by low-profile LHDs and by continuous horizontal transport to shafts and stores using belt conveyors. Conveyors are also used to transport ore to concentrator plants.

The deposit comprises three different rock types, that is dolomite and sandstone separated by a main layer of copper-bearing black shale. The height of extracted seam (different in different mines) is established by channel sampling in the grid from 20x20 to 40x40 m. In each mining area both copper content and proportion of different rock types can substantially vary. The variability of ore excavated in different locations and transported in a complicated network of about 130 km of belt conveying systems, with many mixing and switching points as well as stores, causes the ore quality control to become a significant problem [1].

For improving the enrichment process, the identification of both copper content and proportion of type of rocks supplied are important. In fact, the latter one is the most important and never controlled on conveyors. The Cu content measurements from samples on a conveyor are done on a regular basis

[a] Corresponding author: robert.krol@pwr.edu.pl

on one conveyor linking two mines. The samples are taken in order to make financial accounting between both mines after each shift.

In this paper one month measurements data were used to analyse the variability of Cu content in ore transported by one of KGHM Polska Miedz S.A. mines [2]. The results can be compared with Cu content variability of ore proceeded in the enrichment plants [3] as well as in ore supplied from mining face on divisional conveyor [2].

2 Methodology of ore sampling on a belt conveyor

A sampling station led by staff from the Centrum Badan Jakosci Sp. z o.o. (shortly CBJ) is placed on a transfer belt conveyor linking two KGHM Polska Miedz S.A. (shortly KGHM) mines. The ore is weighed on the conveyor by weightometer and is manually sampled in person by CBJ for metal content determination. These measurements and sampling/assay dictate the commercial payment between the mine, which supplies ore through this conveyor, and the mine, which receives and sends it together with its own ore to the ore processing plant.

The following description come from the QA/QC audit provided to KGHM by Wroclaw University of Science and Technology and SRK Consulting from UK [2].

The ore from the supplying mine was loaded into four bins from which the ore was fed onto the gathering conveyor and then transported ore from four sites of potentially different grade levels. The operations personnel advised that the finer fractions were generally with higher grade. The loading of material onto the analysed gathering conveyor was controlled by supplying the mine operators while the sampling regime was selected independently by CBJ personnel based on an agreed protocol.

The ore in the supplying mine was fed after blasting onto sectional conveyors by loaders through a 400 x 400 mm grille limiting the maximum lump size fed forward onto the analysed gathering conveyor linking both mines. Cumulative sampling was carried out exactly on this conveyor on each shift. It consisted of a dozen of partial samples drawn from equal masses (or periods) according to a special procedure described in the protocol.

The loading station comprised four 300 Mg ore bins fed by a single conveyor. The first three bins were filled via three moveable heavy duty ploughs on the conveyor. The local operator either raised or lowered the ploughs to push ore off the conveyor to discharge into the bins in order to maintain ore levels in the bins. The final bin received the ore from the discharge of the conveyor and did not incorporate a plough. The height of the plough was adjusted allowing partial removal of solids from the conveyor and consequently selective filling of different bins with either finer or coarser material was possible. The ore in the bins exhibited some degree of size segregation. The bins incorporated the level measurement instrumentation.

The bins were discharged via feeders onto the transfer conveyor to feed ore to the receiving mine conveying system via the CBJ weightometer and ore sampling point. The local operator, employed by the supplying mine, selected the plough position and bin discharge point. The procedure indicated that one feeder should be used at any time but also two feeders could be used if required. The first feeder was larger than the others. The operator recorded the timing/usage of the feeders but did not record the bin loading times.

The conveyors in the receiving mine were shut down one hour earlier than the supplying mine conveyors and during this time the bins at the load out point were usually filled. The conveyor tonnage was measured by two weightometers located upstream of the sample point. One weightometer was maintained by the supplying mine, which used the results, and the other was used and maintained by the receiving mine.

The sample cut was 0.3 m wide marked on the conveyor structure. This represented a 30 kg sample. The sampling process, from stopping the conveyor, breaking larger lumps, removing the sample, screening, crushing of oversize, mixing, splitting through labelling – was satisfactory. The CBJ sampling operators, seemed well-drilled in the procedures. The summarized protocol was as follows.

- Based on total planned tonnage to be transferred over the shift, the sampling procedure defined the number of samples to be taken, the sample interval in terms of tonnage and the minimum tonnage between successive samples.
- The sample was 300 mm wide and was taken from the conveyor between two defined marks.
- The conveyor was stopped and samples were taken at tonnages selected by the CBJ operators within the parameters described below. This procedure introduced randomness to the sample timing in order to reduce any potential bias in the type of material fed onto the conveyor at any particular time and to prevent the sampling exercise from being predicted (by loading operators).
- Pieces of ore lying across the sampling width were manually broken.
- The sample was removed and screened at 10 mm. The oversize was removed and crushed to -10 mm. The undersize and crushed solids were mixed in person on a steel plate. The mixed sample was collected and riffled to produce an approximately 4 kg sample. The sample was added to the mixer. All samples were combined and mixed and then riffled to produce the final sample.
  The parameters for selecting sample intervals are given in Table 1.

Table 1. Sampling interval protocol.

The planned supply of dredged material [Mg]/shift	Mass range (layer) [Mg]	Minimum interval by mass [Mg]	Number of incremental samples/shift
to 999	75	37.5	10-13
from 1000 to 1499	100	50	10-15
from 1500 to 1999	125	62.5	12-16
from 2000 to 2499	150	75	13-16
from 2500 to 2999	175	87.5	14-17
over 3000	200	100	15 and more

At the start of the shift, the planned tonnage to be transferred from the supplying mine was used to define the sampling plan. The CBJ sampling team identified the number of samples and the sampling tonnage interval. Then the team then identified the random tonnages at which the conveyor would be stopped. These figures are noted on a log sheet. During the shift the actual tonnage transferred as measured by the CBJ weightometer is noted. For each sample were a unique sample reference number was given which then was noted on the sheet.

The sample size was reported to be approximately 30 kg per cut depending on loading of the conveyor. The particle size was mostly smaller than 300 mm. Larger particles commonly occurred.

The sample size taken from the belt appeared to be too small considering the maximum particle size on the conveyor. Observation indicated that some particles in excess of 250x250x100 mm were possible, which would be 30% of mass contained in a typical 30 kg sample cut. Typically, a sample of this size should not contain particles greater than 40 mm. Increasing the sample width even to 1 m would not significantly improve this situation and would probably result in a unrealistic sample size for manual processing.

The final sample was split into four separate samples: one for i) the supplying mine, ii) receiving mine, iii) assay and one as a control sample.

Similar sampling procedures were used for sampling the ore on other conveyors e.g. irregular control sampling for establishing Cu content in the ore supplied from mining faces on the divisional conveyors. The check of dilution level was the aim of control sampling.

3 Results and discussion

Figures 1 and 2 show the change of the quantity and quality of ore during 120 subsequent shifts. On Saturday and especially on Sunday the analysed conveyor belt was practically not used for ore transportation, and therefore some studies excluded the "zero" shift from the analysis.

Ore quality variations were random, what can be seen from the results of the process test run recorded above and below the median, as well as from the change of the increase and decrease directions (Table 3). In the case of ore quantity, a statistically important data clustering was observed with a 95% confidence level (Table 2).

Out of 120 shifts, ore was transported only during 77 shifts (the quantity of ore was greater than zero). The quantity range was between 560 and 3500 Mg. The median was 3000 Mg. The *P* value for values above and below the median was below 0.025 (0.00978), and therefore a statistically significant clustering exists of the sequence value, with statistical significance of 95%.

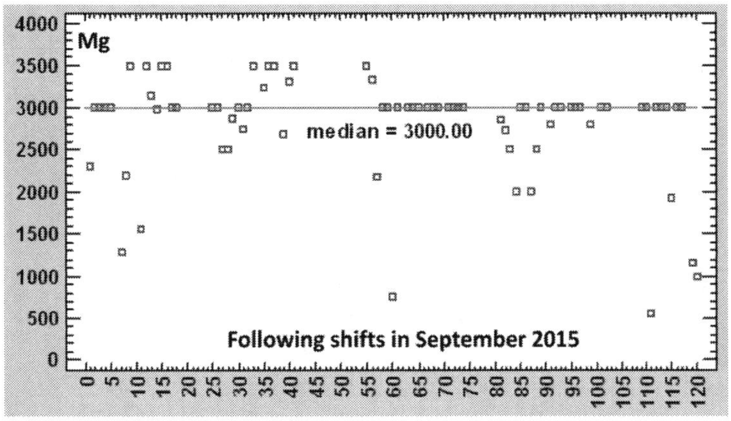

Figure 1. Variations of ore quantity on the conveyor analysed in 120 subsequent shifts in September 2015.

Table 2. Results of the Wald-Wolfowitz test of time sequence for the quantity of run-of-mine material.

Test	*Observed*	*Expected*	*Longest*	*P(>=)*	*P(<=)*
Values above and below the median	11	17.8649	44	0.996549	0.00977954
Increase and decrease of the sequence value	32	29.6667	4	0.254099	0.846738

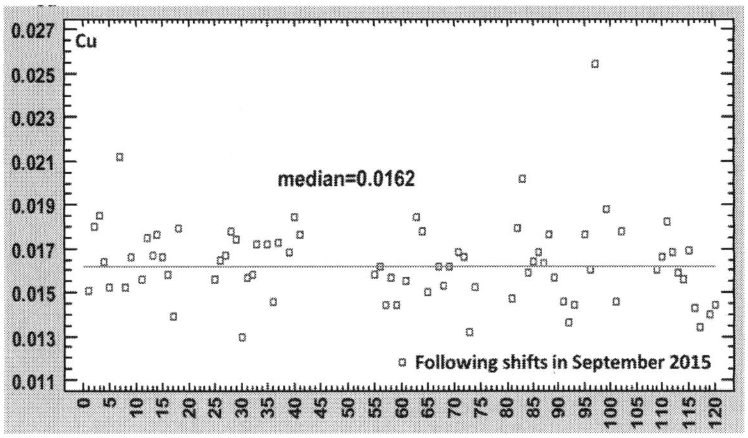

Figure 2. Variations of copper content in the ore transported on the analysed conveyor in September 2015.

Table 3. Results of the Wald-Wolfowitz test of time sequence for the copper content in the ore.

Test	Observed	Expected	Longest	P(>=)	P(<=)
Values above and below the median	35	37.973	7	0.792076	0.281171
Increase and decrease of the sequence value	49	50.3333	3	0.693159	0.409252

Numerous instances of shift output at the level of 3000 Mg indicated that this was the target value and probably such tonnage was usually planned for transportation from one mine to other one using a underground transport. Maintaining control of this value was possible, while maintaining control of quality could be difficult and quality variations might be random.

The ore grade was determined for 77 of 120 shifts, as the discussed conveyor was not used to transport ore during the rest of the shifts. The grade varied from 0.0118 to 0.0254 and the median was 0.0162. As none of the *P* values in Table 3. was below 0.025, non-random patterns were not observed at the significance level of 95%.

A further part of this work included description of the variations in copper quantity and content in the ore transported on the analysed conveyor, depending on the shift and the day of the week, supplemented by the analysis of their significance.

In September 2015, the quantity of the ore transported on the analysed conveyor varied depending on the day of the week (Fig. 3), shift number and day of the week/shift number combined. The variation of the weight of the transported run-of-mine material was significantly smaller if idle shifts were excluded.

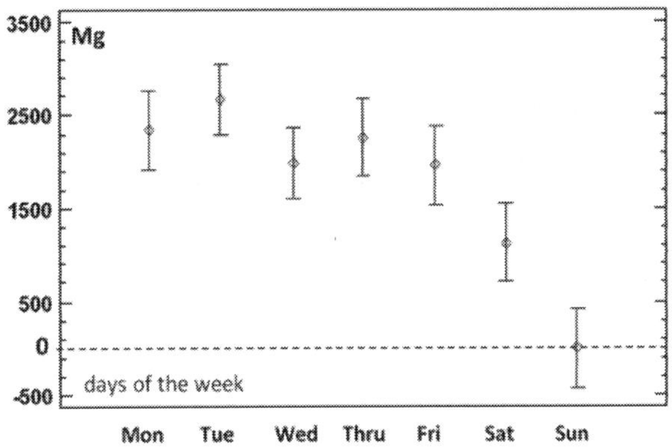

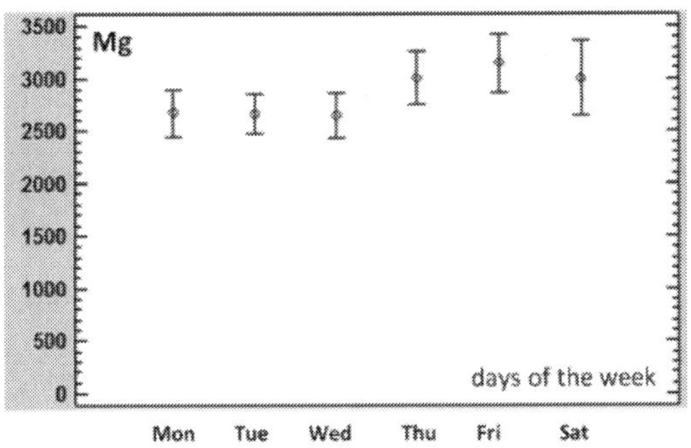

Figure 3. Variation of the average quantity of the transported ore with 95% confidence intervals depending on days of the week (including idle shift on the left and excluding idle shift on the right).

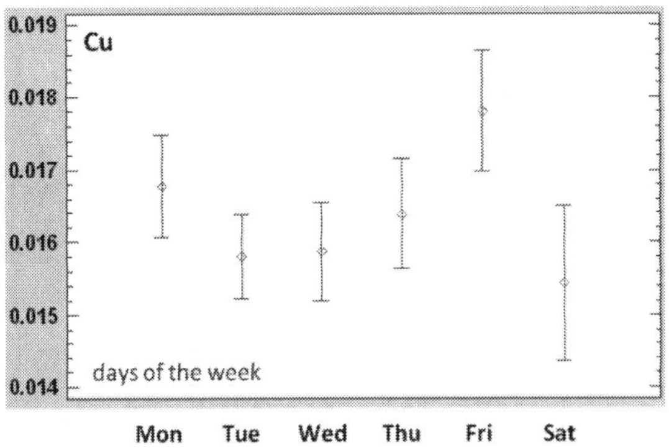

Figure 4. Variation of the average copper content in the ore transported on the analysed conveyor with 95% confidence intervals depending on days of the week.

Such statistical differences, occasionally significant especially if idle shifts with zero transportation were included, were not reflected in grade variations, which showed to be more stable in the discussed subgroups. As far as days of the week were concerned, two homogeneous but not disjoint groups were formed (Mon-Thu + Sat and Thu-Fri + Mon), with significant differences for the following pairs: Tue-Fri, Wed-Fri and Fri-Sat. Such differences may be difficult to explain, without the knowledge on the extent to which the production plans were followed.

Although some variations of copper content may be observed depending on the shift number, they did not seem statistically significant. This was confirmed in the quartile graph (box and whiskers type – Fig. 5, left side). Similar statistically insignificant variation of average grade values may be observed, if they were grouped depending on days of the week (Fig. 5, right side). Although average grade variations seemed to be considerable, especially on Saturday, when represented in the quartile graph, they proved to have a similar variation range for all days of the week. Two non-homogenous but joint groups may be identified (Mon-Thu + Sat and Thu-Fri + Mon). Significant differences can be observed for the following pairs: Tue-Fri, Wed-Fri and Fri-Sat.

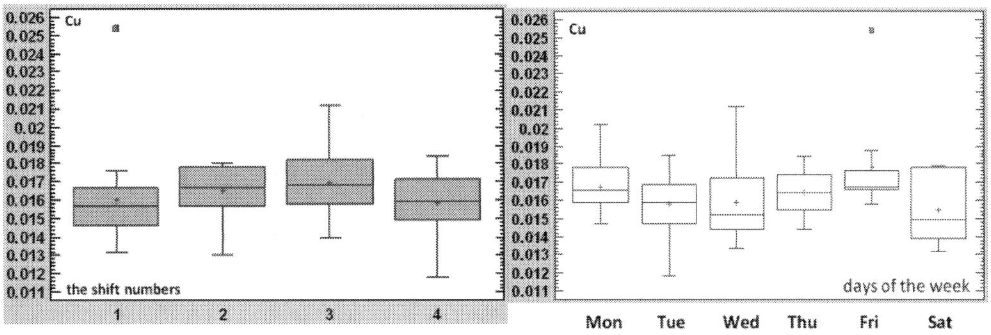

Figure 5. Variation of average copper content in the transported ore shown as a quartile graphs depending on the shift number (left) and days of the week (right).

Another comparison can be made depending on the days of the week and shift number (Fig. 6). The statgraphics Centurion XVII computer software, which was used in the calculations, identified as many as 4 joint homogeneous subgroups based on 17 significantly different pairs.

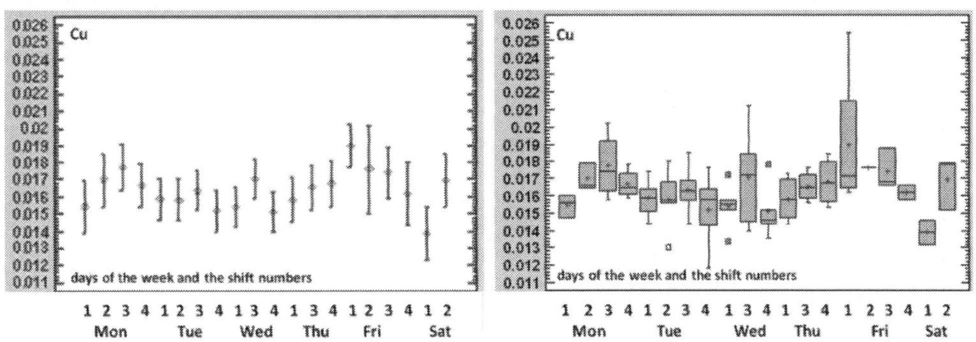

Figure 6. Average copper content in the ore transported on the analysed conveyor with 95% confidence intervals and with quartile graph depending on the day of the week and on the shift number.

These differences can't be logically explained without establishing a connection with production plans and mining face location. However, in the future, a record should be kept and analysed of metal grades in the output stream of mixed ore on the analysed conveyor and/or at the feeding point to Division of Concentrators, as the identified statistical differences may be explained by variation in the quality of ore from different parts of the deposit. Such procedure may serve a base for using statistical methods to forecast variation as related to mining face location (feeding point). The procedure may also be used in the DISIRE project.

If neither the shift number nor the days of the week have a significant influence on metal grade in subgroups, the time sequence of ore grade may be possibly analysed as a random sequence.

4 Conclusions

The analyses of ore transported on the transfer conveyor analysed in September 2015 showed that the quantity of the material in separate shifts was not a random process. The most frequent value and median was 3,000 Mg per shift, what indicated that the quantity of material sent from one mine to another one via the underground transport system was planned and controlled.

The run-of-mine tonnage changed significantly depending on days of the week, number of shifts and days of the week/shift numbers. Eliminating the idle shifts (without ore transport: Wed2, Thu2, Sat2, Sat3, and Sun1-4) allowed to limit the variations. It indicated that during work shifts the attempts were made to follow the plan. Reasons for the lack of transport during the second shift on Wednesday and Thursday are not known. Zero load during two shifts on weekends was due to lack of production.

The preliminary analysis of ore grade changes on the analysed conveyor indicated that grade variations may be considered as a random process. Significant grade differences depending on days of the week and shifts may be the result of conducting mining operations in different parts of the mine.

Detail description of the sampling procedure showed that the Cu content in all ore transported during one shift (circa 3,000 Mg) was estimated based on composite of about a dozen of small subsamples taken randomly during one shift. The subsample size taken from the belt appeared to be too small considering the maximum particle size on the conveyor. Observation indicated some particles in excess of 250x250x100 mm were possible, which would be 30% of mass contained in a typical 30 kg sample cut. Typically a sample of this size should not contain particles greater than 40 mm. Increasing the sample width even to 1 m would not improve this situation significantly and would probably result in a sample size unrealistic for manual processing. An automatic belt sampler should be considered if physical samples are required.

Taking into account variable size of ore lumps sometimes greater than 30 cm in diameter and possible big variation of Cu content in extracted ore: from 0.7% Cu in sandstone and dolomite (cut-off grade) up to even 15% Cu in rich shale layer indicated that both the sample mass as well as frequency of subsampling can hardly be representative.

Therefore, there is a need to introduce another ways of estimating ore quality (both Cu content as well as composition of ore lithology) what has not been measured till now. A belt analyser (neutron technology) could be considered as on application of special pellets with the ore quality information input into ore streams at their beginning in a DISIRE project.

Acknowledgement

This work is supported by the Framework Programme for Research and Innovation Horizon 2020 under grant agreement n. 636834 (DISIRE).

References

1. A. Tasdemir, P.B. Kowalczuk, *Physicochem. Probl. Miner. Process.* **50(1)**, 249 (2014)
2. K. Holodnik, L. Jurdziak, W. Kawalec, Z. Krysa, W. Milczarek, G. Paszkowska, D. Pattinson, R. Simpson. *Raport*, (archive material) (2015)
3. J. Drzymala., P.B. Kowalczuk, *in Proceedings of the 12th International Mineral Processing Symposium*, (Cappadocia, Turkey, 1191, 2010)

Coal waste slurries as a fuel for integrated gasification combined cycle plants

Marcin A. Lutynski[1,a], and Aleksander Lutynski[2]

[1]*Silesian University of Technology, Faculty of Mining and Geology, ul. Akademicka 2, 44-100 Gliwice, Poland*
[2]*KOMAG Institute of Mining Technology, Pszczynska 37, 44-101 Gliwice, Poland*

Abstract. The article summarizes recent development in integrated gasification combined cycle technology and lists existing and planned IGCC plants. A brief outlook on the IGCC gasification technology is given with focus on entrained-flow gasifiers where the low-quality coal waste slurry fuel can be used. Desired properties of coal and ash for entrained-flow gasifiers are listed. The coal waste slurries, which were deposited at impoundments in Upper Silesian Coal Basin, were considered as a direct feed for such gasifiers. The average ash content, moisture content and lower heating value were analysed and presented as an average values. Entrained-flow commercial gasifiers can be considered as suitable for the coal slurry feed, however the ash content of coal slurries deposited in impoundments is too high for the direct use as the feed for the gasifiers. The moisture content of slurries calculated on as received basis meets the requirements of entrained-flow slurry feed gasifiers. The content of fines is relatively high which allow to use the slurries in entrained-flow gasifiers.

1 Introduction

Integrated gasification combined cycle (IGCC) is a promising clean coal technology which has been successfully implemented in several operational, commercial scale and other semi-industrial scale plants (Table 1) [1, 2]. The majority of the plants is fuelled by coal although the main advantage of the IGCC plant is its fuel flexibility. For the IGCC technology to become more competitive and fully commercial, low availability (defined as plant operation time during a certain period of time, usually 1 year) and high capital costs are the main challenges [3]. In general, the cycle in a typical IGCC plant consists of two main stages. Firstly, the fuel is converted to syngas, which is a mixture of hydrogen and carbon monoxide. Secondly, the syngas is converted to electricity in a combined cycle power block consisting of a gas turbine process and a steam turbine process which includes a heat recovery steam generator (HRSG). In fact, the combined cycle technology is similar to the technology used in modern natural gas fired power plants. The IGCC plants have several advantages over either typical PC plants or fluidized bed power plants. The advantages are with lower production of solid waste, fuel flexibility or low NO_x and SO_x emission.

[a] Corresponding author: marcin.lutynski@polsl.pl

Table 1. Existing and planned IGCC plants (after [2]).

Plant (location) initial year of operation	Size (MW)	Feedstock
Buggenum (Netherlands) 1994	253	Coal & biomass
Wabash river (USA) 1995	262	Bituminous coal & petroleum coke
Tampa (USA) 1996	250	Coal & petroleum coke
Pernis (Netherlands) 1997	155	Heavy oil residue
Priolo Gargallo (Italy) 1998	532	Heavy fuel oil, tar
Puertollano (Spain) 1998	335	Petroleum coke
Sarlux (Italy) 2000	548	Residual oil, tar, bitumen
Negishi (Japan) 2003	342	Heavy oil, asphalt
Vresova (Czech Republic) 2005	400	Brown coal
Knox County (USA) 2013	618	Mid-western coal
Kemper County (USA) 2014	582	Lignite coal
Nakoso (Japan) 2014	500–600	Bituminous

As the IGCC technology requires low quality fuel, it is anticipated that coal slurries resulting from the coal beneficiation process can be of great importance as the potential low cost feedstock for the gasification process. In this study, the coal waste slurries resulting from the coal beneficiation process of Polish coal mines were considered as a potential fuel for commercially available IGCC plants.

2 Integrated gasification combined cycle gasifiers overview and fuel requirements

2.1 Gasifier type

There are three common types of gasifiers used in commercially available IGCC systems: fixed-bed, fluidized bed and entrained flow. There is also one technology called the Molten-bath witch is capable of gasification both coking and non-coking coals but its development is not yet advanced enough to be considered in this study. One of the factors for selection of gasifier is the fuel type. In the fixed-bed and fluidized bed gasifiers the fuel (coal) should have low moisture content of <18% [4]. The entrained-flow gasifiers are the most common commercial gasifiers due their high throughput, when compared to the other technologies currently available in the market. Therefore, only the entrained-flow gasifiers could be suitable for coal waste slurries gasification (Fig. 1).

In general, there are five commercially available entrained-flow gasifier. In Table 2, a short summary of commercially available entrained-flow gasifiers is presented.

DOI: 10.1051/e3sconf/20160801056

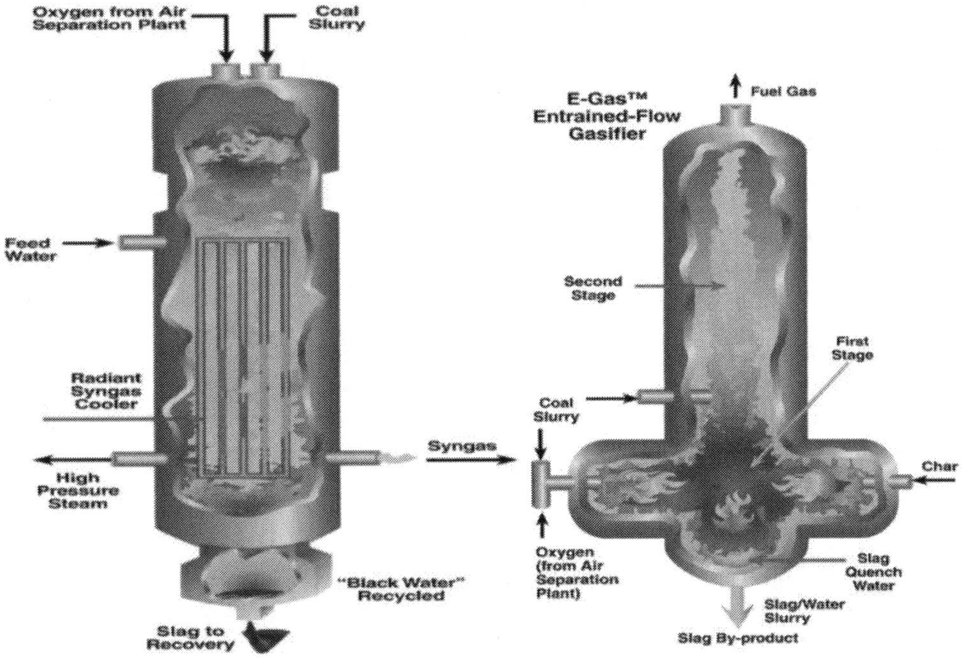

Figure 1. Examples of entrained-flow slurry feed gasifiers. GE Energy Gasifier with Radiant cooling (left) and Conoco Phillips E-gas gasifier [5].

Table 2. Characteristics and features of commercially available entrained-flow gasifiers (after [6]).

Gasification Technology/ Design feature	**Ge energy with Radiant Cooling**	**Ge energy with Water Quench**	**Shell Scgf**	**Conocophillips E-gas**	**Siemens Sfg**
Feed system	60%–70% coal/water slurry; top feed	60%–70% coal/water slurry; top feed	Dry coal (~2% moisture content)/N_2; bottom feed	60 to70% coal/water slurry; bottom feed: 1st stage 80% and 2nd stage 20%	Dry feed/N_2 or CO_2; top feed
Wall	Refractory	Refractory	Membrane	Refractory	Cooling screen (ash content > 1–2 wt%)
Pressure (bar)	30–80	30–80	30–40	~27	≥ 40
Hot syngas temperature (°C)	1260–1500	1260–1500	1360–1650	1st stage 1340–1400; 2nd stage 1038	1400–1600
Syngas quenching method	Radiant cooling then fire tube syngas cooler	Water quench (no syngas cooler)	Recirculated cold syngas quench then water tube syngas cooler	Chemical quench then fire tube syngas cooler	Water quench (no syngas cooler)
Quenched syngas temperature (°C)	~800	~300	~900	~1038	~230

2.2. Desired coal and ash properties

The increased moisture content requires more energy to vaporize the water, while the and increased ash content means a greater mass flow of inerts which must be heated up. Therefore, both the increased moisture and ash content have the effect of higher oxygen consumption and lower cold gas efficiency.

The dry feed gasifiers, the coal is dried before gasification so that the gasifier does not have to process the extra steam from coals with the high moisture content. Obviously, drying of coal to around 2% of moisture consumes a part of the energy from the produced syngas. The slurries of coal with high moisture content achieve lower solids concentration, and therefore a large amount of steam must be vaporized, and then processed in the gasifier. This contributes to reduction of the capacity per gasifier volume because the residence time per unit volume is reduced [3].

The ash content is an important criteria for selection the gasifier. In general, the ash content for slurry feed should not exceed 20%, whereas for the dry feed 40%. The feed particle size in the entrained flow gasifiers should not exceed 100 μm for dry feed and 1000 μm for the slurry feed.

Other criteria for coal and ash properties for the entrained-flow gasifiers are summarized in Table 3. In particular, the ash properties are of great importance in the gasification process. The mineral matter and/or inorganic components may vaporize, form either liquids or solids. Formation of three phase products is a significant difference in the coal burning process. The major problem in the process of gasification is condensation of inorganic compounds which may result in slagging. The slag is formed when the ash particles deposit on the gasifier wall and flow down as highly vitreous product. To some extent the slagging effect is desirable in the entrained-flow gasifiers since a small slag layer on the refractory is desirable, as this layer protects the refractory lining and reduces heat loss. However, at high temperature and reduced viscosity, the molten slag layer can be very corrosive and can penetrate into the air-cooled refractory lining, the desired properties of coal and ash for entrained-flow gasifiers are summarized in Table 3.

Table 3. Summary of criteria for ash and coal properties for entrained-flow gasifiers [4, 6].

Coal and ash properties	**Criteria for entrained-flow gasifier**
Coal rank	Any (dry feed); High (Slurry feed)
Acceptability of fines	Unlimited
Ash contents	20% slurry feeding, 40% dry feeding
Ash Temperature	(FT) <1400 °C is optimal, but 1500 °C is acceptable; >1500 °C flux is needed
Ash composition	silica ratio 100 $SiO_2/(SiO_2+Fe_2O_3+CaO+MgO) < 80$; SiO_2/Al_2O_3 ratio of 1.6–2.0 is optimal for minimum flux amount
Temperature of critical viscosity (Tcv)	<1400 °C is preferred
Viscosity	5–25 Pa·s at 1400 to 1500 °C; Optimum 15 Pa·s

3 Analyses of coal waste slurries properties

Since the IGCC technology, particularly entrained-flow gasifiers, required low quality fuel as the feedstock and the feed can be in the form of a slurry, it was decided to assess coal slurries deposited in impoundments as a potential fuel. The impoundments were filled with the slurries resulting from the process of steam and coking coal beneficiation in Polish coal mines. The small size gangue was difficult to remove using conventional beneficiation methods such as froth flotation. The presence of fines was significantly lowering the quality of coke. Therefore, particles smaller than 1 mm were considered as a waste. The same situation was observed in the case of steam coal as it was impossible to burn small particles in stoker-fired boilers. Due to that fact, the coal slurry was stored on the surface in either impoundments or ponds.

Impoundments are located in the Upper Silesia Coal Basin (Southern Poland) where currently over 30 coal mines are in operation. Formerly, over 100 mines were active, hence the large number of abandoned impoundments could be potentially used as a fuel either directly or after processing. Investigation of coal slurries properties was performed within the framework of development project entitled No N R09 0006 06/2009 "*Identification of energetic potential of coal slurries in the national fuel balance and technological development strategy of their usage*" [7].

For the purpose of this study the following properties of coal slurries were analysed (on as received basis):

- ash content
- moisture content
- content of fine fraction (<0.1 mm)
- lower heating value (LHV).

The properties of coal slurries deposited in 20 impoundments on the as received basis were averaged and are presented in Table 4. Large variations in the properties of coal slurries reflected in the calculated standard deviation for each property.

Table 4. Properties of coal waste slurries considered for the use in IGCC technologies (as received basis).

Property	Average	Standard deviation
Ash content, %	34.2	10.1
Moisture content, %	25.6	5.7
Content of fine fraction (<0.1 mm), %	62.3	14.3
Lower Heating Value, MJ/kg	13.1	3.7

Since the analyses of properties was done on "as received" basis and this state is relevant to the case where slurry feed gasifier is used it was decided to analyse the properties on air dried basis which would be more relevant to dry feed gasifiers. The results of analysis based on air dried basis are shown in Table 5.

Table 5. Properties of coal waste slurries considered for the use in IGCC technologies ("air dried" basis).

Property	Average	Standard deviation
Ash content, %	42.6	12.3
Moisture content (air dried), %	2.9	2.1
Lower Heating Value, MJ/kg	16.4	4.5

4 Discussion

Results presented in Tables 4 and 5 show that the average ash content for both "as received" and "air dried" basis are over the requirements of the entrained-flow gasifiers. Although the water content (moisture) is sufficient (on average 25.6%) to consider the coal slurry as the direct feed for the gasifier, the ash content is significantly too high (above required 20%). If the slurry would be dried to "air dried" state the moisture content would be in average 2.9%, which is still over the required 2% for the dry feed gasifiers.

The lower heating values of coal slurries was on average 13.1 MJ/kg for "as received" slurries and 16.4 MJ/kg for dried slurries. Experience with commercial gasifiers show that best results were obtained for LHV values ranging between 20-30 MJ/kg [8,9]. Therefore, coal slurries cannot be considered as the direct feed for the gasifiers. However, there is a possibility to upgrade coal slurries with commonly known methods such as either gravity separation or flotation [10].

The only parameter, which would allow to use the slurries in IGCC gasifiers, is the content of fines (<0.1 mm) which on average is above 62.3%. Only in two impoundments the content of fines was below 30%, whereas in the rest it was higher tan 60%. The high content of fines makes this feed much easier to process since it does not require comminution.

Other important issue regarding the use of coal waste slurries for IGCC gasifiers is the ash (slag) composition. In this research this issue was not considered although will be taken into account in further studies.

5 Conclusions

The study was a first attempt to assess if the coal waste slurries deposited in impoundments could serve a direct feed for IGCC gasifiers. The following conclusions can be drawn from the study:

- entrained-flow commercial gasifiers can be considered as suitable for the coal slurry feed,
- the ash content of coal slurries deposited in impoundments in the Upper Silesia Coal Basin is too high for the direct use as the feed for the gasifiers,
- the lower heating value of coal slurries is relatively low in comparison to coal used in pilottests,
- the moisture content of slurries calculated on the as received basis meet the requirements of entrained-flow slurry feed gasifiers,
- the content of fines is relatively high, which allows the use of slurries in entrained-flow gasifiers.

One of the most important properties of the fuel for IGCC entrained-flow gasifiers is the ash composition which was not analysed in this study. Further studies focused on this should be carried out to verify if such low quality fuel can be considered for the use in IGCC technology.

References

1. J. Tollefson, R. Van Noorden, *Nature,* **484**, 151 (2012)
2. M. Hossein Sahraei, D. McCalden, R. Hughes, L.A. Ricardez-Sandoval, *Fuel* **137**, 245 (2014)
3. O. Maurstad. *An Overview of Coal based Integrated Gasification Combined Cycle (IGCC) Technology* (MIT 2005)
4. V. Krishnamoorthy, S.V. Pisupati, *Energies,* **8,** 104 (2015)
5. R.W. Breault, *Energies*, **3**, 216 (2010)
6. P. Wang, M. Massoudi, *Energies,* **6,** 784 (2013)
7. Report on *Identification of energetic potential of coal slurries in the national fuel balance and technological development strategy of their usage,* (Silesian University of Technology, 2009)
8. L. Zheng, E. Furinsky, *Energy Convers. Manage.* **46,** 1767 (2005)
9. *Tampa Electric Polk Power Station Integrated Gasification Combined Cycle Project: Final Technical Report,* (US DOE, Tampa, Florida, 2002)
10. J. Szpyrka, M. Lutynski, *Physicochem. Probl. Miner. Process.* **48**, 579 (2012)

E3S Web of Conferences **8**, 01057 (2016) DOI: 10.1051/e3sconf/20160801057
MEC2016

Granulometric composition study of mineral resources using opto-electronic devices and Elsieve software system

Stanislaw Kaminski[1], Piotr Kaminski[1], Dorota Kaminska[1], and Jerzy Trzcinski[2,a]

[1]*Kamika Instruments, Strawczyńska 16, 01-473 Warsaw, Poland*
[2]*Faculty of Geology, University of Warsaw, Żwirki i Wigury 93, 02-089 Warsaw, Poland*

Abstract. The use of mechanical sieves has a great impact on measurement results because occurrence of anisometric particles causes undercounting the average size. Such errors can be avoided by using opto-electronic measuring devices that enable measurement of particles from 10 μm up to a few dozen millimetres in size. The results of measurement of each particle size fraction are summed up proportionally to its weight with the use of Elsieve software system and for every type of material particle-size distribution can be obtained. The software allows further statistical interpretation of the results. Beam of infrared radiation identifies size of particles and counts them precisely. Every particle is represented by an electronic impulse proportional to its size. Measurement of particles in aqueous suspension that replaces the hydrometer method can be carried out by using the IPS L analyser (range from 0.2 to 600 μm). The IPS UA analyser (range from 0.5 to 2000 μm) is designed for measurement in the air. An ultrasonic adapter enables performing measurements of moist and aggregated particles from 0.5 to 1000 μm. The construction and software system allow to determine second dimension of the particle, its shape coefficient and specific surface area. The AWK 3D analyser (range from 0.2 to 31.5 mm) is devoted to measurement of various powdery materials with subsequent determination of particle shape. The AWK B analyser (range from 1 to 130 mm) measures materials of thick granulation and shape of the grains. The presented method of measurement repeatedly accelerates and facilitates study of granulometric composition.

1 Introduction

Measurement using mechanical sieves takes into account the smallest dimension of the particle that has to be inscribed in a square, whose side equals the dimension of sieve mesh. The size of the square sieve mesh is assumed to be the size of the particle. Usually, particles are not spherical. For example elliptical particles are measured on sieves according to the smallest dimension of the ellipse. During sieve measurement, particles are weighted correctly, but measurement of their size is prone to errors. The use of the set of sieves with diverse strength and method of shaking and different time of operation can also have strong influence on the results. Particles of longitudinal shape cause undercounting during determination based on medium sized sieves. Those errors can be avoided by using opto-electronic measuring devices.

[a] Corresponding author: jt@kamika.pl

Addressing this problem generates basic question: is it possible to precisely measure particle size distribution of the material that consists of the particles of various sizes ranging from a few µm to tens of mm? A set of elements of such material is numerically large, and construction of a single measuring device does not allow to measure spectrum of particle sizes this wide. The solution can be to use a system that would consist of a few opto-electronic devices. Each of them would perform independent measurement for the precisely set size range of grains diameters. The particle size distribution would be summed up proportionally to the weight particle size fraction of each range.

2 Principle of measurement

Let us consider how to measure grain size distribution of material consisting of particles from 10 µm to fragments with maximum dimension up to several dozens of mm. Counting by a volume ratio, quantity of smaller particles equivalent to one big fragment is really large. The set of particles is so numerically large that no single device can measure it in the considerable amount of time. This task can be facilitated by a specified method of measurement.

A sample is divided into a few parts (fraction ranges) by being transferred through the sieves. Each fraction is weighted and representative samples for the measurement with opto-electronic devices are prepared. The representative sample for the grain size fraction of fine particles below 1 mm is usually a small part of the whole set and consists of many millions of particles. A medium grain size fraction (from 1 to 10 mm) is represented by tens of thousands of particles. The particle size fraction of thick particles (larger than 10 mm) usually consists of a few thousands of particles and can be measured as a whole. Results of measurement of every grain size fraction are summed up proportionally to their weight with the use of opto-electronic Elsieve method and distribution of volume simulates sedimentary and sieve analysis. The particle size distribution is constructed proportionally to the weight share of divided particle size fraction groups.

Measuring system can consist of any number of devices connected individually to steering and registering computer. A computer of each measuring device is connected to a central computer acting as a server. The software of the measuring system provides access to the results saved on the server. They can be summed up on any computer connected to this network. The presented method significantly accelerates and facilitates study of granulometric composition.

Any system of opto-electronic devices can be used to calculate results of analysis and to obtain the particle-size distribution for all materials. If the user is missing one of the devices, its work can be replaced with traditional sieve method and results (sieve's size and weight of the material on the sieve) can be entered directly into the software.

The presented method of measurement significantly accelerates and facilitates obtaining the particle size distribution curve and automatically determines compatibility of the material with various norms, for example PN-ISO 565: 2000, PN-B-04481: 1988, and PN-EN ISO 14688-2: 2006.

The system of this type is currently used to measure granulometric composition of soils, rocks and various geo materials in the laboratory of Institute of Hydrogeology and Engineering Geology, Faculty of Geology, University of Warsaw [1, 2, 3, 5].

3 Measurement method

The method of measurement by opto-electronic devices relies on determining the size of the smallest particles in order to gradually, in a steady manner, switch to the measurement of changes in the beam or radiation scattered by moving larger particles. There are no optical limits for measurement of single fine and thick particles in the analysers. The beam of infrared radiation not only identifies the particle size, but also allows to precisely count particles in the whole measurement range. Every particle is represented by an electric impulse proportional to its size. A set of particles is primarily measured with the division into 4096 size fractions and transformed (calibrated) into 256 size fractions.

Apart from the software designed for the measurement of particles, analysers are also equipped with a software for optimization of any parameter as a function of granulometric composition of the measured suspension or powder. There is also a software for calculating the grain size distribution by any calibration (for sieve method, hydrometer method or spherical). The measurement results are presented in either colourful diagrams or tables.

4 Selected types of opto-electronic analysers

4.1 IPS L analyser – measurement range 0.2 - 600 μm

The analyser is designed for the measurement of particle size distribution in aqueous suspension irrespective of their chemical and physical properties. Solid and oily particles as well as air bubbles can be measured. It completely substitutes hydrometer method.

4.1.1 Method of measurement

Water or aqueous solutions with suspended particles is introduced from a container, equipped with a mixer, into water passing through the analyser. Flow of water is automatically regulated by switching on a feeding pomp. Measurement of particles size is completely automatic and regulated by a computer. Filter cleans water from the particles. Operation and permeability of the filter is controlled with a manometer. Clean water transported to a chamber flows into the measurement zone, where a needle dosing water with suspended particles is located. Clean water prevents optics in the sensor from getting dirty. Feeding the pomp, that is placed after the sensor, additionally sucks water and is regulated via the computer.

4.2 IPS UA analyser – measurement range 0.5 - 2000 μm

The analyser is designer for the measurement of solid particles size distribution in the air irrespective of their physical and chemical properties under laboratory conditions. With addition of ultrasonic adapter, the analyser can be used to measure the size of moist and aggregated particles (size range from 0.5 to 1000 μm). Construction of the analyser and applied software enable determining second dimension of particle, its coefficient of shape and specific surface area (when the coefficient of intra-particle porosity is known).

4.2.1 Method of measurement

The analyser consists of a measuring sensor and a measuring system integrated with the electronic system that is feeding the particles, providing continuity of the measurement and control of concentration of the particles in the measurement zone. The entire process is regulated by a computer equipped with the technical software system. A small compressor is the executive element of the automatic feeding system. The compressor has specific characteristics, adjusted to stable regulation of the air flow.

An ultrasonic feeder is used to separate 'glued' particles. The feeder has a shape of a concave vessel. Its bottom vibrates with the frequency of 40 kHz and with the amplitude up to a few mm. The moist substance is dried due to vibrations. Even significant humidity of the sample does not disturb the measurement. For feeding various granular materials, regulation of the amplitude and quantity of ultrasonic impulses has around 4000 transitional states between 0 and the maximum operational level of the feeder. Control over the airflow transmitting previously separated particles into measurement zone provides precision of feeding. Regulation of the airflow has around 300 levels of speed. Such a precise control of the feeder enables measuring single particles quickly (up to several thousand particles per second) and avoiding coincidence of particles in the measurement zone.

4.3 AWK 3D analyser – measurement range 0.2 – 31.5 mm

The analyser is destined for the measurement of powdery materials like raw materials (fine aggregates, gravel, coarse sands), coal, plant seeds, food granulates and plastics under laboratory conditions. A software system used to operate this device enables full simulation of the measurement on mechanical sieves. The device can be used during either production or technological processes during optimization of milling and selection of compound of various materials. Additional useful function is the ability to determine the shape of the particle.

4.3.1 Method of measurement

The AWK 3D analyser is equipped with an electronic measuring block, connected with two independent optical tracks to determine the particles size, all together with counter of measurements, that enables to determine the shape of particles in three dimensions. The electronic measurement block is connected to a computer running analysis and controlling the device. The measurement process is controlled automatically, after the data are entered to the computer through the keyboard. The feeding system consists of the feeder with a vibrating slide. Vibrations are caused by a small electric engine with eccentric load. The engine speed (and vibrations caused by it) is controlled with an electronic system. The range of regulation has 225 levels. If there are too many grains fed to the measurement zone, the speed of the engine slows down and if there are not enough grains falling to the measurement zone, speed of the engine accelerates. Particles are fed to the slide from the charge funnel. Size of the funnel and width of the slide are adjusted accordingly to the size of particles measured. There is a small chimney, located between feeder's slide and measuring sensor, that is supposed to stabilize the fall of particles of different shapes. The feeder transports particles from the funnel to the container placed below the device. When the measurement is finished, feeding is automatically terminated and the results, completely simulating measurements according to mechanical sieves, are processed.

4.4 AWK B analyser – measurement range 1 – 130 mm

The analyser is designed for automatic measurement of either thick aggregates or other materials of similar particle size under laboratory conditions. The device can be constructed to measure in 3 dimensions (3D) and have the same properties as AWK 3D – it can determine the particle shape. The device fully simulates measurements according to mechanical sieves and can be used for the optimization of the production process and for determining the shape of particles.

4.4.1 Measurement method

A beam of infrared or laser radiation originating from the sensor is scattered by particles of aggregate flowing through the measurement zone. The measuring system consists of a multi-channel optic system that enables spatial measurement in three directions. Each grain is represented by the corresponding electronic impulse proportional to its size. Set of particles is primarily measured in 4096 size fractions and then transformed (calibrated) into 256 fractions available to the user. A huge measurement surface '*s*' x 300 mm of identical sensitivity is one of the characteristics of this method. Measurement area dimension '*s*' depends on the type of the applied optics and can be 110, 160 or 190 mm. Additionally, the device is equipped with a special feeder that can fit over a dozen of kilograms of material.

4.4.2 Measurement method

The AWK B analyser is equipped with electronic measurement block connected with electric control box and measuring sensor as well as a desktop computer with a special interface. The whole system is controlled with a computer keyboard. When measurement is terminated, its results are automatically saved by computer memory and analyser settings go back to the initial state. The working section of the analyser is built in the shape of a cart. A steering box equipped with an electric motor operator is suspended on the stainless steel frame. The motor operator, with help of a pusher, inclines the slide hanged on the bearing rod. The measuring sensor with the inlet of 200 x 300 mm section is placed on the frame on two supports. The slide is additionally equipped with a vibrator. During the process of measurement, the vibrator gradually reaches maximum vibration, and then the slide is set to the lower position. Through the inlet, the sample slips from the bottom of the slide to right under the analyser. Intensity of the sample slipping is regulated with the use of a computer.

5 Notes regarding the method

AWK 3D and AWK B are the analysers constructed for 3-dimensional measurement of particles. The results of measurements of shape are presented according to the Zingg classification [7]. Information determined additionally, considering indicators of particle size distribution calculated on the basis of results of analysis, are: geotechnical name of the soil according to the classification presented in Polish norms PN-B-04481: 1988 and PN-EN ISO 14688-2: 2006, index of heterogeneity particle size distribution, average diameter of grains, standard aberration, skewness and kurtosis. Those information open the door for wider statistical interpretation of the results of granulation studies. It is possible to obtain other characteristics of particle size distribution, depending on what is required by the user .Previously described devices are presented in Fig. 1.

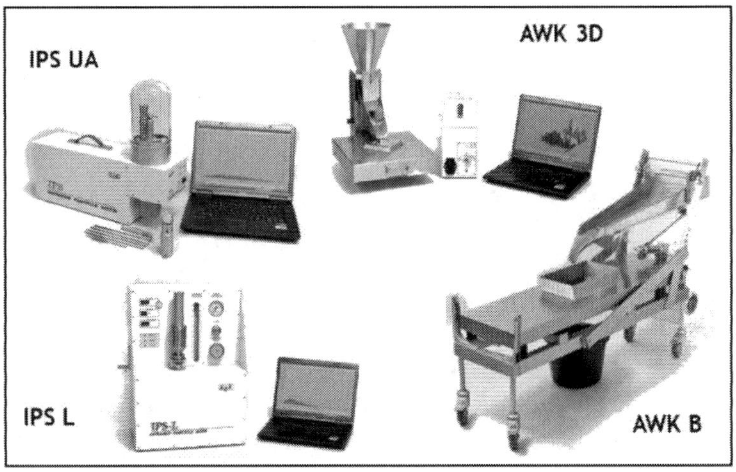

Figure 1. Measuring apparatuses manufactured by KAMIKA Instruments.

6 Description of the software system

The software enables user to:

- sum up results of measurements according to sieve calibration performed on different KAMIKA analysers for the measurement of granulometric distribution,
- determine the curve of granulometric distribution in the range from 0.4 µm to 30 mm,
- automatically sum up results of measurements carried out on mechanical sieves.

DOI: 10.1051/e3sconf/20160801057

6.1 Saving results of sieve analysis

To start working with the IPSL, IPSU, IPSA, 2DiSA, AWK 3D software system, chosen results of sieve analysis should be saved as SI* files. Different analysers may be used to measure different fractions of geo materials because of their measurement range. Results of the sieve analysis carried out on certain analysers should be saved as different file types, as shown in Table 1.

Table 1. Fractions and types of devices used for their analysis.

Fractions	Analyser	File name with results of sieve analysis
Clay	IPS L	*.SI1
Silt	IPS U*	*.SI2
Sand	IPS A**	*.SI3
Gravel	AWK 3D	*.SI4

6.2 Quantity of measurements

First, the quantity of measurements (fractions) that is supposed to be summed up must be chosen. Results from up to 4 analysers can be summed up (Fig. 2).

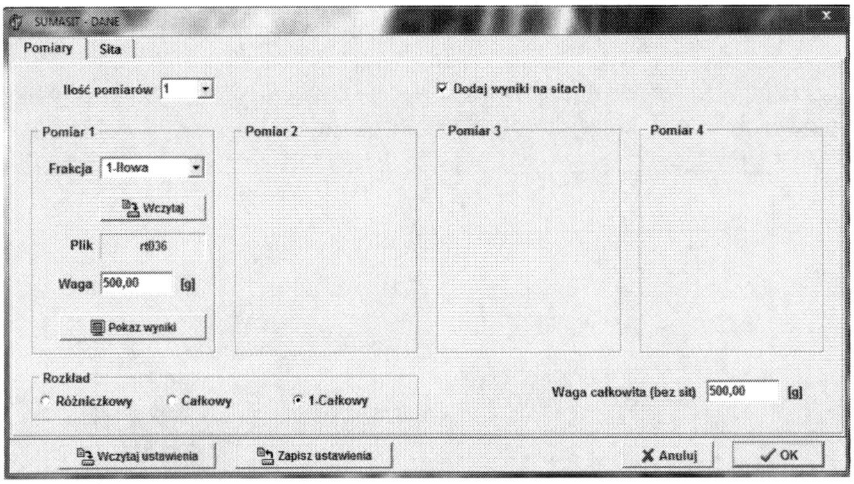

Figure 2. Dialog box for setting number of measurements.

If one measurement is chosen, it can be used for hydrometer analysis, where results from IPS L and measurements on sieves are added. For every result of sieve analysis, that is to be summed up, the type of fraction should be chosen.

6.3 Weight

To sum up results of measurements, weights of particular fractions should entered, for respective measurement results in the corresponding fields in the measurement section. Summed weight of fractions should equal to the weight of the whole sample measured with the use of the analyser. If summed weight of fraction is not equal to the weight of the whole sample, information indicating error will appear.

6.4 Summing up results from mechanical sieves

The 'Sum of sieves' software enables to add results from mechanical sieves to the particle size distribution curve. This solution is helpful, when only a part of the sample can be measured with the Kamika analysers, for example when some particles in the sample exceed measurement range of the devices. The result from up to 11 sieves can be added.

It is important for the sieves to be bigger or equal to the biggest sieve in selected results for the biggest fraction and smaller or equal to the smallest sieve in selected results for the smallest fraction. The results from the sieves cannot be added for the ranges between the ranges of the connected devices. The weight on every sieve is automatically included in the calculation of the particle size distribution curve.

6.5 Results

The calculated grain size distribution curve is presented in Fig. 3.

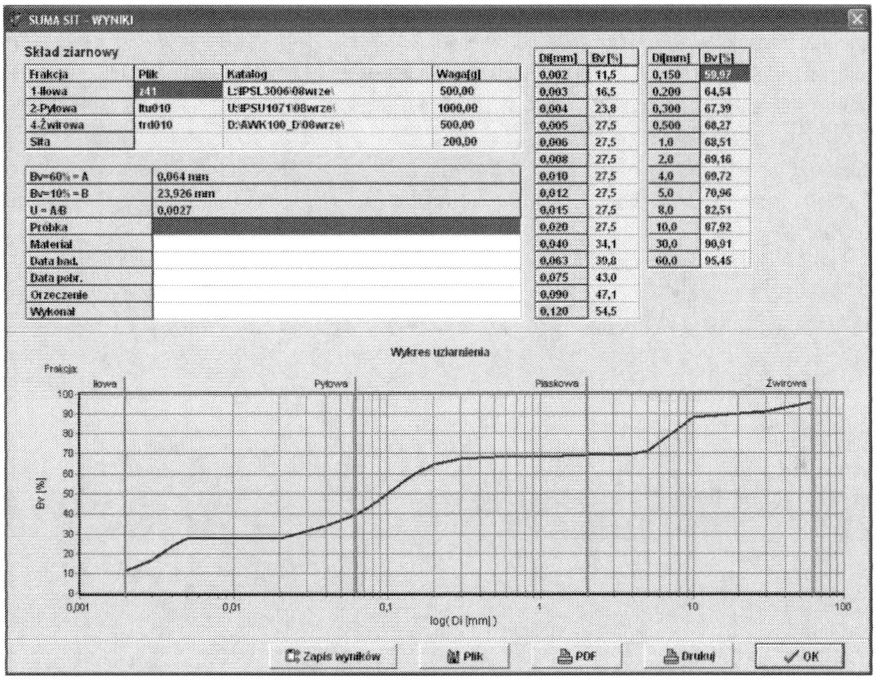

Figure 3. Dialogue box presenting summed results of measurements.

7 Conclusions

Using traditional methods of measurement of granulometric composition of various materials is manually challenging and requires a lot of work, both during studies and in practice. Traditional methods of measurement, such as either sieve or sedimentation analyses can cause errors and inaccuracy connected with the method of measurement. Those measurements can be replaced with the 3D opto-electronic Elsieve analysis of shape and size of particles from 0.5 μm to 100 mm providing also other parameters of the particle size.

When a set of particles is numerically large and varies very much in size, it is not possible to measure it with a single device. This is why the measurement should be performed with a measuring system consisting of a few devices. The whole set should be divided into a few parts (size ranges) and representative samples should be prepared. The distribution of grain size is summed up proportionally to weight share of the respective size ranges.

The measuring system consists of a few analysers individually connected with the computer that controls and registers and is conjugated with the server. The software system provides summing up of the measurements on any computer connected to this network. The presented method of measurement accelerates and facilitates studies of granulometric composition.

References

1. M. Fiolek, *MSc thesis* (Uniwersytet Warszawski, 2014)
2. T. Gotowiec, *MSc thesis* (Uniwersytet Warszawski, 2006)
3. K. Jaglinski, *MSc thesis* (Uniwersytet Warszawski, 2006)
4. S. Kaminski, *ELSIEVE Optyczno-elektroniczna symulacja pomiarów mikroziarn powyżej 0,5 µm według sit mechanicznych,* (2007)
5. S. Kaminski, D. Kaminska, *ABiD*, **12(2)**, 85 (2007)
6. P. Malczyk, *MSc thesis* (Uniwersytet Warszawski, 2015)
7. E. Mycielska-Dowgiallo, in *Badania osadów czwartorzedowych. Wybrane metody i interpretacja wynikow, edited by E.* Mycielska-Dowgiallo, J. Rutkowski (Uniwersytet Warszawski, 1995)
8. PN-88/B-04481:1988. *Grunty budowlane. Badania probek gruntu.* (Polski Komitet Normalizacyjny, Warszawa, 1998)
9. PN-ISO 565: 2000. *Sita kontrolne. Tkanina z drutu, blacha perforowana i blacha cienka perforowana elektrochemicznie. Wymiary nominalne oczek* (Polski Komitet Normalizacyjny, Warszawa, 2000)
10. PN-EN 12620: 2004 *Kruszywa do betonu* (Polski Komitet Normalizacyjny, Warszawa, 2004)
11. PN-EN ISO 14688-2: 2006. *Badania geotechniczne. Oznaczanie i klasyfikowanie gruntów. Czesc 2: Zasady klasyfikowania* (Polski Komitet Normalizacyjny, Warszawa, 2006)

Project DISIRE (H2020) – an idea of annotating of ore with sensors in KGHM Polska Miedz S.A. underground copper ore mines

Witold Kawalec[1,a], Robert Krol[1], Radoslaw Zimroz[1], Leszek Jurdziak[1], Marek Jach[2], Radoslaw Pilut[3]

[1]*Wroclaw University of Technology, Wyb. Wyspiańskiego 27, 50-370 Wrocław, Poland*
[2]*KGHM O/ZWR Polkowice, Kopalniana 1, 59-101 Polkowice, Poland*
[3]*KGHM, M. Skłodowskiej-Curie 48, 59-301 Lubin, Poland*

Abstract. The lithological composition of the copper ore delivered to the mills, when recognised in advance, is considered as the most important factor for the proper settings of grinding/milling equipment that can both decrease the specific energy consumption of ore processing and increase metal recovery. In order to address this problem, the idea of the use of sensors for in-situ sensing with respect to adaptation to harsh environment of copper ore conveying in the underground mines was investigated. The in-situ sensors information carrying pellets, are planned to be used for annotating the copper ore for the needs of ore processing plants control. The improved identification of the lithological composition of ore can be achieved by combination of annotating the stream of transported ore with pellets that keep the information on the original location of the ore (when they are dropped into the transported bulk material), the data of the in-situ lithology derived from the digital, orebody structural and quality block model and the simulation analysis of the transportation system. The idea is being developed within the DISIRE project – a part of the SPIRE initiative, acting under the "Horizon 2020" framework program.

1 Introduction to the DISIRE project

Sustainable Process Industry through Resource and Energy Efficiency (SPIRE) is a proposal for a Public Private Partnership (PPP) aiming to regain competitiveness of the ailing European process industry by "the collaborative efforts of public research and technology organisations and academia as well as industry stakeholders throughout Europe, along and across the value chain" [5, 6]. The SPIRE initiative has been launched in order to address the European efficiency targets set for 2030:
- reduction in fossil energy intensity of up to 30% from current levels,
- up to 20% reduction in non-renewable, primary raw material intensity compared to current levels.

The DISIRE (Distributed In-Situ Sensors Integrated into Raw Material and Energy Feedstock) project belongs to the SPIRE 1 group – Integrated Process Control (IPC) and is aimed to "develop and introduce a technological platform for novel inline measurement techniques and Process Analyser Technology (PAT) sensors based IPC approach" [4, 6]. The project consortium consists of 15 partners

[a] Corresponding author: witold.kawalec@pwr.edu.pl

DOI: 10.1051/e3sconf/20160801058

from Sweden, Spain, Italy, Germany, Poland and Israel. KGHM Polska Miedz S.A. (shortly called KGHM) is the key industrial partner of the Work Package 5 – Non-Ferrous Processes, focused on the improvement of transportation of the copper ore by belt conveyors and processing the ore as well as enhancement of the on-line PAT information on further process stages. Other industrial application work packages deal with ferrous, steel and combustion processes and are supported by the direct research and development activities in the areas of industrial process control, electronics and sensor development as well as data mining.

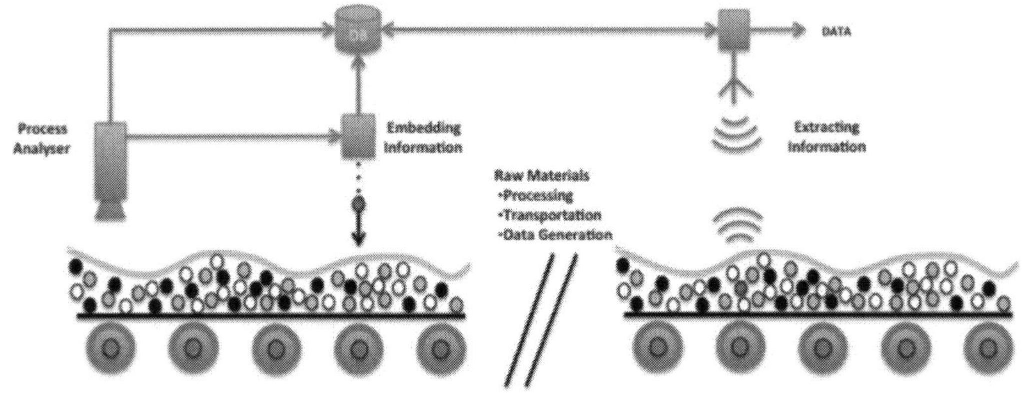

Figure 1. Application of the DISIRE concept in the pelletizing process [4].

The general idea of DISIRE concept is presented in Figure 1. The bulk material, usually undefined on either belt conveyor or a stockyard, is annotated with the help of pellets that are able to carry information. The information embedded into the pellets is obtained from a database supporting the process. As a result, either the transported or transformed bulk material keeps the information on itself, and if needed, on the process that can be read and analysed along its way in order to provide data for improving the consecutive steps of processing.

2 Copper ore identification issues

The problems identified in the copper ore processing at the KGHM processing plants have been addressed in several papers [3]. The authors of the review paper [1] stated that "The ore-treatment process of Polish copper-ores is difficult to proceed from when lithology changes and variability sizes of minerals are connected with an increase in energy consumption. As a result of this, finding new solutions in technology of concentration, new projects solutions are key elements to increasing the Cu concentrate grinding and flotation system productivity" [3]. As presented in Table 1 the lithology of copper ore is KGHM mines is different and changes throughout the years of mining. Authors [1] pointed that "the lithological types of ore treated by KGHM S.A. have not only variable crushing and grinding parameters, but also different floatability, which is the main reason for the creation of unique conditions for treatment during the grinding and flotation operations". The general lithological profile of the copper ore is presented in Fig. 1, but the proportions of thicknesses of each stratum are different from one mining field to another, and in some areas the given lithology can vanish completely.

The Polish resources are ones the most complex in the world, and additionally the most difficult to be processed. The main reasons for rating the Polish copper deposits as among the so-called difficult ones are due to the following factors [1]:

- relatively long flotation time as a result of slow kinetics of sulphide mineral particles,
- three lithological forms occur at the same time, but in different proportions, thus having a significant influence on the grinding and flotation conditions,
- fine and ultra-fine sulfide mineral particles require a fine grinding process.

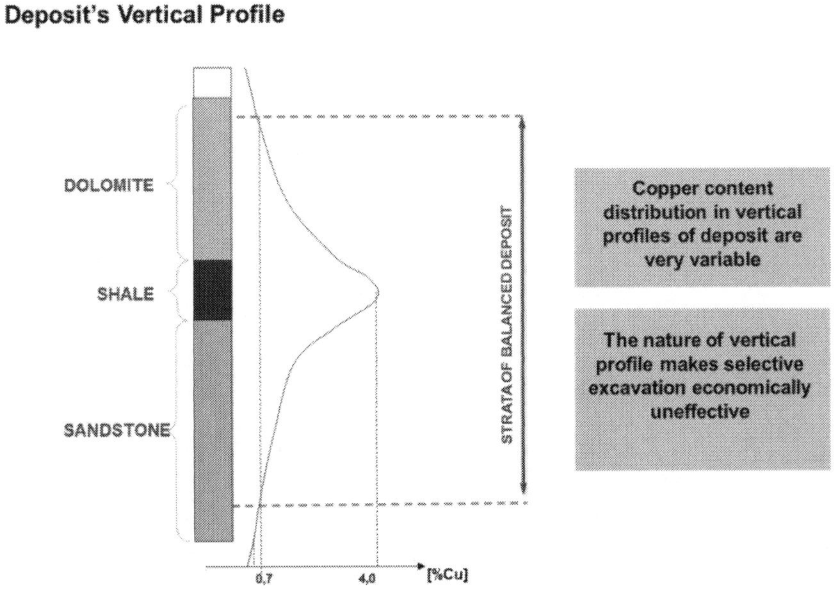

Figure 2. Structure and properties of copper ore deposit.

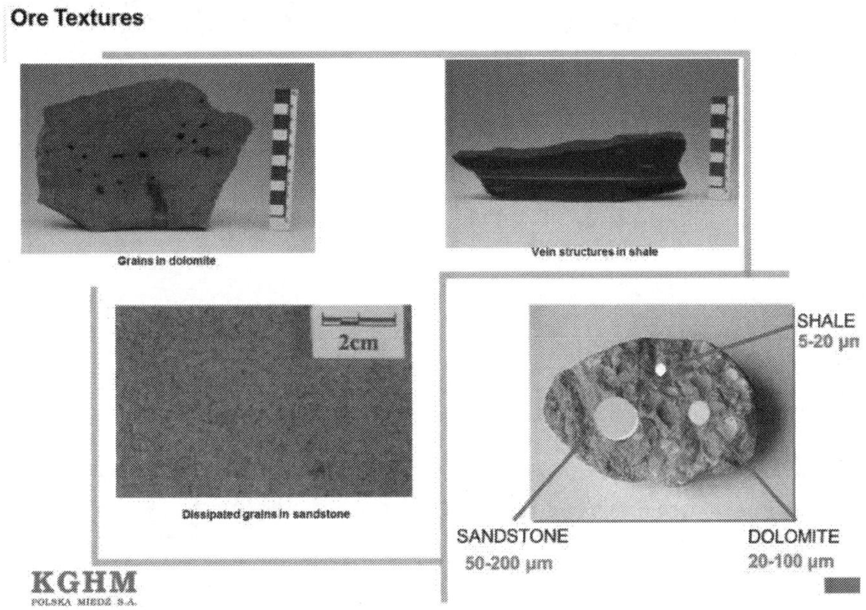

Figure 3. Types of texture of copper ore.

As shown in Figs. 2 and 3, the particles of copper ore are significantly differentiated in size according to the given lithology. Due to the available non-selective mining technology these particles are mixed when supplied to the processing plant. The average proportions of the ore lithology (see table 1) do not reflect the actual differences in mining fields (see Fig. 4).

DOI: 10.1051/e3sconf/20160801058

Table 1. KGHM S.A. deposit's changing of lithology structure [1].

Concentrator name	Sandstone		Shale		Carbonates	
	Years 1985-91	Years 2009-11	Years 1985-91	Years 2009-11	Years 1985-91	Years 2009-11
Lubin	54%	77.545	8.1%	0.94%	37.9%	21.52%
Polkowice	9.9%	22.67%	5.9%	5.28%	84.2%	72.05%
Rudna	43.3%	54.11%	5.5%	11.56%	51.2%	34.33%

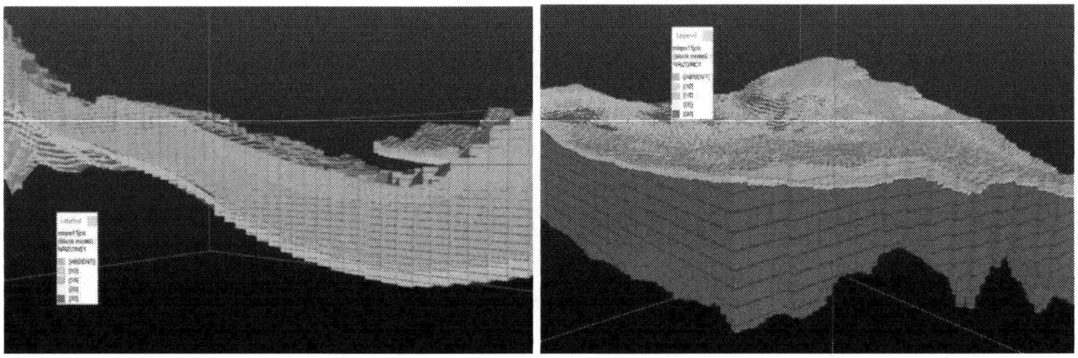

Figure 4. Visualisation of the structural block model of the balance ore from two mining fields of one KGHM Polska Miedz S.A. mine: left panel: carbonates (dolomites, streaky dolomites) and shale in the bottom, right panel: thin layer of shale on the top with sandstone (source: Datamine Studio).

3 Proposal of a PAT solution

3.1 Prerequisites

The DISIRE technology is targeted for the enhancement of existing information systems already implemented at KGHM, that could either provide required information to be embedded into pellets or process and analyse the information extracted thereafter. Is there a sufficient information environment in. mining and processing sites of KGHM?

KGHM successfully implements advanced information systems such as FloVis, MillVis and ConVis for supporting the complex processes of copper ore enrichment [2]. Especially the last one ConVis, designed for optimisation of managing and grinding of ore supplied from various mining fields, needs information on the lithology of ore to achieve the main targets of that is stabilization of ore granulation, optimisation of grinding and maintaining the stable crushed material flow onto mills feeders.

The second reason for considering implementation of DISIRE technology is the fact that KGHM undertook a project to upgrade the existing geological and mine planning systems into the full 3-D geological modelling and mine planning digital environment. The consulting division of CAE Mining (developer and distributer of Datamine software) was commissioned for this work and the training and implementation project was launched in 2015. It is expected that it leads to changes in methods used for data handling and to provide accurate, regularly updated information on the orebody as well as advancing mining fields (Fig. 6).

The last vital element of the existing information environment is managing and control the systems that support underground belt conveying (see Fig.5). These systems (containing also weights on selected conveyors) can provide the information on the actual operation status and load of each object of the transportation system.

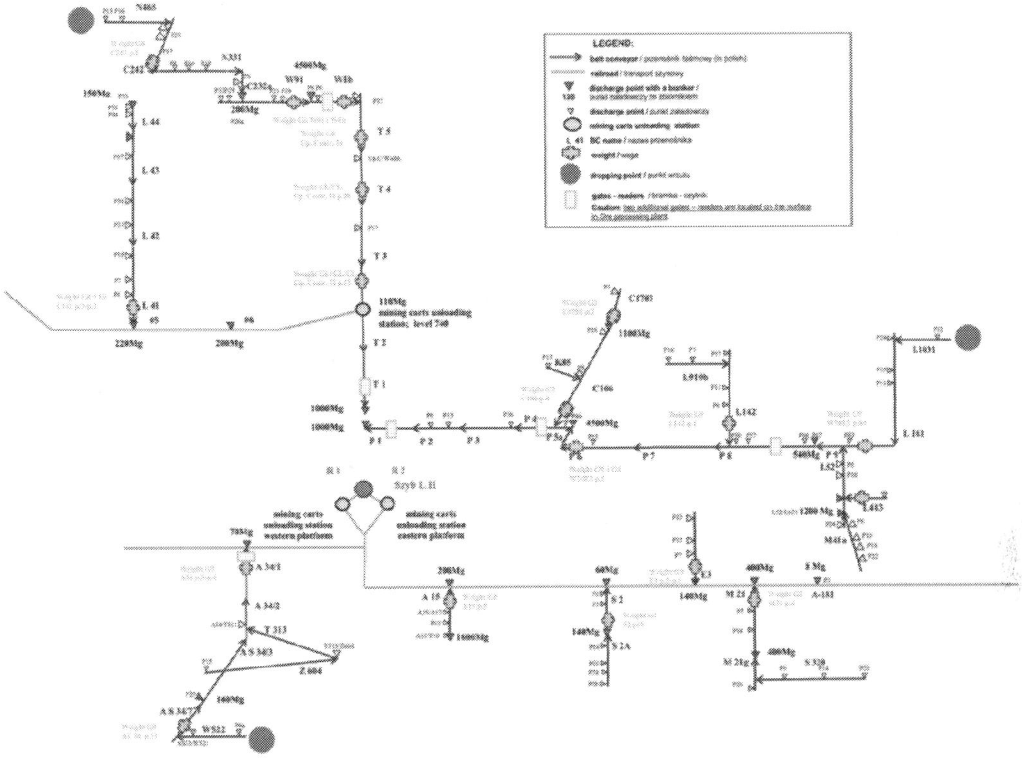

Figure 5. The underground transportation system in the "Lubin" mine of KGHM with the proposed locations of pellet dropping points (red bullet points) and pellet readers (yellow squares, 2 other are located on the surface on the entrance to the processing plant).

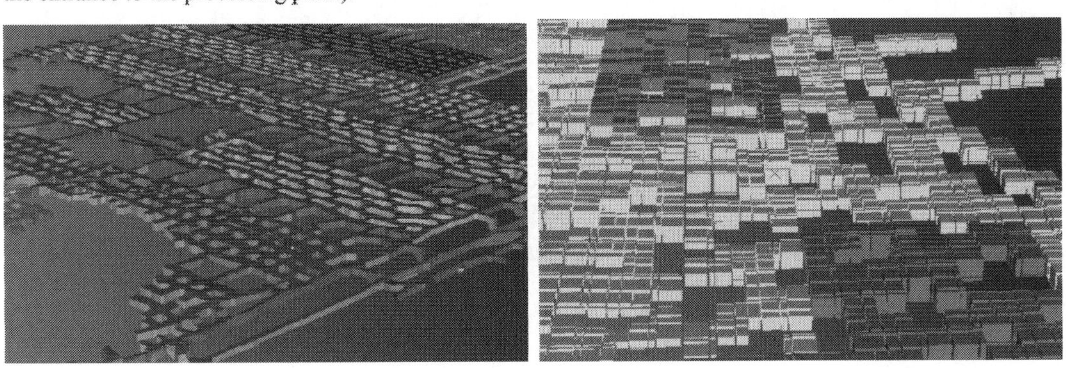

Figure 6. Typical visualisation of a digital 3-D model of underground room-and-pillar developments: left panel: solid wireframes, right panel: block model cells coloured by mining fields numbering (source: *Datamine Studio)*.

3.2 The DISIRE concept

The proposal of the use of DISIRE pellets for tracking the ore in order to provide the suitable information on the processing plants assumes the following:

- pellets will be used to provide the information on lithology of the ore,

- pellets are dropped into the copper ore bulk material while it is being discharged by eiter trucks or loaders onto the division belt conveyor loading point (Fig.5),
- pellets denote the place of mining field and consecutive number of discharging,
- pellets are read instantly after the portion of ore moved by a skip from the shafts the surface (Figs. 1 and 5).

The practical use of DISIRE pellets requires the interaction between the existing, currently being implemented and planned information systems. The 3-D geological modelling and mine planning system should provide the information on the lithology (Fig.4 and 6) on the basis of actual orebody structural model and location of the mining field. It is expected that the system would be on-line examined with annotation of the mining field read from the pellets delivered on the surface to return with the actual lithology composition at the mining faces obtained from the orebody model.

However, the composition of lithology of the ore moved to the surface depends on mixing of the ore from several mining fields. The assessment of this will be achieved with the use of read-in several pellets as well as information on the work of the underground transportation system.

It is not assumed that all pellets can survive transportation from the division belt conveyor to the surface under the difficult operational conditions of underground mines. The in-situ tests will allow to predict the percentage of pellets that are lost but it is obvious that the variance of "losses" will be high. Therefore, the result on lithology, based only on the weighted data from pellets, should be compared with the results of simulations of ore flow within the belt conveyor system.

The Convis system is expected to be fed with the data of lithology of ore that will be soon conveyed to the processing plant. The time delay caused by its transportation can be utilised for the proper adjustment of machinery equipment.

4 Planned digital and in-situ investigations

The DISIRE pellets project has just started. A series of experiments, both theoretical investigations and in-situ tests, are scheduled for the coming months.

The theoretical experiments include simulations of the belt conveyor systems. These are under development by domestic and foreign participants of the research consortium. It is assumed that the dedicated simulation software FlexSim (not exclusively) will be adopted for this purpose.

Another topic of investigation is concerned with interrogation of the modelling and mine planning system which has not been implemented yet. Though the needs of an experiment the query of the system can be simulated, the practical use require safe and regular on-line operations as well as keeping the orebody and mine advancing models updated.

The practical use of lithological information supplied in advance to the information systems supporting the processing plants can be verified by reviewing the procedures of setting the adjustments of processing machinery equipment. This can only be done by the specialists of KGHM processing divisions.

For the in-situ tests of dropping, transporting and reading pellets the dedicated equipment such as sufficiently ruggedized pellets and RFID gates for reading them, is necessary. Completion of the equipment and schedule of tests is under development.

Acknowledgement

This work is supported by the Framework Programme for Research and Innovation Horizon 2020 under grant agreement n. 636834 (DISIRE).

References

1. B. Bazan, E. Kasinska-Pilut, M. Garbacki, A. Bazan-Krzywoszanska, in *Proceedings of the Mineral Engineering Conference MEC2014* (2014)
2. A. Konieczny et al., *Gornictwo i geologia*, **6(2)** (2011)
3. A. Konieczny, B. Bazan, *in Proccedings of MEI Conference Comminution* (2012)
4. DISIRE Newsletter, **2** (2015)
5. P.Tello, R.Weerdmeester, *SPIRE ROADMAP Consultation document* (2013)
6. Desire project home page, http://spire2030.eu/disire (2015)
7. *FloVis - Wizyjny System Optymalizacji Sterowania Procesami Flotacji* http://www.ameplus.pl (2014)

DOI: 10.1051/e3sconf/20160801059

Impact of inorganic salts on behaviour of air bubbles in aqueous solutions

Danuta Szyszka[1,a]

[1]*Department of Geoengineering, Mining and Geology, Wroclaw University of Science and Technology, Na Grobli 15, 50-421 Wroclaw, Poland*

Abstract. This paper comprises an analysis of solutions of such salts as: sodium chloride NaCl, potassium chloride KCl, sodium sulphate Na_2SO_4, ammonium chloride NH_4Cl, and hexylamine hydrochloride $C_6H_{15}N{\cdot}HCl$. The objective of the paper was to determine the Critical Coalescence Concentration (CCC) of four inorganic and one organic salts. The obtained CCC values were similar to those already existing in literature. Also, in this paper the CCC values of investigated inorganic salts were compared in terms of their ionic strength. The results showed that, except KCl, there was a relationship between CCC and ionic strength.

[a] Corresponding author: danuta.szyszka@pwr.edu.pl

Bulk density estimation using a 3-dimensional image acquisition and analysis system

Adam Heyduk[1,a]

[1] *Silesian University of Technology, Department of Electrical Engineering and Control in Mining, 44-100 Gliwice, ul. Akademicka 2, Poland*

Abstract. The paper presents a concept of dynamic bulk density estimation of a particulate matter stream using a 3-d image analysis system and a conveyor belt scale. A method of image acquisition should be adjusted to the type of scale. The paper presents some laboratory results of static bulk density measurements using the MS Kinect time-of-flight camera and OpenCV/Matlab software. Measurements were made for several different size classes.

1 Importance and range of bulk density measurements

Bulk density, also called an apparent density, is defined as the mass per unit volume of loosely piled granular material (eg. coal or coke). The coal bulk density depending on its particle size, impurities, moisture may be in the range of 700 to 1100 kg/m^3. The bulk density varies depending on the way in which this material is stored. Under gravity load or under external pressure, spaces between the particles can be reduced, so the bulk density may increase over time. The bulk density, like the material volume, is particularly important for aggregates used in civil engineering [1, 2] since they act as volume fillers to concrete mixes. However, this parameter is important also for coal [3], especially in the case of measuring the stored coal amount in the landfill. This was sanctioned in national regulation, according to which "*the mass of stored coal is determined by weight or calculated as the product of volume and bulk density*." This is connected with coal excise taxation and with trading CO_2 emission rights. However it should be noticed that, in accordance with [3] the regulation method of calculating the weight of the stored coal on the basis of the volume and bulk density, is only valid for coarse size classes. Therfore, the analysis performed in this paper was restricted to the coarse size classes.

2 Static and dynamic measurements of bulk density

National standards and regulations define the density measurement in a static manner [3] by using the box the with size dependent on the coal particle size (the larger the particle size the greater the box dimensions). The mass of the coal sample is simply defined as a difference of the full box mass and empty box mass. This rule of a discrete measurement is presented demonstratively in Fig. 1. However,

[a] Corresponding author: Adam.Heyduk@polsl.pl

this process needs sampling from the greater volume, so only the small part of the total amount can be measured.

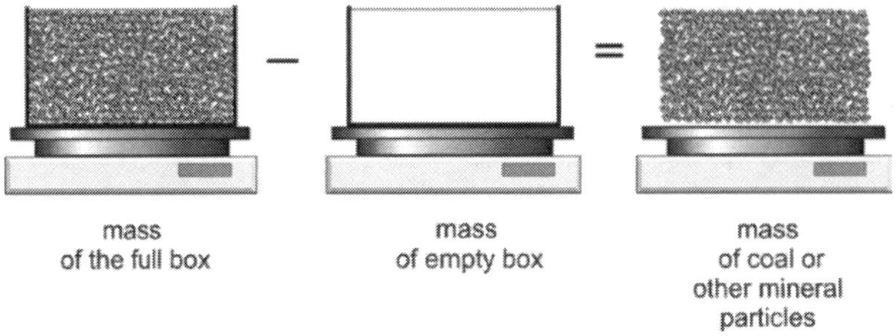

Figure 1. Schematic of discrete (static) bulk density measurement.

It would be desirable to measure bulk density in a continuous way, as it makes possible to measure the whole stream of the granular material. However, this method needs continuous and synchronous measurements of both mass and volume of the whole stream. The result of mass measurement should be divided by the corresponding result of the volume measurement. Mass measurements can be performed by widely used belt conveyor scale, and volume measurement can be performed by recently developed 3-dimensional image acquisition and analysis. There are several methods of 3-dimensional image acquisition. The most applicable in these case are laser triangulation [4] and time-of flight cameras [5-7]. Each of these methods has its own advantages and disadvantages in terms of cost, efficiency and precision. However, the most important difference are the time and space range of these methods. The laser triangulation measures simultaneously the profile of a single line and the time-of-flight camera measures usually a rectangular area (as it has a matrix generator and converter). So, the selection of the method should adjusted to the type of the conveyor belt scale used in a particular application. If there is a single sensor scale measuring only one line, a laser triangulation system [4] presented in Fig. 2, should be used. However, if there is a multi-sensor scale measuring the mass of the greater load area, a time-of-flight camera capturing larger field of vision [5, 6] presented in Fig. 3, should be used.

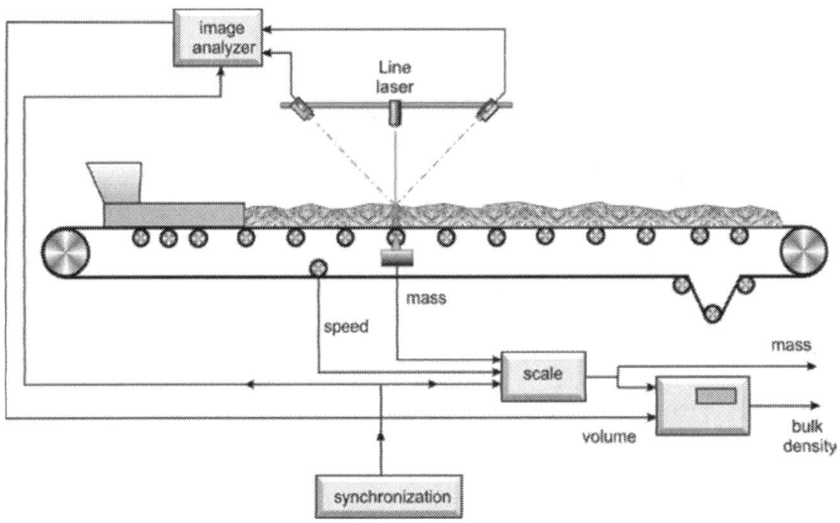

Figure 2. Continuous measurement of bulk density using a single-sensor scale and laser triangulation.

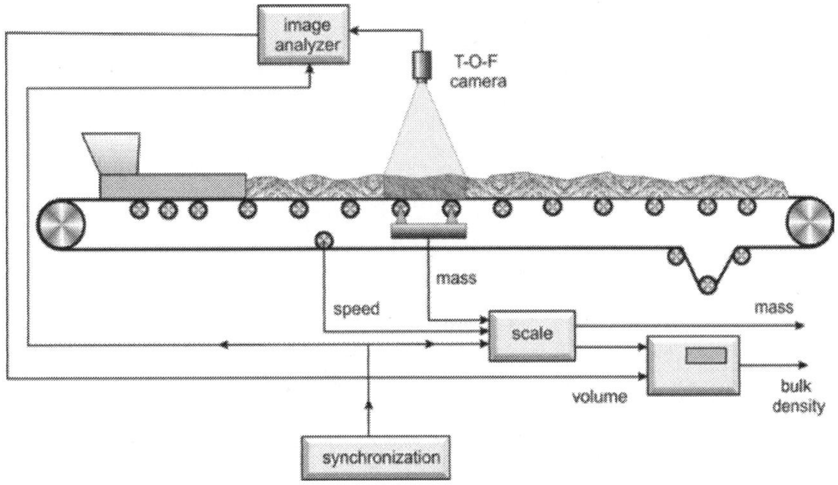

Figure 3. Continuous measurement of bulk density using a multi-sensor scale and time-of-flight measurement camera.

3 Volume estimation methods

Volume estimation algorithms are much simpler then algorithms for calculation of the particle size distribution, because they do not require precise delineation of single touching and overlapping particles (this is usually the most complex, error-prone time-consuming task [4]). There is also no need to analyse the lower, usually invisible from the top, hemisphere of individual particle, because the lower surface of the whole particle population is well known (it is just the upper surface of the conveyor belt). The most important part of the volume estimation algorithm is to calculate a height difference between the surface of the particle stream and surface of an empty conveyor belt, adopted as a reference surface. So, the height map $h_d(x,y)$ of the particulate material stream is:

$$h_d(x,y) = h_s(x,y) - h_b(x,y) \qquad (1)$$

where $h_s(x,y)$ is depth map of the stream surface and $h_b(x,y)$ is depth map of the empty conveyor belt. If for any point $P(x,y)$ the belt area is locally empty:

$$h_s(x,y) = h_b(x,y) \qquad (2),$$

then

$$h_d(x,y) = 0, \qquad (3)$$

and this point P will not be considered in any further volume calculations. Each pixel $P(x,y)$ (where $x \in X, y \in Y$) of the 3-d image represents in a horizontal plane a distinct patch with a constant elementary area A_e. The value of A_e is dependent on image resolution and camera field of view angle. In the 3-d space it can be considered as a vertical stack of voxels (elementary volume elements), as presented in Fig. 4. Therefore, the volume of the mineral particle stream can be calculated as the sum of these voxel vertical stacks:

$$V = \sum_X \sum_Y h_d(x,y) \cdot A_e = A_e \sum_{x \in X} \sum_{y \in Y} h_d(x,y) \cdot \qquad (4)$$

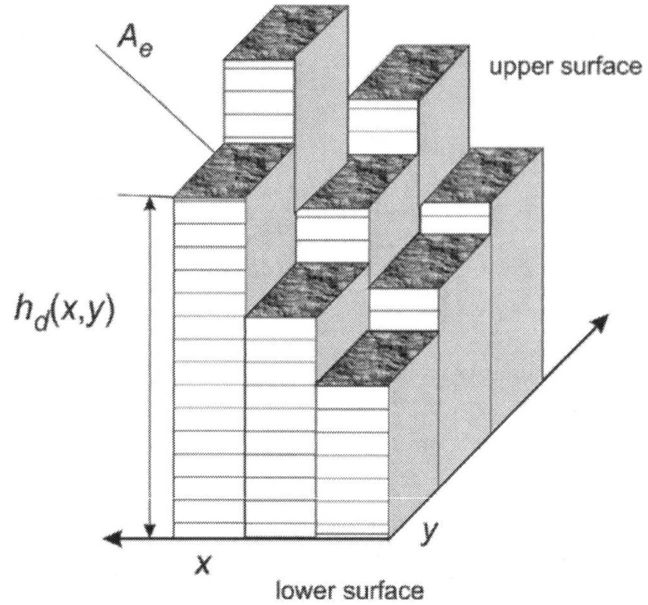

Figure 4. Outline of volume estimation based on the 3-d image analysis according to Eq. (4).

4 Results of laboratory research

Preliminary studies were performed is a laboratory equipped with a section of the conveyor belt, and the MS Kinect v2 time-of-flight camera with vertical precision 1 mm and horizontal resolution of 512x424 pix [7]. The pixel elementary area value A_e was calculated as 0.85 mm^2/pix. It was estimated on the basis of image analysis performed for a flat rectangular reference object with known dimensions. The coal sample mass was measured using the digital scale with 0.001 kg resolution. The experiments were performed for different size classes presented in Table 1.

Table 1. Measurement and calculated results for coal particle size classes.

Size class [mm]	Mass [kg]	Volume [m^3]	Bulk density [kg/m^3]
80 – 160	7.076	0.0106	724
50 – 80	2.936	0.0038	777
40 – 50	3.656	0.0041	888
30 – 40	1.905	0.0020	963
20 – 30	3.936	0.0040	972
10 – 20	4.536	0.0045	1.010
0 – 10	4.638	0.0045	1.021

The results of bulk density estimation from Table 1 are presented graphically in Fig. 5. The smaller the size class. The higher its bulk density.

DOI: 10.1051/e3sconf/20160801060

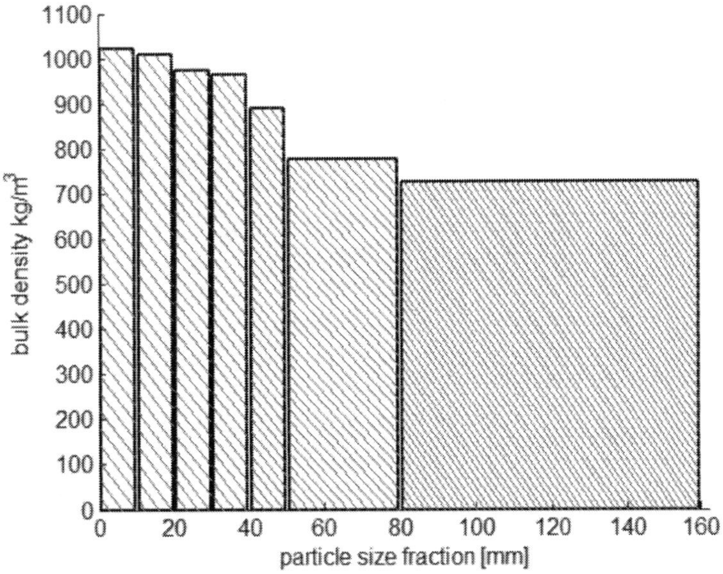

Figure 5. Results of bulk density estimation for different size classes from Table 1.

For comparison purposes, some different acquired (raw) and processed (differential without conveyor belt, ready for volume calculation) depth maps: one set for coarse particles population (size class 80-120 mm), one set for medium particles (30 – 40 mm) and one for the finest particles e (0-10 mm) were presented. In each case the reference depth map of an empty conveyor belt presented in Fig. 6 was used. The map for coarse particles on the conveyor belt was presented in Fig. 7. After subtracting the image of the empty conveyor belt (Fig. 6) we can get an image of the particles alone (Fig. 8). It is ready for volume calculation according to Eq. (4). Similar images for a medium size class (30 – 40 mm) were presented respectively in Figs. 9 and 10 and for the finest size class (0 – 10 mm) in Figs. 11 and 12. All dimensions on these figures were presented in mm.

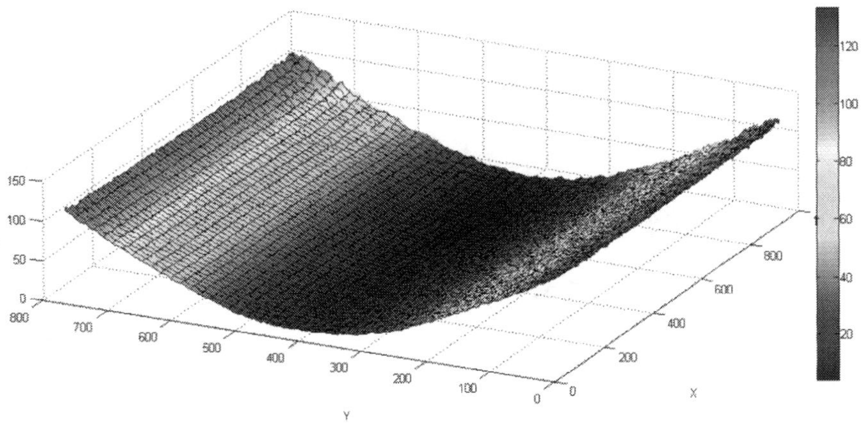

Figure 6. Depth map of an reference empty conveyor belt.

DOI: 10.1051/e3sconf/20160801060

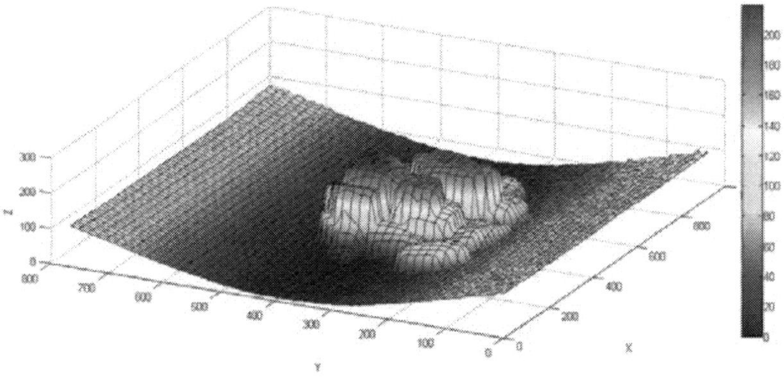

Figure 7. Raw depth map of coarse (80 - 160 mm) coal particles on the conveyor belt.

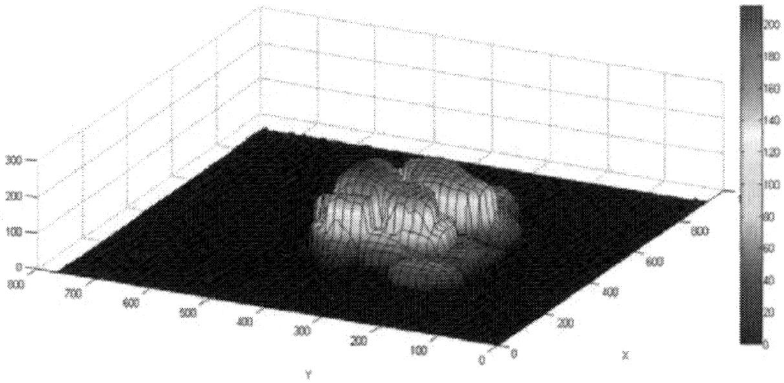

Figure 8. Differential depth map of coarse (80-160 mm) coal particles alone (without conveyor belt), ready for volume calculation.

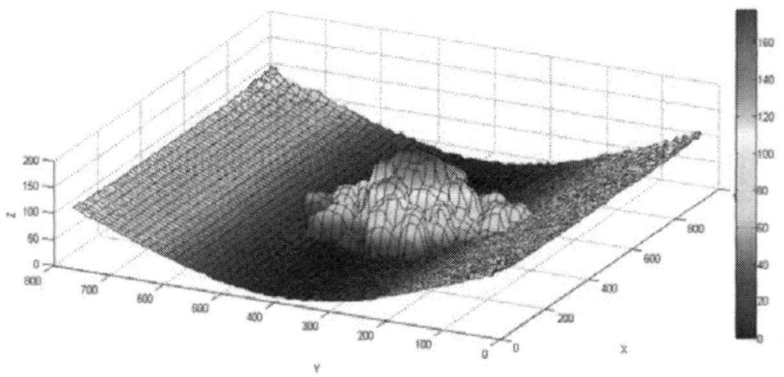

Figure 9. Raw depth map of medium (30 - 40 mm) coal particles on the conveyor belt.

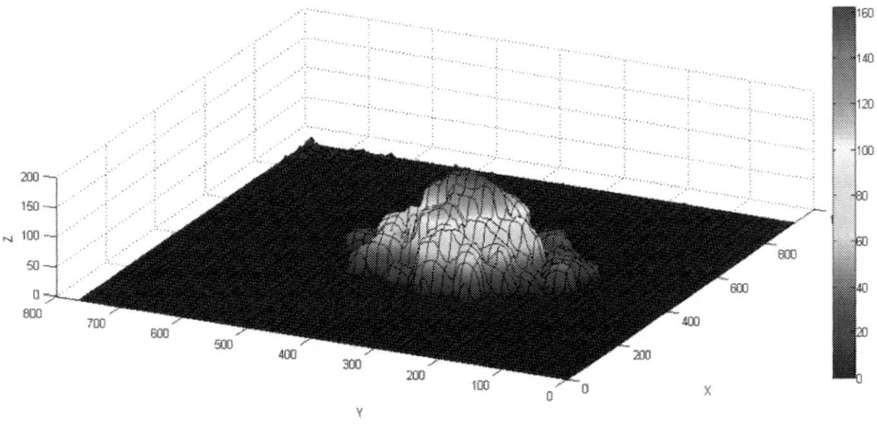

Figure 10. Differential depth map of medium (30 – 40 mm) coal particles alone (without conveyor belt), ready for volume calculation.

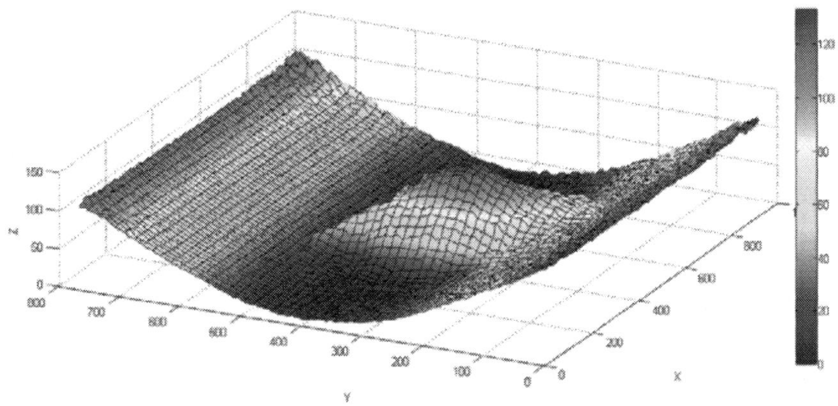

Figure 11. Raw depth map of fine (0 …10 mm) particles on the conveyor belt.

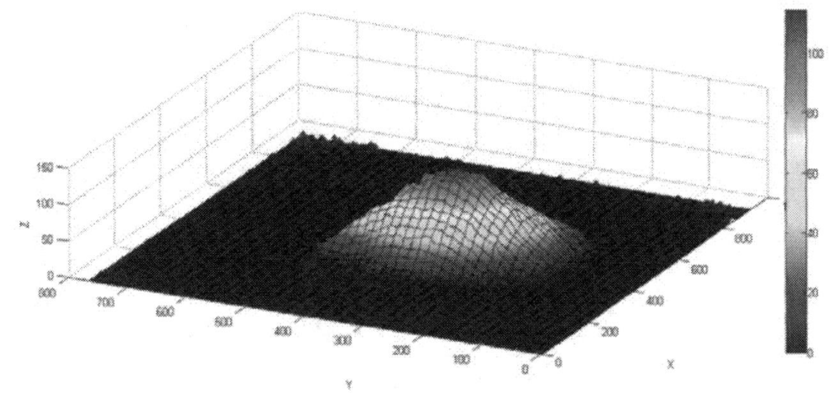

Figure 12. Differential depth map of fine (0…10mm) coal particles alone (without conveyor belt), ready for volume calculation.

DOI: 10.1051/e3sconf/20160801060

5 Conclusions

Modern 3-d image acquisition methods enable continuous measurement of bulk density for mineral particle streams. It can be useful for estimation of the stored bulk material amount e.g in a stockpile. Measurements can be made just on the conveyor belt, using belt the scale and image acquisition system. Volume estimation algorithms are relatively simple, as they do not require precise delineation of individual particles. There must be created a differential image. It can be obtained by subtracting the image of the empty conveyor belt from the image of the particle stream on the belt. The 3-d images, also called either depth maps or height maps, can be obtained by either laser triangulation or time-of-flight measurements. As the bulk density is connected to the particle size, it can be also used, under special conditions, as the rough indicator of the average particle size in a stream.

References

1. H. Zhang, *Building materials in civil engineering* (Science Press, Beijing, 2011)
2. W.F. Chen, J.Y. Liew, *The Civil Engineering Handbook* (CRC Press, Boca Raton, 2002)
3. L. Róg, R. Rompalski. *Eng. Res. Report of CMI*, (2012).
4. A. Heyduk, *Arch. Min. Sci.*, **61**, 15 (2016).
5. M. Hansard, S. Lee, O. Choi, R. Horaud. *Time-of-Flight Cameras, Principles, Methods and Applications* (Springer, London, 2013)
6. F. Remondino, D. Stoppa, *TOF Range-Imaging Cameras* (Springer, Heidelberg, 2013)
7. L.Yang, L. Zhang, H. Dong, *IEEE Sens. J.*, **15**, 42 (2015)

Application of a rapid methodology for preliminary appraisal of kaolinite deposits

Mahir Can Cetin[1] and N. Emre Altun[1,a]

[1]*Middle East Technical University, Mining Engineering Department, Ankara, Turkey*

Abstract. An approach that facilitates the mineralogical-compositional analysis and beneficiation-classification procedure was used for fast assessment of the evaluation possibility of kaolin deposits. The approach was applied on two different kaolin deposits from the Aegean region in Turkey. The kaolin samples were characterized using XRD and XRF analyses to determine the key mineralogical characteristics and major components such as Al_2O_3. The samples were then subjected to the attrition-scrubbing-hydrocycloning procedure to identify the -2 μm matter yielding potential and undesired +10 μm particle contents. Through this fast methodology, it was found that Al_2O_3 content of both deposits were below the acceptable limits and both reserves consisted of alunite as an undesired component in kaolin. The XRD analysis of attrition scrubbing-hydrocycloning products showed the cyclone overflow fractions, which were relatively finer in size, were enriched in terms of kaolinite content. The extent of +10 μm particles in the cyclone overflows was within acceptable limits for both deposits. However, the critical -2 μm matter content was found to be extremely low in the cyclone overflows. The findings confirmed the benefit of this fast methodology to make a quick decision regarding the potential of any kaolin deposit: whether to eithr continue with further verifications through specific tests/analyses or to end the assessment in case of an unpromising deposits, as those investigated in this work.

1 Introduction

Kaolin, white clay composed of mineral kaolinite ($Al_2Si_2O_5(OH)_4$), is primarily used in paper manufacturing and, to a lesser extent, in ceramics, paint, rubber and plastics industries. The content of Al_2O_3 in the kaolin sample and the amount of -2 μm and +10 μm sized particles obtained after beneficiation of kaolin are the three important criteria considered for the preliminary evaluation of a kaolin deposit and for determination of marketable kaolinite product content [1].

Generally, for obtaining products with acceptable quality, Al_2O_3 content should be between 35 to 40% in the kaolin deposit and in some cases 30% of Al_2O_3 may also be considered as sufficient. The amount of -2 μm material directly indicates the quality of the kaolin deposits and the required extent of -2 μm differs with respect to the utilization area. Similarly, the tolerable percentage of undesirable +10 μm sized particles varies with respect to the end use of kaolin. Roughly, the limits for -2 and +10 μm particles with respect to utilization area of are given in Table 1.

[a] Corresponding author: ealtun@metu.edu.tr

Table 1. Acceptable limits of -2 µm and +10 µm material amount from a kaolin deposit.

Particle Size	Acceptable limits in different industries (%)			
	Filler (paper)	Coating (paper)	Ceramics	Paint
-2 µm	25-60	75-95	40-85	15-70
+10 µm	6-25	0-6	1-20	1-30

Depending on the area of usage a vast number of tests would be required to justify the kaolin deposit as a potentially promising reserve. However, these tests are specific, extremely time consuming and expensive. These tests may also require a sophisticated equipment and expertise. In this respect, such tests should be conducted only if the kaolin deposit of question could qualify from the above mentioned three major pre-requisites, i.e. the extent of Al_2O_3, the amount of -2 and +10 µm particles. The mineralogical characteristics and chemical composition may also be evaluated along with these three criteria for a better idea regarding the potential quality of products from a kaolin deposit.

This study aims to introduce a systematics for a preliminary evaluation of the kaolinite deposit. Based on the results achieved from this methodology, the necessity for more specific and detailed tests, related to the anticipated utilization area, could be justified, i.e. if positive indications cannot be achieved from these preliminary tests, further steps and sophisticated evaluations become obsolete.

2 Material and methods

In this work, a methodology is facilitated that consists of the XRD/XRF analyses for mineralogical and compositional diagnosis, and attrition scrubbing-hydrocyclone classification procedure [1], is facilitated for determining the -2 and +10 µm yielding potential of the kaolin deposits. This methodology is tested on two different kaolin deposits from the Aegean region of Turkey. The exact name and location of the deposits were not disclosed due to commercial restrictions and coded as K1 and K2 in the text. Firstly, the K1 and K2 samples were subjected to controlled size reduction to 80% of -75 µm, followed by XRF and XRD analyses and attrition scrubbing-hydrocyclone classification application. Attrition scrubbing is highly effective in breaking up and releasing kaolinite particles from relatively harder phases. Pulps including 10% kaolin samples (by weight) were prepared and subjected to attrition scrubbing for 10 hours. Subsequent classification, using a laboratory scale hydrocyclone, yielded overflow (finer) and underflow (coarser) products. These products were further subjected to a laser size analysis to determine the extent of -2 and +10 µm particles. In addition to the feed, the overflow and underflow products of classification were analyzed in terms of chemical composition and mineralogy and the results were compared to the feed.

Table 2. Compositional analysis results and loss on ignition values of kaolin deposits.

Sample	LOI (%)	Al_2O_3 (%)	Fe_2O_3 (%)	CaO (%)	Cl (ppm)	MgO (%)	K_2O (%)	SiO_2 (%)	Na_2O (%)	TiO_2 (%)
K1	7.96	14.97	0.56	0.22	1310	0.11	0.16	74.55	0.01	0.5
K2	8.35	15.22	0.48	0.17	1293	0.05	0.35	72.80	0.06	0.5

DOI: 10.1051/e3sconf/20160801061

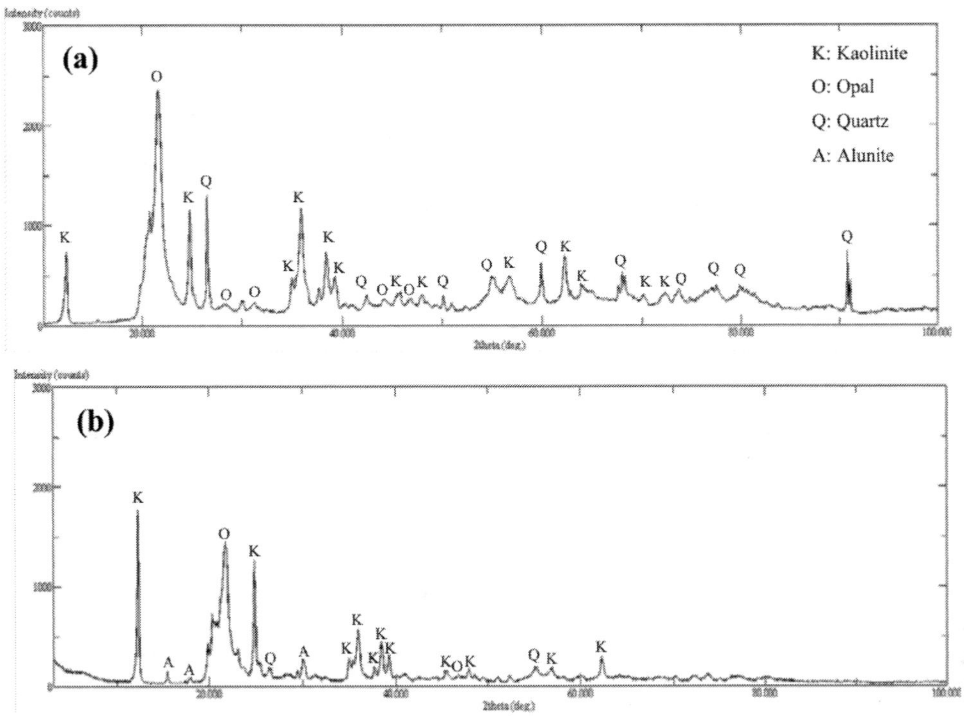

Figure 1. X-ray diffractograms of K1 kaolin: a) original sample b) cyclone overflow.

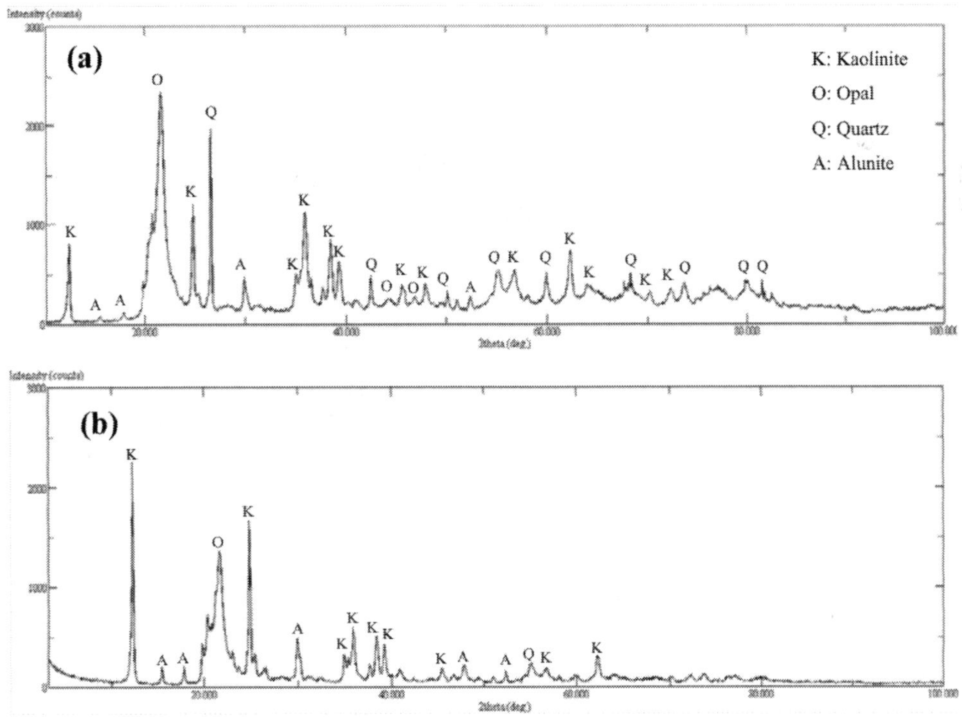

Figure 2. X-ray diffractograms of K2 kaolin: a) original sample b) cyclone overflow.

3 Results and discussion

XRD analysis revealed that K1 sample mostly consists of quartz (SiO_2) and opal ($SiO_2 \cdot nH_2O$) minerals, and additionally contains kaolinite ($Al_2Si_2O_5(OH)_4$ (Fig. 1a). In terms of chemical composition determined by the XRF analysis (Table 2), K1 kaolin is principally composed of SiO_2 (74.55%) and Al_2O_3 (14.97%). The only clay mineral detected in this sample is kaolinite, therefore, the amount of Al_2O_3 could be linked to kaolinite content (Fig. 1a). Similar to K1, quartz (SiO_2) and opal (SiO_2nH_2O) are the major minerals in the K2 sample (Figure 2.a). The only difference is the presence of alunite ($KAl_3(SO_4)_{2.}(OH)_6$) in K2, in addition to kaolinite ($Al_2Si_2O_5(OH)_4$) as the minor minerals. Chemically, SiO_2 (72.80) and Al_2O_3 (15.22) are the most dominant constituents of the K2 kaolin (Table 2). Contrary to K1, Al_2O_3 cannot be directly correlated with kaolinite, since alunite is another clay mineral observed in K2 (Fig. 2a). The mineralogical and compositional analyses of phase showed that Al_2O_3 content in both kaolin reserves were below 30%, indicating that these two reserves were unlikely to be promising deposits even before investigating the -2 μm particles yielding potential of these two reserves.

After attrition scrubbing-hydrocyclone classification, the XRD analysis of the overflow of K1 sample showed that the extent of kaolinite and alunite increase, while quartz and opal minerals were hardly observed (Fig. 1b). This suggests that attrition scrubbing contributes to the release and liberation of clay-rich entities from harder phases and relatively finer overflow product of cycloning became enriched with respect to clay entities. Despite this, the laser size analysis pointed to very low extent of -2 μm material in the overflow product. The -2 μm particles in K1 were only 1.73% (by weight) of the total feed introduced to the scrubbing stage (9.12 % of the cyclone feed). The amount of +10 μm material in the K1 cyclone overflow corresponded to 5.16% (by weight) of the cyclone feed.

From the mineralogical point of view, the K2 sample was quite similar to K1. Attrition scrubbing-classification resulted in relative enrichment of kaolinite and alunite in the cyclone overflow and quartz and opal became almost unrecognizable (Fig. 2b). The extent of -2 μm particles in the cyclone overflow was again quite limited and corresponded to only 2.62% (by weight) of the total attrition scrubbing feed (14.29 % of the cyclone feed). The size fraction +10 μm material in the overflow was 2.39 % (by weight) of the cyclone feed.

Overall, it was seen that the K1 and K2 samples had quite low potentials in terms of yielding -2 μm material, despite favorable +10 μm particles content. The XRD analysis indicated the alunite content in both kaolin samples (Figs. 1 and 2), which was another obstacle against potential utilization of these deposits.

4 Conclusions

The analytical work and lab other studies showed that assessment of kaolin samples by relatively simple mineralogical-compositional analysis and beneficiation-classification procedure, as given elsewhere [1], was a highly applicable and time-saving methodology for preliminary assessment of reserves. The results attained from this methodology allowed the decision for either conducting further tests or leaving the deposit aside without wasting time and financial resources for potentially exploitable kaolin reserves.

Reference

1. A.J. Bloodwoth, D.E. Highley, C.J. Mitchell, *Kaolin.* (Nottingham: British Geological Survey, 1993)

E3S Web of Conferences **8**, 01062 (2016) DOI: 10.1051/e3sconf/20160801062
MEC2016

Control of clay minerals effect in flotation. A review

Hasan Ali Taner[1,a], and Vildan Onen[1]

[1] *Selcuk University, Department of Mining Engineering, 42075 Konya, Turkey*

Abstract. The increased exposure to low grade ores highlights the importance of understanding phyllosilicate gangue mineralogy which consists of common gangue minerals. To improve the flotation performance and ore quality the negative effect of the clay minerals on flotation should be identified. The presence of clay minerals leads to problems, such as changing the froth stability, which are related to swelling behaviour, increase in pulp viscosity, overconsumption of reagents, slime coating and mechanical entrainment. The clay content in the ore is changing from time to time and it is necessary to provide quick solutions to the issues caused by the new ore composition. The objective of this paper is to give an overview how to control the colloidal properties of clay minerals on flotation.

1 Introduction

Clay minerals are the primary natural materials consisting of fine-grained minerals. Clay minerals are commonly found in many mineralization as gangue minerals and they create problems in all steps of mineral processing. Difficulties in processing of high clay content ores are due to their different and complex structures. The presence of clay minerals leads to problems, such as changing the froth stability, which are related to swelling behaviour, increase in pulp viscosity, overconsumption of reagents, slime coating and mechanical entrainment.

There is a strong relationship between the rheological properties of pulp and flotation performance. The presence of clay minerals can cause adverse effects on the pulp rheology. For instance, montmorillonite type clays due to show a considerable amount of yield stress and their impact on the viscosity of the suspension is very strong. As it is known, that the turbulence in the cell increases the possibility of particle-bubble collisions. The influence of enhanced pulp viscosity also has a detrimental effect on gas dispersion, through districting the turbulent region of flotation cells to a small ''turbulent cavern'' surrounding the impeller [1-3]. In addition, it is stated that with increase in the pulp viscosity causes increase in froth stability and decrease in detachment of particle from bubble [4].

Clay minerals are composed of cations (Si, Al, Mg et al.) and anions (O and OH) and contain broken bonds on their corners, edges and surfaces. Broken edge bonds of clay particles are considered to be particularly positive charged at neutral and acidic pH. Clay surfaces have negative charges while edges are positively charged [5, 6]. The anisotropic charges on edges and surfaces of clays allow coating the surface of valuable minerals through the electrostatic attraction. Therefore, it causes

[a] Corresponding author: hasanalitaner@selcuk.edu.tr

inadequate surface hydrophobicity as well reduce the flotation performance and quality of the concentrate [7]. Chemical reagents used in flotation adsorb onto the mineral surfaces. Many clay minerals have a large active surface area supporting adsorption with the ionic and molecular bond structure. Therefore, they also cause more consumption of flotation reagents and increase the operational costs [8].

Flotation is more difficult for fine and ultra-fine particles. One of the challenges in flotation of fine particles is high gangue entrainment. Fine hydrophilic gangue particles easily entrain between bubbles getting into the fine pores and froth layer what results in mechanical entrainment [9]. It is well-known that the froth stability is an important factor determining the flotation performance. Creation and protection of the best froth stability is not a simple process. Clay minerals could prevent adsorption between bubbles and minerals by coating air bubbles [10]. The flotation performance is dependent on ability of the bubbles in collection of hydrophobic particles from pulp. Reduction of capture efficiency of these particles decreases concentration froth stability.

Significant structural differences of clay minerals such as crystallinity, cation exchange capacity, swelling etc. are effective in their behaviour during flotation and these structural differences. The effects can be reduced by:

- removal of clay minerals prior to flotation,
- using of viscosity modifiers,
- removal of clay minerals by washing which are mechanical entrained and not attached to the bubbles,
- operation at low air speed,
- flotation in low pulp densities [11, 12],
- using of special frothers such as cycle-type frothers which are less sensitive during the presence of clay [13],
- using of clay binders as a group of low molecular weight polymers [7, 14].

2 Control mechanism

2.1 Removal of clay minerals prior to flotation

A common method in plants to reduce an unwanted effect of clay minerals prior to flotation is prevention of fine particles entrance into the flotation process and washing of ores. The plants, that do not apply this method, may face serious difficulties, firstly on flotation circuits, and then thickeners and filters. However, the clay fractions may contain large amount of valuable minerals [15]. This method can be applied if the amount of valuable mineral in the discarded size as a slime is at negligible level. However, in the case of comminution to ultra-fine sizes, a deslime process cannot be considered as a solution because of losses of valuable minerals [16]. Figure 1 shows the effect of quartz particles finer than 10 μm on flotation of different iron ores [17].

2.2 The use of viscosity modifiers

The rheological behaviour significantly effects the pulp properties, so this indicates the level of either aggregation or interaction of particles. Therefore, it can be used as a useful control parameter in mineral processing. Enrichment of minerals by flotation could be effected by interaction of particles the rheological data present an approach to directly understand this interaction. The rheology has been used to investigate particle-particle interactions in slurries of oxide minerals [18-20], clays and clay minerals [21-24], coal [25, 26], sulphide minerals [27], and even gangue minerals such as talc [28-30].

Accurate measurement of the rheological properties of pulp is rather difficult, because the solid particles tend to settle during measurement. Some researchers have used different methods such as pumping slurry [31], using capillary viscometers [32] or stirring the suspension while measuring with

 DOI: 10.1051/e3sconf/20160801062

Couette-type viscometer [33], to overcome this problem. However, finding a proper practical way to measure the rheology of such slurries still remains a challenge [30]. It was also demonstrated that the recovery of coarse particles increased with the increased medium viscosity controlled by adding certain amount of glycerol [34].

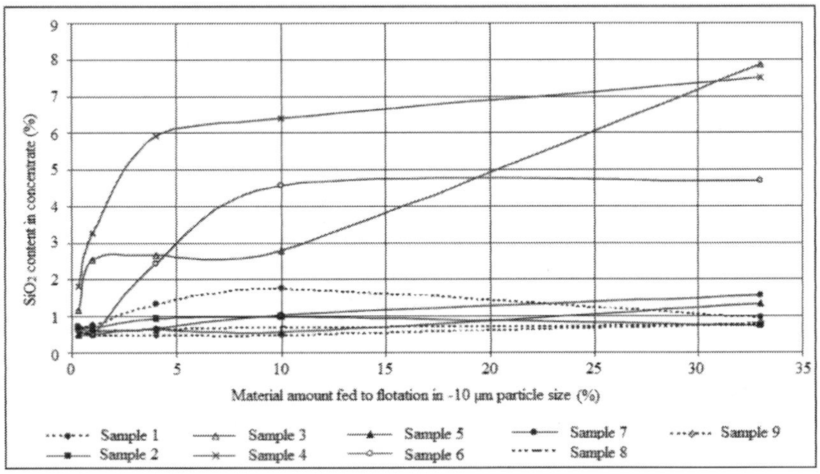

Figure 1. Effect of ultra-fine particles on flotation of different iron ore samples to quartz content in concentrate [17].

2.3 Removal of clay minerals by washing

In almost all scientific researches, it is expressed the mechanical entrainment takes place depending on the amount of water obtained from the concentrate. In other words, unless arrival of water in the cell to concentrate is not prevented, mechanical entrainment of fine gangue minerals will not prevent to entrain to the concentrate. One of the ways to reduce the water come from the concentrate is to increase the fluidity of interfacial water between bubbles. The viscosity of the pulp, which has high amount of fine particles, is higher than that of low amount of fine particles, even though they have the same pulp density. In this case, mechanical entrainment can't be reduced adequately neither to provide low air rate nor deeper froth region in order to flow back the water between bubbles to pulp the region. In such cases, giving the washing water from the top of the cell can be referred to a solution.

2.4 Operation at low air speed

To flow the water between bubbles into the pulp, waiting of froth for an adequate time could be reduced the amount of water transfer to concentrate. This situation is possible by either decreasing the froth flow rate or increasing froth depth, only. However, one of the most important drawbacks of either increasing froth depth or decreasing froth flow rate is to increase probability of bubble coalescence in the froth region.

The most important variable is the air flow rate supplied to the cell. It is expected to decrease mechanical entrainment by two important reasons if the low air rate is used. The first reason is decreasing mechanical entrainment as a result of reducing the froth flow rate because of relative reduction of the froth volume in the froth region. Secondly, when the air flow rate decreases, naturally, stirring in pulp decreases and the flow regime in the cell has lower Reynold number, so the decrease in fluctuation occurs in the pulp-froth interface, and then mechanical entrainment decreases. Therefore, the air flow rate and froth depth are considerable control variables in flotation plants [16].

2.5 The use of depressants and clay binders

In flotation, sodium silicate (Na_2SiO_3), called water glass, has fairly common use to depress silicate and carbonate minerals present as gangue minerals. Also, it is an effective dispersant. Sodium silicate that chemical composition could vary, acts as either a distributor or a depressant depending on the SiO_2/Na_2O ratio, also known as a module number [16].

Aggregation of fine and surface coating of coarse particles are not desired during the flotation. Dispersion is the condition of particles being apart enough from each other in the pulp. Prior to flotation to prevent aggregation of fine and slime coating particles, that is dispersion, one of the polyphosphate compounds can be used. Sodium hexametaphosphate, commercially known as Calgon prevents aggregation and slime coating by forming complex ions with Ca^{2+} and Mg^{2+} ions so minimizing effects in thinning the electrical double layer around these ions. With these effects, they also prevent similar metal ions to activate minerals other than a target [16].

The clay binder plays an important role as a slime depressant that agglomerates clay particles to increase the size, lower the surface area, clean bubble and particle surfaces and reduce adsorption of surfactants (Fig. 2). Selection of appropriate binders and optimizing the dosage is important for the flotation performance and selectivity. Two main adsorption mechanisms of clay binder are dipole-dipole interaction and strong hydrogen bond. Because of amphiprotic characteristic and multiple binding and chelating sites, binding can be very selective [35]. As is shown in Fig. 2, when clay minerals are present as individual particles, they adsorb a large amount of reagents due to their large surface area (clay minerals are being adsorbed on both ore particles and air bubbles). The clay will also be adsorbed on the coal particles and bubble surfaces as a result of electrostatic interactions, what is referred as slime coating. Clay binders were used to agglomerate clay minerals and depress them and also to reduce their surface area and expose fresh coal particle surface and fresh bubble surface.

An experimental laboratory study was conducted to evaluate the performance of clay binders in the fine coal flotation process [14]. According to the results a clean coal product of 8-8.5% ash was produced from a feed of more than 19% ash with 97-98% combustible recovery. Binders significantly enhanced flotation selectivity, reducing clean coal ash from 8-8.5% to 7-7.5%. Also, flotation performance increased 15-30% based on the high rotation speed. Phosphate flotation using clay binders was studied and a significant increase about 5% in phosphate recovery and 1.5% in grade was achieved [36].

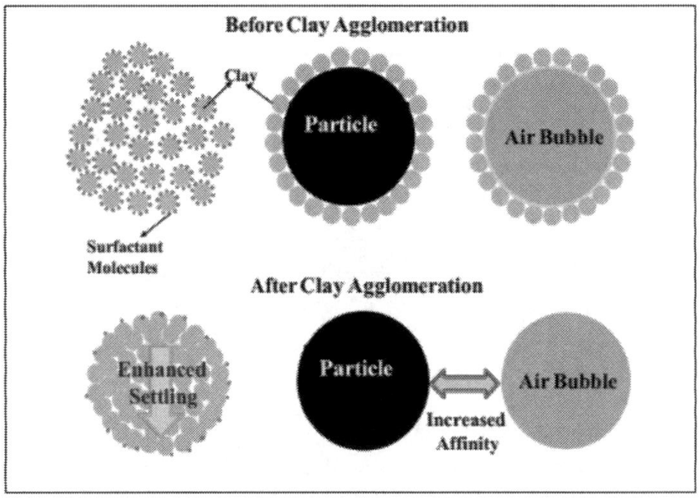

Figure 2. Illustration of how clay agglomeration improves flotation performance by removing slime coating from the mineral particle and bubble surface and increasing clay particle size [14].

3 Conclusions

Clay minerals constitute challenges by decreasing flotation and creating a complex tailing in flotation. Improvement of the flotation performance and product quality require ensuring the effect of clay minerals and understanding of this mechanism.

By taking into account solutions mentioned above to overcome the negative effects of clays, it can be considered to have detailed data related to solve the clay problem in plants. However, clays are still one of the major problems in period from comminution to dewatering stage in plants while enriching high clay content ores. This case can be stated that due to undetailed investigation of effect to mechanism of complex structure of clays and structural differences. Nowadays, this issue is more importance due to necessity of enrichment the low-grade ores.

References

1. C. W. Bakker, C. J. Meyer and D.A. Deglon, *Miner. Eng.*, **22(11)**, 944 (2009)
2. N.Z.P. Shabalala, M. Harris, L.S. Leal Filho, D.A. Deglon, *Miner. Eng.*, **24(13)**, 1448 (2011)
3. E. Forbes, K.J. Davey and L. Smith, *Miner. Eng.* **56**, 136 (2014)
4. D. Xu, I. Ametov and S.R. Grano, *Int. J, Miner. Process.*, **101(1-4)**, 50 (2011)
5. Y. Akın, M.S. Celik, *End. Hammaddeler Semp.* (İzmir, 1995)
6. A. Evcin, T. Kavas, *5.Endüstriyel Hammaddeler Semp.* (İzmir, 2004)
7. M. Zhang and Y. Peng, *Miner. Eng.*, **70**, 8 (2015)
8. W.J. Oats, O. Ozdemir and A.V. Nguyen, *Miner. Eng.*, **23**:413-419 (2010)
9. H. Ozer, *Yüksek Lisans Tezi*, (Süleyman Demirel Üniv. Fen Bilimleri Ens,. Isparta, 2007)
10. W.W. Wen and S.C. Sun, *Trans. of AIME*, **262**:174 (1977)
11. S.M. Bulatovic, *Miner. Eng.*, **10(9)**, 895 (1997)
12. T. Ozgen, *Yüksek Lisans Tezi,* (Hacettepe Üniv. Fen Bilimleri Ens., Ankara, 2014)
13. S.M. Bulatovic, *H. of Flot. R.: Chem., Th. and Prac.: Flot. of Sulp. O.,* (Elsevier: Boston, 2007)
14. D. Tao, P.G. Dopico, J. Hines and D. Kennedy, *Pro. of the Int. Coal P. C.*,(USA, 2010)
15. İ. Bayraktar, *14. Cevher Hazırlama Kongresi,* Inv. papers, Kuşadası, (2014)
16. E.C. Cilek, *Mineral flotasyonu,* SDU. Müh.-Mim. Fak., Kitap Yayın, 59 (2006)
17. N.P. Lima, *M.Sc thesis*, In Portuguese (2001)
18. M.L. Taylor, G.E. Morris and R.S.C. Smart, *J. Colloid Interface Sci.,* **262**, 81 (2003)
19. F.J. Rubio-Hernández, M.F. Ayúcar-Rubio, J. F. Velázquez-Navarro and F. J. Galindo-Rosales, *J. Colloid Interface Sci.*, **298**, 967 (2006)
20. S. Farrokhpay, G.E. Morris, D. Fornasiero and P. Self, *Powder Tech.*, **202(1-3)**, 143 (2010)
21. P. Mpofu, J. Addai-Mensah and J. Ralston, *Int. J. Miner. Process.*, **71**, 247 (2003)
22. A. J. McFarlane, J. Addai-Mensah and K. Bremmell, *Miner. Eng.*, **18(12)**, 1173 (2005)
23. A. Nosrati, J. Addai-Mensah and W. Skinner, *Chem. Eng., Sci.* **66**, 119 (2011)
24. Z. A. Zhou, O. Worku, L.X. Wang, Z. Xu and J.H. Masliyah, *CIM J.*, **3(1)**, 54 (2012)
25. R. W. Lai, L.G. Mc Mahan, A.G. Richardson and S.H. Chiang, *Advances in Coal and Min. Proc. Using Flot., Soc. for Mining, Met. and Exp. Inc.* (Littleton, Collorado, 1989)
26. H. Dincer, F. Boylu, A.A. Sirkeci and G. Atesok, *Int. J. Miner. Process.*, **70**, 41 (2003)
27. T. H. Muster and C.A. Prestidge, *Miner. Eng.*, **8(12)**, 1541 (1995)
28. G.E. Morris, D. Fornasiero and J. Ralston, Int. J. Miner. Process., **67**, 211 (2002)
29. E. Burdukova, D.J. Bradshaw and J. S. Laskowski, *Can. Metall. Q.,* **46**, 273 (2007)
30. S. Farrokhpay, *Miner. Eng.*, **36**, 272 (2012)
31. C. Boudrias-Chapleau, C. Bazin and D. Cotnoir, *In: Pawlik, M. (Ed.), Proc. of 8th UBC-McGill-UA Inter. Sym. on the Fund. of Min. Proc.: Rheology and Proc. of Fine Partic.*, 43 (2011)
32. S.K. Kawatra and A.K. Bakshi, *Int. J. Miner. Process.*, **47**, 275 (1996)
33. J. Gustafsson, M. Toivakka and K.K. Koskinen, *Annu. Tr. Nord. Rhe. Soc.*, **13**, 277 (2005)
34. D. Xu, I. Ametov and S. R. Grano, *Miner. Eng.*, **39**, 89 (2012)

35. L. Zhang, MSc thesis., (University of Kentucky, 2013)
36. J. Kou, X. Guangyang, T. Daniel, G.D. Pablo, H. John and K. Dennis, *Phosphate flotation enhancement using clay binder* (2012)

World production and possible recovery of cobalt from the Kupferschiefer stratiform copper ore

Paulina M. Pazik[1,2,a], Tomasz Chmielewski[2], Hylke J. Glass[1], and Przemyslaw B. Kowalczuk [2]

[1]*Camborne School of Mines, University of Exeter, Tremough Campus, Penryn, Cornwall, TR10 9EZ, UK*
[2]*Wroclaw University of Science and Technology, Faculty of Geoengineering, Mining and Geology, Wybrzeze Wyspianskiego 27, 50-370 Wroclaw, Poland*

Abstract. Cobalt is recognized as a strategic metal and also E-tech element, which is crucial for worlds development. An increasing demand for cobalt forces for searching of new resources that could be explored in European countries. There are many examples of cobalt recoveries, mostly from laterite and sulphide deposits. However, the accurate choice of the technology depends on many factors. The Kupferschiefer stratiform copper ore located in Poland is the biggest deposit of cobalt in Europe. Although KGHM Polska Miedz S.A. recovers many precious metals from this ore, cobalt is not recovered yet. This metal occurs as an accompanying element, mostly in the form of cobaltite (CaAsS), with the average content of 50-80 g/Mg. In this paper a possible recovery of cobalt from the Kupferschiefer ore, with the use of hydrometallurgical methods, was investigated.

1 Introduction

In 2014, the European Commission published a report on critical raw materials for the European Union, defining cobalt as a strategic metal. It is due to rising consumption of cobalt and its very low production in European countries. Cobalt is also classified, by the National Environmental Research Council (NERC), as an E-tech element what means that this element is essential for a technological development [1].

Cobalt is a transition metal appearing in the periodic table between iron and nickel. It a silvery-white, shiny, ductile and malleable metal with density of 8.9 g/cm^3 [2]. Cobalt has melting and boiling points at 1495 and 3100 °C, respectively [1]. Cobalt exhibits ferromagnetic properties and can be magnetized to produce permanent magnets since it has the Curie point at 1121 °C, which is higher than any other metal. As a result cobalt exhibits its magnetic properties even at high temperatures [3, 4]. Due to its heat resistance, strength and magnetic properties, cobalt has wide variety of applications. The most common uses are magnets, battery and alloys [5]. In comparison to either nickel or iron alloys the cobalt ones exhibit superior thermal weariness resistance and weldability [4]. Cobalt chrome alloys have the best combination of properties like good strength, corrosion resistance and good tolerance by the human body [6].

[a] Corresponding author: paulina.pazik@pwr.edu.pl

Almost all of cobalt that is used in EU comes from resources outside Europe. It was estimated that in 2012 an about 95% of cobalt was imported came from Russia and 3% from the United States [1]. The largest World resources of cobalt are located in the Central African Copper Belt in the Democratic Republic of Congo and Zambia, where the resources are estimated to be around 15 million tonnes. An additional one billion tonne may exist in manganese nodules and crusts on the ocean floor [1]. These resources are characterized as not yet economically exploitable, however there are many studies which qualify them as a strategic resource with huge future potential. Almost half of world cobalt production comes from Democratic Republic of the Congo. Figure 1 shows world cobalt mining in 2015. In Europe there are known only some small cobalt resources like the Kupferschiefer copper ore extracted by KGHM Polska Miedz S.A. [1].

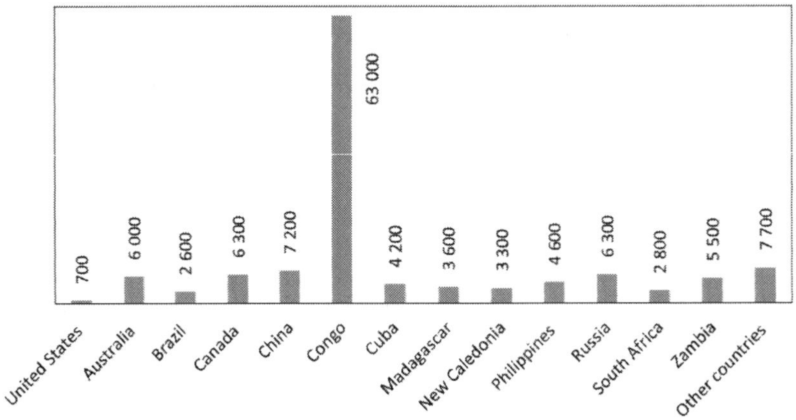

Figure 1. World cobalt mining in 2015, based on [5].

Cobalt is the 33th element in terms of spreading in the earth curst. In nature, cobalt is a dispersed element and its average content in the continental crust is 20 ppm [3]. Cobalt is commonly found in sulphide minerals, associated with copper, nickel and iron. The most common minerals of cobalt are cobaltite (CoAsS), safflorite ($CoAs_2$), glaucodot ((Co,Fe)AsS) and skutterudite ($CoAs_3$). Typically, cobalt does not form independent deposits, but it appears as an important component of different types of ores such as copper, silver, bismuth and uranium [4].

Most of metallic cobalt is obtained as a by-product of nickel and copper extraction. Production of cobalt may require combination of chemical, thermal and electrolytic methods. The right technology of cobalt production depends on its concentration, type of ore, other metals production and final form of cobalt products [7]. In this paper possible recovery of cobalt from the Kupferschiefer ore, with the use of hydrometallurgical methods was investigated.

2 Cobalt extraction and production

Overall world production of cobalt can be divided in two groups. Approximately half of this output is produced as a by-product of nickel extraction, where the other half is produced from copper mines [8]. Cobalt production depends on the ore type wherein its occurred.

2.1 Laterite ores

In this type of ore, cobalt mostly occurs as a goethite (Fe,Ni,Co)OOH. Most of laterite ores are smelted to produce ferronickel, however cobalt cannot be removed during this process. On the other hand cobalt can be recovered from the limonite layer of laterite ores by a leaching process [8]. The most common way of cobalt recovery is high-pressure acid leaching in sulfuric acid (HPAL). After leaching cobalt can be recovered with the use of sulphide precipitation, re-dissolution and solvent extraction. Then, it can be recovered by either electrowinning or hydrogen reduction [8].

2.2 Sulphide ores

Flotation is the most efficient method for recovery of cobalt from sulphide ores. Flotation concentrates are further either directly processed by smelting and converting or leached [8]. Leaching can be carried out in either hydrochloric acid, ammonia solutions or sulfuric acid [8]. After that cobalt can be recovered by either electrowinning or hydrogen reduction.

A large source of cobalt comes from copper-cobalt ore types like the Central African Copper Belt in the Democratic Republic of Congo and Zambia. The influence of rain, air and vegetative acid make some of this deposits weathered, which can be easily dissolved in sulfuric acid. According to Fisher [7] in some cases flotation is inefficient, therefore a large amount of this metal is lost in flotation tailings from which cobalt can be recovered. The example of such process is the Kolwezi Tailings Project, in which the tailings have a great potential to produce large amount of cobalt with very low cost. The processing would require leaching of using sulfuric acid and after that solvent extraction and electrowinning would be applied to produce metallic cobalt [9].

The problem in pyrometallurgical processing of cobalt is that this metal follows iron during smelting, therefore some of the copper smelter operations generate slags with high cobalt content [7, 17]. These slags could be a great potential of cobalt resource.

3 Cobalt at KGHM Polska Miedz S.A.

3.1 Mineralogy

In 2007 Bachowski et al. [10] analysed the prospective of extension the resources of KGHM Polska Miedz S.A. It was established that in the context of high prices of some metals, cobalt, molybdenum and platinum group metals would add a great value to this company.

The Kupferschiefer copper ore mined by KGHM Polska Miedz S.A can be divided into three groups. The first one consists of such elements as Cu, S, Pb and Fe with contents of each above 0.1%. The second group includes such elements as Zn, F, As, Ni, Co, V, Ag, Mo, U which average content oscillates in the range of 10-1000 g/Mg. The third group consists of Sb, Bi, Th, Sn, Cd, Se, Hg, Ge, Re, Au, Pd i Pt and the average content of each element is less than 10 g/Mg [11]. According to Pieczonka [12] the most important cobalt forms of occurrence are minerals from cobaltite (CoAsS)–gersdorffite (NiAsS) series, but this element also occurs as isomorphic impurities in other minerals. Additionally, it was found that cobalt also concentrates in the organic matter especially in copper-bearing shale [12, 14].

Cobaltite (CaAsS), as the main cobalt-bearing mineral occurring in the ore mined by KGHM Polska Miedz S.A., appears in all types of lithological fractions. The largest concentration of this mineral was found in the sandstone and shale fractions [11]. The cobalt concentration in the vertical profile of copper ore is shown in Fig. 2. Cobaltite mostly creates small xenomorphic inclusions (10-20 μm) in the copper-bearing minerals such as chalcocite, bornite and digenite [11, 12]. Larger concentration of cobalt is observed in the epigenetic veins of ore. Grotowski and Henc [14] provided information on behaviour of cobalt in industrial enrichment circuits and established that cobalt mostly occurred as independent mineral with fine grade.

Due to small size of cobaltite crystals it is very difficult to identify this mineral. Cobalt belongs to the group of common accompanying elements, where its average content in the Kupferschiefer deposit is 50-80 g/Mg, particularly 140 g/Mg in the Lubin ore. On this basis, cobalt can be classified as the secondary group of elements, with the local maximum level of 1.7% [11].

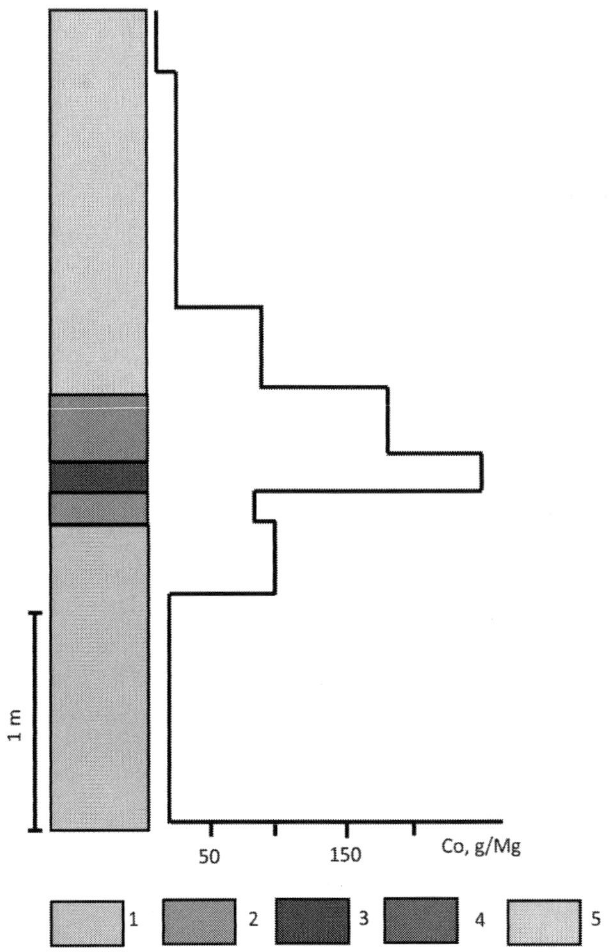

Figure 2. Cobalt concentration in vertical profile of the Kupferschiefer ore, where 1-sandstone, 2-dolomitic sandstone , 3-clay-organic shale, 4-dolomitic shale , 5-dolomite, based on [11, 13].

3.2 Mineral processing and metallurgy

The complex mineralogical structure and unique chemical composition of ore mined from sedimentary deposits are reasons of copper, silver, cobalt, molybdenum, nickel and other minerals loses in industrial processing [14]. On the other hand, copper-bearing shale, which contains the highest content of cobalt and high content of organic carbon causes many metallurgical problems at KGHM Polska Miedz S.A. It was found that the content of copper in the excavated ore decreased, however the content of cobalt increased [14]. Kubacz and Skorupska [15] reported that the content of shale fraction in the Lubin deposit exceeded 25% and it was expected to increase in next years. This situation creates additional technical as well as ecological issues, especially in the smelting process. The increased content of carbonate and organic coal as well as metals disseminated in the ore, especially in the black shale fraction, makes this material difficult to float, but first of all it causes remarkable metals losses [14]. According to Chmielewski [16] the application of modern hydrometallurgy techniques should be considered as a complimentary process for processing, particularly for the shale fraction. It would be also beneficial for the cobalt recovery, since it can be recovered by leaching.

3.1.1 Cobalt recovery from converter slag

In the smelting process, metals accumulating as by-products of pyrometallurgical processes are mainly Pb, Zn, Re, As, Se, Co, Hg, Sb, Ge and In [17]. These metals accumulate in the form of either dusts or slag as a residue from a dedusting process of gases produced during metallurgical processes. The exception is cobalt, which concentrates in the converter slag (1-2%) [17]. Some technologies of recovery of cobalt from these slags were proposed. It was found that separation of this metal from the metallic phase was relatively easy for either fire or electrolytic refining. However, separation from the silicate phase required much higher costs. In order to separate cobalt, the slag could be reduced at a high temperature in an electric furnace to the metal alloy, and then cobalt could be easily separated [17]. The possible way of cobalt recovery is shown in Fig. 3.

Converter Slag ⇨ Reduction in an electric furnace ⇨ Leaching in sulfuric acid ⇨ Iron removal ⇨ Solution purification ⇨ Electrolysis ⇨ Metallic cobalt

Figure 3. Unit operations of cobalt recovery from converter slag [17].

The alloy received in an electric furnace contained 5-7% Co, 4-5% Cu, and 85% Fe. Pure cobalt can be recovered from the alloy by leaching in sulfuric acid. The pollutants from a leaching solution could be recovered by a multi-stage purification process. The final step of this process was electrolysis of cobalt from the purified solution which allowed to produce metallic cobalt [17].

3.1.2 Cobalt recovery from flotation concentrates

Cobalt recovery from the flotation concentrates in the oxygenated leaching process has been extensively examined [16, 18, 19]. The results are presented in Fig. 4-10. These studies showed the influence of different factors on the effectiveness of cobalt leaching. The recovery increased with the increase of temperature of solution (studied temperatures were 50, 70, 90 °C) (Fig. 4). The same trend of process occurred with increasing the concentration of iron(III) ions (Fig. 5) and oxygen flow rate (Fig. 6) [18]. Higher oxygen flow rate was beneficial for regeneration of iron(II) ions to iron(III), which in the studies were used as oxidants. As it was expected, the redox potential of slurry decreased in the initial phase of process (Fig. 7). It was due to the dissolution rate of easy-to-leach copper minerals like bornite and chalcocite. In means that the oxygen flow rate is the crucial factor and it has to be controlled in the first phase of process for improving leaching and reducing the redox potential drop.

The effect of chloride ions addition to the leaching solution on recovery of cobalt was also investigated (Fig. 8). Chloride ions are present in the mine water at KGHM Polska Miedz S.A., where its value varies from 3 to 5 g/dm^3 [18]. Even a small addition of chloride ions cause increased in the leaching recovery of cobalt, which would be beneficial for industrial use of this process.

The most important finding in leaching experiments was that the efficiency of cobalt recovery highly depended on the copper recovery (Fig. 9). A strong relationship between cobalt and copper recoveries by oxidative leaching was found. Cobalt started to leach rapidly after recovery of an about 90–95% of copper. Therefore, it was concluded that cobalt was finely disseminated in the copper-bearing minerals.

Figure 10 shows that the recovery of cobalt depended on the sulfuric acid concentration (Fig. 10). Higher concentration enhanced the recovery of cobalt from 0 to 45% after 10 hrs. This results show that leaching can be a sufficient method for recovery of cobalt from the Kupferschiefer ore. Further studies are needed to improve the process.

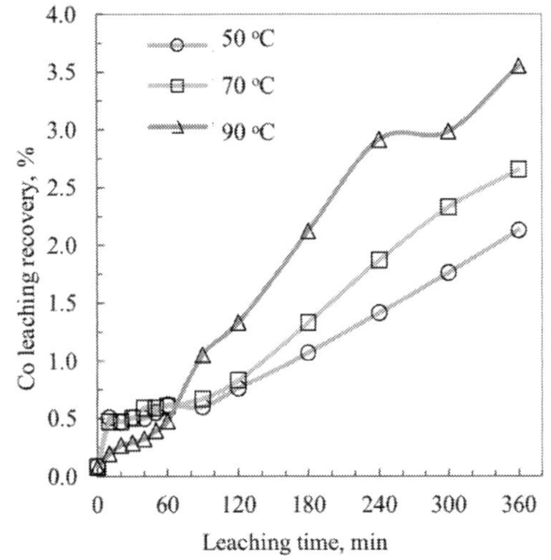

Figure 4. Effect of temperature on Co atmospheric leaching recovery (solid-to-liquid ratio 1:8, initial H_2SO_4 concentration 50 g/dm^3, O_2 flow rate 30 dm^3/h, based on [18]).

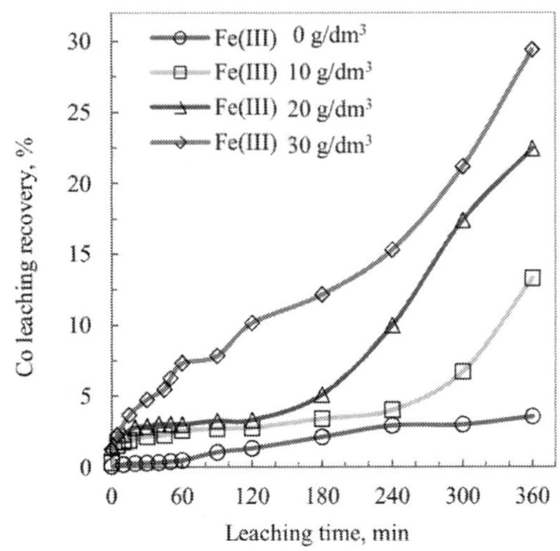

Figure 5. Effect of Fe(III) ions concentration on Co recovery in atmospheric leaching (solid-to-liquid ratio 1:8, O_2 flow rate 30 dm^3/h, Fe(III) concentration 50 g/dm^3, temperature 90 °C, initial H_2SO_4 concentration 50 g/dm^3, based on [18]).

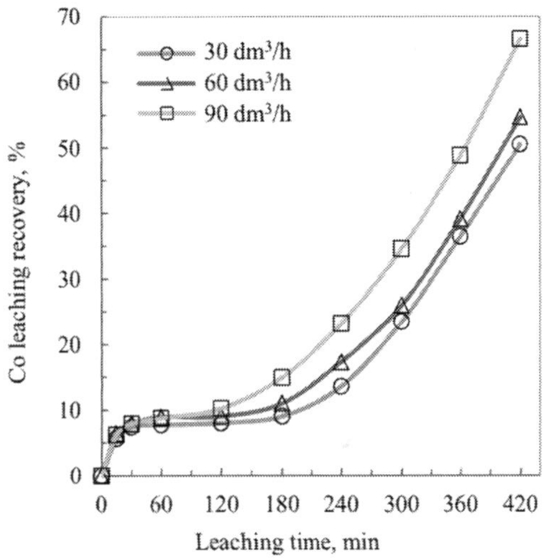

Figure 6. Influence of oxygen flow rate on Co recovery in atmospheric leaching (solid-to-liquid ratio 1:8, Fe(III) concentration 30 g/dm^3, temperature 90 °C, initial H_2SO_4 concentration 50 g/dm^3, based on [18]).

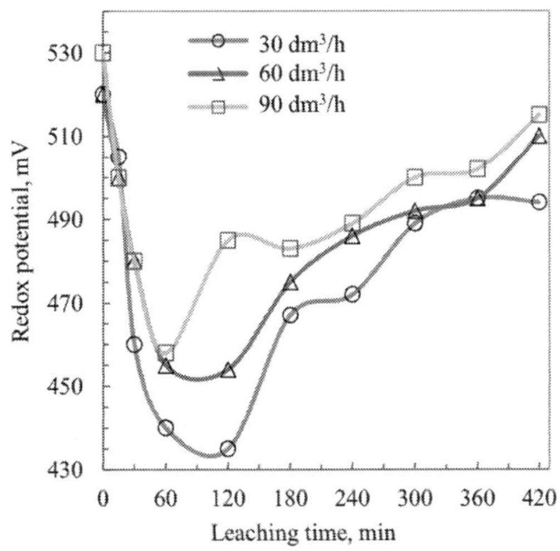

Figure 7. Effect of oxygen flow rate on the redox potential (solid-to-liquid ratio 1:8, O_2 flow rate 30 dm3/h, Fe(III) concentration 50 g/dm^3, temperature 90 °C, initial H_2SO_4 concentration 50 g/dm^3, based on [18]).

 DOI: 10.1051/e3sconf/20160801063

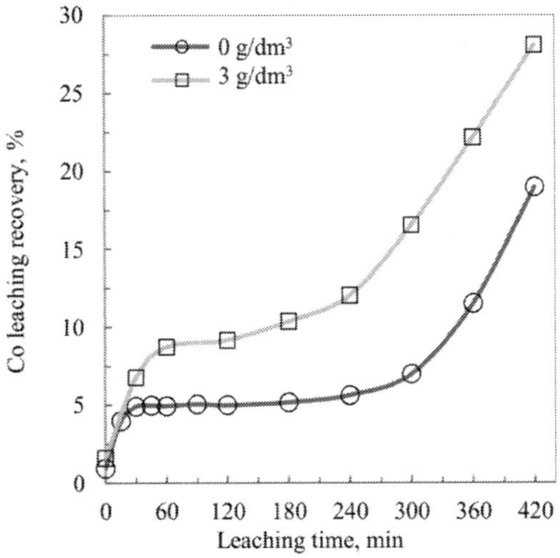

Figure 8. Influence of chloride ions addition on Co atmospheric leaching (solid-to-liquid ratio 1:6, O_2 flow rate 60 dm^3/h, Fe(III) concentration 30 g/dm^3, temperature 90 °C, initial H_2SO_4 concentration 50 g/dm^3, based on [18]).

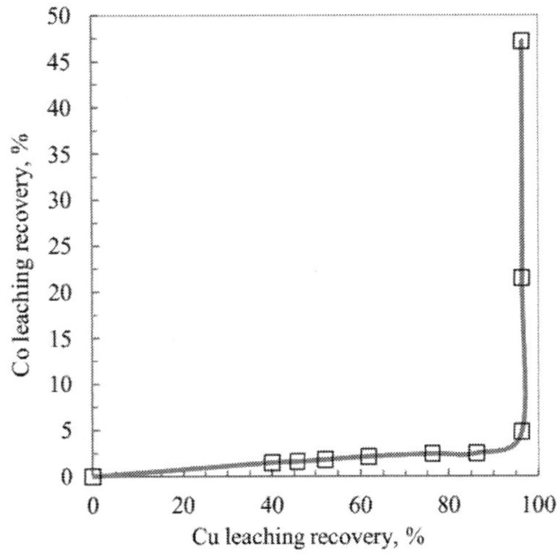

Figure 9. Relationship between Co and Cu recoveries in atmospheric leaching (solid-to-liquid ratio 1:6; Fe(III) concentration 30 g/dm^3, O_2 flow rate 60 dm^3/h, temperature 90°C, initial H_2SO_4 concentration 100 g/dm^3, based on [18]).

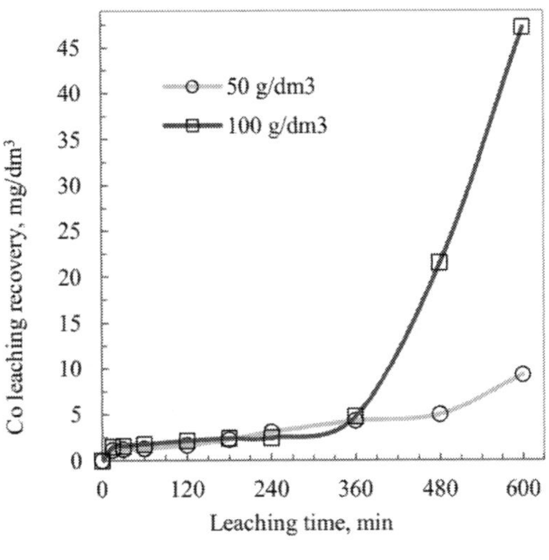

Figure 10. Influence of initial sulphuric acid concentration on Co atmospheric leaching recovery (solid-to-liquid ratio 1:6, Fe(III) ,concentration 30 g/dm^3, temperature 90 °C, O_2 flow rate 30 dm^3/h, based on [18]).

Wodka et al. [19] investigated pressure leaching of black shale middlings in the presence of H_2SO_4 solution, at high temperature and high pressure of oxygen. There are many advantages of application of pressure leaching like the high rate of reaction, high recovery of noble metals, high selectivity or elimination of lower gases and dust emission [19]. Pressure leaching was carried out at higher temperatures from 100 to 180 °C. The results showed that cobalt leaching at temperature below 120 °C did not occur (Fig. 12). Cobalt started to leach at higher temperatures. With increasing temperature above 140 °C the recovery increased up to 40 – 50 mg/dm^3. Additionally, the influence of initial sulfuric acid concentration and oxygen partial pressure on the cobalt recovery was also investigated. It was shown that no changes in the cobalt recovery at different acid concentrations were observed, while higher pressure provided better results [19].

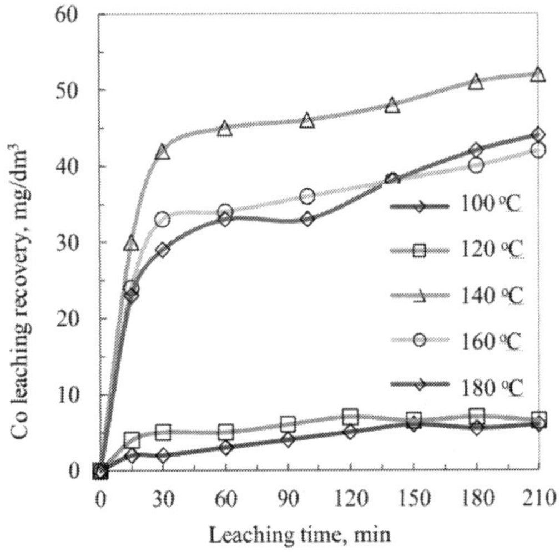

Figure 11. Influence of temperature on pressure leaching of Co (solid-to-liquid ratio 1:10, oxygen partial pressure 0.5 at, initial H_2SO_4 concentration 50 g/dm^3, based on [19])

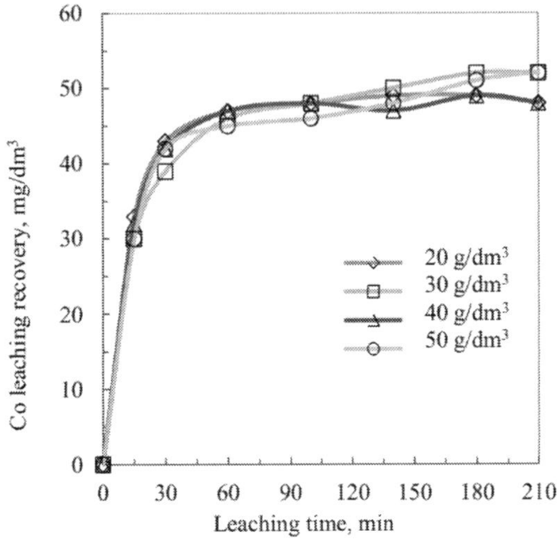

Figure 12. Effect of sulfuric acid concentration on Co pressure leaching (solid-to-liquid ratio 1:10, oxygen partial pressure 0.5 atm, temperature 140 °C, initial H_2SO_4 concentration 50 g/dm^3, based on [19]).

Recovery of valuable metals from pregnant leach solutions (PLS) has to be followed by purification and separation, which can be done with the use of solvent extraction. Rotuska and Chmielewski [14] studied the copper(II), cobalt(II), and nickel(II) recovery from PLS after atmospheric leaching of the Lubin shale fraction. Cobalt was recovered by extraction of copper with extractants such as LIX 84, LIX 612 and LIX 984. After that copper free raffinate zinc was extracted with use of di-(2-ethylhexyl)phosphoric acid (D2EHPA). At that point cobalt and nickel were separated from the zinc free raffinate with di-(2,2,4-trimethylpentyl)phosphinic acid (Cyanex 272), where all extractants were dissolved in kerosene [14]. The results showed that cobalt could be sufficiently extracted from PLS. However, the applied extractants were not enough selective, what resulted in transition of other metals to the organic phase. Additionally, it was proved that the recovery of copper, cobalt and nickel was better from the solutions obtained from leaching of shale without use of iron salts. Therefore, there is a need for further studies in this area to find the most optimal conditions of extraction and reextraction processes for the Kupfeschiefer cooper ore.

4 Conclusions

Although process of cobalt recovery is very complex, there are many examples that it could be possible and profitable. Considering the world demand for this metal and its many applications, cobalt is classified as a strategic metal. There is a possibility to recover cobalt from the Kupferschiefer ore mined by KGHM Polska Miedz S.A. Due to unique mineralogy of this ore, it is difficult to find an efficient technology for recovery of cobalt. The previous studies showed that it is possible to recover cobalt with the use of leaching technology, both for flotation concentrates as well as converter slag, where currently cobalt is cumulating in metallurgical processes.

Acknowledgement

This work was partially financed by the Polish Statutory Research Grant 0401/0121/16.

References

1. European Comission, *Report on Critical Raw Materials for the EU* (2014)
2. E. Lewicka, *Gospod. Surowcami Min.* 23 (2), 5 (2007)
3. G.F. Nordberg, B.A. Fowler, M. Nordberg., *Handbook on the Toxicology of Metals*, (Elsevier, 2015)
4. C.Y. Cheang, N. Mohamed, *Sep. Purif. Technol.*, **162**, 154 (2016)
5. Cobalt Development Institute, *Cobalt Facts*, http://www.thecdi (2016)
6. S. Hannis, T. Bide, A. Minkd, *Raport of BGS* (United Kingdom, 2009)
7. K.G. Fisher, *in Preccedings of The Southern African Institute of Mining and Metallurgy, 6th Southern African Base Metals Conference* (2011)
8. F. Crundwell, M. Moats, V. Ramachandran, T. Robinson, W.G. Davenport, *Extractive Metallurgy of Nickel, Cobalt and Platinum Group Metals* (Elsevier, 2011)
9. Kolwezi Tailings Project, Raport (SRK Consulting, Democratic Republic of the Congo, 2008)
10. C. Bachowski, J. Kudełko, H. Wirth, *Biul. P.I.G.*, **423**, 189 (2007)
11. J. Dabski, Dunaj A., Markiewicz M., Paździora J., Rydzewski A., Siewierski S., *KGHM Polska Miedz S.A. Monograph*, Edited by: Banaszak A., Banas M. (1996)
12. J. Pieczonka, *Prawidłowosci w rozmieszczeniu minerałów kruszczowych w złożu rud miedzi na monoklinie przedsudeckiej* (Wydawnictwo AGH, Krakow, 2011)
13. KGHM Polska Miedz S.A., http://kghm.com/en (2016)
14. A. Grotkowski, T. Henc, *in Proceedings of. International Conference Metale towarzyszące w przemysle metali nieżelaznych* (Wrocław, 2008)
15. N. Kubacz, B. Skorupska, in Proceedings of VIII International Conference on Non-ferrous Ore Processing (KGHM Cupru, Wroclaw 2007)
16. T. Chmielewski, *Physicochem. Probl. Miner. Process*, **41**, 337 (2007)
17. A. Rapacz, in Proccedings of Konferencja Naukowa, Złoza rud miedzi — geologia, mineralogia, stan badan, perspektywy (Lubin, 1998)
18. K. Gibas, K. Borowski, T. Chmielewski, K. Wejman, *Physicochem. Probl. Miner. Process.* **51(1)**, 191 (2014)
19. J. Wodka, T. Chmielewski, B. Ziółkowski, *Physicochem. Probl. Miner. Process.* **41(1)**, 349 (2007)

Technology of Polish copper ore beneficiation – perspectives from the past experience

Andrzej Wieniewski[1], and Bozena Skorupska[1,a]

[1]*Institute of Non-Ferrous Metals, ul. Sowinskiego 5, 44-100 Gliwice, Poland*

Abstract. The paper describes the main types of existing copper deposits in the world and the most common enrichment technologies. The characteristic elements of the current flowsheet of the Polish ore beneficiation were discussed together with the reasons for using them. The paper presents a perspective flowsheet based on more than 50-years of experience in beneficiation of Polish copper ore. The main elements of proposed technology are: leaving in justified cases the division of ore into fractions of increased sandstone and carbonate ores content, flotation in the milling circuit as part of the effective preparation of the ore to rougher flotation, intensive rougher flotation, classic cleaning system, processing of the middlings with grinding system in new type mills, classification and flotation with outlet of final tailings.

1 Introduction

Polish copper ore deposits are of sedimentary type, quite commonly found in the world. However, they have different mineralogical-petrographic and physicochemical properties. The similarities occur only with the ores from the region of Lusatia and Mansfeld (Germany), currently not under exploitation. There are also some similarities with respect to the White Pine (US) and in a lesser extent to Dzheskazgan (Kazakhstan) ore deposits.

Under these conditions, during the technology development of Polish copper ores in the sixties of the last century, it was difficult to fully follow the experience of foreign facilities. These technologies were developed by the Institute of Non-Ferrous Metals and tested in the Experimental Department of KGHM Polska Miedz S.A., under which BP Bipromet and ZBiPM Cuprum performed technical design.

Currently, for the same reasons, not all the achievements in the beneficiation technology in the world can be applied in the Polish enrichment plants. Limitations are also imposed by existing technical and organizational requirements in individual enrichment plants.

In the world, among the total amount of produced copper, about 90% is recovered from the ore containing sulphidic minerals. The main method of beneficiation is flotation, while flotation enrichment flowsheets are relatively simple. In most cases, the particle size of the ore directed to beneficiation is coarse (degree of comminution $K_{80\%}$ = 120-200 μm), flotation time is short and circuits of middlings are simple. The concentrates of rougher flotation to are usually reground to liberate copper sulphides from other easy-floating minerals such as pyrite.

[a] Corresponding author: bozenas@imn.gliwice.pl

Each beneficiation flowsheet has its own specification resulting from the type of enriched ore, but it is also a reflection of the technique state and technology period in which arose [1, 2].

In the world, from the point of beneficiation process, the copper ore deposits can be divided into the following main types [3, 4]:

- porphyry ores occur very often, and it is estimated that more than 60% of the copper production comes from this deposit type. They are mainly copper-containing ores and copper and molybdenum ores. Examples of porphyry deposits are Chuquicamata and La Escondida (Chile), Bingham Canyon (US), Grasberg (Indonesia)
- deposits of copper-bearing pyrite occur among volcanic and volcanic-sedimentary rocks. The composition of ores is variable, usually polymetallic with plenty of either pyrite or pyrrhotite. Examples are deposits in North America
- stratified deposits in sedimentary and metamorphic rocks. The economic significance of these deposits is high. The Cu deposits occur among the marine clay formations, marly less carbonate and sandy. Copper-bearing shales and copper-bearing sandstones are present in this deposits. Minerals create a sandstone binder, microscopic grains scattered in the clay-carbonate rocks and veins. Examples of stratified deposits are Nchanga (Zambia), Kupferschiefer (Poland), White Pine (US)
- other sulphe ores of polymetallic type. They include for example Cu-Zn-Pb, Cu-Mo, Cu-Ni, Cu-Pb ore etc.

Figure 1 shows an example of the most common porphyry copper ore type beneficiation flowsheet. Porphyry ores, although generally contain small amount of copper below 1%, belong to easy-to-float ores, and the obtained concentrates are about 30% of Cu. Technological flowsheets for porphyry copper ores are simple. A typical flowsheet is shown in Fig. 1 [5]. It includes the following circuits:

- ore preparation for flotation
- flotation in the milling circuit – rougher flotation
- classification and regrinding of rougher flotation concentrate
- cleaning flotation with the outlet of the second component of the final tailings from the I cleaning flotation.

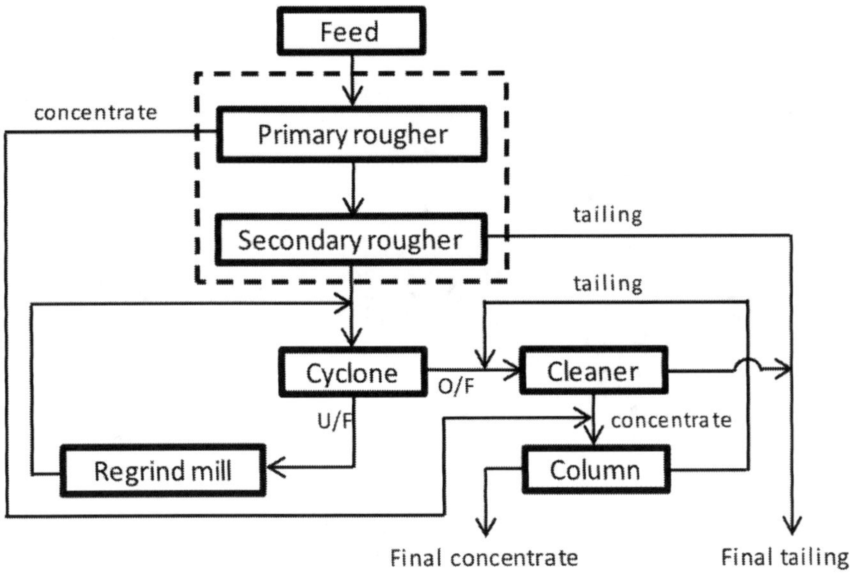

Figure 1. A typical flowsheet used in the flotation process of porphyry copper [5].

In the case of copper-molybdenum ores the flowsheet is more complex, especially when the collective concentrate of copper and molybdenum is heat treated before separation of copper from molybdenum.

2 Beneficiation technology of Polish copper ores - past experience and perspective

2.1 Characteristics of the ores

The copper ore exploited and processed at KGHM Polska Miedz S.A. plants has a complex chemical composition and unfavourable mineralization. It is characterized by high variability in susceptibility to enrichment. The main reason is that the ore occurs in three lithological varieties, namely carbonate, shale and sandstone, with different physical and chemical properties. The shale ore, the richest among all lithological variations (fractions), is characterized by the worst beneficiation. Its share in the deposit is steadily increasing, but is variable, even within the same mining area. This directly affects the basic parameters of the deposit, i.e. the share of individual lithological fractions, copper content, and therefore the characteristics of the exploited ore beneficiation. Illustrations of the problem are made based on groove samples from the prospective mine areas, maps of the metal recovery in concentrate in the study area of deposits [6] (Fig. 2).

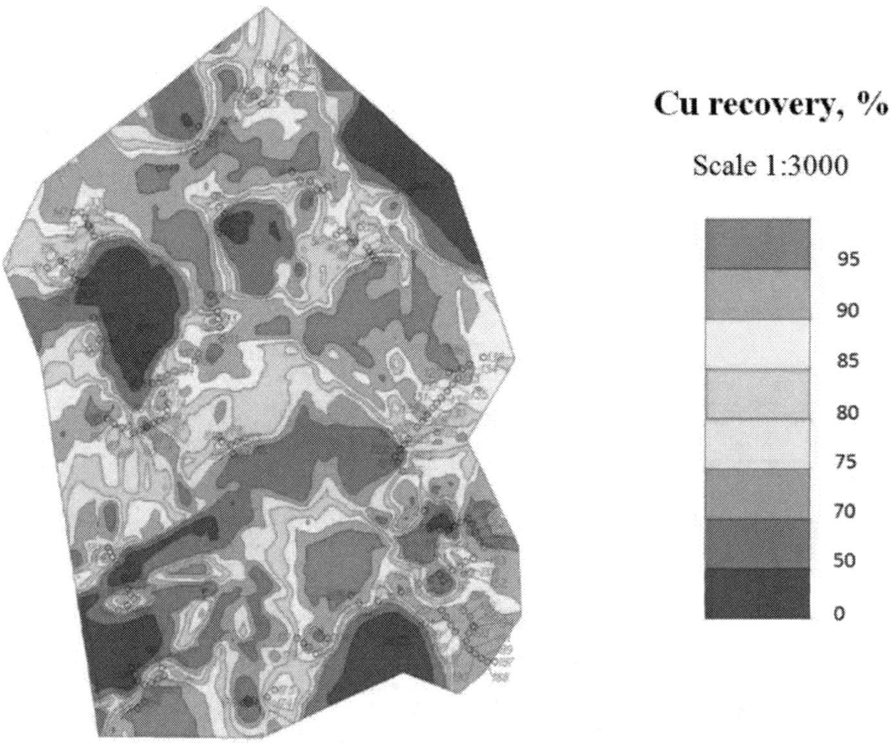

Figure 2. Values of the copper recovery possible to obtain for the studied mining area.

Skorupska et al. [6] described variations in copper content in the area of the deposit. The attention was also paid to a frequent lack of correlation between the content of copper in the deposit and ore beneficiation. It is found that with the progress of exploitation and depletion of ore resources, the metal content in the ore was decreasing and the properties of ore were deteriorating (Table 1).

Table 1. Copper (Cu) and organic carbon (C_{org}) contents in the ore directed to beneficiation in last years.

Year	Lubin concentrator		Polkowice concentrator		Rudna concentrator	
	Cu, %	C_{org}, %	Cu, %	C_{org}, %	Cu, %	C_{org}, %
1995	1.34	0.76	1.75	1.14	1.90	0.64
1996	1.36	0.65	1.86	1.10	2.06	0.71
1997	1.37		1.94	0.62	2.10	0.69
....	....	......	......	....	....	....
2000	1.17	0.87	1.96	1.07	2.20	0.76
2001	1.14	1.30	1.98	0.98	2.14	0.74
2002	1.32	1.40	1.99	1.22	2.36	1.14
2003	1.30	1.06	2.00	2.00	2.31	0.98
2004	1.28	1.62	2.03	1.66	2.23	1.49
....	....	....	....	....	....	....
2010	1.03	0.85	1.82	0.95	1.85	0.75
2011	0.96	0.69	1.80	0.92	1.84	0.70
2012	0.91	0.76	1.81	0.85	1.79	0.74
2013	0.91	0.84	1.75	0.80	1.80	0.73

2.1.1 Problem of organic components

Organic components appearing in the Polish copper ores, mainly in a shale layer, are the main cause of difficulties occurring in the flotation process. Their presence in ores essentially determine the course of the enrichment process and the results of this process. An example of the impact of organic carbon on the technology results expressed as the recovery of copper in concentrate, is shown in Fig. 3 [6].

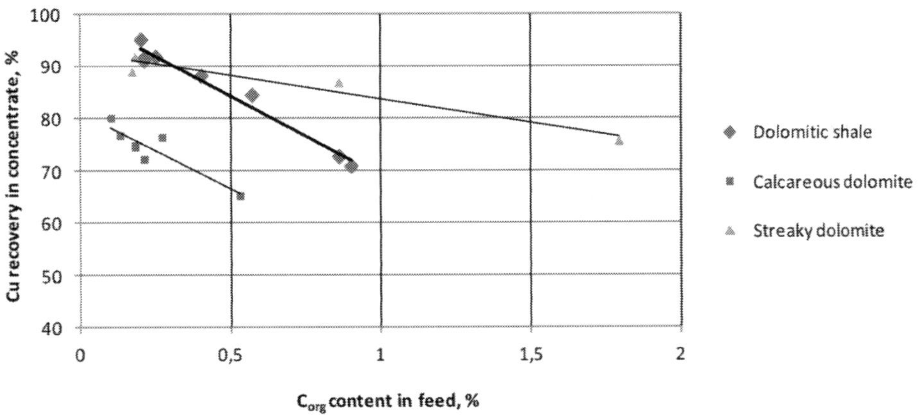

Figure 3. Example of the impact of organic carbon content on the recovery of copper in concentrate [6].

In the view of the negative effect of organic compounds in the flotation process for years it is tried to find ways to reduce their negative impact on this process. One of the methods were attempts to float these components before rougher flotation [7]. This type of flotation, so called preflotation, exists in several enrichment plants in the world, i.e. Mt. Isa, Red Dog in Australia for graphitic carbon preflotation, before flotation of copper, lead and zinc minerals. In the case of Polish copper ores, this kind of flotation did not bring satisfactory results and has not found practical application. Besides the negative impact on the flotation process, the problem is also with the content of organic components in the produced copper concentrates. Currently, this problem is partially solved by the use of segregating flotation and production of two concentrates with different organic carbon contents directed to various metallurgical processes [8]. After elimination of a shaft furnace, in which the concentrate with a high content of organic components is processed, necessary will be roasting the part of the concentrate prior to directing it to metallurgical processes.

2.2 Characteristics of beneficiation technology

The complex lithological, mineral and chemical composition of Polish copper ores is a cause of problems during processing these deposits and force to use more complex technological systems. Moreover, different particle size distributions of minerals in the individual lithological variations cause the need of different grinding degree to liberate minerals during preparation for flotation. Therefore, a flowsheet for mixed ore enrichment was introduced, covering separation into fractions with increased share of sandstone and dolomite-shale ore and subsequent enrichment in separate lines [9]. A block diagram of mixed ore beneficiation processed in the Rudna area is shown in Fig. 4 [10].

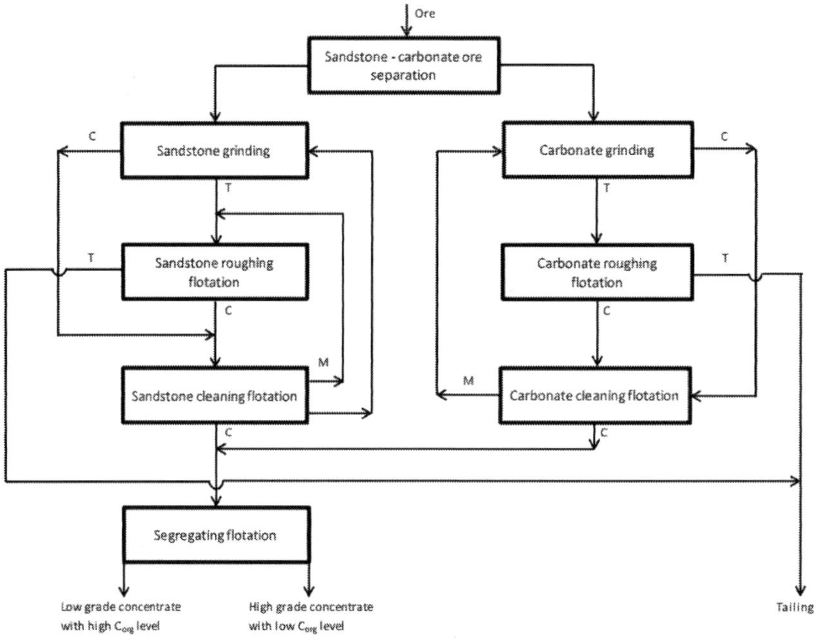

Figure 4. Processing of mixed sandstone-carbonate ore.

The flowsheet includes:

a) beneficiation of dolomite-shale ore, which consists of:
   - a second grinding stage, due to high requirements for grinding of carbonate- shale ore sufficient to liberate sulphide minerals
   - flotation in the milling circuit
   - rougher flotation and three-stage cleaning
   - regrinding the middlings; in the case of beneficiation in the Lubin area flotation in the circuit of middlings regrinding is also possible

b) beneficiation of sandstone ore, which consists:
   - classification and regrinding
   - flotation in the milling circuit
   - rougher flotation along with two-stage cleaning

c) combined concentrate derived from the enrichment line of sandstone and carbonate fraction in the Rudna area is subjected to segregating flotation, by which the concentrate so-called low calorie with high copper content and relatively low organic carbon content and high calorie with lower copper content and a high organic carbon content is obtained.

Table 2. Comparison of enrichment indexes in different years.

Polkowice concentrator

Year	Feed		Concentrate		Recovery
	Cu, %	C_{org}, %	Cu, %	C_{org}, %	Cu, %
1995	1.75	1.14	26.94	6.57	88.58
1996	1.83	1.10	26.70	6.37	88.65
1997	1.94	0.62	25.74	7.74	88.80
....	....	....	....	....	....
2001	1.98	0.98	27.24	7.28	87.40
2002	1.99	1.22	27.20	8.47	87.62
2003	2.00	2.00	27.12	8.04	87.73
....	......	......	......	......	......
1010	1.80	0.95	24.13	8.07	87.55
2011	1.80	0.92	23.70	8.57	87.44
2012	1.81	0.85	24.58	8.25	87.42

Rudna concentrator

Year	Feed		Concentrate		Recovery
	Cu, %	C_{org}, %	Cu, %	C_{org}, %	Cu, %
1995	1.90	0.64	28.40	7.26	88.97
1996	2.06	0.71	28.57	7.45	89.31
1997	2.06	0.69	28.07	7.52	89.98
....	....	....	....	....	....
2001	2.14	0.74	28.88	6.18	90.12
2002	2.36	1.14	30.42	6.82	89.74
2003	2.31	0.98	30.55	6.33	89.91
....	....	....	....	....	....
2010	1.79	0.73	26.26	7.92	89.95
2011	1.84	0.75	26.41	7.93	89.71
2012	1.79	0.70	26.11	7.92	90.12

Lubin concentrator

Year	Feed		Concentrate		Recovery
	Cu, %	C_{org}, %	Cu, %	C_{org}, %	Cu, %
1995	1.34	0.76	18.97	7.33	89.86
1996	1.36	0.65	18.67	6.74	89.98
1997	1.37	0.84	18.45	7.66	90.05
....	....	....	......	......	.....
2001	1.14	1.30	17.13	7.25	86.70
2002	1.32	1.40	17.33	8.30	87.15
2003	1.30	1.06	17.79	8.67	87.01
....	....	....	......	......	......
2010	1.02	0.95	14.50	7.61	86.69
2011	0.96	0.69	13.90	7.33	86.98
2012	0.91	0.76	14.05	7.88	88.98

Currently produced copper concentrates contain from 14 to 14.5% copper (Lubin area) to 26.4% (Rudna area). In the past, the contents were at a higher level, on average, for all plants, about 26 to 27% (from 2001 to 2003 year). The higher copper content in the produced concentrate favoured lower content of organic carbon. Table 2 shows a comparison of enrichment indices in different years for three plants Rudna, Polkowice and Lubin. In spite of processing the ore of difficult beneficiation the enrichment indices are at good level and the overall evaluation, carried out by SGS company from Lakefield (Canada) in 2007, of the used technology can be positive.

2.2.1 Flotation in a milling circuit

A special role in the technology of Polish copper ore beneficiation belongs to flotation in the milling circuit [11]. The main task of this flotation is to remove from the classification-grinding system the concentrate to cleaning flotation. It prevents overgrinding of liberated copper sulphides. Otherwise, these minerals accumulate in the hydrocyclones underflow due to the higher specific gravity in relation to the waste rock. Then, they pass unnecessary fragmentation, sometimes to the level preventing effective flotation. This makes that in greater amount they get to final tailings. The effects of flotation in the milling circuit is well illustrated by the analysis of the copper content in the material recycled to the mill as a function of particle size distribution (Fig. 5) [12].

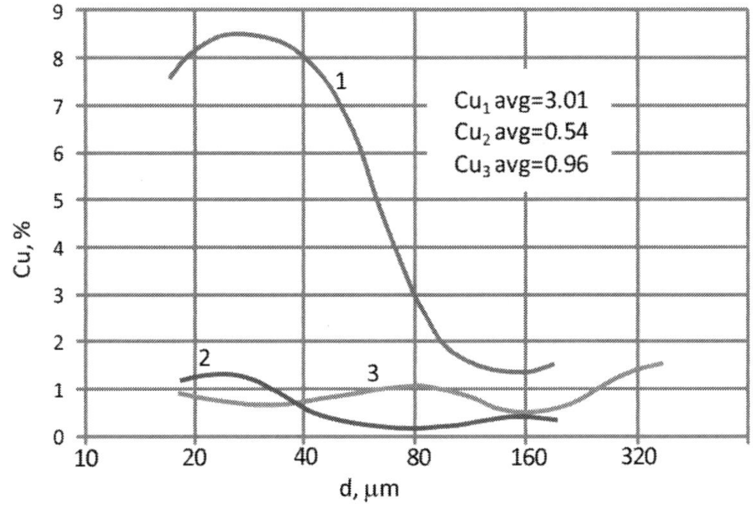

Figure 5. Copper content in size distribution, where 1 is the product recycled to mill without using flotation in the milling circuit, while 2,3 are the products recycled to mill using flotation in the milling circuit.

Flotation in the milling circuit can be carried out in conventional flotation machines. It can also be carried out in a special construction machines. Machines of special construction such as IZ- 9S, IZ-16S types worked in the nineties of the last century in all areas of ore enrichment plants [13]. The feed for these type of machines was underflow product of either ball mills or hydrocyclones. Therefore, they worked with a high density of the slurry and under the relatively coarse particle size.

The aim of flotation, as it was in the classic flotation in grinding circuit, was to remove the liberated copper sulphides in the primary beneficiation phase. The obtained concentrates were of high quality corresponded to the quality of the final concentrate. It had a favourable effect on circulation of middlings and stable production of the final concentrate.

It is worth to emphasize that the low organic carbon content in concentrates was obtained in this flotation. In the case of the concentrate from the Polkowice concentrator, content of organic carbon was only 1.2-2.0% with a 36-47% of copper content in the concentrate. This was a factor for stabilizing an organic carbon content on the lower level in the produced final concentrate. Table 3 presents the test results of flash flotation carried out in flotation machines IF-19S in the area of Rudna and Polkowice in years 1998-1999.

Table 3. Flotation results in milling circuits in machines IF-19S and IF-9S.

Product, component content %	**Sandstone fraction Rudna concentrator**	**Carbonate fraction Rudna concentrator**	**Polkowice concentrator**
Feed, Cu %	0.9-1.20	2.45-3.48	2.5-2.92
Concentrate Cu %	34.34-44.2	31.61-41.9	35.58-47.04
$C_{org.}$ %	-	-	1.2-2.0
Tailing Cu %	0.54-0.74	1.71-2.73	2.32-2.64

In the years 2001-2003 (Table 1) high grade concentrates were produced. Despite the high organic carbon content in the feed the content of organic carbon in the produced concentrates was at an acceptable level. It should be associated with a positive role of flotation in the milling circuit. Despite achieving good technological results because of operational difficulties, flotation in these machines was eliminated. Currently, flotation in the milling circuit is performed in the standard flotation machines of IF type. An important consequence of the correct operation of flotation in the milling circuit should be decopperization of the feed for rougher flotation. As a result, the copper content in the feed in this flotation should be significantly lower than in the raw ore. This ensures a low copper content in the rougher flotation tailing, and thus a high recovery in the whole process of beneficiation.

2.2.2 Rougher flotation

Rougher flotation in the technology of Polish ores beneficiation, differently than porphyry ores, is characterized by long flotation time. The system of rougher flotation has a significant impact on the final indexes of enrichment, because it directly determines the losses of copper to the tailings. It also affects the quality of the final concentrate. This quality increases as copper content is increasing in the concentrate from rougher flotation. Minimizing the losses of copper in the tailings, especially in the coarse classes can be achieved by optimizing for instance classification the feed for rougher flotation in hydrocyclones.

2.2.3 Role of middlings in a beneficiation process

One of the causes of the technological problems of the beneficiation process are circulating in current circuits so-called middlings (I cleaning tailings). The question of middlings and the scale of the problems created by them is still unsolved. It is estimated that in the middlings circulate about 30-35% of the ore by weight carrying 30-35% of copper for the most part in the form of not liberated copper sulphides, and about 30-40% organic carbon [14].

In recent years, in all enrichment plants at KGHM Polska Miedz S.A., despite decreasing of copper content in the ore directed to beneficiation, there is increased copper and organic carbon content in the middlings. This phenomenon is a disadvantage, especially in respect of the rougher flotation operation. The basic chemical characteristics of middlings from Rudna and Polkowice areas is presented below based on the available data from 2015:

Rudna area:

- sandstone middlings – Cu – approx. 6%, $C_{org.}$ – approx. 5%
- carbonate middlings – Cu – 13.5-14%, $C_{org.}$ – approx. 6%

Polkowice area:

- middlings of II technological line – Cu – approx. 6%; C_{org} – approx. 8%.

Current processing of middlings consists of their regrinding in the cylpebs mills, and then classification in hydrocyclones and recycling to rougher flotation. Multiple recycling of middlings results in accumulation of copper in them, but also other adverse components of ore, including organic carbon, clay minerals etc. This has a negative impact on the work of rougher flotation and achieved final indices of the enrichment process.

3 Beneficiation technology - future from the past experience

Due to the unavoidable deterioration of mined ores the natural tendency will be lowering of enrichment indexes. Therefore, an essential element is the analysis of the possible measures to inhibit these negative trends. In recent years, a lot of modernization activities in enrichment plants were made, e.g. exchanged flotation machines, introduced a number of solutions in the field of flotation machine control (FloVis system), continues modernization of classification systems in hydrocyclones.

In addition to the modernization efforts including implementation of new devices also should take action to adapt flowsheets to changing characteristics of ore and metallurgical requirements. The perspective technology should take the best of ore beneficiation technology in the world, but also take into account more than 50 - years of the Polish plants working experience.

The future technology should include the following main elements (Fig. 6):

a) preparation of ore for flotation taking into account solutions that have the most favourable industrial references. The traditional three-stage crushing in either cone crushers or SAG mills was used [15]. The use of SAG mills could afford a marked reduction in the number of operations and unit processes compared to multi-stage grinding scheme in cone crushers with simultaneous possibility of obtaining a finer feed to ball mills. The way of ore preparation for flotation is of particular interest of KGHM Polska Miedz S.A.

b) processing in plant a mixed ore with a relatively constant proportion of sandstones and carbonates is worth to consider to keep separation into fractions with a higher share of sandstones and carbonates, with their separate processing. This allows to adjust the operating parameters of individual systems (grinding, flotation, middlings processing) to the diverse requirements in terms of beneficiation of separate fractions

c) it is necessary to increase the intensity of flotation in the milling circuit. It is an essential part of fulfilling the requirement of preparing the feed for rougher flotation (decopperization of the feed) and stabilizing production of the final concentrate with reduced content of organic carbon. This flotation may be performed in the standard flotation machines by mounting the froth crowder to reduce the yield of the concentrate

d) change of the middlings processing by grinding in mills of a new type i.e. Vertimill®, IsaMill or other (magnetic), and then flotation in open cycle with tailings outlet - the second component of the final tailings. Vertimill® and IsaMill for many years were used in industries for very fine grinding. Currently, the mills are used in non-ferrous metal industry, generally for concentrates regrinding. It is believed that the use of fine grinding mills would enable improvements in liberation of sulphides from fine-mineralised ore and improve the flotation results

e) implementation of cleaning in two- and tree-stages system

f) changing in the way of dewatering the final concentrate by carrying out the process at increased temperatures, enabling to reduce the water content to a level of 8-8.5% and elimination of the drying process.

Thinking about future technologies it cannot be forgotten the importance of the development of machinery construction and processing equipment, new flotation reagents, including activators, containing in its composition new functional groups for improved adsorption of either existing or new collectors on the surface of copper sulphide. Processing is an important element in mining-metallurgical process for copper production. The final indices of the enrichment process, i.e. recovery and quality of concentrate, should be related not only to beneficiation of ore, but also take into account the economic balance of the whole production line including a certain level of stock market prices of copper and silver.

DOI: 10.1051/e3sconf/20160801064

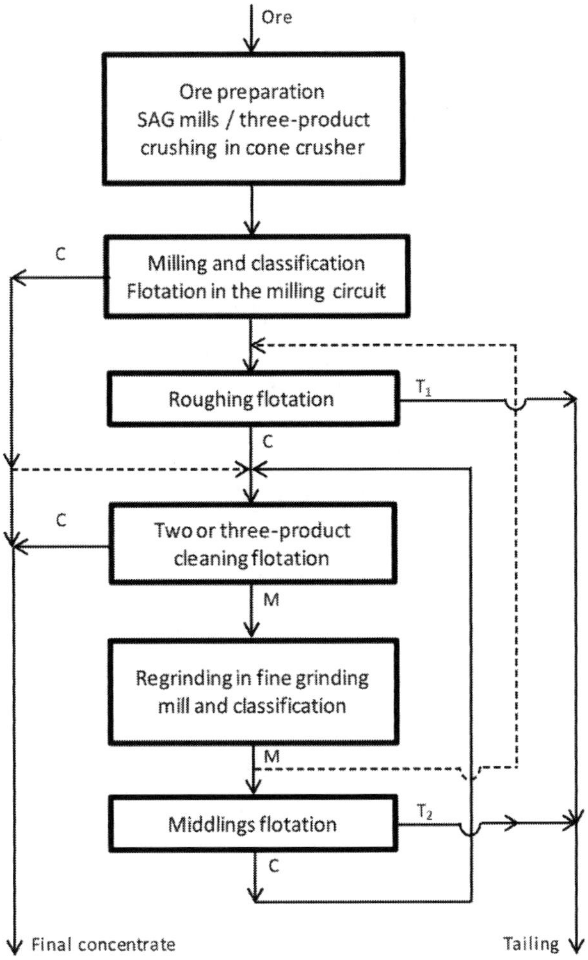

Figure 6. Proposed flowsheet for mixed ore processing at KGHM Polska Miedz S.A.

4 Summary

A beneficiation technology of Polish copper ores is developed to their specific properties. In different periods of application high indices of enrichment were achieved. Gained experience of the past allows to develop the technologies of the future. This does not mean omitting new solutions, particularly in the field of machinery and equipment, process control and the introduction of new flotation reagents. It is important to treat processing as an important element of the mining-metallurgical process for copper production. Metallurgy sets specific requirements to produced concentrate, the achievement of which constitute a separate optimizing tasks. Future processing should consists of technologies that use more than fifty years potential of knowledge in the field of concentrates production for certain pyro- and hydrometallurgical processes ensuring comprehensive recovery of contained in them metals.

References

1. S. M. Bulatovic, D.M. Wyslouzil, *Miner. Eng.*, **11(4)** (1998)
2. W. Blaschke, Z. Blaschke, *Inzynieria Mineralna*, **8(2)** (2008)

3. A. Bolewski, *Surowce mineralne swiata*, (Wydawnictwo Geologiczne, Warszawa 1997)
4. S.M. Bulatovic, *Handbook of flotation reagents* (2007)
5. B.A. Wills, T. Napier-Munn, *Mineral processing technology* (2005)
6. B. Skorupska, A. Wieniewski, N. Kubacz, *Zastosowanie metod geostatystycznych do oceny zmiennosci wzbogacalnosci rudy* (Monograph, 2013)
7. R. Bortel, J. Gramala, Z .Grzebieluch, *IMN Report No. 1489* (1970)
8. B. Skorupska, A. Wieniewski, N. Kubacz, *IMN Report No. 6956* (2011)
9. A. Wieniewski, S. Sobierajski, *Stan aktualny oraz dalsze zamierzenia w zakresie flotacji rudy miedzi w Zakladach Wzbogacania KGHM Polska Miedz S.A.* (Lubiatow, 1996)
10. A. Wieniewski, *in Proccedings of V Miedzynarodowa Konferencja Przerobki Rud Metali Niezelaznych ICNOP* (1999)
11. A. Wieniewski, J. Dziadek, *Ph.D. thesis* (1983)
12. W. Madej, A. Wieniewski, S. Zienkiewicz, *IMN Report No. 3* (1977)
13. E. Szczerba, J. Komorowski, B. Skorupska, *in Proccedings of V Miedzynarodowej Konferencji Przerobki Rud Metali Niezelaznych ICNOP*, (1999)
14. A. Luszczkiewicz, A. Wieniewski, *Gorn. Geoinz.*, **3(1)** (2006)
15. A. Grotowski, A. Szubert, W. Mizera, B. Mizera, *Cuprum report* (2009)

Role of classification in grinding using the electromagnetic mill. A case study

Marta Wolosiewicz-Glab[1,a], Dariusz Foszcz[1], Tomasz Gawenda[1], and Szymon Ogonowski[2]

[1]*AGH University of Science and Technology, Faculty of Mining and Geoengineering, Department of Environmental Engineering and Mineral Processing, Poland*
[2]*Silesian University of Technology, Faculty of Automatic Control, Electronics and Computer Science, Institute of Automatic Control, Poland*

Abstract. In the process of classification, one of the methods used to evaluate the effectiveness classifiers is to plot a separation curve, which determines the size of seprated particles and characterizes the process accuracy. The article presents an assessment of the impact of parameters on grinding and classification using an electromagnetic mill. The results allow a detailed assessment of the effectiveness and efficiency of the device and facilitate the optimization of the grinding process by establishing an appropriate control algorithms.

1. Introduction

The quality of the products obtained from the preparation processes depends primarily on the correct selection of equipment and auxiliary devices depending on the type of raw material processed. Assuming a specific technological process, usually one takes into account the characteristics of individual machines performance, which depends, among other factors, on the physico-mechanical feed properties and quantities of the designable and steerable machines' parameters [1, 2]. It turns out, however, that obtaining products with the appropriate parameters is also affected by the conduct of the technological process that is the number of grinding stages, technological cycles, flow control of material to the selected devices etc. [1, 2]. For example, in the processes of crushing and grinding during the production of crushed aggregates closed technological cycles are often used. It is so called a grinding process in a closed circuit. There are also technological systems with open flow of material. Such systems have the advantage over closed cycles in terms of performance, but often the obtained products have lower quality [1, 2].

Figure 1 shows the technological cycles of grinding with screening grain classification but systems of this type can be also used for other devices, such as mill-separator [1, 2].

[a] Corresponding author: wolosiewiczm@gmail.com

DOI: 10.1051/e3sconf/20160801065

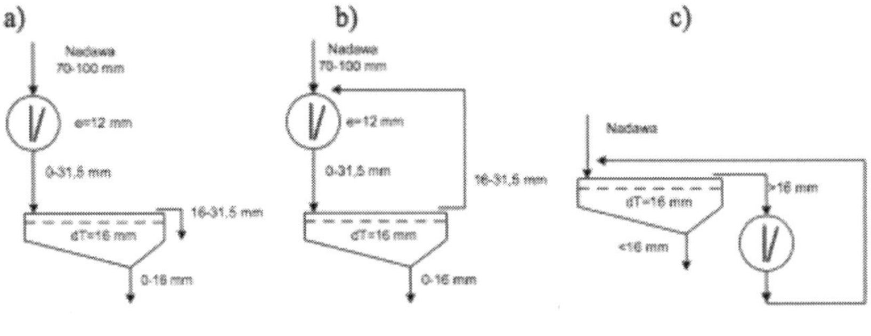

Figure 1. Technological cycles: a) open cycle, b, c) closed cycles [1, 2].

2. Inertial-impingement separator used in electromagnetic mill

An inertial-impingement separator with a cyclone has been designed, basing on computational flow dynamics (CFD) for explicit performance parameters and particle size of individual streams. Figure 2 shows the chosen sorted grain sizes that result from the CFD simulation of the flow.

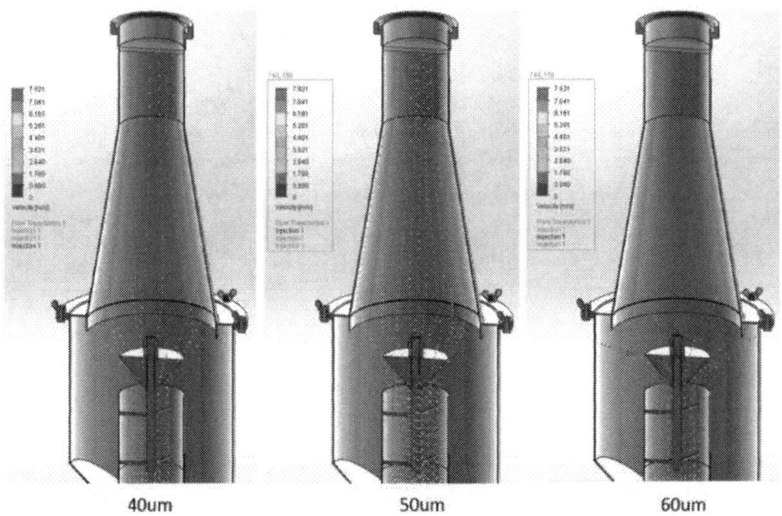

Figure 2. Results of flow simulation CFD for different sizes of feed grains led to separator (courtesy of Pro-mill - cyclone contractor).

The separator is designed to allow a change of the sorted grain size by changing the position (mounting height) of the impactor or replacement with another element with different opening angle. Figure 3 shows the design, fabrication and impingement component of this type of separator. As a result of shredding and other physical and chemical processes, a mixture of solid particles of different sizes and shapes is obtained. Very often a specific grain size or range of sizes are required. The classification can be done by [3]:

- in the gas stream – pneumatic classification,
- in the stream of liquid – hydraulic classification,
- sieve – screening.

In the system with an electromagnetic mill, the inertial-impingement separator has been designed, shown in the system functional layout in Figure 4.

 DOI: 10.1051/e3sconf/20160801065

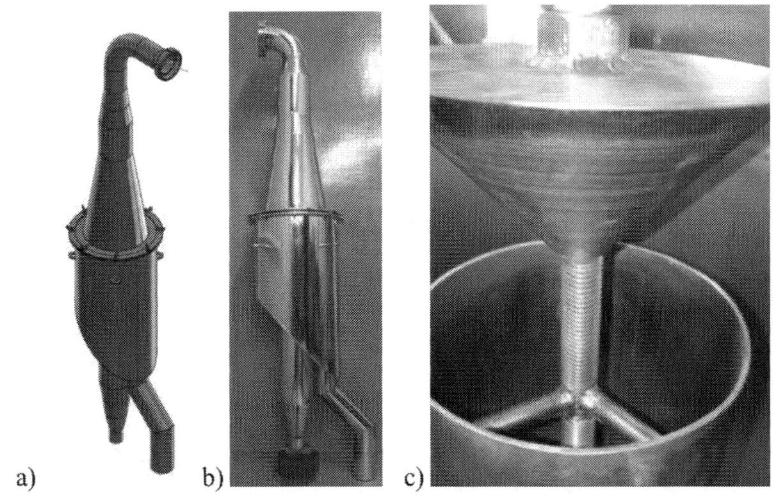

Figure 3. Inertial-impingement separator: design (a), fabrication (b), impingement element (c) (courtesy of Pro-mill - cyclone contractor).

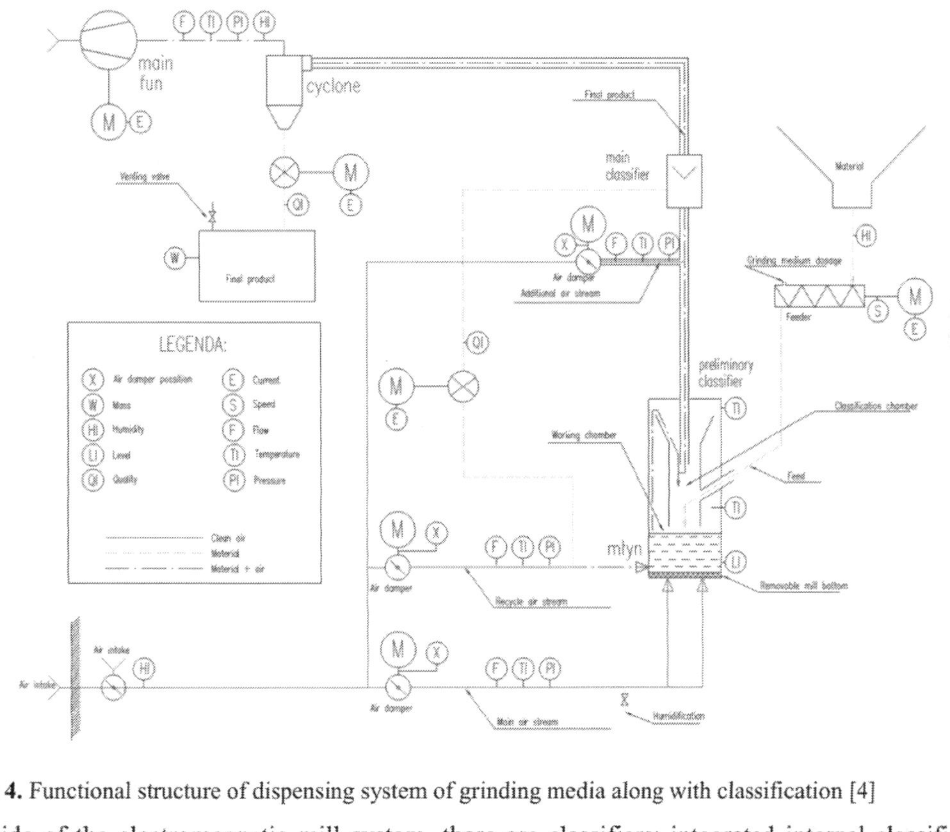

Figure 4. Functional structure of dispensing system of grinding media along with classification [4]

Inside of the electromagnetic mill system, there are classifiers: integrated internal classifier and inertial-impingement. They create a closed system. Over the initial classifier, the inertial-impingement separator was placed from which the recycle flow is led back into the mill, feeded from the bottom of the working chamber (Fig. 5).

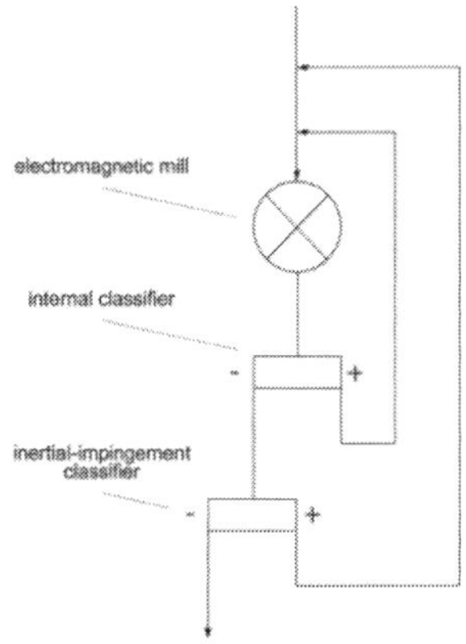

Figure 5. Functional structure of classifiers in electromagnetic mill.

A properly selected classification device should give a sharp separation effect curve, which means that the larger particles should not blend with the fine product as well as the fine particles corresponding to the final product should not blend into recycle flow coming out from the classifier [1, 2].

Multiplicity of the circuit is defined as the ratio of the amount of material recycled to the quantity of new feed. Circuit multiplicity reaches high values in the milling processes. In practice, for grinding in mills cement or limestone it is equal to 2-2.6. In case of the material processed in crushers, these values are much lower [5].

In the case of separation of small grains blends (approximately <1 mm) and very small (<0.1 mm), the classification process is often carried out in either aqueous or air medium. During the processes of grading, different forces are acting (centrifugal, resistance of the medium, Coriolis, gravity, buoyancy, gradient resulting from changes in grains concentration) on the particle [1, 2]. Differences in classifiers' design determine the values and directions of these forces as well as the quantitative and qualitative characteristics of the implemented process.

In case of an ideal material separation with sieves or classifiers, grains characterized by a distribution parameter value (e.g. size or weight) lower than a predetermined, are in one product and the other in the second. Under real conditions, the separation is never perfect, and part of the grains go to the wrong product. To characterize the accuracy of the separation effect, we introduce separation values or separation efficiency for the classes or fractions. It is defined as the ratio of mass of the end product to the weight of that class in the feed [6, 7]:

$$T(r_i) = \frac{Q_{P1} \cdot a_{i_{P1}}}{Q_N \cdot a_{i_N}} = \gamma_{P1} \cdot \frac{a_{i_{P1}}}{a_{i_N}} \tag{1}$$

$$\tau(r_i) = \frac{Q_{P2} \cdot a_{i_{P2}}}{Q_N \cdot a_{i_N}} = \gamma_{P2} \cdot \frac{a_{i_{P2}}}{a_{i_N}} \tag{2}$$

where:

$T(r), \tau(r)$ – separation curves,
r – the average value of the characteristics under which the separation takes place,
i – number of fractions (i=1, 2,...,k),
a_{i_N} – i-th fraction in the feed
$a_{i_{P1}}$ – i-th fraction in the first product
$a_{i_{P2}}$ – i-th fraction in the second product
Q_N – feed weight
Q_{P1} – first product weight
Q_{P2} – second product weight

γ_{P1} – first product flow off
γ_{P2} – second product flow off.

Each classifier has a specific separating ability, which operates at a predetermined separation boundary or otherwise, at a given sorted grain. The sorted grain quantity is the particle diameter expressed in eg. μm (micrometers), for which the probability to be found in iether lower or upper fraction is equal to 50% [6].

The separating ability of the classifier or separation effect curve is a function of the separation depending on the diameter of the grain. It is most commonly presented in the form of a curve in a cordinate system in which the abscissa indicated the grain size and the ordinate the degree of separation in [%]. The separation effect curve is not a curve of particle size distribution, it is a set of points representing the infinitely narrow grain class with an average grain size *d* splits into upper and lower product [6].

Probable error E_p, is defined as [8]:

$$E_p = \frac{\rho(T=75)-\rho(T=25)}{2} \quad (3)$$

where: $\rho(T = 75) = \rho_{75}$ and $\rho(T = 25) = \rho_{25}$[9] and represent readings from the distribution curve of the density values for the ordinates 75% and 25%.

Probable error E_p is a threshold density, which may or may not be exceeded with the same probability by the individual grains with disperse of elementary fraction of specific density. The E_p value is smaller, then the separation is more accurate. For perfect separation E_p=0.

Imperfection I, a benchmark for different separation density distribution, expressed as:

$$I = \frac{E_p}{\rho_r} \quad (4)$$

where $\rho_r = \rho_{50}$ is a reading from the distribution curve of the density value for the ordinate 50% [3].

3. Electromagnetic mill. Separation effect curves for inertial-impingement classifier

Figure 6 shows the separation curve for the electromagnetic mill with inertial-impingement classifier and particle size distribution of up to 5 mm. Values obtained for the separation curve were:

- probable error: 0.1
- imperfection: *I*=0.05.

In an ideal material separation for the sieves or classifiers, grains having a distribution parameter value (eg. size or weight) less than a predetermined, are placed in one product and all other in the second. In terms of actual separation, it is never perfect, and part of the grains goes to the wrong product. To characterize the accuracy of separation effect, the number of separation effectiveness for classes or fractions has been introduced and presented in Fig. 6.

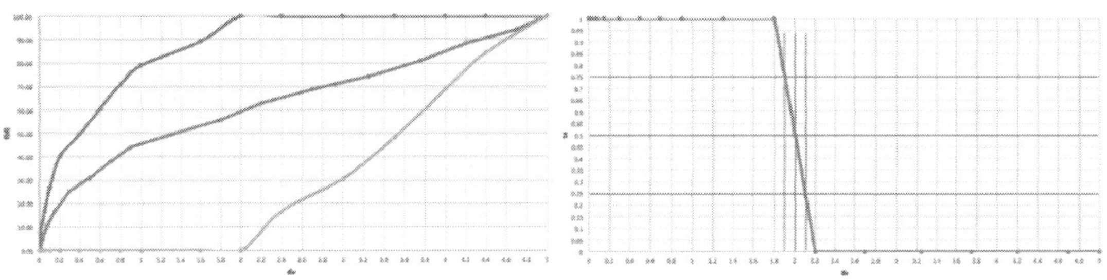

Figure 6. a) Feed, recycle, overflow curves; b) separation effect curve.

Figure 7 shows the separation curve for the electromagnetic mill with inertial-impingement classifier and particle size distribution of up to 5 mm. Values obtained for the separation curve were:

- probable error: 0.3625
- imperfection: I=0.18125.

From Figure 7 one can see that the imperfection, relative to the ideal system, has risen, therefore the classifier works less efficiently and there is a risk of recycle which will reduce the efficiency of the mill.

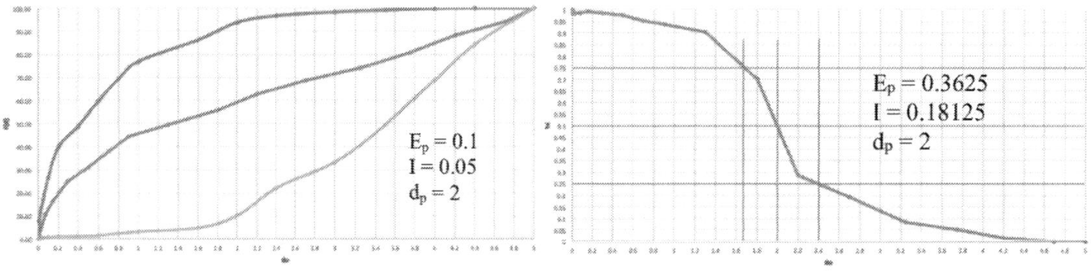

Figure 7. a) Feed, recycle, overflow curves; b) separation effect curve.

Figure 8 shows the separation curve for the electromagnetic mill with inertial-impingement classifier and particle size distribution of up to 5 mm. Values obtained for the separation curve were:

- probable error: 0.925
- imperfection: I=0.4625.

From Figure 8 one can see that imperfection reaches the highest value among all presented in the paper separation effect curves, and the recycle gets a larger share in the feed than the overflow, which causes a significant reduction in efficiency of grinding system.

A significant amount of recycled material caused by the mismatched parameters of the classifier is a significant burden for the system of the mill, the separator and the transport equipment.

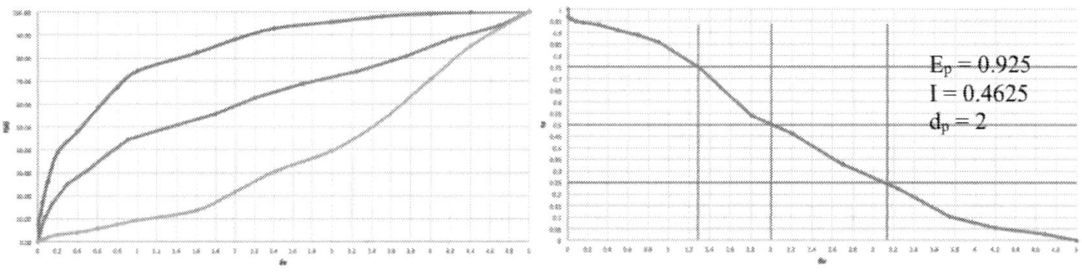

Figure 8. a) Feed, recycle, overflow curves; b) separation effect curve.

Figure 9 shows the separation curve for the electromagnetic mill with inertial-impingement classifier and particle size distribution of up to 5 mm. Values obtained for the separation curve were:

- probable error: 0.65
- imperfection: I=0.325.

Figure 10 depicts the content of the particles with the diameter larger than d_p in the overflow. It is easy to notice that the content of such grains decreases when the imperfection increases.

Thanks to the simulations, we can obtain the information related to the parameters of operation of the system what enabled the introduction of appropriate control in order to improve the performance of grinding and classification. Matching appropriate parameters of the classifier allows to optimize system operation. Such matching is possible and much easier due to the separation effect curve.

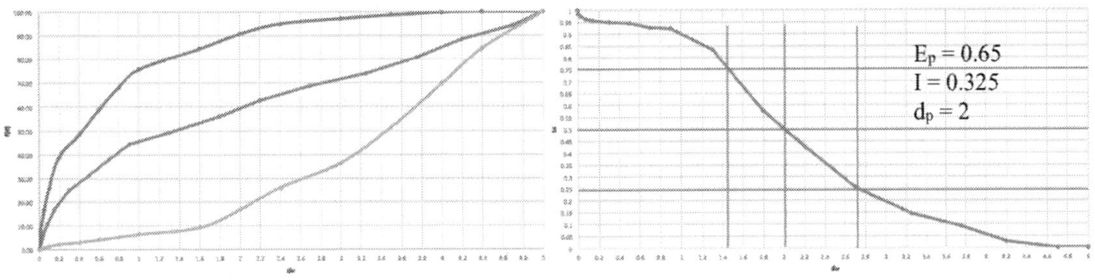

Figure 9. a) Feed, recycle, overflow curves; b) separation effect curve.

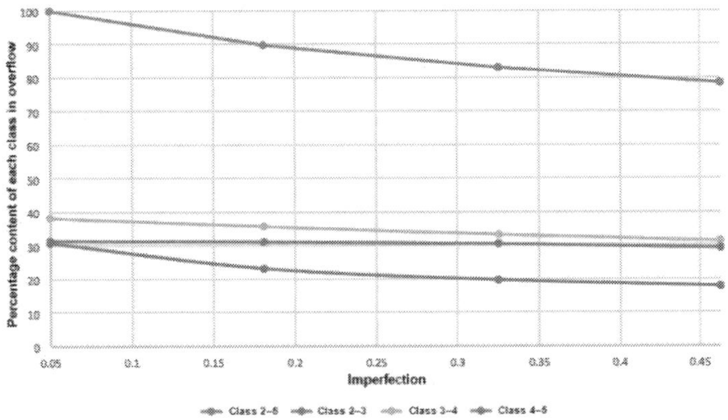

Figure 10. Content of grain classes in overflow. Only size fractions with the diameter of grain higher than d_p.

4. Summary

When designing a classifying system for the electromagnetic mill, the goal is to obtain the most efficient operation of the mill. With a separation effect curves, it is possible to make simulation of the grinding and classification process and determine the size of recycle flow which would allow for introduction of appropriate control algorithms that guarantee the correct operation of the mill and the entire system. With the calculations, one obtaines characteristics of the recycle value. In the review of the classification devices for fine-grained products, to predict the final products of the electromagnetic mill, one should use the separation effect curves, which provide opportunities to optimize the operation of the grinding-classification system. One can read from them the sorted grain size and get information about the accuracy of the implemented process. Made in the article comparisons of the efficiency of the device with different parameters, permit us to determine the impact of the classification efficiency on the level of recycle that will determine the efficiency of the whole grinding and classification system. The paper shows how to determine the optimum operating conditions of abruptly-inertial classifier, which was built specifically for the designed electromagnetic mill. The simulations allow to determine the functional dependencies for the actual classification process and enable the development of algorithms for control system in order to obtain the best product quality in the real system. This will be a subject of further research.

Acknowledgement

Development of the article has been funded in the framework of the project: PBS3/B3/28/2015.

DOI: 10.1051/e3sconf/20160801065

References

1. T. Gawenda, *Monograph No. 304 (*Wydawnictwo AGH, 2015)
2. Z. Naziemiec, T. Gawenda: *in Proccedings of VII Conference Kruszywa mineralne: surowce – rynek – technologie – jakosc (*Oficyna Wydawnicza PWR, 2007)
3. J. Warach, *Aparatura chemiczna i procesowa* (Oficyna Wydawnicza Politechniki Warszawskiej, Warszawa, 1996)
4. M. Pawelczyk, Z. Ogonowski, Sz. Ogonowski, D. Foszcz, D. Saramak, T. Gawenda, Patent No. P.413041 (6.07.2015)
5. T. Gawenda, *Monograph No. 304 (*Wydawnictwo AGH, 2015)
6. Z. Naziemiec, D. Saramak, *Min. Sci.*, **22(1)**, 119 (2015)
7. Z. Blaschke, M. Brozek, E. Mokrzycki, Z. Ociepa, T. Tumidajski, Górnictwo Cz. V. *Zarys technologii procesów przeróbczych* (Skrypt AGH, no 931, 1983)
8. W. Stepinski, *Wzbogacanie grawitacyjne* (PWN, Warszawa, 1964)
9. A. Surowiak, *Ph.D. thesis*, (AGH University of Science and Technology, 2006)

Design of an electromagnetic mill. Its technological and control system structures for dry milling

Marta Wolosiewicz-Glab[1,a], Dariusz Foszcz[1], Tomasz Gawenda[1], and Szymon Ogonowski[2]

[1]*AGH University of Science and Technology, Faculty of Mining and Geoengineering, Department of Environmental Engineering and Mineral Processing, Poland*
[2]*Silesian University of Technology, Faculty of Automatic Control, Electronics and Computer Science, Institute of Automatic Control, Poland*

Abstract. A modern industry could not operate without complex processing of mineral raw materials, supported by innovative technologies. That is the reason to design and create a new electromagnetic mill. The paper presents a design of its technological structure and control system. It is presented that, in relation to traditional solutions, the electromagnetic mill is a fully automated system, which affects the effectiveness of its operation. The effectiveness and efficiency of processes depend on a proper selection of the physical parameters of the system and process parameters. The device is characterized by low energy consumption.

1. Introduction

Specified requirements for systems of grinding and classification, regarding the reduction of energy consumption, grinding, optimal particle size, particle shape and their surface properties showed that there is a need to design a modern fine grinding system with recirculation of grains which do not meet quality requirements. To meet the requirement of a competitive system, it should have a universal application, possible for all kinds of raw materials and configurable by appropriate measuring and control system and a dedicated application HMI/SCADA. An electromagnetic mill with the control system allows either dry or wet milling of raw materials in a closed cycle, with the possibility and very wide range, selection of speed of movement of the grinding media with different parameters. The ability to control the size of the dosing feed and grinding media, the ongoing analysis of particle size and status of the mill and the flow of recycle will enable much more effective mechanical activation of generated particles with specific physical and mechanical parameters (size, shape, surface area and energy, surface properties), and also significant increase in the grinding efficiency and energy savings. Heat losses in the working chamber will be used to simultaneously dry and heat the material to be processed and will increase the milling efficiency. The grinding process is usually only one stage of processing of materials, and the parameters of the ground product are closely related to the requirements of technological processes in the next stages. Not always the greatest possible degree of fineness of the material in the mill is required. The best results in the following process after grinding is being obtained for the certain particle size range. In addition, the parameters of the grinding process

[a] Corresponding author: wolosiewiczm@gmail.com

are often subject to changes resulting from the technological situation of the whole production line. Sometimes it is necessary to change the parameters of mills due to the deteriorating performance of the flotation process of the milled material. Currently, the electromagnetic mill available on the market neither has a feeding-receiving system nor a dedicated control system.

2. Electromagnetic mill – design

The electromagnetic mill is a device in which the ferromagnetic grinding media are moved by deliberately generated rotating electromagnetic field, as an energy carrier. The basic elements of the mill are: inductor of rotating magnetic field and placed in its axis tube, forming a working chamber [1]. The electromagnetic mill was a little-known grinding machine used in laboratories rather than in industry. This limitation of applications results from the very low productivity and low efficiency of this type of mills. Low technical-motor parameters were influenced by improper design assumptions led to very large losses in the inductor windings of a rotating magnetic field. One of the advantages of the electromagnetic mill can be the possibility to enhance many processes by carrying out grinding in a region where, due to the electromagnetic field, ferromagnetic rods (grinding media) with a suitably chosen length to diameter ratio, rotate [1, 2].

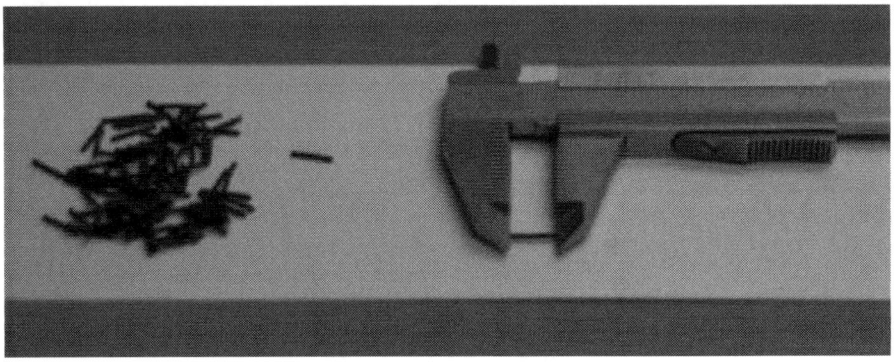

Figure 1. Grinding media used in electromagnetic mill (author's photo).

Figure 2. Magnetic inductor developed by ELTRAF company.

In contrast to conventional mills (ball, rod, or cylpebs) the mill housing is stationary while the grinding (or mixing) takes place in the working chamber, inside which move small ferromagnetic elements - called grinding media. Their movement is caused by the action of the vortex electromagnetic field [2, 3]. The effectiveness and efficiency of the process carried out in the mill are dependent on the proper selection of the physical parameters of the structure and process parameters such as the flow rate of feed (residence time), particle size of the input, speed and intensity of the rotating electromagnetic field, temperature, humidity, and other parameters of the workpiece (e.g. lithological composition) [1].

A basic condition for industrial use of electromagnetic mill is its continuous operation. Grinding media should be selected for the particle size of the particulate material. In the case of large variation in the material grain size, a blend of grinding media of different sizes can be applied. Any change in the size of the grinding media can be adapted to the physical properties of the ground material and the desired final fineness of the ground product [2, 3]. Defined requirements for grinding and classification systems related to the reduction of energy consumption of grinding, optimum grain size, grain shape and their surface properties showed that there was a need to design a modern fine grinding system with recirculation of grains which do not meet quality requirements. To make this solution competitive, system should be universal for possible all kinds of raw materials, configurable using a measuring and control system as well as a dedicated HMI/SCADA software. Such requirements impose a need for mills with a unique operating principle and the wide possibilities of parameterization of its work. As part of the SCADA system, the following optimization problems will be solved:

- maximization of the processing throughput with the limitation on the quality of the product,
- minimizing the energy consumed in grinding at a specified throughput, and
- maximization of the degree of fragmentation.

Figure 3. Classifiers close-up.

 DOI: 10.1051/e3sconf/20160801066

Decision variables of these tasks will be the defining points of the mill work (gear motor output, product temperature, etc.). Arguments of the criterion functions arise from the evaluation of the quality of the process, such as the parameters of the particle size distribution, average or borderline grain size, the total value of the electricity for the processing, etc. To determine the values of decision variables from these arguments, the suitable polynomial models and artificial neural networks will be identified [4]. Such requirements impose a need for mill with a unique operating principle and the wide possibilities of parameterization of its work. The SCADA system will record parameters i.e. the quality parameters defined as the design criterion function optimization tasks and the process parameters, either measured or adjusted during grinding for a single charge. On this basis, the SCADA system on a regular basis will perform principal component analysis (PCA). On the basis of the principal component scatter, the criterion indication of the deregulation process of batch grinding will be developed. There will be developed automatic decision-making procedures in the event of deregulation.

3. Design of the technological system

The processes of mechanical comminution are characterized by enormous energy intensity, that is why in this direction intensive research on modern design solutions of machines and systems for shredding and classification are carried out. Shredding processes are widely used in various fields of material processing to obtain finished products, e.g. mineral aggregates with a suitable grain size, sorbent limestone, cement. The most important characteristics of the milling process include:

- producing much larger quantities of the smallest particle size fraction,
- a much higher energy consumption per unit of mass-processed raw material (technological efficiency)
- higher particles fineness than in the processes of crushing,
- a large part of abrasive action in most of these processes,
- high wear of working elements per unit of processing,
- higher than in other processes probability of producing a certain amount of undersize particles.

Figure 4. Electromagnetic mill and classifiers.

The final product coming out from an accurate classifier is separated from the transporting air in the cyclone and goes into the final product tank. In addition, the installation is equipped with a cooling system of the working chamber in the form of a set of fans with controlled output. The concept of the installation is shown in Fig. 5 which depicts the chosen measuring points parameters such as speed, pressure, humidity and air temperature, temperature and humidity of the material, mass of material in individual tanks, position of the flaps and engine speeds.

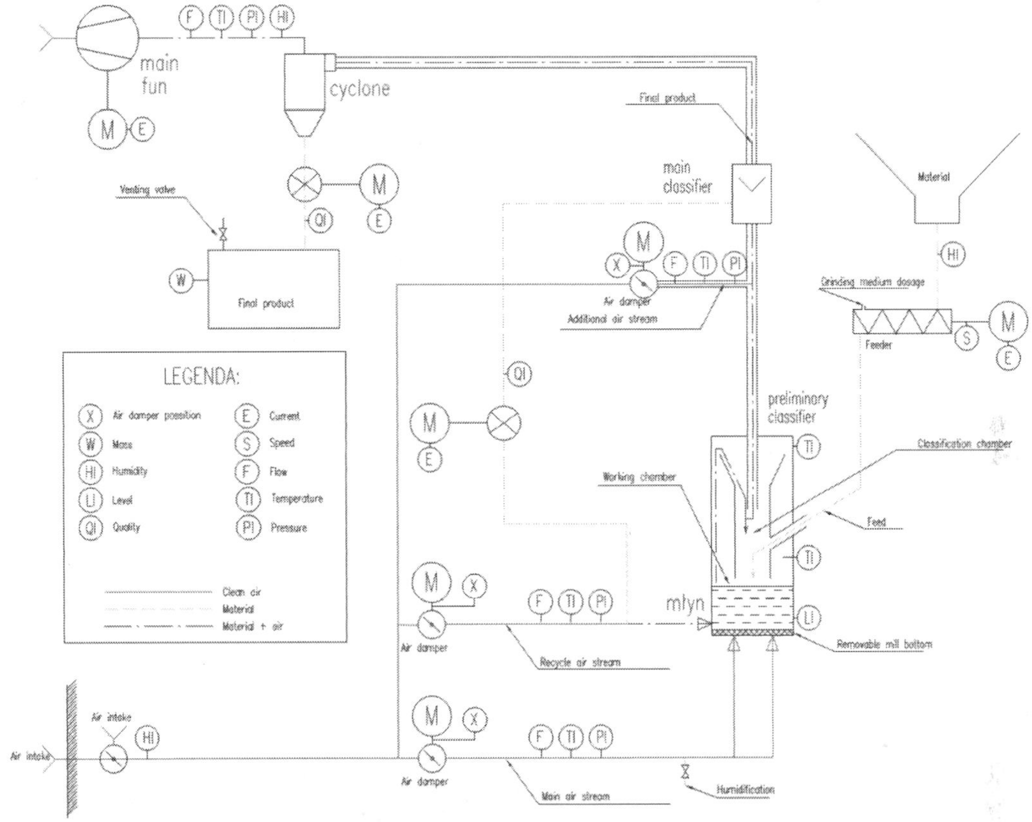

Figure 5. Functional structure of dispensing system of grinding media along with classification system [5].

4. Design of the control system

The installation of the electromagnetic mill is a complex system with many control goals defined for each subsystem. Obtaining the output with required particle size, which is the main objective of the control system, is carried out during the operation of the mill under nominal conditions, which means after startup procedures, constituting a separate group of tasks performed by the control system. Additional supervisory objectives of the control system in the electromagnetic mill is to obtain the required temperature and humidity of the product. The vertical arrangement of the mill causes the control system of air streams fed to the mill system is designed to keep the ground material in the working chamber. For a better tightness of the transport system it was decided to use a vacuum system with a single pulling fan as a major source of air flow as shown in Figure 6.

Therefore, the control system uses flaps regulating the flow of a stream of primary air in the pipes, recycle and secondary air, as actuators of control systems, keeping ground material in the working chamber. This results in a significant complication of the control system, because any change in the position of any flap, associated with the reaction of the local control system changes the flow of the remaining air streams [4, 6].

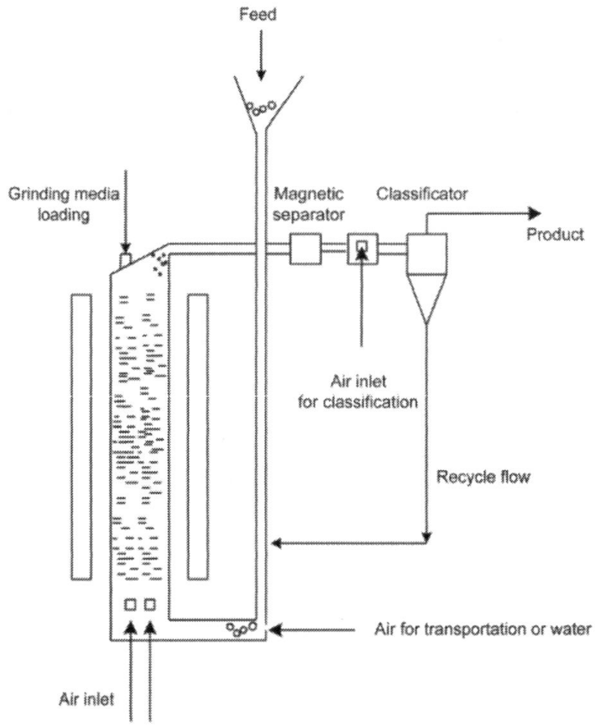

Figure 6. Dry and wet milling in vertical layout [5].

Figure 7. Flaps regulating flow of stream of primary air, recycled air and secondary air .

Another task of the control system is to stabilize the mixture of grinding media and ground material as well as to control the rotating electromagnetic field. The initial studies showed that changing mentioned process parameters significantly affected the fineness of the product. All parameters mentioned above are interrelated, meaning that the control object is multi-dimensional and has a strong internal cross-coupling feedback. From the point of view of the design of the control system, a plurality of the above-mentioned quantities (measured, adjusted and state) were divided into control, disturbance and output signals (Fig. 8). This approach not only allows the systematization of each quantities in the system, but also modelling the object for the control purposes [7].

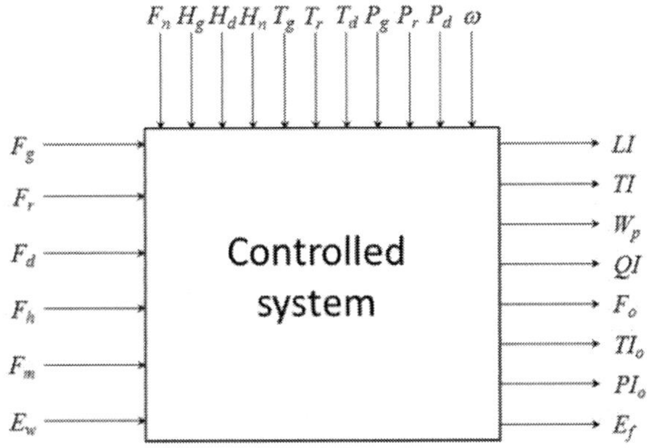

Figure 8. General structure of the electromagnetic mill installation as a controlled system.

Control quantities are: primary air speed F_g, recycle air speed F_r, additional air speed F_d, volumetric flow of water mist to feed F_h, weight stream of grinding media F_m and voltage control of cooling fans E_w. Disturbance variables are: weight feed stream F_n, humidity of air in primary stream H_g, air humidity in the additional stream H_d, initial feed humidity H_n, temperature of air in the primary stream T_g, temperature of air in the recycle stream T_r, air temperature in an additional stream T_d, air pressure in primary stream P_g, air pressure in recycle stream P_r, air pressure in an additional stream P_d, frequency of rotating electromagnetic field ω as well as grain composition of feed and susceptibility to crushing of the material during the shredding process. The output quantities are: working chamber filling level Ll, average temperature of preliminary classifier TI, weight of product in product tank W_p, product quality QI, speed of exhaust air F_o, temperature of exhaust air TI_o, pressure of exhaust air PI_o, load current of inverter generating rotating field E_f.

The classification was the starting point for the design of the functional structure of control systems implementing the objectives described above. These objectives are carried out through the implementation of specific tasks in the control system. These tasks, however, are carried out within the framework of unequal time horizons. Moreover, they are interconnected by the multidimensionality of the plant, and through cross-compliance control purposes. It was essential to implement the control system with hierarchical structure [4], which general overview is presented in Fig. 9.

For each of the subsystems, specific tasks were formulated, carried out by their local control systems. These tasks were defined by the states of the subsystems to be controlled. Basic tasks of direct control are [7, 10]:

- stabilization of speed of the air flow in the working chamber in open and in recycle circuit,
- stabilization of speed of the air stream at the input of accurate classifier,
- control of grinding media dispenser,
- adjust the degree of filling of the working chamber,
- feed material humidity control,
- stabilization of product moisture,
- stabilization of the temperature in the working chamber.

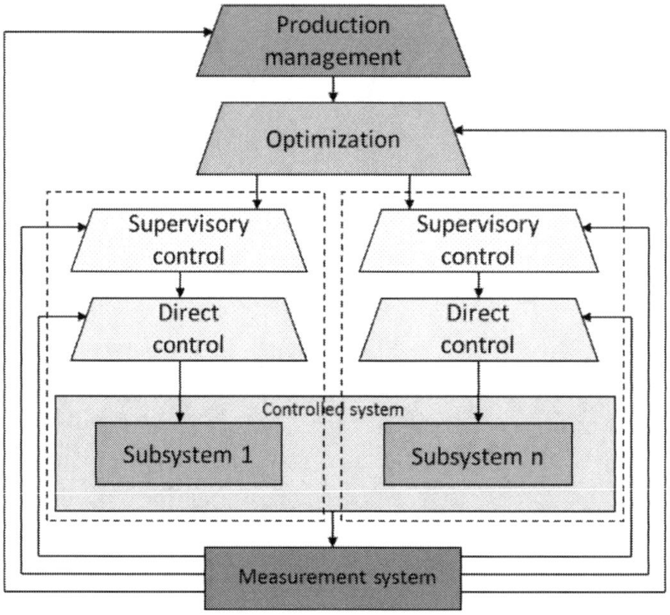

Figure 9. General structure of the multi-layered control system.

The tasks of the supervisory control layer concern, *inter alia*, the determination of the set point correction for air flow regulators depending on such parameters as humidity and temperature of the feed and air; set point correction control of the primary stream regulator, depending on the recycle state; determining a set point for additional flow controller; startup sequences management; the pre- and post-startup control of actuator elements in the installation [11].

The task of optimization control layer is to determine the operating points of the electromagnetic mill. The most important operating point of the mill is its load – the mass flow of the feed. For the supervisory control layer, as well as for the direct layer, the feed stream is the leading quantity, which is the basis for design of the functional structure of control systems in these layers. In general, the task for optimization layer is to determine the operating parameters of the mill, which minimize energy consumption with the restrictions arising from the required product quality, expressed by the granulation. For the optimization task, also other constraints such as temperature and humidity of the product and the geometric properties of grain, can be taken under consideration:

The primary objective of the measurement system in the electromagnetic mill is to measure all physical quantities necessary to implement the mill control algorithm. The measurement system should enable the measurement of physical quantities associated with all media streams present in the system.

Implementation of the control system assumes the use of SIMATIC S7-300 controller with digital and analog I/O modules for direct and basic supervisory control. All analog measurement and control signals were brought to the 4.20 mA standard, which provided a lower susceptibility to disturbances. Direct control with the use of inverters will be executed in an additional PLC system with SIMATIC S7-1200 controller, which also provides the MODBUS communication standard [12]. The communication between the controllers will take place in the PROFINET network. The tasks of the supervisory layer are implemented in PLC controllers and in SCADA IFIX system installed on the operator panel SIMATIC IPC. For the purpose of the data processing for soft-sensing system [12] using basic measurement values and additional quality measurements (eg. the vision system 6), PC computer was installed, connected with the panel via Ethernet network. Additionally, for the registration of data from acoustic [11, 10] and vibration sensors [9] Dataloger (DLAS), was directly connected to the operator panel.

5 Conclusions

The article presents the construction of an innovative electromagnetic mill, which compared to traditional solutions provide a significant reduction in energy consumption and higher technological performance. The device is equipped with a control system, either measuring the grinding quality or analyzing the operating status of the mill. Direct control of the grinding process is implemented in industrial controllers (PLC) and is supported by the optimization algorithms that run in the supervisory control system (SCADA). Designed during the project low-cost way of the real time measuring of particle size of the feed and products, as well as flow classification, can be used in any field of mineral processing, not only related to the grinding in the electromagnetic mill, in which there are similar processes (crushing and grinding, screening, hydraulic and air classification, flotation and gravity enrichment), but also in cement and lime industry and in energy sector.

Acknowledgement

Development of the article was funded in the framework of the project: PBS3/B3/28/2015

References

1. M. Wolosiewicz-Glab, D. Foszcz, *in Proceedings of 5th European Young Engineers Conference* (Warsaw University of Technology, 117, 2016)
2. M. Wolosiewicz-Glab, D. Foszcz, *in Proceedings of 5th European Young Engineers Conference* (Warsaw University of Technology, 117, 2016)
3. M. Wolosiewicz-Glab, D. *Logistyka*, **4**, 9930 (2015)
4. Z. Ogonowski, *Appl. Energy*, **88**, 1586 (2011)
5. M. Pawelczyk, Z. Ogonowski, Sz. Ogonowski, D. Foszcz, D. Saramak, T. Gawenda, Patent No. P.413041 (6.07.2015)
6. Z. Ogonowski, *in Proceedings of 19th International Conference on Process Control* (High Tatras, Slovak Rep., 2013)
7. Sz. Ogonowski, Energy Build., **42**, 1510 (2010)
8. P. Tatjewski, Advanced Control of Industrial Processes: Structures and Algorithms (Springer–Verlag, London, 2007)
9. P. Kadlec, B. Gabrys, S. Strandt, Comp. Chem. Eng., **33(4)**, 795 (2009)
10. A. Konieczny, M. Kurzydlo, Sz. Ogonowski, D. Foszcz, *in Proceedings of the Mineral Engineering Conference MEC 2014* (Istebna, 2014)
11. S. J. Spencer, J. J. Campbell, K. R. Weller, Y. Liu., *IPMM*, **'99(2)**, 939 (1999)
12. M. Kurzydlo, M. Pawelczyk, Vibroeng. *Procedia*, **6**, 18, (2015)

E3S Web of Conferences **8**, 01067 (2016) DOI: 10.1051/e3sconf/20160801067
MEC2016

A phenomenological model for improving understanding of the ammonium nitrate agglomeration process

Alvaro Videla Leiva[1,a], and Catalina Polanco Palacios[1]

[1] *Mining Engineering Department, Pontificia Universidad Catolica de Chile, Chile*

Abstract. Ammonium nitrate is intensively used as explosive in the mining industry as the main component of ANFO. The ammonium nitrate is known to be a strong hygroscopic crystal matter which generates problems due to the creation of water bridges between crystals leading later to nucleation and crystallization forming an agglomerated solid cake. The agglomeration process damages the ammonium nitrate performance and is undesirable. Usually either organic or inorganic coatings are used to control agglomeration. In the present work a characterization method of humidity adsorption of the ammonium nitrate crystal was performed under laboratory conditions. Several samples were exposed into a defined humidity in a controlled chamber during 5 hours after which the samples were tested to measure agglomeration as the resistance force to compression. A clear relation was found between coating protection level, humidity and agglomeration. Agglomeration can be then predicted by a phenomenological model based of combination of the mono-layer BET adsorption and CNT nucleation models.

[a] Corresponding author: avidela@ing.puc.cl

Aggregates from mineral wastes

Ireneusz Baic[1,a], Wieslaw Koziol[1], and Lukasz Machniak[2]

[1]*Institute of Mechanized Construction and Rock Mining, Division in Katowice, 40157 Katowice, Poland*
[2]*AGH University of Science and Technology, 31059 Cracow, Poland*

Abstract. The problem concerning the growing demand for natural aggregates and the need to limit costs, including transportation from remote deposits, cause the increase in growth of interest in aggregates from mineral wastes as well as in technologies of their production and recovery. The paper presents the issue related to the group of aggregates other than natural. A common name is proposed for such material: "alternative aggregates". The name seems to be fully justified due to adequacy of this term because of this raw materials origin and role, in comparison to the meaning of natural aggregates based on gravel and sand as well as crushed stones. The paper presents characteristics of the market and basic application of aggregates produced from mineral wastes, generated in the mining, power and metallurgical industries as well as material from demolished objects.

1 Introduction

Limited resources of mineral raw materials and competitive methods of utilization of areas where they are deposited (e.g. environment protection, cities, etc.) cause that more attention is paid to rational management of natural resources and optimization of secondary raw materials utilization. The EU raw material policy assumes maximum utilization of secondary raw materials for production of goods of quality not worse in comparison to goods made of natural resources, as one of the major goals [1]. Also national regulations [2], recommend utilization of all raw materials, including secondary and waste raw materials. The paper presents the condition of production and application of aggregates other than natural, included in the common group of alternative aggregates.

2 Classification of aggregates

Natural aggregates (gravel and sand as well as ground material) represent ca. 1/3 of raw materials utilized in the world and in terms of volume, this is the largest group of mined minerals. World's production of aggregates is estimated at 20–40 milliard Mg (no detailed data). Both gravel and sand as well as ground aggregates are used mostly in the construction industry, for the production of concrete, roads and other applications: asphalt, cements and glues, prefabricated products, land macro-levelling, sporting infrastructure, mining backfills, embankments, etc. Fine sands of high silica content, are used almost in all contemporary electronics (mobile phones, PCs, TVs, solar panels, etc.), in technical

[a] Corresponding author: i.baic@imbigs.pl

 DOI: 10.1051/e3sconf/20160801068

engineering, and lately for the construction and maintenance of golf course, for gas and oil mining using hydraulic fracturing method from shale deposits, etc. The wide range of application of natural aggregates, especially in many Arabic countries and other highly industrialized areas with very high constructions and specialist engineering building industry make that in some countries there are significant shortages of natural aggregates and the necessity to import them from remote deposits. Prices of imported aggregates in those countries are high (50–100 USD/Mg), i.e. exceed current hard coal prices. Problems concerning covering current demand for natural aggregates cause the increase of interest in alternative raw materials and technologies of their production in order to satisfy the demand and limit costs, especially costs of transport from remote deposits.

Standards concerning aggregates, harmonized with European standards, beside the requirements, set the following division of aggregates together with their definitions. The following types of aggregates can be distinguished:

- natural aggregates - aggregates from natural sources (deposits), which, except mechanical processing, are not subjected to any other processing. Natural aggregates can be divided into ground as well as gravel and sand
- artificial aggregates - mineral origin aggregates, obtained as a result of industrial process, covering thermal or other modification
- recycled aggregates - aggregates produced as a result of processing inorganic material previously applied in the construction industry.

However, the above mentioned classification adopted according to the PN-EN standard is not unequivocal and raises many reservations, mostly due to a large group of aggregates produced of waste, accompanying and by-product raw materials, which beside the mechanical processing (crushing, sieving) are not subjected to any other modification. Can these aggregates be included to the artificial aggregates group?

Within 2010–2011 in the European Union, except narrowing the aggregates classification, also normative Annex A has been enclosed, which includes a list of raw materials and their origin, which can be the base for aggregates production. Due to the type and origin of raw materials for aggregates production, one can proposed a division to primary raw materials for production of natural aggregates and secondary raw materials for production of artificial and recycled aggregates.

Primary raw materials cover all petrographic types of rocks (magmatic, sedimentary, metamorphic) included in the PN EN 932-3 standard. However, the following is the source of secondary raw materials: construction industry and recycling, combustion of solid municipal waste, energy industry, metallurgy, non-ferrous metals industry, founding, coal and rock mining, dredging works and other [3].

In the construction industry, except for artificial and recycled aggregates, separate group of aggregates can be noticed - "waste aggregates" which include mineral origin aggregates from any technological process, subjected only to mechanical processing. This group includes, among other, aggregates produced from metallurgical and power plant slag as well as self-burnt coal associated shales.

Considering the above, it is proposed to adopt the classification of aggregates to two large groups of aggregates, i.e. natural and alternative aggregates. Similar division is also adopted in the case of fuels: fossil fuels (natural) and alternative fuels. The scope of application of all alternative aggregates is set in the PN-EN standard and it is identical for natural aggregates, thus the basic criterion of aggregates application is their properties.

Aggregates are manufactured mostly from various kinds of waste raw material, generated in the metallurgical, mining, power and other industries. Raw material origin has a significant influence on physical and mechanical parameters of produced aggregates.

3 Production and application of alternative aggregates

3.1 Aggregates from waste raw material from mining and metallurgical industry

Aggregates are acquired from waste or accompanying raw materials from current production or kept on stockpiles. Volumes of stored waste (over 1.6 milliard Mg - mostly from mining industry), which are annually announced by the Central Statistical Office, should not be perceived as the volumes of available resource base.

In the case of coal waste (deposited), red shale (burnt) is the most desirable for production of aggregates, the "operating volumes" of which on all dumps or reclamation area amounts ca. 20–25 million Mg, which is not even 5% of waste in this group. In turn, in the case of the copper ore mining wastes, their application is not possible at this moment (e.g. for production of mining bentonites, heavy concrete, cement, porous aggregates, cellular concrete, glass powder - component of bituminous compounds) because they are used as construction (embankment) material and to seal the object used for rendering harmless the waste deposited in the Zelazny Most area. Infringement of this construction stability would be a serious environmental hazard due to the type of waste being rendered harmless. About 25-30% of the waste present in this object could be potentially used, maintaining the structure and current function [4].

In case of aggregates produced of waste stored in metallurgical and open-heart furnace dump yards, the main problem is the depleting resources or their complete utilization (in case of some objects) for the production of aggregates. In 2004, the amount of stored gravels from melting processes was ca. 22 million Mg, in 2013 it was ca. 4 million Mg [5, 6]. Similar problems apply to aggregates produced from waste collected on non-ferrous metals metallurgy waste yards (copper and zinc). Some minor quantities of copper derivative gravels are available in Legnica Steelworks, and gravels deposited in Glogow Steelworks have been fully utilized.

At the moment, the ongoing production provides resource basis. Production of waste from current operations is closely related to standing of a given sector, and its volume is correlated with the size of production (mining). In the case of [4]:

- hard coal mining waste - the volume is ca. 35–40% of the coal mining volume,
- copper ores mining waste - the volume is ca. 90% of the ore mining volume,
- metallurgical and open heart furnace gravels – the volume is ca. 30% of steel production.
- copper ores gravels – the volume is ca. 190% of copper production.

From the mentioned data it results that current production provides about 30 million Mg of coal derivative waste, ca. 2.5 million Mg of metallurgical and furnace gravels and about 1 million Mg of gravel from copper metallurgy. In relation to gravels from steel and copper metallurgy, one can say about their total depletion from aggregates production. However, in case of coal derivative waste, only a small part is used for economic purposes. The example could be technological line for aggregates production in the Katowicki Holding Weglowy (Wujek Mine). This mine started production of aggregates from waste raw materials used in the road industry and other engineering works.

Aggregates are produces from waste raw materials and depending on their quality and origin they can be used in numerous construction applications:

- aggregate for concrete,
- aggregate for soil gradation improvement,
- backfill of engineering objects consisting of stone and coarse grain soil,
- filling and / or stabilizing material for auxiliary base course,
- construction of embankments (below the ground freezing zone, in the freezing zone, in dry locations or when isolated from water or represent backfill for stone and coarse grain soil),
- freeze resistant layer, draining layers,
- backfill of:
  - engineering objects,

  - excavations for installations,
  - abutments and supporting structures,
- land macro-levelling,
- land reclamation,
- construction of flood banks,
- reinforcement of unsurfaced roads,
- maintenance of roads during winter (grip improvement aggregate).

Due to low quality of aggregates produced from hard coal mining waste, their application is limited to simple engineering works.

3.2 Aggregates from energy industry waste raw material

Despite relatively large volumes of ashes and slag produced by power plants and CHPs (ca. 20 million Mg/year), two companies are engaged in the production and distribution of waste based aggregates in the energy sector: EDF Ekoserwis Sp. z o.o. from Wroclaw and Epore Sp. z o.o. from Bogatynia. Aggregates are produced mostly basing on the furnace slag which is a by-product generated during combustion of hard and brown coal in the furnace.

The slag based aggregate is not homogeneous and its bulk density is 700–1100 kg/m^3. Low quality of aggregates for concrete production purposes and cases of defective elements made of slag-concrete are the grounds to limit the application of slag based aggregates for reinforced elements as well as medium and large structural non-reinforced elements. At the moment, aggregates are mainly applied for:

- base and auxiliary courses of roads,
- drainage layers of roads,
- base courses for yards and walkways,
- levelling backfill for cobblestone,
- sanitary and sewerage construction, backfill of cables and systems,

The second group is aggregates generated during thermal treatment of ashes in energy and heating sector. At the moment, the only company offering such aggregate is LSA Sp. z o.o. from Bialystok. The main orientations of the aggregate application are:

- production of light structural concrete as well as structural and insulation concrete,
- production of abrasive elements (products), ceiling elements, stack materials, insulation materials and products for small architecture,
- production of thermal insulating mortars,
- acoustic and thermal insulation,
- geo-technology (reduction of buildings settlement, reduction of horizontal thrust and transfer of structure loads, stabilization of foundations, replacement of the base),
- road industry (road base courses),
- gardening (drainage layers under crops or a component for hoeing of soil, controlling penetration of water and air),
- construction and road filtering and draining layers,
- filtering and water accumulation layers under green (green roofs) and sporting areas,
- engineering construction,
- reinforcement of unsurfaced and forest roads,
- formation of road embankments,
- land levelling,
- soil stabilization,
- ceramic product manufacturing,
- aggregate for light concrete,

- reclamation and macro-levelling of degraded elements.

3.3 Recycled aggregates

Recycled aggregates are produced as a result of construction works such as: demolition works, repairs, modernizations. Quality of recycled aggregates depends on the quality of recycled materials (concrete, brick, asphalt). There are two methods of producing recycled aggregate: "in situ" at the site of demolition, construction, modernization, repair and "ex situ" at the disposal plants. Significant savings, on transportation costs, can be achieved in case of in situ production of aggregates. Turnover with these raw materials is very difficult to register. Based on historical data, share of construction materials in total volume of waste from demolition, construction and repair works is ca. 30% (40% scrap, 30% soil and earth), in turn, the degree of recovery of waste is at the level of ca. 80% [7]. Assuming the increase of recovery index to 90% as well as the option to recycle waste of code 10 12 08 (rejected ceramic waste, brick, tiles and construction ceramics) and 10 13 82 (rejected products), one can say that practical level of recycled aggregates production may be at the level of ca. 2.0 to 3.0 million Mg. This value should be increased by aggregates from asphalt recycling, the volume of which is estimated at the level of ca. 5 million Mg per year.

Places of recycled aggregates production and at the same time their economic utilization should be linked mostly to large urban agglomerations, where most of demolition works are focused. Main directions of utilization are related to land levelling, ballast (lower layers) under cobblestone pavement, backfilling of foundations in building construction industry, replacement of soil, lower level of road base courses (concrete aggregates).

4 Summary

Poland is a major producer and user of natural aggregates. Current production is at the level of ca. 210–240 million Mg/year, i.e. ca. 5.5–6.3 Mg/year per citizen. Over 70% of the production of natural aggregates is related to gravel and sand aggregates located within the whole country and mined in all Provinces (leaders in this list are Podlasie, Warminsko-Mazurskie, Malopolskie – year 2014 data) and the remaining part applies to aggregates, produced of solid rocks, which are located mostly in Southern Poland, Dolnoslaskie Province (ca. 44%), Swietokrzyskie Province (ca. 34%), Malopolskie Province (over 11%) and Silesian Province (ca. 5%).

The production of natural aggregates is complemented by alternative aggregates, produced mostly from mineral wastes. Production of alternative aggregates matches the message concerning basic goals and principles of waste management, i.e.: reduction of waste to minimum, minimization of their impact on the environment, recovery in accordance with the environment protection. The name "alternative aggregates" seems to be fully justified due to adequacy of this term for the origin and roles of these raw materials in comparison to natural aggregates.

Alternative aggregates are used in building construction industry and in horizontal structures. Intensification of interest in these aggregates is observed when major constructions are located near the area of resource base for their production. Projects requiring significant volumes of construction works (geotechnical applications), alternative aggregates, especially light ones, due to three-times less bulk density, in some cases, have the edge over natural aggregates. Their application may reduce the cost of construction investment, including transport costs.

In developed countries, with vast ecological awareness, production of alternative aggregates represent significant share in the total production of aggregates. These are Great Britain (68 million Mg) and the Netherlands (18 million Mg) where the share of alternative aggregates is ca. 25% of total production. Share of over 15% in total production can be noticed in Belgium (16 million Mg), Germany (the largest production, almost 100 million Mg). Average share of artificial aggregates for 28 EU states is 10%.

In Poland, according to the UEPG data [8], production of alternative aggregates in 2013 amounted 30 million Mg, which is 11.7% of the total production. Within the closest future, one can expect reduction of alternative aggregates production (mostly secondary raw materials) due to depletion of the resource base. Complementation of lost capabilities can be provided by utilization of recycled aggregates as well as common utilization of mining waste from current production and production of light aggregates from new raw materials such as sewage sediment, bottom sediments. Mining aggregates and some recycled aggregates, in relation to metallurgical aggregates can be characterized by worse utility parameters, thus they cannot fully substitute them.

The last specific group of natural waste aggregates is small fractions of gravel-sand aggregates (sands) as well as ground aggregates (sand generated during production of grit). In relation to large production capacities of mines, less demand in relation to 2010–2012 and low prices of aggregates, some fractions of aggregates (especially fine ones) are at the moment difficult or even impossible to sell. At the same time, the quality of gravel and sand deposits deteriorates. For example, within 8 years 2007–2015, the share of industrial resources of gravels in Poland was reduced from 8.8 to 4.2% and in the share of sands (fraction below 2 mm) increased from 25.7% to 37.4%. The most significant deterioration of deposits quality was in the northern and central part of Poland. Very often, these aggregates are returned to pits as useless waste, even though, after re-qualification and enrichment, they could be a valuable raw materials applied outside road industry.

References

1. W. Koziol, A. Cieplinski, L. Machniak, A. Borcz, *Nowoczesne Budownictwo In zynieryjne*, **5** (2015)
2. Ministerstwo Gospodarki, (Warszawa, 2014)
3. S. Goralczyk, D. Kukielska, *Prace Naukowe Instytutu Gornictwa PWR. Studia i Materialy* **43** (2013)
4. L. Machniak, W. Koziol, *Kruszywa*, **4** (2014)
5. Glowny Urzad Statystyczny, *Ochrona Srodowiska 2005* (Warszawa, 2006)
6. Glowny Urzad Statystyczny, *Ochrona Srodowiska 2014* (Warszawa, 2015)
7. *Krajowy Plan Gospodarki Odpadami 2014* (Warszawa, 2010)
8. UEPG, *Estimates of Aggregates Production 2012* (Brussels, 2013)

AUTHOR INDEX

AUTHOR INDEX